Jan J. W. Lisman

Cosmogony, Theogony and Anthropogeny in Sumerian Texts

Alter Orient und Altes Testament

Veröffentlichungen zur Kultur und Geschichte des Alten Orients
und des Alten Testaments

Band 409

2013
Ugarit-Verlag
Münster

Cosmogony, Theogony and Anthropogeny in Sumerian Texts

Jan J. W. Lisman

2013
Ugarit-Verlag
Münster

Jan J. W. Lisman

Cosmogony, Theogony and Anthropogeny in Sumerian Texts

Alter Orient und Altes Testament, Band 409

www.ugarit-verlag.de

Herstellung: Hubert & Co, Göttingen
Printed in Germany
ISBN 978-3-86835-095-1

Printed on acid-free paper

to

Ine
Anne-Marie
Marie-Claire
Jan-Willem

'Wisdom means
searching for the nuances,
testing your own opinions,
and judging moderately
because you know you might be wrong.'

Rev. A. Tollefsen

Table of Contents

Preface - Acknowledgements

The first step is the hardest, but well begun is half done. What was that first step? During my stay for one year in Heidelberg I was looking for an interesting subject for my undergraduate thesis. That subject simply fell into my lap on the occasion of a Blockseminar in Oberflockenbach: 'Der Ursprung der Welt in griechischer und altorientalischer Sicht', presided over by Prof. J. Halfwassen and Prof. S. Maul. My contribution to this seminar concerned the cosmogony in Sumerian texts.

By courtesy of Prof. W. van Soldt (promotor) I obtained the permission to expand this topic into 'At the beginning... Cosmogony, theogony and anthropogeny in Sumerian texts of the third and second millennium BCE' for the writing of my PhD thesis. My research was also supervised by Dr. D. Katz as co-promotor. In the final stage of my study Dr. A. Jagersma gave valuable comments, for which I am very grateful to him. The present book is based on this PhD thesis.

This study is built on the work of various scholars, including: Kramer, Jacobsen, van Dijk, Lambert and Pettinato. It is my wish that my study will contribute to and will be the subject of further discussion on these topics.

When I was given the opportunity to study Assyriology with a specialization in the Sumerian language and culture, I experienced the stimulating support of all my teachers, some of whom I would like to mention by name. Mr Theo Krispijn in Leiden, Prof. S. Maul and Prof. H. Waetzoldt in Heidelberg, each of them with an accent on different aspects, initiated me into the knowledge of the Sumerian language and showed me the beauty of the Sumerian culture, which has greatly appealed to me since my youth. I owe much gratitude to these teachers.

I would like to thank Helen Richardson-Hewitt, who corrected the English of this book. For any errors that remain the author alone is responsible.

Most of all I am deeply grateful to Ine, who has always been on hand to listen to me, to hearten and inspirit me, even in periods when I seemed to be 5000 years and just as many kilometres away.

Prologue

Sumerian Beginnings

ἀρχὴ ἥμισυ παντός

Many people were and are searching for the beginnings, for the ultimate source of our world and of our existence. Nowadays physicists are able to come close to the beginning of our universe, the moment of the 'big bang'; the development of life from atomic and molecular level to living cell to living organisms is being unravelled; details of the evolution of man are becoming more and more clear. But what were the ideas about the beginnings in the remote past? To discover this we have to look at the oldest texts that are available to us: Sumerian texts. In 2004, during a seminar about 'Der Ursprung der Welt in griechischer und altorientalischer Sicht' [1], I was confronted for the first time with the Assyriologist's publications about the Sumerian beginnings.

In the very early stages of Sumerology, in the middle of the 20th century, some scholars tried to describe in broad outline a picture of these beginnings according to the line of thought of the Sumerians. The first one to summarize the Sumerian literature in this respect was Samuel Noah Kramer in 1944. In his *Sumerian Mythology. A Study of Spiritual and Literary Achievement in the Third Millennium B.C.* there is a chapter on 'Myths and Origins', in which among other things the creation of the universe and the creation of man are discussed. Thorkild Jacobsen reviewed Kramer's book, and added several new ideas [2]. In 1964 van Dijk, in his article 'Le motif cosmique dans la pensée sumérienne', resumed the thread of the Sumerian cosmogony [3]. This article set the standard: van Dijk's ideas, which he repeated and supplemented in the subsequent years, have been accepted by nearly all scholars writing about Sumerian beginnings.

Some revised translations, though only about cosmogony, as part of my contribution to the above-mentioned seminar, were in fact the direct motivation for the present study. It is hoped that these new translations might shed fresh light on the current theories about the Sumerian cosmogony.

At the start of this study, two terms have to be clarified: 'Sumerian' and 'Beginnings'.

'Beginnings' in this context has three aspects: 1. cosmogony – how and when the universe came into being –; 2. theogony – how and when the gods appeared –; 3. anthropogeny – the origin of man [4].

The term 'Sumerian' may refer to the language in which the sources were written or to the people inhabiting Sumer, whose existence could no longer be traced after the Ur III

---

1 Blockseminar in Oberflockenbach, Germany (10-12 July 2004), organised by Prof. J. Halfwassen (Philosophy) and Prof. S.M. Maul (Assyriology), University of Heidelberg.

2 Jacobsen 1946.

3 See chapter 1 of this study for an extensive discussion of van Dijk's publications, and of other relevant publications.

4 When the term 'religion' is used or when is referred – directly or indirectly – to religion in this study, it must be understood as Veldhuis (2004, 11-17) discussed this concept in his introduction to the Sumerian composition 'Nanše and the Birds'.

period when 'Sumer ceased to exist as a political, national, and cultural entity' [5]. Or as Lambert says: '(...) Sumerian culture declined outside the scribal schools.' [6]

The following considerations have been the starting-points for the present study.

1. It would appear that no study exists in which all the extant texts relevant to the Sumerian Beginnings defined above are discussed. The intention of this study is to present as complete as possible a discussion of all the extant Sumerian texts in which 'beginning' is mentioned, from the time when the first written documents appeared (~ 3000 BCE) up to and including the late Old Babylonian period (~ 1500 BCE). The Sumerian literary texts used in this study date from the Early Dynastic period onwards. Some of the Sumerian literary texts from the Old Babylonian period contain older traditions, and in this respect they still may represent a great number of Sumerian ideas.
2. Because of the influences of other ethnic groups on the Sumerians, already discernable from the first half of the third millennium BCE, it is scarcely possible to speak of 'genuine Sumerian'. Nevertheless, by comparing the ideas about Beginnings as these can be inferred from Sumerian texts with those from texts written in Akkadian, an attempt will be made to map out as far as possible Sumerian versus Akkadian motifs. The most recent Akkadian text included here dates from the Middle Babylonian period.
3. It proved to be necessary to revise the existing translations. Therefore all the relevant texts in which 'beginnings' are mentioned – both Sumerian and Akkadian texts – have been re-edited in this study.
4. Another major weakness of former studies is the absence of a diachronic survey of the Sumerian ideas with respect to these beginnings. There are too many examples in the Assyriological literature in which data from such widely distant periods as the third and the first millennium are compared without any comment, as if they had appeared synchronically and as if there had not been any development in whatever respect during that long period of 1500-2000 years. In this study I will try to describe a diachronic survey of the development of the Sumerian ideas about Beginnings, from the Early Dynastic period onwards. These results will be compared with the Akkadian concepts concerning 'origins' as they are found in texts of the third and second millennium BCE.
5. A point of interest will be to use the structural analysis method for the interpretation of the mythological stories with respect to the beginnings. Therefore the literature of Lévi-Strauss and of Doty has been consulted and used as far as possible throughout the analyses and discussion of the texts [7].

The structure of this study is as follows.

In chapter 1 – Introduction – two topics will be reviewed. The first one describes the history of Mesopotamia from the ʿUbaid period onwards. The aim is to try to answer the questions about precisely when the first Sumerians were present in South Mesopotamia and which other ethnic people might have influenced their community, in order to have a clear idea of how 'Sumerian' can be defined. The second topic concerns the current theories

---

5 Katz (2003, *xv*), who has given a more comprehensive definition of the term 'Sumerian'.

6 Lambert 2011, 72.

7 Lévi-Strauss 1969; *idem* 1977; Doty 2000. This last scholar also critically reviewed the methods of Lévi-Strauss. The critical discussion of Civil (1980) concerning the limits of textual information has been taken into consideration, too. The results of this approach will be discussed in ch. 4.10.

about Sumerian Beginnings. The studies of van Dijk, Pettinato and Lambert, scholars who have contributed a great deal to the theories about Sumerian Beginnings, will be discussed in detail.

Chapter 2 contains the translations of those (parts of) Sumerian texts which are relevant for the Beginnings. The concerning texts are:

Period	*Text*
Early Dynastic / Presargonic	IAS 114 IAS 136, 113 and 203 IAS 174 Ukg 15 (AO 4153) Barton Cylinder
Ur III	NBC 11108
Ur III / Old Babylonian	Debate Tree - Reed Debate Grain - Sheep
Old Babylonian	Gilgameš, Enkidu and the netherworld Enki and Ninmaḫ Song of the Hoe
Middle Babylonian / Kassite	KAR 4

A preliminary summary and conclusions with respect to the Sumerian Beginnings will be given at the end. In an Excursus attention will be paid to the cosmic 'marriage' of **an** and **ki**, and to the possibility that this 'marriage' might have been celebrated as a ritual.

Literary texts are not the only source for the study of the Sumerian ideas about origins. The god lists are a useful complement. They consist of an enumeration of gods, the order of which is theologically or lexically [8] determined. God lists are known to have existed from the Early Dynastic period onwards. In chapter 3 – 'God lists or Lists of divine names' – the relevant parts of the god lists, from the Fara period to the end of the Middle Babylonian time, will be analysed. The relatively late list **an** = *anum* (second half of the second millennium BCE) has been included, because this list is '(...) an explanatory list that seeks

8 It is outside the scope of this study to make a structural analysis and theoretical interpretation of the complete lists, analogous to the one performed by Gantzert (2011) for the Emar lexical lists; only a small section of each of these god lists will be used in this study.

to clarify the offices and relationships of the numerous members of the pantheon. Unlike most lexical texts, this series is not even Akkadian in its composition, but rather, Sumerian.' [9]

This chapter will be concluded with an Excursus about the gods Enlil and Ninlil and one about Enki, in which their historical background and their origin are the main topics, and one Excursus in which the power relations between Enlil and Enki and the spheres of influence of both gods will be examined.

In chapter 4 the results obtained from the texts (ch. 2) and the god lists (ch. 3) will be combined to give a diachronic survey [10] of the Sumerian Beginnings, from the Early Dynastic period up to the end of the Old Babylonian period. A comparison will be made with some Akkadian mythological texts about Beginnings: *atra-ḫasīs* and *enūma eliš*.

Finally, in the Epilogue the Sumerian Beginnings will concisely be compared with the main themes from stories about the same topics as they are known not only in the Ancient Near East but also in other parts of the world, in the past and at present.

The Appendix contains the editions of the texts that are discussed in chapter 2, with a score and a commentary, as well as an Excursus about a grammatical question, viz. about the animate versus the inanimate class of **an** and **ki**. Finally a survey will be given of the relevant parts of all the god lists that are used in this study.

***

---

9 Litke 1998, 6.

10 Michalowski (1998, 239) argued in favour of studying a subject like the present one from the oldest times onwards to more recent periods, and not in the backwards direction: 'Mesopotamian civilization, or better, civilizations, have been read backwards for over a hundred years. (...) The backward reading may have been beneficial – within reason – in the case of languages, but the consequences have been less fortuitous in other areas, particularly in the study of religion. This reverse evolutionary trajectory is particularly problematical in the case of Mesopotamian religion since, as we now know, some of the crucial later documents that have been used in the debate – *Enūma Elish* in particular – were composed as cultic and religious polemics, and represent a radical break with past beliefs.'

Chapter 1

Introduction

'Interest in cosmogony marks a hope that despite
the perpetual changes evident to traditional cultures
and the diversity of beliefs apparent to moderns,
some realities may be permanent and some things
may be dependable enough to build a way of life on them.'
[Lovin and Reynolds, *In the Beginning*; 1985, 6.]

This introduction contains two main subjects which constitute the background and basis for my study. In the first place it concerns the history of Mesopotamia, and especially the question: is it possible to draw a distinction between the Sumerian and the Semitic/ Akkadian cultural heritages, and if so: what is the relation between the Sumerian and the Semitic/Akkadian culture and especially between Sumerian and Semitic/Akkadian Beginnings? The second subject concerns the existing scientific theories or views with respect to Sumerian Beginnings. In this overview the contributions of several scholars – van Dijk, Pettinato, Lambert – will be discussed in detail.

1.1 *Mesopotamian history*

1.1.1 *Archaeological evidence*

Describing the development of the Mesopotamian ideas with respect to cosmogony, theogony and anthropogeny from earliest times, the terms 'Sumerian' and 'Semitic/ Akkadian' will be used. This section will be devoted to examining to what extent the conceptions about 'the beginnings' may be attributed to these different ethnic entities, instead of the terms Sumerian and Semitic only being used to point to the fact that the texts have been written in the Sumerian or Semitic/Akkadian language. Some of the most intriguing questions are: since when were the Sumerians present in South Mesopotamia? Were they the original inhabitants or did they enter an already populated land? A brief summary of some main points concerning the ethnicity of the inhabitants of Mesopotamia and the history of that country will be given [11].

The ˁUbaid period (ca. 6000 - ca. 4000 BCE)

The ˁUbaid period is when the first settlement in South Mesopotamia could be proven. The ˁUbaid culture developed in South Mesopotamia, and spread widely beyond the southern alluvial plain into the neighbouring regions of Syria, South-East Anatolia and Iran [12]. Several opinions about the cause(s) of the spread of the ˁUbaid culture have been put forward; migration from the South into the other areas has not been excluded. Whatever the cause, it appears, as far as this discussion is concerned, that there must have been contact between southern Mesopotamia and the other areas with ˁUbaid culture. In their summary

---

11 For the ethnicity aspects: see van Soldt (ed.) 2005; for a detailed treatment of (aspects of) the history, see e.g. Akkermans and Schwartz 2004; Algaze 2005; Algaze 2007; Charvát 2002; Maisels 1999; Nissen 1988; Nissen 1999; Rothmann 2001 (with a new chronological framework for Greater Mesopotamia for the 4th and 3rd millennia BC: Table 1.1, p. 7).

12 Akkermans and Schwartz 2004, 154-157; Stein and Özbal 2007.

Stein and Özbal state: 'A contextual analysis comparing different regions shows that the ˤUbaid expansion took place largely through the peaceful spread of an ideology, leading to the formation of numerous new indigenous identities that appropriated and transformed superficial elements of ˤUbaid material culture into locally distinct expressions.' [13]

Excavations in Eridu and Uruk have shown that in those places where in ancient times the presence of temples could be demonstrated, the building activity goes back to the beginning of the ˤUbaid period. It is likely that the oldest buildings there would also have been shrines or temples.

The Uruk period (ca. 4000 - ca. 3100 BCE)

The ˤUbaid period was followed – without interruption and with a gradual transition – by the Uruk period [14]. The Uruk culture was almost as widespread as the ˤUbaid culture in roughly the same areas; the direction of this spread was from its homeland – the south of Mesopotamia – to the other regions, e.g. Syria and western Iran [15]. The expansion of the Uruk culture has been studied especially by Algaze [16]. The Uruk expansion was predominantly based on trade between southern Mesopotamia and the rest of Great Mesopotamia (including Syria, South-East Anatolia, Iran). There was an intensive contact between the southern alluvial plain and the other regions. In some instances one may speak of a process of colonization on the part of South Mesopotamia. Algaze concluded [17]: '(...), it may be inferred that the expansionary processes of Mesopotamian societies of the Uruk period were firmly rooted in earlier developments.' Hesse, noting that in the second half of the 4th millennium Upper and Lower Mesopotamia no longer developed in parallel, ascribed this asymmetry to the natural advantages of the South, which promoted agricultural production and (inter)regional trade [18].

By the final quarter of the 4th millennium the expansion of the 'Uruk world system' ended. A factor that may have contributed in an important way to this collapse may have been a change in the climate, viz. a decrease in rain fall, leading to a less strong economic point of departure for the alluvial community [19]. In the same period there were also demographic changes in South Mesopotamia. Pollock has re-examined the data of Adams in order to describe settlement patterns and demography in two main areas of southern Mesopotamia: the Nippur-Adab region and the Uruk area [20]: 'The corrected figures result in far fewer sites and hence total hectares occupied than conventional observations would suggest (figs. 6.11 and 6.12).'[21] The Nippur-Adab region appeared to have been more

---

13 Stein and Özbal 2007, 329.

14 Maisels 1999, 116. For a recent view on the Uruk period, see the article of Algaze 2013.

15 Akkermans and Schwarz 2004, 181.

16 Algaze 2005; Algaze 2007. After the first edition of his 'Uruk world system' in 1993, Algaze received several criticisms, which he has incorporated in its second edition of 2005.

17 Algaze 2005, 121.

18 Hesse 2010, 76-82.

19 For the climate change: see Nützel 1976. For more detailed descriptions of this collapse, see e.g: Akkermans and Schwarz 2004, 207-209; Algaze 2005, 104-107; Charvát 1981, 686; Charvát 2002, 176-183, 238; Nissen 2001, 174-175.

20 Pollock 2001, 210-221. This re-examination was done with the correction method of Dewar: 'His method is based on a model that estimates the rates at which sites were founded and abandoned and, from these, the average number of contemporary occupations.' (Pollock 2001, 211).

21 Pollock 2001, 212.

stable, both demographically and with respect to settlement longevity, than the Uruk area. Pollock wrote that her conclusion 'contrasts with conventional interpretations that suggest that the proportion of population living in towns and cities in the Nippur-Adab area by Late Uruk (LC 5) times was substantially greater than in the Warka area (Adams 1981:75, table 4). (...) However, this revised evaluation of the settlement data does not bear out the contention that the Nippur-Adab area lost a substantial portion of its settled population, probably to emigration, by Late Uruk (LC 5).'[22] The growth of the population in the Uruk area may have had several causes: people coming from the Nippur-Adab region; people moving from the Susiana plain; a substantially higher rate of biological reproduction in the Uruk area than in the Nippur-Adab region.

Jemdet Nasr (ca. 3100 - 2900 BCE) - Early Dynastic period (ca. 2900 - ca. 2350 BCE)

Steinkeller also discussed the Uruk expansion [23]. He agrees with Algaze that the main cause of the end of this expansion may be 'the growth of native political structures in the peripheral areas impacted by the Uruk colonization, (...). As can be inferred from various types of data, during this transitional phase (i.e. Jemdet Nasr and the beginning of ED I; *JL*) there was a major influx of Semitic peoples into Syria and Upper Mesopotamia, probably in several waves and over an extended period of time. One of those peoples, probably the ancestors of the Akkadians, migrated into the Diyala Region and northern Babylonia, (...).' In northern Mesopotamia the political configuration known as 'the Kiš civilization' was formed [24]. During the Jemdet Nasr and ED I period there was an ever-diminishing presence of Sumerians in North Mesopotamia and in the Diyala region; finally the Sumerian presence was confined to the territory south of Nippur.

In the beginning of the Early Dynastic period, the Semitic Kiš civilization extended its political influence from Kiš and neighbouring cities – e.g. from Abu Ṣalābīḫ – to Akšak, Mari and Ebla [25]. The focal point was not Uruk anymore (as in the previous period), but Nippur. Evidence of cultural contacts between the Kiš civilization and the Sumerian one can be inferred from the following. Some of the Kiš rulers had Akkadian/Semitic names, some of them Sumerian names [26]. In Abu Ṣalābīḫ some of the scribes had Sumerian names. Similar lexical lists and incantations were found in Ebla, Abu Ṣalābīḫ and Fara/Šuruppak [27]. Enlil, the supreme god of the Sumerian pantheon [28], was present in the ED god lists from Fara and Abu Ṣalābīḫ. In the 'archaic City List' from Uruk, Enlil's name represented the city of Nippur [29]. According to Steinkeller [30], the rise of Enlil to the position at the head of the Sumerian pantheon was concomitant with the rise of Kiš and its region to power.

---

22 Pollock 2001, 215.

23 Steinkeller 1993, 115.

24 Gelb 1981; Gelb 1992; Steinkeller 1993, 117-121. Archaeological evidence has established that Kiš was not founded before the Jemdet Nasr period (McCarthy 2010, 838).

25 Gelb 1981, 54; Charvát 2002, 213-214; Steinkeller 2002, 257.

26 Steinkeller 2003; The Sumerian King List, ETCSL 2.1.1, ll. 40-94 and 160-178.

27 Biggs 1988; Krebernik 1984; Pomponio 1983; Westenholz 1988.

28 About Enlil's origin, see this study, ch. 3: Excursus 1 'Enlil and Ninlil'.

29 Englund 1998, 91-94 and fig. 25-27. The sequence of the cities in this list is not obvious. Englund supposes that the order may reflect a mythological or cultic hierarchy; but an economic or political meaning, reflecting a 'league of cities' could not be excluded.

30 Steinkeller 2002, 257.

Akkad period - Ur III - Old Babylonian period (ca. 2350 - ca. 1600 BCE)

The period after the Early Dynastic time, the Sargonic or Akkad period, is the first period in which the south of Mesopotamia – Sumer – was dominated by Semitic people, i.e. the Akkadians [31]. After the Ur III period a Sumerian ethnic identity is no longer identifiable; only the Sumerian language continues to be used in the literature.

1.1.2 *Linguistic evidence*

Although archaeological remains from the prehistory are not labelled 'Sumerian', they give at least some basis for drawing tentative conclusions. Another approach to the 'Sumerian question', – the question from what time onwards the Sumerians inhabited the South of Mesopotamia – is to try to gather linguistic evidence. If the Sumerians were not the first inhabitants of South Mesopotamia, then there might be traces of one or another pre-Sumerian substrate language in the Sumerian language. This idea was first proposed by Landsberger in 1944, and thereafter several other scholars attempted to find evidence along similar lines [32]. The most obvious difficulty in this respect, of course, is the relatively late appearance of the first written texts, viz. in the late Uruk period, in relation to the period in which the first settlements in South Mesopotamia had been attested. However, the hypothesis of a pre-Sumerian substrate language proved to be not very successful [33]. In the view of Michalowski: 'On comparative grounds, it is more probable that this language (*i.e. the Sumerian one; JL*) represents but a remnant of a much broader linguistic continuum, areal if not genetic, that had occupied much of Western Asia before the Semitic spreads.' [34] In the concluding words of Rubio: '... there was no identifiable single substratum (proto-Euphratic, Indo-European, or otherwise) that would have left, in a sort of primeval age, its vestiges in the Sumerian lexicon.' [35] Steiner simply asserted that the question of a possible pre-Sumerian substrate language has to be answered negatively [36].

1.1.3 *Presence of the Sumerians in South Mesopotamia*

To summarize: the results from archaeological research were as follows. The continuity of the cultural remains, e.g. religious buildings and temples, in Uruk and Eridu; the direction of the cultural flow during the ʿUbaid and the Uruk periods from South Mesopotamia to the other areas; and the absence of disturbances indicative of an influx of foreign people in the south of Mesopotamia [37]. The tentative conclusions of linguistic studies also point to the early presence of the Sumerians in South Mesopotamia [38].

---

31 Liverani 1993; Westenholz 1999.

32 For a summary, see Rubio 1999; Whittaker 2005.

33 Michalowski 2000; Rubio 2005; Steiner 2005; Black and Zólyomi 2007, ch. 3 and 4.1.

34 Michalowski 2000, 180.

35 Rubio 2005, 331.

36 Steiner 2005, 350.

37 The fact that the extent of some geographical names in Mesopotamia, being neither Sumerian nor Semitic, fits rather well with the extent of the ʿUbaid culture, and that consequently the ʿUbaid culture cannot be Sumerian (Gelb 1960, 263-264), does not necessarily mean that the Sumerians entered into the south of Mesopotamia only *after* the ʿUbaid culture. Kienast (1985, 108) supposes that the Sumerians entered the area *ca.* 3500 BCE.

38 See also the reviews of Englund (1998, ch. 4, 56-81) and Glassner (2003, ch. 2, 29-47).

Therefore it is my suggestion, based on both lines of evidence, that Sumerians had been present in this area already since the ʿUbaid period. Cultural contacts between the Sumerians and other ethnic people have always existed. From the Kiš period onwards, the Semitic influence on the Sumerian culture had been increasing, with a culmination during the Akkad time. After the Ur III period, which was the final heyday of the Sumerian culture, the dominant sphere was the Akkadian/Semitic one. This short overview may be helpful in my attempt to identify the Mesopotamian ideas about beginnings.

1.2 *Existing theories about the Sumerian Beginnings*

In this section the theories of several scholars about the Sumerian Beginnings will be discussed. It is not my intention to give a complete survey of the literature in this respect, as the knowledge of the Sumerian language at least in the first half of the 20th century was too inadequate. This does not mean that there are no studies available to provide a firm basis for our thinking about Sumerian Beginnings, see e.g. those of Kramer and Jacobsen [39]. My survey will start with the studies of van Dijk, who has admitted to being indebted to both these scholars. Next the contributions of other authors about the Sumerian Beginnings, in so far as they differ from the theories of van Dijk, will be reviewed and discussed. Only those studies will be discussed which contain new ideas and contributions to clarify my thoughts about the beginnings which have been crystallized in the Sumerian literature [40].

1.2.1 *Van Dijk: the cosmic and the chthonic motif; emersio and formatio*

Until now, the ideas and theories of van Dijk about the Sumerian Beginnings – cosmogony, theogony, anthropogeny –, which he for the first time published in a comprehensive article in 1964, are still held by most scholars [41]. Van Dijk based his studies on several publications by Kramer and Jacobsen [42]. This chapter will not contain a detailed summary of the publications of van Dijk; I will focus the attention in the first place on the two main points in the theory of van Dijk, viz.:

1. the cosmic motif *versus* the chthonic motif;
2. anthropogeny via *emersio* and via *formatio*, respectively.

1.2.1.1 *The cosmic motif*

For a better and correct understanding of van Dijk's concept of the cosmic motif, we will summarize his ideas in this respect [43]. According to van Dijk's definition, the cosmos is the universe: heaven and earth and all the inhabitants thereof. It follows that cosmic means: 'ce qui appartient à l'univers considéré comme une totalité en interdépendance'. If one considers the parts that comprise the universe while leaving aside this interdependence, then one cannot speak of heaven, stars and earth as *cosmic* entities, but only of celestial,

---

39 Kramer, *Sumerian Mythology*, 1944; Jacobsen 1946.

40 E.g. the publications of Römer (1969a) and Seux (1987) do not contain new ideas and thus these are not discussed here.

41 The most important articles in which van Dijk unfolded his theories are published in 1964, 1969, 1971 and 1976. In this section, only van Dijk's conclusions are rendered, without comments.

42 Kramer, *Sumerian Mythology*, 1944; Jacobsen 1946. Other references to Jacobsen: *passim* in van Dijk's publications.

43 Van Dijk 1964, 5-6 (quotation: p. 5).

astral and earthly entities, respectively. In the Sumerian way of thinking, the interdependence of the male heaven and the feminine earth seems to play a significant role in the cosmogony as well as in the religious idea. These considerations brought van Dijk to examine whether a number of items in which 'heaven' or An, the god of heaven, and 'earth' or events happening on or with earth show some interdependence, might be reduced to one leading principle, for which he coined the phrase 'le motif cosmique' [44]. That principle of the the cosmic motif appears to consist of the male heaven and the feminine earth. Among the examples given by van Dijk to support this interdependence are the 'noces cosmiques', the cosmic marriage between heaven and earth, which can be found in the oldest Sumerian literature. The goal of van Dijk's study is to search in various texts for this cosmic motif, i.e. the interdependence of the two principles of the Sumerian cosmogony: heaven and earth. For this study he wants to analyse:

- the 'cosmogonic introductions' of the god lists;
- the position of the expression **u$_4$-ri-a** 'in illo tempore' in the Sumerian mythology and literature;
- texts which express the idea of a cosmic marriage.

Van Dijk starts with the analysis of the god lists; here he distinguishes two forms of genealogy:

- vertical genealogy: the ancestry line from father to son/daughter;
- horizontal genealogy: the several spouses who appear via syncretism.

In this respect the Nippur list (SLT 122-124) is a good example for van Dijk. The following scheme is his interpretation of the beginning of this list [45]:

An
∞
Antum (Ki) ——— Uraš
↓
Enlil (Nunamnir)
∞
Ninlil (Šulpa'e) Ninmaḫ Ninḫursaga Nintu Ninmena Maḫ Bēlit-ilī

Van Dijk explains this scheme, in which Šulpa'e, the spouse of Ninḫursaĝa at Adab, has been added in the horizontal line by force of tradition as follows. It is evident, he says, that this list is intended to rank all the mother goddesses in one line with Ninlil; this happens via the syncretism of the spouses of Enlil. Also by way of syncretism, Antum (= **ki**) and Uraš have been placed on a horizontal line. At the same time it is clear that the theological idea that the marriage between **an** and **ki** forms the origin of the pantheon, was known in several cities but in a different way. The theogony based on **an** x **ki** originates from Uruk, the one based on An x Uraš originates from cities dedicated to Gula-Baba-Ninisina. In this theogonic introduction, Namma has been passed over without comment.

44 Van Dijk gives several examples in which An is the protagonist, e.g.: An grants kingship, An founds several cities, An engenders plants, stones, sons and daughters, demons, but without any literature reference.

45 Scheme and text: van Dijk 1964, 8-9. Van Dijk wonders – indicated by a question mark in the scheme, but not rendered here – if these mother goddesses (Ninlil etc.) are children of An.

1.2.1.2 *The chthonic motif*

In opposition to the cosmic motif, van Dijk introduced the chthonic motif [46]. This motif, he said, can be associated with Eridu and the goddess Namma. In the theogonic system of Eridu, Namma has a prominent position, as mother of heaven and earth and as mother of Enki:

An ∞ Nammu
↓
Enki

This mere fact means that the Eridu system greatly differs from the previous one, the cosmic sytem. The Eridu system may be rendered schematically as follows [47]:

Apsû + Mummu + Ti'āmat
↓
Laḫmu ∞ Laḫamu
↓
Anšar ∞ Kišar
↓
An ∞ [Nammu] [48]
↓
Nudimmud (= Enki)

From these schemes van Dijk concluded that the Eridu system is *chthonic*; it only seems to become cosmic as a result of integration into another system. In this way the theogony and cosmogony of *enūma eliš* become cosmic only from the moment that Marduk slays Ti'amat and separates heaven and earth. Marduk has derived this last function from Enlil, not from Enki. Van Dijk offers two possible solutions for this situation:

1. The Sumerian religion had formed such a solid unity that Enki and Enlil could have acted as a couple and taken each other's place;
2. The separation of heaven and earth originally does not belong to the Eridu cosmogony; it has been introduced into it by way of syncretism as a heterogeneous element.

But then van Dijk encounters a problem. He asks how Enlil could have separated heaven and earth, while he in all probability has been born from the union of both? He states that on several points the Eridu theology differs profoundly from the cosmic system. In the Eridu cosmogony the primary origin is the **abzu** as a male principle.

This last remark reveals why van Dijk named the one system cosmic and the other one chthonic: in his concept the male principle – An and Apsû, respectively – determines the name of the system. Obviously, the female principle **ki**, being chthonic, is not relevant for the determination of the character of the beginnings.

Van Dijk has also a simple explanation for this bipolarity in the Sumerian ideas. The chthonic concept, at home in the south of Sumer, is connected with a farming community.

---

46 Van Dijk 1964, 9-12 (Nammu is the old spelling for Namma).

47 Van Dijk 1964, 10; he does not mention explicitly that this scheme has been derived from *enūma eliš*.

48 Van Dijk remarks that Namma has been passed in silence in *enūma eliš*.

The farmers are dependent on irrigation, not on the scanty rainfall. This last one is more or less sufficient for the pastoral people. In this way, says van Dijk, it is easy to explain why the cult of the sky god An and that of the shepherd Dumuzi are both at home in Uruk; and also why the god of the crafts, Enki, is connected with Eridu. He concluded this argument as follows [49]. The Eridu system may belong to an agricultural and sedentary people, the cosmic system to a pastoral people. Immigrants to Mesopotamia mostly were pastoral people who became sedentary. Therefore it is not surprising that in the end the cosmic system ousted the chthonic system, according to van Dijk [50].

1.2.1.3 *Combination of the cosmic and the chthonic system*

The god list TCL XV 10

According to van Dijk [51], the cosmic doctrine, present in the god list of Nippur (SLT 122-124) and the chthonic doctrine of Eridu have been combined (van Dijk: 'ont été syncrétisés') in the god list TCL XV 10. Its introduction can be divided into two parts [52]:

1. A vertical genealogy, listing the generations of gods who fill the embryonic universe, that finally brings forth the god of heaven An.
2. A horizontal genealogy which results in the creation of Enlil and Enki.

In scheme:

Enki (**en-ki-e-ne**) ∞ Ninki (**nin-ki-e-ne**)

↓

Enmešarra ∞ Ninmešarra

↓

An (= Anšargal; Enurulla)

∞

↓ ^{d}Uraš ^{d}nin-ì-li ↓ Enlil

↓ ^{d}Nammu ↓ Enki

Van Dijk calls the Enki-Ninki gods 'êtres androgynes' [53]. But these chthonic gods, he says, who inhabit the earth, were already well-known before the existence of the list TCL XV 10.

---

49 Van Dijk 1964, 11-12: 'Si l'on admet l'hypothèse que le système d'Eridu est celui des agriculteurs et du peuple sédentaire et le système cosmique celui des pasteurs, il ne faut pas s'étonner que celui des pasteurs ait réussi à évincer celui des agriculteurs: l'afflux et le renouvellement de la population en Mésopotamie sont en effet toujours et en tout temps venus des pasteurs devenus sédentaires.'

50 As we shall see in ch. 4, this hypothesis is not justified.

51 Van Dijk 1964, 12.

52 Vertical genealogy means: successive generations.
Horizontal genealogy concerns members of the same generation, in this case: Enlil and Enki are half-brothers.

53 Van Dijk 1964, 12. Moreover van Dijk called these gods 'dema-gods' (van Dijk 1957-1971, § 2a, 535; *id.* 1964, 12 note 21; *id.* 1969, 178; *id* 1971, ch. 4, 449-452). According to the definition given by Jensen (ref. in Cavigneaux-Krebernik 1998-2001, 446b) or the definition in the Encyclopaedia Britannica it is not justified to classify the Enki-Ninki gods as 'dema'-gods,

They were introduced into this list in a logical way: these androgyne beings lived in an embryonic universe before the birth of the heaven [54]. The same gods appear in the Sumerian literature with a chthonic character as ancestors of Enlil (e.g. in 'The Death of Gilgameš'). As a consequence, according to van Dijk, the text of TCL XV 10 refers to a religious tradition older than that in which these chthonic gods have evolved into Enlil's ancestors. The Fara god lists SF 23-24 confirm this view, though the enumeration there of the Enki-Ninki gods is not followed by a concluding addition such as 'ancestors of Enlil'.

The doctrine reflected by the introduction of TCL XV 10 seems to be the following [55]:
1. The embryonic universe in which chthonic gods were living pre-existed.
2. The universe was conceived of as a city: **uru-ul-la** 'city of ever'.
3. From that universe An – **en-uru-ul-la** 'Lord of the city of ever' – has come into being.
4. Heaven and earth were joined in a cosmic sacred marriage.
5. At a certain point heaven parts from earth.
6. From the union between heaven and earth originate the great gods via *emersio*.

Van Dijk then continues, that the necessary complement to this doctrine must be that the insemination of the earth by the heaven results in chthonic gods, vegetation and man. He states that Namma is in fact a stranger in this concept, introduced by syncretism, and that she seems to be some kind of 'Mother Earth'. It is also his opinion that the introduction of Namma is not a new speculation, but an attempt to integrate older myths into a homogeneous system. In her quality of Mother Earth, Namma becomes the spouse of An in the cosmogony of the list TCL XV 10, just like Uraš and Ki. Van Dijk also remarks that the story 'Enki and Ninmaḫ' differs in this respect, because of the fact that An, as a principle of life and as partner of Namma, is absent.

The god list an = *anum*

Next van Dijk analyses the god list **an** = *anum* [56]. In his opinion this list has lost the logic of TCL XV 10. To give preference to the 'system of Eridu' [57], the gods of the embryonic universe have been included under Enlil. The cosmic system has been lost. In fact there is, according to van Dijk, now the following new system in **an** = *anum*:

1. Heaven and earth are androgyne and the origin of themselves. The list does not recognize the separation of heaven and earth after the embryonic universe. According to van Dijk, the same phenomenon is present in the first line of the text KAR 4 [58]: **u$_4$ an-ki ... tab gi-na- ... eš-a-[ba]** 'the day that heaven and earth were founded together'. According to van Dijk this is a new theological concept, because the original one surely was: **u$_4$ an-ki-ta ... bad-a-ta ...** 'since the day that heaven removed from earth...'. Van Dijk continues, that one has to

because the Enki-Ninki gods do not meet the qualities of dema-gods ['Dema deity: any of several mythical ancestral beings of the Marind-Anim of southern New Guinea (...);' The New Encyclopaedia Britannica, vol 4, 1; 15th ed. 1991].

54 Van Dijk refers in this respect to Eliade (1954, 78-87) in order to support his proposition.

55 Van Dijk 1964, 13-14.

56 Van Dijk 1964, 14-16.

57 See ch. 1.2.1.2.

58 'KAR 4, 1, qui introduit une glosse: **u$_4$ an-ki ... tab gi-na- ... eš-a-[ba]** «le jour où le ciel et la terre furent fondés ensemble». C'est un théologoumenon nouveau, car l'original portait à coup sûr: **u$_4$ an-ki-ta ... bad-a-ta ...** dès le jour où le ciel s'éloignait de la terre...' (van Dijk 1964, 12).

conclude that the extension of the god lists has not been the foundation for the Sumerian cosmic doctrine, but has moved away from it. Therefore the following was necessary:

- Uraš had to be classified in the vertical genealogy, together with the ancestors of An.
- ^{d}**en-uru-ul-la** (= **an**) had to be classified in the same vertical genealogy, together with a – newly created – ^{d}**nin-uru-ul-la** at his side. In this way the concept of a 'cité de jadis', a city of former days, got completely lost.

2. The 'ancestors of An' in TCL XV 10 have become 'the ancestors of Enlil' in **an** = *anum*. In this way the last list places Enlil at the same level as An by composing two parallel genealogies instead of two genealogies in juxtaposition. The resulting theogony is not the outcome of doctrine development, but it is a completely new doctrine. Therefore the most pure and most logical content of the cosmic motif is not found in the most developed theogony. It seems, says van Dijk, as if the authors of **an** = *anum* no longer knew the Sumerian doctrine. Comparison with the god lists from Nippur and TCL XV 10 makes clear that this doctrine was not an invention of its authors; that doctrine was incorporated into these lists.

1.2.1.4 *The expression* **u_4-ri-a**

According to van Dijk, the expression **u_4-ri-a** is closely linked with the cosmic motif. He has made several statements about this expression:

- Sumerian cosmogonic events centre around the expression **u_4-ri-a** [59].
- The cosmogony with a cosmogonic motif precedes **u_4-ri-a** '*in illo die*', the moment of the separation of the heaven from the earth [60].
- The expression **u_4-ri-a** is even more than the deluge the prototype of all destructive violence. The day of 'earth in labour' is the day of violence *par excellence* [61].
- **u_4-ri-a** is the day of the birth of humanity [62].

1.2.1.5 *Anthropogeny via emersio and formatio*

In connection with the expression **u_4-ri-a**, van Dijk discussed the origin and the birth of mankind. As already mentioned: '«Ce jour-là» est le jour de la naissance de l'humanité.' In the Sumerian tradition he distinguished two different explanations for the birth of man: *emersio* and *formatio*.

His definition of *emersio* is: Man emerges from the earth after the earth has been fertilized by the heaven [63]. Evidence for this *emersio* can be found in the following texts:

---

59 Van Dijk 1964,16: 'Les événements de la cosmogonie sumérienne se centrent autour de l'expression **u_4-ri-a** «ce jour-là»'.

60 Van Dijk 1976, 128.

61 Van Dijk 1964, 21: 'Plus encore que le déluge, «ce jour-là» «**u_4-ri-a**», est le prototype de toute violence destructive. Le jour de la «terra parturiens» est le jour de violence par excellence.'

62 Van Dijk 1964, 23: '«Ce jour-là» est le jour de la naissance de l'humanité.'

63 Van Dijk 1964, 23: 'L'homme surgit de la terre après que celle-ci a été fécondée par le ciel'.
Van Dijk (1971, 489): 'Aber das Motiv der Erschaffung des Menschen ist ein Leitmotiv in der religiösen Vorstellungswelt der Sumerer. Auf dieses Motiv gründet sich eigentlich die ganze Anthropologie der Sumerer: die Menschen sind geschaffen, um die harte Arbeit der Götter zu übernehmen, um die Götter zu versorgen.'
Comments: It is true that man has been created to take over the heavy burden of the gods, but the creation of man can only be found in the literature of the second millennium; in the third millennium texts the creation of man is not a topic at all.

1. 'Enki's Journey to Nibru', ll. 1-3 [64]. In these lines, according to van Dijk, man has been created by An; consequently this story is 'cosmic' and probably reflects the theology of Uruk.
2. 'The Song of the Hoe', ll. 18-20 [65]. This is only an aberrant version from the former one, in which Enlil replaces An.
3. The *emersio* motif is also shown in l. 10 of 'Gilgameš, Enkidu and the Netherworld': 'als der Same der Menschheit (in die Erde) gelegt war' [66].

Van Dijk's definition of *formatio* is: Namma, 'a form of the earth', brings forth man, i.e. she gives life after having formed his image of clay [67]. The example, of course, is the story 'Enki and Ninmaḫ' or 'le récit d'Eridu' as he called it. In his comment on this story, van Dijk wrote [68] that, if Namma is a manifestation of Mother Earth, then man originates – also in the Eridu theology – from the womb of the earth via modum partus. This creation story differs profoundly from those in 'Enki's Journey to Nibru' and in 'The Song of the Hoe': it is completely chthonic, and the birth from Namma is preceded by a formatio by Enki, the chthonic god *par excellence.*

Finally van Dijk discusses shortly two lines from 'The Sumerian Flood Story': **^{d}en-líl ^{d}en-ki ^{d}nin-ḫur-saĝ-ĝá-ke$_4$ saĝ-gi$_6$-ga mu-un-dím-eš-a-ba** 'Lorsque Enlil, Enki et Ninḫursaga eurent créé les hommes' [69]. In his opinion these lines indicate to what extent the stories about the creation of man have been syncretized.

---

64 Van Dijk 1964, 23-24. His explanation of these lines is based, inter alia, on a misreading in the first line: **a-ri-a** instead of **u$_4$-ri-a**. This **a-ri-a** should be a derivation from the verb **a-ri** 'fertilize', with An as subject, as is proved by the second line **an ù-tu-da**. Based on these suppositions van Dijk says: 'L'homme est né de la terre fécondée par le ciel comme l'étaient les plantes dans cette «année d'abondance»'.

Translation of van Dijk:	*My translation*:
1. Lorsque le destin fut fixé de tout ce qui était engendré (par An),	On that remote day, when the fate has been determined,
2. qu'An eut engendré l'année d'abondance,	the year that he (= *Enki*) has brought about abundance,
3. que les hommes brisèrent la surface de la terre comme les herbes,	(and) that he has made it (= *abundance*) break through the earth for the people like green plants,

65

Translation of van Dijk:	*My translation*:
19. (Celui qui..) mit l'individu humain dans le moule;	he placed the first of humankind in a brick-mould.
20. devant Enlil, son peuple de Sumer surgit brisant la surface de la terre.	In favour of his Land he (*the first one*) split open the earth towards Enlil.

For an account of my translation and explanation of these texts: see ch. 4.3.3.

66 Van Dijk 1971, 488. My translation of this line is: '(after) the name of mankind has been established on it'.

67 Van Dijk 1964, 23: 'Nammu (une forme de la terre) donne naissance à l'homme, c.à.d., elle donne la vie, après avoir formé son image d'argile'.

68 Van Dijk 1964, 30.

69 Poebel 1914, PBS V 1, *i*: 13-14; Civil 1999, 140, ll. 47-48; this study: ch. 2.1.8.2, ll. *i*: 11'-12'. Van Dijk (1964, 31) by mistake has interpreted the first sign **AN** as being the determinative **diĝir** of Enlil; these lines read: **an ^{d}en-líl ^{d}en-ki ^{d}nin-ḫur-saĝ-ĝá-ke$_4$ saĝ-gíg-ga mu-un-dím-eš-a-ba** 'After An, Enlil, Enki (and) Ninḫursaĝa had formed the black-headed people'.

In his article about Sumerian religion, van Dijk discussed also the creation of man [70]. With reference to 'Enki and Ninmaḫ', he wrote that there is no question of a god who has been slain and whose blood had to be mixed with clay, as happens in the Akkadian creation myths of man. In 'Enki and Ninmaḫ' man is created in the womb of Mother Earth. Discussing the anthropology of the Sumerians, van Dijk writes that the relation between god and man is twofold. On the one side man is a slave of the god; on the other side there is a family relationship: man has been created from the semen of the god or even from his blood, the immortal principle of life.

1.2.1.6 *Summary and conclusions of van Dijk*

At the end of his study of 1964, van Dijk summarizes his theories and encounters some difficult questions [71].

What has been called 'le motif cosmique', a cosmic sacred marriage, has become the formative element of the systematic pantheon. There has been 'une autre tendance formative' of the theological idea, which can be found in the Inanna hymns of Enḫeduanna [72]. This other trend has influenced the cosmic system of the Babylonian pantheon, the result of which is evident from the list **an** = *anum*. Then van Dijk remarks that the Nippurian theologians, so devoted to Enlil, have not succeeded in eliminating this other theological idea, which probably had originated in Uruk [73].

According to van Dijk, the only serious alternative for the cosmic motif originates from Eridu. The pluralism of the Sumerian idea is based on two principles:

1. The chthonic motif: the **abzu** and Mother Earth form the principle of life. The *formatio* of man corresponds with this system. As a hypothesis this doctrine may be attributed to the sedentary and agricultural population.
2. The cosmic motif: the interdependence of the universe is the principle of life; 'Ciel et Terre se fertilisent mutuellement.' With this system corresponds the *emersio* of gods and man. The cosmic hierogamy has been derived from it. This doctrine may stem from the milieu of the desert-dwellers, whose life depends on pastureland. In this way it might be explained that the cosmic religion and the cult of Dumuzi are found together in Uruk.

About these two motifs says van Dijk, that both are opposed to each other, and that the one does not form an answer to the other. They do not constitute a harmonious system in themselves [74]. Moreover he finds it astonishing that the theologians have had a preference

---

70 Van Dijk 1971, 489. About the relation god - man: 'Das Verhältnis Gott-Mensch ist latreutisch, d.h. ein Sklavenverhältnis. Auf der anderen Seite ist es auch ein "Familienverhältnis": der Mensch wird aus dem Samen des Gottes geschaffen, oder sogar aus seinem Blut, dem unsterblichen Lebensprinzip.'

71 Van Dijk 1964, 57-59.

72 Van Dijk has not indicated in which lines exactly he has found this idea. One possibility is line 115: 'how supreme you are now over the Anunna, the great gods.' (Translation 'The Exaltation of Inana', ETCSL 4.07.2). But it cannot be excluded that he alluded to the general tenor of the whole text.

73 'Les théologiens de Nippour si dévoués à la cause d'Enlil n'ont pas réussi à évincer cette pensée théologique fondamentale qui n'était probablement pas le fruit de leur école mais de celle d'Uruk.' (van Dijk 1964, 57).

74 'Ces deux motifs sont en opposition l'un à l'autre; l'un n'est pas la réplique de l'autre: ils ne forment pas d'eux-mêmes un système harmonieux.' (van Dijk 1964, 58).

for the cosmic motif. Perhaps the reason therefore is – according to van Dijk – that the nomadic population of Mesopotamia always had dominated the sedentary population [75].

Finally the chthonic motif has become integrated with the cosmic doctrine. The syncretism of both systems may be symbolized by the marriage of An (Heaven) with Namma (Mother Earth in the Eridu theology; the mother of heaven and earth).

1.2.2 *Pettinato: anthropogeny* [76]

In his monograph 'Das altorientalische Menschenbild und die sumerischen und akkadischen Schöpfungsmythen', Pettinato tried to answer two principal questions: 1. Why was man created? and 2. What is the human being? Therefore he analysed several Sumerian and Akkadian texts [77]. His conclusions are the following.

The cuneiform texts agree on the first question: Man has been created in order to work, in order to relieve the gods of their heavy task [78].

In order to answer the second question, the creation texts have to be examined because, as Pettinato writes, we cannot expect from the Mesopotamians that they give a definition of man like the Greek philosophers [79].

The manner in which man was created may tell us what he was in the eyes of the Mesopotamians. There appeared to be two concepts of creation in the Sumerian myths. Pettinato follows the study of van Dijk and uses his terminology: *emersio* 'sprouting forth' which happens in two phases [80], and *formatio* 'forming'.

The first phase of the creation by *emersio* entailed that mankind, just like plants, sprouts forth from the earth, after An has impregnated Mother Earth. The second phase of this creation is the introduction of civilization.

On the basis of the texts used, Pettinato made some conclusions [81]. The Sumerians never have given a definition of the human being; instead they describe his activities and his place in the world order. For the Sumerians the human being is an 'ens rationale et sociale'. After the first creation phase man behaves like an animal; only in the second phase does he get the instruments for the establishment of culture. Finally he becomes 'human' by

---

75 Van Dijk did not explain or prove this statement on the nomadic domination.

76 Pettinato 1971; philological notes and comments with respect to Pettinato's translations have been made at the respective translated texts in this study (Appendix Text editions). For reviews of 'Pettinato 1971': Cooper 1973a; Hruška 1974; Jestin 1972; Kümmel 1973-1974; Lambert 1972.

77 The most important texts are: 'Enki and Ninmaḫ'; KAR 4; 'The Song of the Hoe'; 'The Debate between Sheep and Grain'; 'Lugal-e'; 'The Sumerian Flood Story'; '*atra-ḫasīs*'; '*enūma eliš* '.

78 Pettinato 1971, 21: 'Der Mensch wurde erschaffen, um zu arbeiten, und zwar um die Götter von ihrer schweren Arbeit zu entlasten.'

79 Pettinato 1971, 29: 'Von den Sumerern und den Akkadern dürfen wir nicht erwarten, daß sie eine Definition des Menschen nach dem Vorbild der griechischen Philosophie geben! Deshalb müssen wir die Schöpfungstexte befragen, denn nur aus der Art der Schöpfung können wir entnehmen, was der Mensch für die Babylonier war.'

80 Pettinato (1971, 31-35) refers for the *emersio* to the following texts: 'Enki's Journey to Nippur', 'The Song of the Hoe', KAR 4, 'The Debate between Grain and Sheep', 'How Grain came to Sumer', 'The Sumerian Flood Story'.

81 Pettinato 1971, 35: 'Nach dem ersten Akt der Schöpfung stand das menschliche Wesen auf derselben Stufe wie die Tiere, erst nach dem zweiten Akt, durch das Geschenk der Hilfsmittel zum Aufbau der Kultur, und nachdem die Götter ihm den Lebensodem, hier im Sinne vom griechischen *noũs* zu verstehen, verliehen hatten, wurde es zum "Menschen" '.

a gift from the gods: that what the Greek name νόος 'spirit'. This concept is likely to represent the mentality of the sedentary population. Some texts [82] are very reminiscent of the description of the nomadic Amorites by the sedentary Sumerians in their literature. The Amorites repeatedly are called 'those who do not know grain', or 'those who do not know houses or cities' [83].

The creation by *emersio* is connected with the Nippur cosmogony [84]. Pettinato follows van Dijk's theory and summarizes it as follows. According to the Nippur cosmogony, heaven and earth were coupled. Enlil, the air, caused their separation. Then earth received An's semen and she bore the gods, men and animals. For the birth of man by An's seed Pettinato refers to the lines 1-3 of 'Enki's journey to Nibru'.

Creation by *formatio* is linked with the Eridu cosmogony [85]. The central god is Enki, the son of An/heaven and Namma, the female freshwater. Enki creates everything without the agency of An. The vegetation emerges due to the periodical union between Enki and Mother Earth (= Ninḫursaĝa). Man has been created by Enki and Namma with the aid of clay; the role of Ninmaḫ in this process is not completely clear, but the relation between creation and birth is made understandable by her presence [86].

According to Pettinato the creation by *emersio* is absent in the Akkadian texts. In *atra-ḫasīs* and in *enūma eliš*, man was created by *formatio*, with the elements clay and divine blood and flesh (*atra-ḫasīs*) or with divine blood alone (*enūma eliš*). In the opinion of Pettinato, the text KAR 4 is an attempt to harmonize both creation traditions – *emersio*

---

82 Pettinato refers here to 'The Debate between Grain and Sheep', ll. 20-25, and 'How Grain came to Sumer', line 1.

83 Pettinato 1971, 36, notes 120 and 121, respectively.

84 Pettinato 1971, 62-63: 'Nach der Nippur-Kosmogonie waren Himmel und Erde zunächst verbunden wie in einer Ehe. Enlil, die Luft, veranlaßte die Trennung von Himmel und Erde, und erst dann gebar diese die Götter, die Menschen und die Tiere, nachdem sie aber den Samen Ans empfangen hatte. Daß die Menschen von dem Samen Ans gezeugt wurden, erfahren wir aus den ersten drei Zeilen des bereits Zitierten E'engurra-Preisliedes (= Enki's Journey to Nibru, ETCSL 1.1.4; *JL*):

'Als allen Gezeugten das Schicksal bestimmt wurde,
als in einem Jahr des Überflusses, das An geschaffen/gezeugt,
die Menschen wie Pflanzen die Erde durchbrochen hatten.'

Dasselbe, wenn auch indirekt, läßt sich aus dem $^{\hat{g}iš}$**al**-Mythos herleiten: Auch hier trennt Enlil den Himmel von der Erde, legt den Samen der Menschheit in eine Spalte, woraus dann die Menschen hervorsprießen.

Nach dieser kosmogonischen Vorstellung kommt alles wahrhaftig von oben; und die Menschenschöpfung ist im Grunde genommen eine Geburt der Muttererde, so daß das Verhältnis zwischen Schöpfung und Geburt verständlich ist.'

85 Pettinato 1971, 63: 'Die Eridu-Kosmogonie kreist dagegen rund um Enki, der im Apsû wohnt. Enki ist zwar der Sohn Ans, des Himmels, und der Nammu, des weiblichen Süßwassers, doch er vollbringt die Schöpfung ohne Zutun seines Vaters. Die Vegetation entspringt aus der periodischen Vereinigung Enkis mit der Mutter Erde (= Enki & Ninḫursaĝa; JL). Der Mensch seinerseits ist von Enki und Nammu mit Hilfe der Muttererde erschaffen worden (= Enki & Ninmaḫ; JL). Dabei ist die Rolle Ninmaḫs, der Muttererde, nicht ganz eindeutig, doch ihre Anwesenheit und ihr Zutun machen das Verhältnis zwischen Schöpfung und Geburt wiederum verständlich.'

86 See also my discussion of the role of Namma and that of Ninmaḫ and their respective tasks in 'Enki and Ninmaḫ' in ch. 4.4.3.

and *formatio* – [87]; the blood of the gods has not been used in the creation of man, as in the Akkadian tradition, but it has the function of 'the seed of the humanity' which emerges from the earth.

Comments on Pettinato's study

Pettinato's study has been reviewed by several scholars. The main conclusions of the most important reviewers are given below.

Cooper criticized what is, in his opinion, the inaccurate use by Pettinato of the terms Sumerians and Akkadians [88], instead of speaking about 'Sumerian and Akkadian literature'. As none of the Sumerian texts discussed by Pettinato antedate the Old Babylonian period, it is anachronistic to speak of Sumerians at that time. Although many texts may go back to earlier times, it is highly questionable, according to Cooper, if even then it is possible to speak about two fundamentally different population groups [89]: 'Thus, any differences between the traditions of Sumerian and Akkadian creation texts should be seen for what they are, and not projected onto fictitious or, at best, shadowy population groups.'

Lambert wonders why there should be two stages of man's creation [90]: 'first man was made and secondly he was civilized by the express intervention of the gods' (*Pettinato, pp. 30-9*). Becoming civilized is not a matter of creation [91]. Moreover Lambert has the same objection as Cooper: 'The other chief ideological difficulty arises from the attempt to distinguish between Sumerian and Akkadian concepts of creation.'

Hruška also does not agree with Pettinato's theory in making a sharp division in a Sumerian and an Akkadian tradition with respect to the creation of man; it is an oversimplified picture of the Mesopotamian civilization and culture [92].

---

87 Pettinato 1971, 61: 'Hier tritt nämlich, als Element aus dem der Mensch besteht, das Blut der Götter auf, jedoch nicht in der Funktion eines Bildungselements, wie in der akkadischen Überlieferung, sondern als 'Samen' der Menschheit, die aus der Erde hervorsprießt.'

88 Cooper 1973a, 583.
Cooper (1973, 585) also criticized the 'attitude toward life and work' of Sumerians and Akkadians, one of the topics of Pettinato's study: 'The reviewer's position, then, can be summarized as follows: there is no basic difference in attitude toward life and work or their ultimate value in the Sumerian and Akkadian creation myths. They both recognize the difficulty of human existence as well as its importance in the cosmic order. Nothing, in any case, justifies speaking in terms of an optimistic, positive, sedentary Sumerian on the one hand, and a pessimistic, negative, originally nomadic Akkadian on the other.' The present author fully agrees with Cooper's opinion.

89 Cooper 1973a, 583a, referring to Pettinato 1971, 17. Cooper (1973b, 242) supposed that already since ED III a displacement of Sumerian as a spoken language was in progress.

90 Lambert 1972, 135a.

91 I agree with Lambert's view. In some texts there is mention of people behaving like animals ('The Debate between Grain and Sheep', ll. 20-25 [ch. 2.1.6]; 'How Grain came to Sumer', ll. 1-2 [ETCSL 1.7.6]). They are indicated as **nam-lú-ulu$_3$** and **ùĝ**, respectively, thus as real man. Perhaps the reference to the uncivilized status of mankind has been caused by confrontation of the Sumerians with people like e.g. the Guti, who in the Sumerian opinion may have been 'barbarians', people with a civilization standard lower than their own.

92 Hruška 1974, 274a: 'Obwohl die einzelnen keilschriftlichen Quellen zur Menschenschöpfung in zwei Sprachen geschrieben sind und ihr Inhalt mit der Zeit ab und zu Änderungen und Neuerungen aufweist, kann man doch nicht von zwei grundverschiedenen Überlieferungen von der Erschaffung des Menschen sprechen.'

Kümmel, in his review, raised also other objections. First of all he mentions methodological shortcomings. Pettinato applies an unreflected idea of myths to stories of man's creation, which stories are all very different. There is no attempt to determine the so-called 'Sitz im Leben' [93]. Moreover Kümmel pointed to dubious interpretations, such as the positive appreciation of labour by Sumerians and the negative one by Akkadians, and also to an oversimplified division into 'Sumerian' and 'Akkadian' texts [94].

1.2.3 *Lambert: cosmogony – theogony*

On several occasions Lambert has written about the Mesopotamian cosmogony [95]; his contribution to the Reallexikon gives a good overview of his ideas, which will be summarized and commented upon. In the light of the absence of an extent Sumerian story about the beginnings, Lambert remarked that 'one-sentence myths and allusions have as much importance as lengthy epic-style narratives.' [96]. Then he continues: 'As in most philosophy and science, ancient Mesopotamian thinkers tended to assume that everything known went back to a single element in the beginning. Three such elements are attested in Sumerian and Babylonian texts, (1) Earth, (2) Water and (3) Time. A fourth element, Heaven, occurs, but not with the same status as the three.'

1. *Earth.* This prime element occurs most often in the ancestry of Enlil in the form of a list of **en-nin**-pairs, leading from Enki and Ninki to Enlil and Ninlil. All in the list save Enlil and his spouse Ninlil can be summed up as 'the Enkis (and the Ninkis)'. Thus the purpose of the list was to trace Enlil's descent from **ki** 'earth'. A similar origin for Anu and Antu is given in **an** = *anum* I 4-23, where Uraš ('Earth') and Nin-uraš ('Lady Earth') head a nine-pair ancestry.

A more complex attestation of Earth as a prime element occurs in the Mother Goddess. One of her many Sumerian names is Ninḫursaĝa. As responsible for the births of both gods and men one could expect the Mother Goddess to have a cosmogonic function. In **an** = *anum* I 25 she is described as Anu's spouse, under her Akkadian name Bēlet-ilī, which makes her Earth, to match An/Heaven. In the single-line ancestry of Anu (i.e. TCL XV10; *JL*) Uraš ('Earth') seems to be equated with Bēlet-ilī.

---

93 Kümmel 1973-1974, 26-27. 'Hier wird ein völlig unreflektierter Mythos-Begriff unterschiedslos angewandt auf die Berichte von der Menschenschöpfung in so verschiedenen Texte wie 'Enki und Ninmaḫ', der sumerischen Sintfluterzählung, theologisch-lehrhaften Dichtungen (Lobpreis der Hacke, **u₈** und **ašnan**), dem Epos von Atraḫasīs und gar auf einen Text esoterischer Geheimlehre wie KAR 4. Dabei fehlt jeglicher Versuch, die innere Gesetzmäßigkeit literarischer Gattungen und ihre literarisch-historische Entwicklung, den "Sitz im Leben" des jeweiligen Schöpfungsberichts im literarischen Kontext, zu bestimmen.'

94 Kümmel's criticism (1973-1974, 28-29) that several texts that Pettinato used (viz. 'The Song of the Hoe', 'KAR 4', and 'The Sumerian Flood Story') should not mention the reason for the creation of mankind, is not correct for some texts. This reason can be found in 'The Song of the Hoe' l. 31 'she makes them provide for the daily rations of the gods'; and in 'KAR 4' *obv.* l. 21: 'Let the work assignment of the gods be its job (= *the job of mankind*)'. The tablets with 'The Sumerian Flood Story' are too damaged to be able to say that this text is missing the fundamental reason for man's creation.

95 Lambert 1974; 1975a; 1980-1983a; 2008; 2010.

96 Lambert 1980-1983a, 219a.

2. *Water*. As a prime element, the term 'water' does not normally occur, but Sea, River, and other terms are used. The best known example occurs in *enūma eliš*, where a bisexual ancestry is starting with Apsû and Ti'amat. The watery beginning is taken over from a single-line ancestry of Anu found in an Old Babylonian forerunner to **an** = *anum*, TCL XV 10. It traces his origin back through Uraš ('Earth', and apparently identified with Bēlet-ilī) to Namma, apparently meant also in the name Amatuanki. She is a single prime element, presumably derived from the traditions of Eridu, in contrast with the tradition of bisexual beginnings, which no doubt came from Nippur.

3. *Time*. A third, and much less common, prime element from which everything in the universe evolved, was Time: eternal Time, for which there are various terms. It too was expressed in a bisexual pair Dūri Dāri 'Ever and Ever'. This first appears in an incantation – dating to Old Babylonian times in the Elamite language [97]– atypically serving as the ancestry of Enlil. It is probable that another cosmogonic pair, (Ḫ)alma and (Ḫ)al(l)ama, also express eternal Time. They occur in Anu's ancestry in **an** = *anum* I 20-21 written ^{d}ALAM.

4. *Heaven*. Heaven as a prime element is not on the same level as Earth, Water and Time, since it usually appears only with Earth, or is the god An who is acting outside the physical elements. Rarely however is the god An a prime element.

5. *Heaven and Earth*. Another widely held view had creation begin from the combined operation of heaven and earth. Here again creation is modelled on animal reproduction: Father Heaven copulates with Mother Earth presumably by sending down his fertilizing rain into her bosom. Of course agriculture in southern Mesopotamia does not depend on rainfall, so the origin of this myth must be sought outside, and perhaps for this reason the 'marriage of Heaven and Earth' is not a major item of mythology, but survives only in allusions and with its original function disguised or lost [98].

General comments on Lambert's ideas

The first objection against this overview, and in general against Lambert's main articles about Mesopotamian cosmogony, is the absence of any diachronic study: he refers to Sumerian and Akkadian texts, thereby mixing up texts of both languages and of all periods. Presented in this way, it may seem as if all the prime elements mentioned can be found in texts of every period. Another conclusion must be that Lambert's *cosmogony* is only formulated in terms of *theogony*.

1.2.4 *Other scholars*

Until now I have discussed in this chapter the theories with respect to the Mesopotamian ideas about cosmogony, theogony and anthropogeny of those scholars who have published

---

97 Lambert 2008, 32.

98 Lambert (1975a, 47) says about this cosmic marriage: 'The idea that life arose from the marriage of heaven and earth is a widespread one, and often implies the shooting up of plant life as the result of father heaven sending down his rain into the bosom of mother earth. This idea occurs only rarely and marginally in Sumero-Babylonian texts. It is not the centre of any major text.'

in detail about these subjects: van Dijk, Pettinato and Lambert. All other authors writing about the 'Sumerian Beginnings' refer to one or more of these scholars [99]. In an exceptional case a minor detail is added, without deviating from the outlines. An example of such 'sophistication' has been given by Luginbühl [100]. With respect to anthropogeny, Luginbühl added a subdivision in the 'formatio' type, viz. the '*sacrificatio*' type. *Sacrificatio* is the creation of man with the aid of divine blood; it is always connected with *emersio* (in the text KAR 4) or with *formatio* (in *atra-ḫasīs* and in *enūma eliš*).

In his articles about 'Mesopotamian Beginnings', Dietrich based his theories on those of van Dijk and Lambert [101]. His goal is to make some refinements, especially of van Dijk's concept.

***

---

99 E.g. Chiodi (2008).

100 Luginbühl 1992, 30-31: 'Der Typ "sacrificatio" bezeichnet die Menschenschöpfung aus dem Blut eines Gottes oder einer Gruppe von getöteten Göttern. Er erscheint niemals isoliert, sondern verbindet sich mit dem Typ 'emersio', so im KAR 4-Mythos, oder "formatio", so im Atraḫasis und Enuma Eliš.'

101 Dietrich 1984; 1991; 1993 a-b; 1994; 1995; 2007. Dietrich's translations of the respective texts were based on those of van Dijk ('GEN'; NBC 11108), Pettinato ('Enki & Ninmaḫ'; 'The Song of the Hoe'; KAR 4), and Lambert-Millard (Atra-ḫasīs). His theories are also not free from 'sophistication'. For example: He discerns two phases in the primaeval time. The first phase is an undivided cosmos in which the gods exist 'gewissermaßen präexistent' within a primaeval city – **uru-ul-la** –, which phase he called 'die Embryonale Welt'. In the second phase – 'die Jetzt Welt' – heaven and earth are separated. The term **u$_4$-ri-a** characterizes the change from phase 1 into phase 2. Because of the threefold formulation – **u$_4$-ri-a** / **ĝi$_6$-ri-a** / **mu-ri-a** –, is Dietrich's opinion: ' ... legt es sich nahe, dort, wo die Dreiergruppe "Tag - Nacht - Jahr" auftritt, eher an einem "Schöpfungszeitraum" als an einen "Schöpfungstag" zu denken.' (Dietrich 1995, 57).

Chapter 2

The Sumerian story of the Beginnings: the texts

'The faculty in man, then, which is susceptible to and also generates myths is more than merely an archaic stage of cognitive development or a primitive curiosity about how things work; it is rather an alternative mode of consciousness, with an a-priori, instinctive impulse toward this different, sacred mode of comprehension.'
[S. Larsen, *The Shaman's Doorway*; 1998, 28]

The Sumerian texts dealing with cosmogony, theogony and/or anthropogeny (or the creation of man) will be discussed separately in this chapter, not in a philological way [102], but with respect to their content. What does a specific text tell and what does it not? Who are the protagonists? What are the similarities and what the differences between the several texts? Does the main point or the content of 'the Beginning' change in the course of time (from Early Dynastic times until the Old Babylonian period)? In addition the most important items of the Sumerian texts will be summarized in a table.

Finally the conclusions that can be drawn from these texts will be formulated. Since the texts collected and edited in this study cover an era of about a millennium [*ca.* 2500 BCE - Kassite period], it is justifiable to investigate whether changes in the Sumerian ideas about 'origins' can be demonstrated.

2.1 *The texts*

2.1.1a *IAS 114* [*Abū Ṣalābīḫ; ca. 2500 BCE*]

column i

	UGN-orthography [103]	'Normal' orthography	
... Unknown number of lines broken			
1'	**[ ] ⌜UD⌝ ki LAGAB**	**[ ] ⌜an⌝ ki niĝin$_2$**	[ ] **an**/heaven and **ki**/earth together
2'	**ĝiškim NAM$_2$-⌜LU$_3$⌝**	**ĝiškim nam-⌜ĝar⌝**	have placed a sign.
3'	**^{d}GAL-ki ^{d}nin-ki**	**^{d}en-ki ^{d}nin-ki**	Enki and Ninki
4'	**7 àm-ta-tud**	**7 àm-ta-tud**	have brought forth the seven.
5'	**a UNUG šim-LAGAB**	**a ki šim-gin$_7$**	Water for the earth, like something fragrant,
6'	**rig$_7$-dug$_3$ ŠA-NAM$_2$**	**rig$_7$-dug$_3$ na$_5$-nam**	is really a gift.

---

102 In the Appendix Text editions the scores of the texts are given (as far as possible and applicable), and also philological analyses and comments. In this chapter only the main or reconstructed Sumerian text of each composition is present.

103 UGN = UD-GAL-NUN

7’	**a pú-šè**	**a pú-šè**	Water for a well
8’	**rig$_7$-dug$_3$ ŠA-NAM$_2$**	**rig$_7$-dug$_3$ na$_5$-nam**	is really a gift.
9’	**^{UD}GAL-NUN**	**^{d}en-líl**	Enlil
10’	**AMA-a**	**tud-a**	has been brought forth
11’	**^{d}GAL-UNUG ^{UD}nin-ki**	**^{d}en-ki ^{d}nin-ki**	by Enki and Ninki.
12’	**^{d}GAL-UNUG AMA-a**	**^{d}en-ki tud-a**	Enki has been brought forth
13’	**UD-GIŠGAL *ù* ki-EREN-TUKU**	**an-uru$_{16}$ *ù* ki-dilmun**	by the mighty An and the luxuriant earth.
14’	**^{UD}ŠEŠ-KI AMA-a**	**^{d}nanna tud-a**	Nanna has been brought forth
15’	**^{UD}KIŠ-NUN**	**^{d}en-líl**	by Enlil
16’	**^{UD}nin-KID**	**^{d}nin-líl**	and Ninlil.

Before discussing the text of this tablet from Abū Ṣalābīḫ it is useful to repeat that duplicates of it were found in Fara [104]. If Deimel is right that about 9 lines are missing in the first column of VAT 12680 [105], then in IAS 114 about 4-6 lines may be broken [106]. Each line of the remaining text contains only a small part of the story; sometimes even one sentence is divided over three lines (e.g. 9’-11’; 14’-16’). This means that just a small part of the beginning of the text is missing, perhaps no more than a traditional introduction in primaeval times.

In the cosmogonic introductions of the other third millennium texts **an** and **ki** are shouting together, they lie together, or they have intercourse (see table 1). In the text IAS 114 **an** and **ki** are cooperating to bring forth omens. This very notion can also be found in a much later, Neo-Assyrian text [107], which has been designated as ‘SB instruction to the interpreter of omens’ [108]. In this text we read: *šamê u erṣetim ištēniš giskimma ubbalūni* ‘Heaven and Earth all together bring forth omens’ [109]. This Neo-Assyrian text continues:

---

104 SF 37 (= VAT 12680) and SF 38 (= VAT 12766) [Deimel 1923, 33-35].

105 See also the photograph of VAT 12680 at CDLI no. P010619.

106 The dimensions of the tablet IAS 114 (AbS-T 247) are given by Biggs (1974, 104) as 20 cm in height and 24.5 cm in width, but the top and right side are broken. The left edge is intact, indicating that there are no columns missing at the beginning. Another tablet (AbS-T 230, IAS 61) measures 23.2 h. x 24 cm w., with the height complete and the width nearly complete. Of tablet AbS-T 228 (IAS 116) the full dimensions are known: 26.5 h. x 26.8 w. From these data it may be concluded that the original height of IAS 114 could have varied between 24 and 27 cm. In other words: probably no more than 4 - 6 lines (each with a height of 1.1 mm) are missing in column *i*.

107 Virolleaud 1911, 111, lines 39-40; there are several duplicates.

108 CAD G, 98 *ad* ‘*giskimmu*’.

109 Instead of the translation *ištēniš* ‘alike’ (in CAD G, 98 *ad* ‘*giskimmu*’) the present author prefers ‘all together’, ‘jointly’, meanings given in CAD I-J 279.

[a-ḫi]-en-na-a ul **BAR-MEŠ** *šamê u erṣetu itḫuzū* 'they cannot be separated from each other: Heaven and Earth are interconnected'. To express this 'interconnection' the Gt-stem of the verb *aḫāzu* [110] has been used. This fact, and also the other cosmogonic introductions which tell about **an** and **ki** – their shouting or lying together, their intercourse – may give a more profound meaning to **LAGAB** = **niĝin₂** in line *i* : 1' of IAS 114: probably it points to an intimate relationship between **an** and **ki**. The introduction of **ĝiškim** in the very beginning may be explained by the fact that this text is also dealing with extispicy; several lines tell about performing an extispicy (viz. in column *iii*) [111].

The identity of 'the seven' [112] who have been brought forth by Enki and Ninki (ll. *i*: 3'-4'), the pair that also produced Enlil (ll. 9'-11'), has not become clear from the text. As will be clear from the discussion of the god lists [113], the group of seven probably belongs – at least in later times – to the ancestor group of Enlil. These seven must not be interpreted as 'brothers and/or sisters', but as a genealogical tree, at the head of which are Enki and Ninki. According to this text, the youngest descendant of this tree is Enlil. The lines 9'-11' mention: 'Enki and Ninki have brought forth Enlil'. The verb used to express this is **tud**, which may indicate that Enki and Ninki are the parents of Enlil; if my hypothesis about the seven as an 'ancestor group' is correct, then Enki and Ninki – as primary ancestors – are the 'ultimate parents' of Enlil.

The lines *i*: 5'-8' about 'water being a gift' may sound rather enigmatic placed between the production of an omen and the birth of Enlil. A tentative explanation may be that in this line we are told that water from now on could be applied in a useful way: for the earth, so that it would become fertile, for people, who could draw it from a well [114]. In a more detailed way the use of water for 'cultivation' was described in Ukg 15, *i*: 2-4 and NBC 11108 *obv.* 2.

The next passage concerns the birth of Enki. The mention of this birth shows also a remarkable detail, insofar as **an** and **ki** are Enki's parents. In particular the mention of **ki** as his mother is a unique feature. **an** and **ki** are not mentioned as the parents of Enki in any other extant text, except for IAS 114. Only the epithet of Enki, **ᵈam-an-ki**, reminds us of this ancestry [115]. In the Old Babylonian text 'Enki and Ninmaḫ' there is mention of Namma as the mother of Enki; this is one of the rare attestations of 'the mother of Enki'. The god list **an** = *ᵈa-nu-um* does not mention the parents of Enki, only giving Namma [116] as his mother. From the OB period onwards Namma has become the mother of **an-ki**. It is not unlikely that Namma's motherhood of Enki originated from the same time, in any case after the ED period; in this way Namma as primaeval mother replaced **ki** as the mother of Enki, the lord of the **abzu** [117].

---

110 Gt-stem *aḫāzu* (= *itḫuzu*), AHw 19: 'für die Dauer nehmen;' CAD A I 173: 'to be interconnected'. The verb *aḫāzu* has, as one of its meanings, 'to marry'.

111 At this moment the complete text of this UGN-tablet has not yet been translated to the best of my knowledge.

112 The number 'seven', being a holy number, has not to be taken too literally. See also e.g. the Barton cylinder *ii*: 9-10.

113 See chapter 3 on 'God lists' in this study.

114 Water in the beginning, being the primaeval sea, i.e. Namma, will be discussed in the chapters 3 and 4.

115 See also ch. 2.2.3, note 256, about the name **ᵈam-an-ki**.

116 Litke 1998, I:27.

117 In chapter 4 the position of Namma as primaeval mother will be discussed in detail.

2.1.1b *IAS 136, 113, 203* [*Abū Ṣalābīḫ; ca. 2500 BCE*]

IAS 136

iii	UGN-orthography	'Normal' orthography	
1'	**^{UD}GAL-NUN**	**^{d}en-líl**	Enlil,
2'	**an UNUG-ta bad**	**an ki-ta bad**	who has separated heaven from earth,
3'	**ki an-ta bad**	**ki an-ta bad**	who has separated earth from heaven.

IAS 113

ii	UGN-orthography	'Normal' orthography	
5	**^{UD}KIŠ-[NU]N**	**^{d}en-líl**	Enlil,
6	**GAL nu-nám-NAGAR**	**en nu-nam-nir**	Lord Nunamnir,
7	**GAL du$_{11}$-TUKU DU$_{6}$-GAG-GAG**	**en du$_{11}$-ga nu-gi$_{4}$-gi$_{4}$**	the lord who does not revert to an order,
8	**UD UNUG-ta LAGAB**	**an ki-ta bad**	who has separated heaven from earth,
9	**ki UD-ta LAGAB**	**ki an-ta bad**	who has separated earth from heaven,
10	**UD UNUG-ta LAGAB**	**an ki-ta bad**	who has separated heaven from earth.

IAS 203

ii	UGN-orthography	'Normal' orthography	
3'	**[^{UD}GAL/KIŠ]-NUN**	**^{d}en-líl**	Enlil,
4'	**[U]D [k]i-ta [LAGAB]**	**an ki-ta bad**	who has separated heaven from earth,
5'	**ki UD-ta LAGAB**	**ki an-ta bad**	who has separated earth from heaven.

In these few lines from third-millennium UD-GAL-NUN-texts, originating from the north of Sumer – Abū Ṣalābīḫ –, is said that Enlil separated heaven and earth. The mention of it is unique, insofar this separation by Enlil was not mentioned in any other third-millennium text written in the normal orthography which was published until now [118].

---

[118] In ch. 4.1.1 will be tried to find an explanation for this unique mention.

2.1.2 *Ukg 15 (AO 4153)* [*Lagaš; ca. 2400 BCE*]

Obverse column iii

1	**u$_4$-⸢da⸣ im-ma**	One day, in a bygone year,
2	**ul-[lí-a] im-m[a]**	in the long-ago, in a bygone year,
3	**u$_4$ nu-zal**	the sun did not get up early,
4	**ì-ti nu-è-è**	the moon did not appear all along.

column ii

1	**an en-nam šul-le-éš al-gub**	An, as **en**, was standing there as a youthful man.
2	**an ki téš-ba šeg$_{12}$ am$_6$-gi$_4$-gi$_4$**	An/heaven and Ki/earth, in their unity, were shouting.
3	**u$_4$-ba en-ki nun-ki nu-se$_{12}$**	Then Enki and Ninki did not exist;
4	**^{d}en-líl nu-ti**	Enlil did not exist,
5	**^{d}nin-líl nu-ti**	Ninlil did not exist.

column i

1	[...]	[...]
2	**[k]a-<a>-muš ḫa-mu-ni-se$_{11}$-se$_{11}$**	He has lowered the inlets of the irrigation channels in it,
3	**ki-e dilmun-na dalla ḫa-mu-ak-e**	in order to make earth appear in luxuriance:
4	**kiri$_6$ duru$_5$-am$_6$ te-me-nam**	a garden, moist and cool;
5	**ki-buru$_3$ a šè-ma-si**	water has filled the holes in the earth.

The text of this tablet is unique: there is no other copy of it known today. In the editorial part of this text (Appendix) it is explained why I presume that in fact this copy is a 'miscopy': starting with column *i* at the left side, the story is told in reverse.

The tablet originates from the ED IIIb period [119]. This date is e.g. supported by the spelling **AN** for **am$_6$** and **MUNUS-ḪUB$_2$** for **dilmun**.

Despite the time gap between the Early Dynastic and the Old Babylonian time, we will have a brief look at the study of Veldhuis concerning the elementary education at the scribal schools, the Eduba. This study is mainly focused on the Old Babylonian period. The form of tablet AO 4153 corresponds with type IV, one of the four main types of tablets

---

[119] CDLI, no. P315470.

distinguished by Veldhuis [120]. Alas the size of this tablet is not mentioned in the literature, but a 6-8 cm diameter seems possible. This type IV tablet has been used for the teachers' examples and the pupils' copies. Although the way AO 4153 is inscribed, in particular the subdivision in columns, does not correspond to that described by Veldhuis for these type IV exercise tablets, the miscopying, the abrupt ending of the text and the fact that the rest of the tablet is not inscribed, strongly suggest that a pupil was active in this case. Biggs also labels certain tablets from Abū Ṣalābīḫ as exercise tablets: these too have a 'lentil' form, with three or four columns on one side [121].

Before I will try to answer the question of what kind of text we are dealing with, a closer look at the text itself is necessary. Column *iii* places the whole scene in primaeval times, in a bygone time when there was no sun and no moon. Column *ii* shows the existence of **an** and **ki**. Only with respect to **an** are some details told. The first characteristic feature is that **an** is **en**, 'Lord' or 'high priest'; the second one is that **an** is called **šul**, a youthful man; in other words: **an** is in the prime of his life. The second line of column *ii* introduces the other protagonist in addition to **an**, viz. **ki**. Both are shouting. I agree with Michalowski who says about this shouting: 'This is, of course, but a metaphor for sexual union.' [122]. In this respect the words **téš-ba** are very important. For **téš** we find in the ePSD 'unity'. In text editions and translations, **téš-ba/bi** is mostly translated as 'together'. But in my opinion, the deeper meaning of what one would like to express in this Sumerian sentence is better represented by a translation like 'in their unity', i.e. it is a matter of sexual intercourse. This event is designated by van Dijk as 'les noces cosmiques' [123].

Let us now reconsider the meaning of the epithet of **an**, given in line *ii*: 1, viz. **en**. The meaning of the adjective **EN** = **uru$_{16}$** 'exalted' is, of course, derived from the meaning of the noun **en** 'priest, lord'. The adjective might be very appropriate in this case; however, the noun seems to be more preferable due to the context. Cooper and Steinkeller have discussed the sacred marriage [124]. Steinkeller wrote about the royal figure from Uruk art, the man in the net kilt: '(…), the visual contexts in which he appears make it quite certain that he occupied a top supervisory position over the matters related to administration, economy, war, and cult, and therefore, that his office combined equally secular and religious powers. (…), it is generally believed that the royal figure of Uruk art bore the title **en**, (…).' [125] Then Steinkeller describes the development of this archaic **en**-ship into two different and separate powers, viz. a political (**lugal**) and a ritual one (**en**). Further he shows that in Early Dynastic times a high priestly official functions as the consort of the goddess Nanše. In the light of this fact he concluded as a virtual certainty that such too was the role

---

120 Veldhuis 1997, 38-39.

121 Biggs 1974, tablet nos. IAS 318-321, 458-471. A great number of these exercise tablets have a diameter in the range of 6-8 cm. Jagersma informed me that the 'lentil' form of the clay tablets in Veldhuis' typology differs from the tablet form of e.g. Ukg 15. The form of tablet Ukg 15 is the form that was in use for the tablets up to and including the early Akkadian period. Also all pre-Sargonic administrative texts and letters from Lagaš had the same form (personal communication).

122 Michalowski 1998, 240, note 4.

123 Van Dijk 1964, ch. 3, 34-57.

124 Cooper 1993; Steinkeller 1999.
The dissertation of Lapinkivi (2004) will only be mentioned in this note. Apart from the criticisms of George (2006) and Katz (2006) that do not encourage to refer to this study, the basic useful information for my study has been found in the articles of Cooper and Steinkeller.

125 Steinkeller 1999, 105.

of the **en** of Inanna. Steinkeller emphasized that in the Ur III- and Isin-period '(...), the prerogative of active participation in the symbolic nuptials with Inanna was a specific attribute of the **en**-ship of Uruk and not a general feature of Babylonian kingship.' [126] It is very likely that in the text of AO 4153 *ii*: 1-2 we are dealing with the primaeval 'cosmic marriage' [127] of **an** and **ki**. Based on the evidence brought together by Cooper and Steinkeller about the **en**-ship and sacred marriage, the epithet **en** for **an** seems more appropriate here than **uru$_{16}$**. The terminology used in this respect might indicate that the sacred marriage has been used as an image for the 'primaeval cosmic marriage' [128]. Strictly speaking this text does not mention explicitly 'marriage': the only thing that can be said is that there is 'shouting', which may be explained as 'courting'.

If there were consequences from this courting, these are not told to us, at least not in a direct way. The next three lines (*ii*: 3-5) confirm once more that this cosmic courtship of **an** and **ki** is taking place in primaeval times: Enki and Ninki, Enlil and Ninlil do not exist. The last two lines about Enlil and Ninlil may seem somewhat superfluous: as Enki and Ninki, the ancestors of Enlil do not exist, it is not surprising that their offspring – in this case Enlil – also does not exist. However, these lines may be understood as a confirmation that really no gods were present. A striking detail in the enumeration of the gods that are absent is their order: 1. Enki and Ninki, 2. Enlil and 3. Ninlil. As we shall see in the chapter about the god lists, this is precisely the order in which the gods appear: Enki and Ninki are the ancestors of Enlil. It seems as if these lines say: at the moment of the courtship and intercourse of **an** and **ki** these gods are *not yet* present, but as a consequence of this event they will and in this order.

Column *i* then proceeds with an act – the irrigation of the earth – so that earth can appear in luxuriance and is embellished. It does not need much imagination to suppose that in line *i*: 1 An is mentioned as the subject of this operation: An is lowering the inlets of the irrigation channels. The lines 4-5 explain what demonstrates the luxuriance of the earth: a cool and moist garden with water in wells or low-lying areas. Because of the climate of Mesopotamia it is quite understandable that a moist and cool garden must be an ideal situation, almost 'heavenly'. But on reconsideration of the text of column *i* a metaphor becomes visible. As there are no other protagonists mentioned than **an** and **ki**, it is obvious that An himself has lowered the inlet of the irrigation channels for **ki**. The lowering of canal inlets as such has not been a topic in the Sumerian literature [129]. The effect of the lowering of the inlets is inundation. Another 'inundation' is described in a rather expressive way: Enki is filling the Tigris by an ejaculation, a clear sexual act [130]. The inundation of the earth by 'lowering the inlets of irrigation channels' may also be regarded as a metaphor for the sexual union of **an** and **ki**. In the Dumuzi-Inanna songs, which originate from a later date than AO 4153, there are examples of comparable metaphors for intercourse [131].

---

126 Steinkeller 1999, 131. The **en**-ship here only refers to the ritual powers.

127 When 'marriage' (with ' ' marks) is written, then this means that the text does not use an expression for marriage explicitly, but there is at least courting or intercourse.

128 See also the Excursus '**an** x **ki**: the cosmic marriage as model or motif for a ritual?' at the end of this chapter.

129 A search in the literature was negative in this respect. What could be found were some texts with offerings for the mouth of a river/canal: ITT 4, 07738, *obv.* 4 (Lagaš II) [CDLI P217855]; from the Ur III period: TCTI 2, 03184, *rev.* 1 [CDLI P132438]; TAD 49, *obv.* 12 [CDLI P131091]; ITT 5, 06925, *obv.* 4 [CDLI P111632]; ITT 2, 00893, *obv.* ii: 9 [CDLI P110763].

130 'Enki and the world order' [ETCLS 1.1.3], lines 250-259.

131 Sefati 1998, 90-92: 'Metaphors from the Domain of Farming'.

Irrigation therefore may be considered as a metaphor for having intercourse and the consequent ejaculation. The result for **ki** has been described as follows: **kiri$_6$-(...)-àm**, in other words: **ki** has been transformed into a garden. Or: vegetation is the result of the irrigation, that is: the intercourse of **an** and **ki**. And as if this is not enough: 'water has filled the holes (or: the low-lying areas) of the earth'. This may be a suggestion for the origin of wells, but also in a broader sense: for that of lakes and the sea.

As cosmogonic texts about the Beginning mostly are known as introductions to a main text, this justifies the suspicion that the text of AO 4153 also is an introduction, but an introduction to what kind of text? This question cannot be answered until now because of the uniqueness of this text.

To sum up the conclusions that may be drawn from this short 'beginning':
- in the beginning **an** and **ki** are present;
- there are no gods present;
- there is a courtship between **an** and **ki**, presumably followed by an intercourse;
- as a result of this intercourse, water reservoirs and vegetation come into being;
- there is indirect evidence for the origin of gods.

2.1.3a *The Barton cylinder* [*Nippur; ca. 2300 BCE*]

column i

1	**u$_4$-rí-a u$_4$-rí-šè**	On that remote day, until that remote day,
2	**na-nam**	it was indeed;
3	**ĝi$_6$-rí-a ĝi$_6$-rí-šè**	in that remote night, until that remote night,
4	**na-nam**	it was indeed;
5	**mu-rí-a mu-rí-šè**	in that remote year, until that remote year,
6	**na-nam**	it was indeed.
7	**u$_4$ na-du$_7$-du$_7$**	Then a gale was really blowing unceasingly,
8	**nin na-ĝír-ĝír**	there were really flashes of lightning continuously.
9	**èš-nibru^{ki}**	Near the sanctuary of Nippur
10	**u$_4$ na-du$_7$-du$_7$**	a gale was then really blowing unceasingly,
11	**nin na-ĝír-ĝír**	there were really flashes of lightning continuously.
12	**an-né ki-da**	An/heaven is shouting $^{(l. 13)}$ together with Ki/earth;
13	**gù am$_6$-dab$_6$-e**	
14	**ki an-da ⌞gù⌟ am$_6$-dab$_6$-e**	Ki/earth is shouting together with An/heaven.
15	**[...]**	[...]

[*about 7 lines broken*]

column ii

1	**igi-⌜zi⌝-gal-an-n[a]**	With the true, great Queen of heaven,
2	**nin-gal-^{d}en-líl**	the older sister of Enlil,
3	**^{d}nin-ḫur-saĝ**	Ninḫursaĝ,
4	**igi-zi-gal-an-na**	with the true, great Queen of heaven,
5	**nin-gal-^{d}en-líl**	the older sister of Enlil,
6	**^{d}nin-ḫur-saĝ-ra**	Ninḫursaĝ,
7	**ĝìš mu-ni-du$_{11}$**	he has had intercourse;
8	**ne mu-ni-sub$_{5}$**	he has kissed her;
9	**a-maš-imin**	the seed for a set of septuplets
10	**š[à] mu-ni-ru**	he has poured into her womb.
11	**ki muš-ĝír-da**	Earth chatted cheery (l. 12) with the **muš-ĝír**-snake:
12	**BAL-bal am$_{6}$-da-za**	
13	**^{d}íd-maḫ**	'Exalted Divine River,
14	**níĝ-tur-zu a mu-da-de$_{6}$**	your small things have brought along water;
15	**[p]a$_{5}$-ra diĝir-íd-da-ke$_{4}$**	in the canals, the god of the river
16	**[...] mu [...]**	[...] has ? [...]'
17	**[...]**	[...]

[*about 6 lines broken*]

2.1.3b *IAS 174* [*Abū Ṣalābīḫ; ca. 2500 BCE*]

column i

	UGN-orthography	'Normal' orthography	
... Unknown number of lines broken			
1'	**[ ] NUN-ni-⌜sub$_{5}$**	**[ne ?] mu-ni-⌜sub$_{5}$**	He kissed her.
2'	**ki-ná**	**ki-ná**	Together with her (l. 3')
3'	**NUN-GAL-ak**	**mu-da-ak**	he made up the bed (l. 2'),
4'	**ĝiš$_{3}$ ḪI-NUN-ŠID**	**ĝiš$_{3}$ ḫe-mu-du$_{11}$**	to have intercourse with her.
5'	**a MAR**	**a šà**	He has poured (l. 6') the seed
6'	**NUN-ŠA-ru**	**mu-na$_{5}$-ru**	into her womb (l. 5').

7'	**[a⌟-maš-imin**	**[a⌟-maš-imin**	The seed for a set of septuplets
8'	**[x]-[ru⌟**	**[x]-[ru⌟**	[he has] poured [into her womb].

... *Unknown number of lines broken*

column ii

... *Unknown number of lines broken*

1'	[ ]	[ ]	....
2'	**BUR$_2$-BUR$_2$**	**BUR$_2$-BUR$_2$**	She murmured (ll. 2'-3').
3'	**mu-za**	**mu-za**	
4'	**MAR-^{d}nin-gal**	**šà-^{d}nin-gal**	Inside Ningal
5'	**ušum mu-lu**	**ušum mu-lu**	stirred the snake.
6'	**DAL-DAL**	**DAL-DAL**	She babbled (ll. 6'-7').
7'	**NUN-za**	**mu-za**	
8'	**⌞KU$^?$⌟ a nin-[ ]**	**⌞KU$^?$⌟ a nin-[ ]**	....

... *Unknown number of lines broken*

After Alster and Westenholz edited the text of the Barton cylinder in 1994, we also tried to fathom the text of this cylinder [132]. Column *i* describes primaeval times, the beginning, that was depicted as 'tempestuous'. In the midst of this commotion **an** and **ki** are shouting to each other. This reminds us of an analogous situation between **an** and **ki** in Ukg 15 (AO 4153) *ii*: 2, but with an important difference: where in the last text it was explicitly mentioned that **an** and **ki** were united (**téš-ba**), the Barton cylinder makes no announcement of the particular situation between **an** and **ki**, except perhaps in the following broken (~7) lines.

The female consort of the sexual union is called: **igi-zi-gal-an-na** [133], **nin-gal-^{d}en-líl** and **^{d}nin-ḫur-saĝ**, 'the true great queen of **an** (An, Heaven), the older sister of Enlil, and the divine lady of the mountains', respectively. The most simple and logical explanation of the 'Barton' text is the following. This text is telling about 'the beginning', when **an** and **ki** celebrate a cosmic courtship and intercourse. This ceremony begins in col. *i*:12-14. The epithets **igi-zi-gal-an-na** and **^{d}nin-ḫur-saĝ** point to the direction of the 'bride', **ki**: 'the true great queen of **an**, and 'the divine lady of the mountains'. Then it is nearly inevitable, as a logical continuation of the story of column *i*, that **an** as the male consort participates in the kissing and intercourse mentioned in column *ii*. Thus line *ii*:11 is also more comprehensible: the text just continues with the partner of **an**, viz. **ki**, who is talking with some kind of snake.

---

132 Krispijn and Lisman: to be published. Contrary to the idea of Alster and Westenholz (1994, 38), column *iii* does not describe the produce of plentiful fruits, but describes the construction and decoration of a statue, presumably of a god.

133 The epithet **égi-zi-an-na** seems to be almost exclusive for Ninḫursaĝ (Steinkeller 2005, 303).

Alster and Westenholz comment on line *ii*:7 as follows [134]: ''He' is hardly 'Earth', since it is most likely that Heaven and Earth represent the original male and female couple, and that the male counterpart of Ninḫursaĝ mentioned in *ii*:7 was the son of Heaven and Earth, presumably Enlil himself.' It is impossible that 'he' is 'Earth', because in intercourse or 'marriage' scenes in which **ki**/Earth is involved, **ki** is always the female consort. It is also not very likely that Enlil is Ninḫursaĝa's consort in this primaeval situation [135]. Another example of **an** being the partner of **ḫur-saĝ** (= Ninḫursaĝa) is to be found in 'The Debate between Grain and Sheep' lines 1-2, as a consequence of which the Anunna-gods were engendered. Further, with respect to Enlil, to the best of my knowledge there is no mention anywhere in the Sumerian mythological literature of Enlil being the son of **an** and **ki** [136]. In line 12 of 'The Debate between Winter and Summer' [137] Enlil is indeed the sexual partner of Ninḫursaĝa: 'He (= Enlil) copulated with the great Mountain Ranges (= Ninḫursaĝa) , the Mountain (= Enlil) gave her its share'. But the Sumerian lines in the Barton cylinder *ii*:1-7 are constructed in such a way, that we have to conclude that the subject of these lines has already been mentioned before, because the subject is not explicitly mentioned here, merely 'he'.

A possible parallel to the text of the Barton cylinder col. *ii* has been found in the fragment IAS 174, written in the UGN-orthography [138]. The fragment stems from the ED IIIa-period, and is therefore about 200 years older than the Barton cylinder [139]. In common with this cylinder, the name of the male consort who is kissing and making love has also been broken in IAS 174. The begetting of septuplets is not unique to this story: e.g. in IAS 283 *ii*: 10-13 [140] is reported that someone had intercourse with Ezina who then gave birth to septuplets.

If text IAS 174 is really telling the same story as the Barton cylinder, then it is inevitable that **an** and **ki** again are the protagonists in the first, cosmic 'marriage'. Both are not mentioned as such in the remaining fragment. IAS 174 col. *ii*: 4' mentions **^{d}nin-gal**. If both texts are parallel to each other, then the **^{d}nin-gal** in IAS 174, who has contact with an **ušum**-snake, and **ki**, who is talking with the **muš-ĝír** in the Barton cylinder, must be the same figure. Moreover, **nin-gal-^{d}en-líl** = Ninḫursaĝa = **ki** (Barton cylinder, col. *ii*) [141]. The

---

134 Alster and Westenholz 1994, 33.

135 Should this be Enlil, then it is more likely that the lines *ii*:2 and 4 would have been formulated as: **nin-gal-a-ni**, and not as now **nin-gal-^{d}en-líl**.

136 Recently, Andersson (2012, 68) stated that Enlil was the son of An, but without any reference to a text. Enki, who is the sexual partner of Ninḫursaĝa in the story entitled 'Enki and Ninḫursaĝa', also does not fit as the possible partner for Ninḫursaĝa in the 'Barton' story which tells about the primaeval marriage between **an** and **ki**. For a detailed analysis of 'Enki and Ninḫursaĝa': see Katz 2007a, 2008. In the second part of this story there is the report of a 'marriage' between Enki and Ninḫursaĝa, but without the birth of septuplets.

137 **ḫur-saĝ-gal-gal-la ĝiš bí-in-dug$_4$ kur-re ḫa-la ba-an-šúm** (For the published sources: see Vanstiphout 1987a; composite line in ETCSL 5.3.3, l. 12). For the discussion of **ḫur-saĝ** and **kur**: Steinkeller 2007, 223-232 (for this line: 229-230).

138 Preliminary comments on the text of IAS 174 has been given by Krebernik 1984, 277-278 *sub* *MAR; and Krebernik 1993-1997, § 4.3, 508b).

139 CDLI, no. P010156.

140 IAS 283 *ii*:10. **u$_4$ ^{d}ezina-ra** 11. **ĝiš mu-du$_{11}$** 12. **ne mu-ni-sub$_5$** 13. **dumu imin-áš mu-tud** (Biggs 1974): 'Then he had intercourse with Ezina, he kissed her; she gave birth to seven children.'

141 Another example of the identification of **ḫur-saĝ** (= Ninḫursaĝa) with **ki** can be found in 'The Debate between Winter and Summer' [ETCSL 5.3.3], line 12: **ḫur-saĝ-gal-gal-la ĝiš bí-in-dug$_4$**

lines IAS 174 *ii*: 4'-5' make that of the Barton cylinder *ii*:11 a bit more comprehensible: as **ušum** is living inside Ningal, thus in the earth [142], the **muš-ĝír** presumably does also. But who or what does this snake represent?

Serpents are widespread symbols of e.g. fertility and water [143]. As shown by stamp and cylinder seal impressions, serpents played a very important role in the region outside the real Sumer, viz. in Elam and in the Transtigridian region [144]. Some examples of the serpent as fertility symbol, from the post-ˁUbaid and Jemdet-Nasr period, are given by Amiet: a marital scene is accompanied by a serpent [145]. In historical times several serpent gods [146] can be discerned. Van Buren suggested that, in particular, entwined serpents might be the symbol of the mystery of the *hierós gámos* [147]. With respect to the copulating couple, von Wickede has a more explicit meaning [148]: erotic scenes on stamps were probably introduced at the end of the ˁUbaid period and in this respect he has no doubt about the reproduction of a *hierós gámos*. Amiet is more careful and does not corroborate the hypothesis that the sacred marriage would have been depicted [149]. It seems that the motif of 'entwined serpents' was developed at a later date than that of 'the marriage couple plus single serpent' [150]. It may be hypothesized that the entwined serpents of a later time were developed from this copulating couple, which has faded away in due course, and that they eventually have replaced that couple symbolically. In this respect Frankfort refers to a remark by Campbell Thompson [151], who ' (…) once suggested to me that it (i.e. entwined snakes, *JL*) might represent two snakes copulating.' This suggestion was confirmed to Frankfort in a letter from Parker [152]. In any case, both texts under discussion here – the Barton cylinder and IAS 174 – are dealing with the primaeval intercourse of **an** and **ki**,

---

kur-re ḫa-la ba-an-šúm 'He (= *Enlil*) copulated with the great Mountain Ranges (= *Ninḫursaĝa*), the Mountain (= *Enlil*) gave her its share', and line 14: **^{d}en-líl-le ki ĝìš dug$_4$-ga-ni am-gin$_7$ mur im-ša$_4$** 'While Enlil copulated with earth, there was a roar like a bull's'.

142 Van Dijk (1983, 15, and note 48) supposes that the **ušum** may be the same mythical motif as 'le serpent enlacé qui se mord la queue et qui se trouve dans l'océan qui entoure le monde.' Here he refers to examples of glyptic art, which do not seem to justify his supposition that these serpents are present in an ocean surrounding the world.

143 See e.g. van Buren 1935-1936; Mundkur 1978; Egli 1982, 144 ff; von der Osten-Sacken 1992, 73, 79-80; *idem* 2009; Wiggermann 1997, 47-48; Pientka-Hinz 2009.

144 Von der Osten-Sacken 1992, 79; Wiggermann 1997; Koch 2007, *passim*. The 'two spadelike staffs' resembling 'the identifying sign later associated with Marduk' (Hole 1993, 33-34) most probably represent snakes.

145 Amiet 1980, pl. 2, nos. 45 ('post-ˁUbaid') and 54 ('Jemdet-Nasr'). These pictures were originally published by Tobler 1950.

146 Wiggermann 1997.

147 Van Buren 1935-1936, 65.

148 Von Wickede 1990, 260: 'Siegelthemen mit erotischen Szenen (*hieros gamos*) werden vermutlich gegen Ende der 'Obēd-Zeit eingeführt und in der Gaura-Periode sowie in der späten Uruk-Zeit fortgeführt.'

149 Amiet 1980, 134: 'A la vérité, nos documents ne permettent guère de vérifier cette hypothèse.'

150 Van Buren 1935-1936; von Wickede 1990; McCaffrey 2013, 240-241.

151 Frankfort 1934, 12.

152 Parker, Assistant Keeper of Zoology at the British Museum of Natural History, wrote to Frankfort (1934, 12): 'I do not think there is any doubt that the symbol in which you are interested may well represent two snakes pairing. As a general rule the male seizes the female by the back of the neck and the two bodies are more or less intertwined. (…) Vipers (…) are said to have the body completely intertwined.'

where is also the report of a serpent, albeit not of 'entwined serpents'. The tentative conclusion may be that the *hierós gámos* may have been mirrored in the 'marriage' of **an** and **ki**. The alternative hypothesis is that a story from olden times about the cosmogonic beginning, i.e. a 'marriage' between **an** and **ki**, may have been the inspiration for the *hierós gámos* between a goddess and the **en** [153].

The relationship between a serpent, the reproduction of a serpent, the reproduction of water as undulating lines and the original cuneiform sign of water, is well known [154]. In the Barton cylinder, the **muš-ĝír** is addressed as **^{d}íd-maḫ**, 'Exalted Divine River'. What does this **^{d}íd-maḫ** represent? Jacobsen has pointed out [155], that the terms **engur** and **abzu**, being primarily the body of sweet water below the earth, feeding rivers (**íd**) and wells, are always clearly distinguished from **a-ab-ba**, the sea. Therefore the serpent **muš-ĝír**, alias **^{d}íd-maḫ**, does not represent any cosmic ocean, but a great river – with diverging branches, a kind of delta – or the sweet waters extending below the surface of the earth [156]. This corresponds very well with that what has been said about the snake in IAS 174 *ii*:4'-5': **MAR-^{d}nin-gal ušum mu-lu** 'Inside Ningal stirred the snake'. The 'small things' (Barton cyl. *ii*:14) are the (smaller) rivers that brought their life-giving waters to Sumer/Mesopotamia.

Wiggermann [157] tries to correlate the Late-Babylonian so-called 'Mappa Mundi' with the drawing on Fara tablet VAT 12772. On the 'Mappa Mundi' is an ocean [*marratu*], called 'the cosmic river' by Wiggermann, which surrounds the earth. The drawing on the Fara tablet, about two millennia older than the 'Mappa Mundi', confirms, according to Wiggermann, 'the existence of these notions [i.e.: those as rendered on the Mappa Mundi, *JL*] at a much earlier period.' [158]. The drawing on the tablet VAT 12772 [159] consists of the sign **KUR** ('mountain') in the centre, surrounded four times by the sign **GANA$_2$** ('field'), forming some kind of a cross. Perpendicular to each right end of the **GANA$_2$**-sign are two long straight lines (representing canals?). In the quadrants there are meandering lines, presumably representing rivers, which intersect each other. In some quadrants it seems as if there are more than two rivers rising from one central point (the photograph of the tablet at

---

153 See also the 'Excursus: **an** x **ki**: the cosmic marriage as model or motif for a ritual?' at the end of this chapter.

154 Porada (1992, 232): 'In early Iranian art the undulating bodies of serpents can be recognized as having been identified with watercourses.' Porada (1992, 235 note 21) refers to a relief on an Early Dynastic vase from Khafadja, in which 'On one side a deity holds two large undulating serpents, and on the other side, the same undulating outlines obviously signify watercourses.' For this vase: see Strommenger 1962, plate 38; Frankfort (1970, 40-41) shows also a drawing of the representations. Stevens (1989) has given a survey of the iconography of snake representations in Mesopotamia, from pre-ʿUbaid times up to and including the Jemdet-Nasr period, but without any interpretation.

155 Jacobsen 1946, 139-140, note 21.

156 It is remarkable that Amiet, referring to the same article of Jacobsen 1946, says: ' (…) l'Apsû, des eaux qui ceinturent le monde comme un serpent et s'écoulent sur terre sous forme de fleuves.' (Amiet 1980, 181). Amiet's interpretation of Jacobsen's explanation about **engur-abzu** *vs* **a-ab-ba** seems to be not suitable.

157 Wiggermann 1996, 208-209.

158 Wiggermann 1996, 208: The 'Mappa Mundi' and the like are the result of 'the tension between empirical geography and mythological world view.'

159 This tablet is an excercise tablet and contains an excerpt from the standard profession list on the obverse; the drawing is on the reverse. The text is a duplicate of SF 33, 34, 35 and 75 (Deimel 1923, 24* *ad* 11; Biggs 1974, 37 and 39).

CDLI [160] gives more details than its picture in Deimel 1923). My interpretation of this drawing is that it represents Sumer/Mesopotamia, with its main characteristics: mountains, fields and big canals fed by rivers [161]. In an iconographic respect this drawing has a cross form, and in an abstract form it is shown on the tablet IAS 2 from Abū Ṣalābīḫ [162]. This theme has been encountered many times from Jemdet Nasr-periods on [163], whereas the *swastika*-shape of the cross alternates with the *sauwastika*-form [164]. If my interpretation of the drawing on VAT 12772 is correct, then it is also clear that the picture on IAS 2 is just symbolically representing Sumer/Mesopotamia; nothing points to a picture of the 'cosmic world'. Or it may be a Sumerian interpretation of the old symbol that presumably represents the movement of the sun [165].

In their edition of the Barton cylinder, Alster and Westenholz hesitated about the identity of **ᵈíd-maḫ** [166]; they comment: ' **ᵈíd-mah** is here perhaps an epithet of Ninhursaĝ'. On the other hand, a few years later Alster wrote [167]: 'The existence of the primeval river god Nāru can be inferred from anthroponyms from the Pre-Sargonic and Sargonic periods. The earliest reference to the primeval river in mythological context is the name **Id-mah**, "Mighty River" (written with the divine determinative) in a Sumerian myth' (with reference to the Barton cylinder) [168].

---

160 VAT 12772: CDLI P010672; Deimel 1923, Tafel 8.

161 Mander (1995, 23) gives nearly the same explanation for the drawing on VAT 12772, although he calls the design a *crux ansata*. Then he compares the motif of crossing rivers on this tablet with the 'warp and weft' motif, about which Mander says (p. 24): 'We may safely conclude that a kind of «warp and weft» motif is peculiar to the Mesopotamian cosmogonic thought throughout all of its history; so when the mythical tales, in their symbolic language, depicted this motif as lying on a horizontal plane, we must understand it as related to the whole cosmic reality.' The present author cannot subscribe to this conclusion, and also not to the ideas of Mander about Uttu, the goddess of weaving: ' (…) her presence is critical in establishing the cosmic order' (p. 25, with reference to 'The Debate between Ewe and Wheat').

162 Biggs 1974, 31, fig. 29 (= IAS 2, *rev.*).

163 Legrain 1936, 12 *sub* Geometrical designs, a. Human motives.

164 Goblet d'Alviella 1912, 39; ch. 3. The swastika is a very old symbol, wide-spread in Eurasia with its origin probably in India (the Indus civilization). The right-hand swastika is considered as symbolizing the (daily) movement of the sun; the left-hand swastika (= sauwastika) represents the night.

165 Examples of iconographically related pictures from the earliest times in Mesopotamia are given by Legrain (1936). A rich collection of archaic seal impressions from Ur show e.g. a combination of irrigated areas within the *swastika*-from (no. 412, lower register). This seal impression has a close relationship with the design on a tablet from Abu Ṣalābīḫ (Biggs 1973, 31, fig. 29).

166 Alster and Westenholz 1994, 33, *ad ii*:13. See also my comments on this in the present edition of the Barton cylinder.

167 Alster 1999, 870b. There was no indication how he has decided to change his interpretation of **ᵈíd-maḫ** from the suggestion of 'perhaps being an epithet of Ninḫursaĝ' to 'the earliest reference to the primeval river' shortly thereafter.

168 The CAD N part I, 374b, '*nāru* personified or deified' refers to Gelb (1957, 191) for **ᵈíd** in personal names. Gelb gives only a few examples of **ᵈíd** in personal names for the Sargonic period. There is no remark of a '*primaeval*' river god.

2.1.4 *NBC 11108* [*Ur III, ca. 2000 BCE* [169]]

Obverse

1	**a[n]-uru$_{16}$-né an mu-zalag$_{2}$ / ki mu-kikki kur-šè igi m[u]-[íl$^{?}$]**	The mighty An lighted heaven, earth he darkened, he looked at the netherworld.
2	**buru$_{3}$ a nu-bal ninda nu-ĝar ki-daĝal / ⌜uru$_{4}$⌝ nu-ak**	From the depths no water was drawn; bread was not put down. Cultivation of the wide earth did not happen.
3	**išib-maḫ-^{d}en-⌜líl⌝-lá nu-ù-ĝál / [š]u-luḫ-kù-ge šu nu-ù-⌜ma-du$_{7}$⌝**	The supreme purification priest of Enlil was not there, a holy purification rite was not perfected.
4	**[igi-z]i-an-na-ke$_{4}$ šu nu-ù-tag / [zà$^{?}$]-mí nu-di**	The priestess of An did not play an instrument, did not sing a song.
5	**[an k]i téš-bi-a mu-lug**	An/heaven lived together with Ki/earth,
6	**[nam-dam-š]è ⌜nu-ù⌝-tuku**	(but) as wife he had not taken her.
7	**⌜u$_{4}$⌝ nu-⌜zalag⌝ ĝi$_{6}$-àm mu-lá**	The day did not dawn, the night had spread over her (= **ki** earth).
8	**an-né da-ga-an-na / kiri$_{3}$-zal mu-ni-íb-guru$_{17}$**	An was wearing luxuriance in the residence,

Reverse

9	**ki-gub-ú-šim-ma / ní nu-mu-[gíd]-gíd-e**	(but) on a verdant place he still did not lie down.
10	**me-^{d}en-⌜líl-lá-ke$_{4}$⌝ kur-kur-ra / ⌜šu nu-ù-du$_{7}$⌝**	The ritual of Enlil had not been completed in the lands.
11	**[k]ù-⌜in-nin⌝ [nin]-⌜an⌝-na-ke$_{4}$ / ni[dba] ⌜šu nu-mu⌝-ši-te$^{?}$**	The holy lady, the lady of heaven, did not receive$^{?}$ the food offerings.
12	**[diĝir-ga]l a-nu[n]-[ra a]d$^{?}$ [nu]-um-di-di**	For the great gods, the Anunna, one was not singing$^{?}$:
13	**diĝir-an-⌜na⌝ ⌜an⌝-[ki]-a / nu-ù-⌜ma⌝-su$_{8}$-⌜su$_{8}$⌝-ge-éš**	the gods of heaven still did not stand in heaven and on earth.

When van Dijk published the text of this tablet in 1976, he asked himself what kind of text he was dealing with. It had the form and size of small incantations, but as van Dijk did not know of any non-canonical incantation that was divided over several tablets, he was inclined to believe that this fragment belonged to a sacred tale, a real cosmogony [170].

Until now there has been no evidence of a single pure Sumerian cosmogonic text. This tablet NBC 11108 fits very well with the 'type III: single column tablets', as described

169 The tablet was written in the ductus of the Ur III period (van Dijk 1976, 128).

170 Van Dijk 1976, 128: 'L'analyse de texte nous inclinera à croire plutôt que le fragment appartient à un hiéroslogos, à une vraie cosmogonie.'

by Veldhuis [171]: 'Type III tablets are one-column tablets with a single extract from some school texts. The length of the extract is between 10 and 15 lines. The same exercise continues from obverse to reverse.' As with all other cosmogonic texts, this one of NBC 11108 is probably no more than an introduction, but it is not known to which text this applies. It is not unlikely that it might be a text related to one or more rituals, because of the mention of the absence of several rites (ll. 3, 4, 10-12) in primaeval times.

The first sentence is very clear about the presence of the sky god An, heaven, earth and the netherworld [172]. What might be the meaning of the contrast 'heaven-lighted' and 'earth-darkened'? The clue may be found in the next lines (2-4) and in the lines 10-12, that mention more about the earth. From these lines we learn that there are no cultural or ritual activities at all on earth. A salient feature is the order in which these facts are told to us: 1). There is no bread and no water because the earth had not been cultivated; thus no offerings to the gods are possible. 2). The lines 3-4 tell us that there are no priests or priestesses present to fulfil the necessary rites. 3). The final conclusion is seen in the lines 10-12: no rites could be fulfilled for Enlil, Inanna could not receive food offerings, and for the Anuna there is no song. The information that the ritual for Enlil had not been completed is rather surprising from the perspective of line 13: the great gods, the Anunna, had not (yet) arrived. After all, Enlil – who belongs to the Anunna – is not yet present, why then the remark about the absence of Enlil's ritual? Or is it another way of saying that Enlil is not yet present? The same reasoning, *mutatis mutandis*, applies to Inanna. This construction of arguments shows a gradually build-up to a climax – not unusual in Sumerian texts –, until the last line gives the final clue: the great gods of heaven, the Anunna, are not even there. This is a nice example of building up a certain tension during the telling of a story.

In between, as if to postpone the end of the story and to increase the tension even more, attention is drawn to the relation between **an** and **ki**. Line 7 not only partly repeats the content of line 1, but also splits up the statements about **an** and **ki**. This line, literally in the centre of the story, enhances the dramatic expression of it. Therefore it seems that the lines 5-6 and 8-9 on both sides of this pivotal line form the most important information. This information concerns **an** and **ki**, in the phase of being in unity, but before the primaeval marriage. In these lines **an** has been conceived the one time as animate, the other time as inanimate [173], but it seems that **an** is very close to becoming seen in his ultimate form, the sky god An. The opening sentence (line 1) does show the mighty An already as animate, but also functioning in a clearly primaeval situation; this situation – the darkness of the earth – is repeated in line 7, in the middle of the preparation of the cosmic marriage between **an** and **ki**. The audience has been kept in suspense: **an** was prepared, but the *moment suprême* had not been reached. All the sentences with negations, describing what does not yet exist, are in fact suggesting what the situation will be in a positive sense after the cosmic marriage between **an** and **ki** has eventually been celebrated, i.e. the present situation that is well-known to the audience. This marriage is considered as the starting-point, the *sine qua non*, for everyone and everything: gods, people and culture. But until

---

171 Veldhuis 1997, 37-38; citation on p. 37. The labels for the tablet types are basically those proposed by Civil in MSL XII (1969, pp. 27 ff.) and Civil 2000a, 2308 (first published in 1995). Jagersma doubts if the typology of Civil/Veldhuis can be used for the texts of the Ur III period (pers. comm.).

172 The comment of Katz (2003, 13-14, note 36) that the **kur** is invisible or non-existent (based on van Dijk's edition of this text) has to be revised (see the collation of Sjöberg 2002, 240; and also note 1070 of this study).

173 See Excursus: 'The animate *vs* inanimate class of **an** and **ki**' in the Appendix Text editions.

that moment there was nothing but **an** and **ki**. It must be noticed that this is the only text discussed in this study in which is alluded to a marriage between **an** and **ki**: in line 6 – but in a negative way – the expression **nam-dam-šè tuku** 'to take as spouse' was used.

Let us return to the question: what might be the meaning of the contrast between 'lighted heaven' and 'darkened earth' in line 1? Perhaps a (partial) explanation for the darkness of the earth may be that it is a metaphor for the absence of everything which makes earth a living entity. Another conspicuous fact is that first of all the 'darkness of the earth', i.e. the absence of any culture and any living being, is mentioned (ll. 2-4) before, as the final climax, the absence of the Anunna (ll. 12-13). These lines seem to say: the absence of all mentioned above is understandable and explicable, for the gods of heaven are not yet present! All that is dependent on the performance of the primaeval cosmic marriage, which takes up the central place in this short story.

On the other hand: in heaven, where there is light, there is activity due to the preparations of **an** for his marriage with **ki**. The light then symbolizes these activities. The tentative conclusion about the identity of **an** – sky god An or heaven – may be that before the moment of the real performance of the cosmic marriage **an** always had been considered as inanimate, yet comprising all the later elements evolving from this primary unit.

2.1.5 *The Debate between Tree and Reed* [*Ur III* [174]]

1	**ki-ùr-gal-e ní pa bí-íb-è bar-dul-le-eš nam-sa$_7$**	The large surface of the earth introduced herself; then she has embellished herself as with a *bardul*-garment.
2	**[ki]-daĝal-e kù ^{na4}za-gìn-bi bar-ba àm-mi-íb-si**	The vast earth has filled her exterior with precious metals and lapis lazuli.
3	**$^{[n]a4}$esi ^{na4}nir$_7$ ^{na4}gug sù-du-áĝ-ĝá šu-tag ba-ni-ib-du$_{11}$**	With diorite, **nir**-stone, cornelian and **suduaĝa** she has adorned herself.
4	**[ki]-ú-šim-e ḫi-li gú bí-íb-è nam-nun-ba mu-un-gub**	The earth, the fragrant vegetation, covered herself with attractiveness. She stood in her magnificence.
5	**[k]i-kù-ki-sikil-la an-kù-ra ní-bi mu-na-ab-sa$_7$**	The pure earth, the virgin earth, has beautified herself for the holy An.
6	**an-an-maḫ-e ki-daĝal-la dùb im-ma-ni-ib-nir**	An, the exalted heaven, had intercourse with the vast earth.
7	**a-ur-saĝ-ĝiš-gi-bi-da-ke$_4$ šà-ga ba-ni-in-de$_5$**	He poured the seed of the hero's Tree and Reed into her womb.
8	**ki-šár-áb-zi-dè a-du$_{10}$-ga-an-na da bí-íb-ri**	The whole earth, the fecund cow, took the good seed of An under her care.

174 The text of this debate most likely dates from the Ur III-period (van Dijk 1953, 35-36).

9	**ki-ú-nam-ti-la-ke$_4$ šà im-ḫúl ù-tu-ba mu-un-gub**	The earth, life-giving vegetation, innerly happy, devoted herself to the production of it (*i.e. the vegetation)*.
10	**ki-kiri$_3$-zal-e ḫé-ĝál im-gùr kurun$_2$ lal$_3$ ir sù-ud**	The earth, full of joy, bore abundance, while juice and syrup gave out their smell.

The introduction to 'The Debate between Tree and Reed' places the scene in primaeval times. It specifies especially the preparation of **ki** for her wedding with **an**. This preparation has been described extensively and in detail, more than in any other cosmogonic introduction, but it presents no new information about the primaeval circumstances. The embellishment of **ki** forms a sharp contrast with the inactivity of **an** in this respect; in NBC 11108 line 8, at least 'An was wearing luxuriance in his residence'.

2.1.6 *The Debate between Grain and Sheep* [*Ur III - OB*]

1	**ḫur-saĝ-an-ki-bi-da-ke$_4$**	When Heaven [l. 2] had caused the mountain range of **an-ki**
2	**u$_4$ an-né diĝir-^{d}a-nun-na im-tu-dè-eš-a-ba**	to bring forth the Anunna gods
3	**mu ^{d}ezina$_2$ nu-ub-da-tu-da-aš nu-ub-da-an-sig$_7$-ga**	– because Ezina had not been engendered together with them, had not grown up with them;
4	**kalam-ma gu-^{d}uttu nu-ub-da-an-dím-ma-aš**	because in the Land the thread of Uttu had not been fashioned together with them,
5	**^{d}uttu-ra temen nu-mu-un-na-si-ga-aš**	(and) for Uttu no poles (*for the loom*) had been driven in the ground –,
6	**u$_8$ nu-è-a sila$_4$ nu-šár-ra**	when the ewe had not appeared, so lambs were not present in great numbers,
7	**ùz nu-è-a máš nu-šár-ra**	when the goat had not appeared, so kids were not present in great numbers,
8	**u$_8$-e sila$_4$-min-bi nu-ub-tu-ud**	the ewe did not give birth to her two lambs,
9	**ùz-e máš-eš$_5$-bi nu-ub-tu-ud**	the goat did not give birth to her three kids;
10	**mu-^{d}ezina$_2$-^{d}kù-sù-u$_8$-bi-da-ke$_4$**	the names of Grain, the holy blade, and of Ewe,
11	**^{d}a-nun-na diĝir-gal-gal-e-ne nu-mu-un-zu-uš-àm**	did even the Anunna, the great gods, not know!

12	**še-muš$_5$ u$_4$-30-àm nu-ĝál-la-àm**	The 30-days *šeĝuš*-barley [175] was not there!
13	**še-muš$_5$ u$_4$-40-àm nu-ĝál-la-àm**	The 40-days *šeĝuš*-barley was not there!
14	**še-muš$_5$ u$_4$-50-àm nu-ĝál-la-àm**	The 50-days *šeĝuš*-barley was not there!
15	**še-di$_4$-di$_4$ še-kur-ra še-á-dam-kù-ga nu-ĝál-la-àm**	The small barley, the foreign barley, (and) the barley from the holy fields around the dwelling-places were not there!
16	**túg níĝ mu$_4$-mu$_4$-bi nu-ĝál-la-àm**	A garment (or) something to dress was not there!
17	**^{d}uttu nu-ub-tu-ud men nu-íl**	Uttu had not been born. A cap was not worn.
18	**en-^{d}niĝir-si en-kal-kal nu-ub-tu-ud**	Enniĝirsi, the very mighty lord, had not been born;
19	**^{d}šakan$_2$ bar-rim$_4$-ma la-ba-ra-è**	Šakan had not appeared in the dry area.
20	**nam-lú-ulu$_3$-u$_4$-ri-a-ke$_4$-ne**	Mankind of those faraway days
21	**ninda gu$_7$-ù-bi nu-mu-un-zu-uš-àm**	had not got to know eating bread,
22	**túg níĝ mu$_4$-mu$_4$-bi nu-mu-un-zu-uš-àm**	had not got to know a garment (or) something to dress in!
23	**ùĝ ĝeš-gen$_6$-na-a šu-bi mu-un-ĝen**	The people moved on all fours;
24	**udu-gin$_7$ ka-ba ú mu-ni-ib-gu$_7$**	like sheep they ate grass with their mouths,
25	**a-mú-sar-ra-kam i-im-na$_8$-na$_8$-ne**	the water from the gardens they were drinking.
26	**u$_4$-ba ki-ulutim$_2$-diĝir-re-e-ne-kam**	At that time, near the place of residence of the gods,
27	**é-bi-du$_6$-kù-ga u$_8$ ^{d}ezina$_2$-bi mu-un-se$_{12}$-eš-àm**	in their abode, the holy hill, Ewe and Grain were living.
28	**èš-ninda-gu$_7$-diĝir-re-e-ne-ka mi-ni-ib-de$_5$-de$_5$-ge-eš-a**	After they [*the Anunna*] had gathered them in the dining hall of the gods,
29	**ḫé-ĝál-u$_8$-^{d}ezina$_2$-bi-da-ka**	of the abundance of Ewe and Grain
30	**^{d}a-nun-na-du$_6$-kù-ga-ke$_4$-ne**	the Anunna of the holy hill
31	**i-im-gu$_7$-gu$_7$-ne nu-mu-un-dè-si-si-eš**	were eating, but all along they could not satisfy themselves.
32	**ubur-kù-ga-ne-ne ga-bi níĝ-dùg-ga**	The milk, that sweet substance of their pure udder,

[175] *šeĝuš*- or *šigūšu*-flour was used for offerings to strange ghosts; that kind of offering was considerably less expensive than the offerings made to family ghosts or to gods who were invoked in ghost-expelling rituals (Scurlock 2000, 1891).

33	**^{d}a-nun-na-du$_6$-kù-ga-ke$_4$-ne**	the Anunna of the holy hill
34	**i-im-na$_8$-na$_8$-ne nu-mu-un-dè-si-si-eš**	were drinking, but all along they could not satisfy themselves.
35	**ubur-kù-ga níĝ-dùg-ga-ne-ne-šè**	For the sake of the sweet substance of their pure udder
36	**nam-lú-ulu$_3$ zi-šà im-ši-íb-ĝál**	they have inspirited mankind.
37	**u$_4$-ba ^{d}en-ki-ke$_4$ ^{d}en-líl-ra gù mu-un-na-dé-e**	At that time Enki spoke to Enlil:
38	**a-a-^{d}en-líl u$_8$-^{d}ezina$_2$-bi-da-ke$_4$**	'Father Enlil, now that Ewe and Grain
39	**du$_6$-kù-ga um-ma-da-an-se$_{12}$-eš-a**	have resided together at the holy hill,
40	**du$_6$-kù-ta ga-àm-ma-da-ra-ab-e$_{11}$-dè-en-dè-en**	let us send them down together from the holy hill'.
41	**^{d}en-ki ^{d}en-líl-bi inim-kù-ga-ne-ne àm-dug$_4$-ga**	At the holy command of Enki and Enlil
42	**u$_8$ ^{d}ezina-bi du$_6$-kù-ta im-ma-da-ra-an-e$_{11}$-dè**	Ewe and Grain set off down together from the holy hill.

In the edition of this text [176] I have spent ample discussion on the translation of the lines 1-2. I have argued there that **ḫur-saĝ-an-ki-bi-da** represents the mother goddess, later known as Ninḫursaĝa [177], who – impregnated by **an** – brings forth the Anunna-gods [178], and that she may be identified with **ki**. The expression used in 'Lugalbanda and the Anzud bird', viz. **^{d}a-nun-na diĝir-ḫur-saĝ-ĝá** 'the Anunna, the gods of the mountain range'[179], may be an allusion to the origin of the Anunna gods as expressed in the text under discussion now. In the two other examples in which a god, viz. Enlil, has copulated with **ḫur-saĝ** [180], the addition of **an-ki-bi-da** after **ḫur-saĝ** is missing. A possible explanation for this absence may be that the copulation of Enlil with Ninḫursaĝa did not happen in primaeval times, i.e. when **an** and **ki** were united. In 'The Debate between Grain and Sheep' the male protagonist during the copulation is not Enlil but **an**. The addition of **an-ki-bi-da** after **ḫur-saĝ** is an indication that the intercourse of **an** with Ninḫursaĝa occurred in primaeval times when **an** and **ki** formed a unity. The **an** in line 2 of this debate was considered to be inanimate [181].

In this text **ḫur-saĝ-an-ki-bi-da** and **du$_6$-kù** are mentioned. Wiggermann wrote about these hills [182]: 'The **ḫur-saĝ-an-ki-bi-da-ke$_4$** 'Mountain of Heaven and Earth' is,

---

176 See Appendix Text editions, 6. The Debate between Grain and Sheep.

177 The name Ninḫursaĝa was given to the mother goddess by Ninurta: 'Ninurta's exploits: a *šir-sud* (?) to Ninurta' (also called: 'Lugal-e') [ETCSL 1.6.2], ll. 390-396.

178 This may answer the question of Vanstiphout (2009, 20, note 27): '[W]ere the gods already there when heaven and earth were separated, or did they come into being at the moment of separation?' The formulation 'the mountain range of **an-ki**' in l. 1 suggests that heaven and earth are not yet separated while the Anunna were brought forth.

179 'Lugalbanda and the Anzud bird', line 82 (Wilcke 1969, 100; ETCSL 1.8.8.2).

180 'Gudea cyl. B' xxii, frgm. Sii 3-5 (Jacobsen 1973, 282); 'Debate between Winter and Summer' [ETCSL 5.3.3], l. 12 (see Appendix Text editions no. 6, comments on the lines 1-2).

181 See Excursus 'The animate *vs* inanimate class of **an** and **ki**' in the Appendix Text editions.

182 Wiggermann 1992, 295 *ad* 5a.

according to all commentators, identical with the **du$_6$-kù** from which Laḫar and Ashnan descend later in the text.' In a review of Kramer's 'Sumerian Mythology', Jacobsen discussed **ḫur-saĝ** and **du$_6$-kù** [183], and his conclusion was that for the Sumerians **du$_6$-kù** was located in the eastern mountains. For van Dijk [184] the **du$_6$-kù** is a holy hill above the Weltberg (=**ḫur-saĝ**). Hruška supposed three layers between heaven and earth with their links: **ḫur-saĝ-an-ki(-bi-da)**, **du$_6$-kù** and **kur**, respectively [185]. Maul, writing about the **du$_6$-kù**, also did not identify **du$_6$-kù** with **ḫur-saĝ** [186]. From these examples may be concluded that there is no support for Wiggermann's statement.

As has been made plausible in my commentary on the edition of this text, **ḫur-saĝ-an-ki-bi-da** represents the primaeval mother goddess Ninḫursaĝa – who has never been named with the epithet **du$_6$-kù** –, and this **ḫur-saĝ** does not seem to be identical to a range of mountains as a place to live. Therefore **ki-ulutim$_2$** (line 26), **é** and **du$_6$-kù**, evidently a place where the gods were residing, are not identical to **ḫur-saĝ**.

In the lines of this debate shown above, there are numerous negations. About negations in Mesopotamian texts, Michalowski has written that these devices in creation stories accentuate the idea that the world is not static [187]. We may add to this that the negations also describe what we may expect in the future; they hold a promise for the presence of things and situations that are absent at the moment of negation.

In the third-millennium texts human beings were absent. In the prologue to this debate mankind is put on the scene, but this is mankind still at a very primitive stage: they behave like animals. Finally the Anunna, in their own interests, 'inspirit mankind' (line 36). This 'inspiration' is a transformation, or better, civilization, of mankind in such a way that people are able to practise agriculture and cattle breeding. In this way the satisfaction of the needs of the gods, i.e. their needs concerning food and drink, was guaranteed. The real creation of man is not described in this prologue; it is only an adaptation of already present living beings.

Vanstiphout discussed the 'Remnants of a Cosmogonic Mythology?' and 'The Intention of the Introductions' with respect to the introductions at the Mesopotamian debate poems [188]. After he said that the general tenor of the introductions is specific to the genre, he remarked (p. 289): 'It is never a piece of regular creation or even procreation activity, (...).' But as we have seen in the debates 'Tree-Reed', 'Grain-Sheep' and 'Summer-Winter' there is explicit mention of procreation activity or even copulation [189]. Another remark of Vanstiphout is (p. 289): 'No cosmic system is mentioned or even alluded to.' But in the debate 'Tree-Reed' we find **an** x **ki**, and in the debate 'Grain-Sheep' **an** x **hursaĝ** (**hursaĝ** represents **ki**), the primaeval cosmic system. Vanstiphout also wrote: 'The conclusion can hardly be escaped that the objective of the introduction will *not* have been to give a cosmogony in any serious way' (p. 290), and 'The mythological aspect is artificial, relatively unimportant, and possibly secondary.' (p. 291). These arguments have been

---

183 Jacobsen 1946, 141b.

184 Van Dijk 1960, 134. The term 'Weltberg' was introduced by Jacobsen (1946, 141a) for Kramer's interpretation of **ḫur-saĝ-an-ki-bi-da** in the lines 1-2 of 'The Debate between Grain and Sheep' (Kramer 1944, 39).

185 Hruška 1996, 166.

186 Maul 1997, 116.

187 Michalowski 1991, 134.

188 Vanstiphout 1990, 289-291.

189 For 'The Debate between Summer and Winter': see Appendix Text editions, 6. The Debate between Grain and Sheep, *ad* Comments ll. 1-2.

countered, in my opinion, by the texts and interpretations of the introductions of the 'Debate' texts which are discussed in this study.

2.1.7 *Gilgameš, Enkidu and the Netherworld* (or: *Gilgameš and the Ḫalub-tree*) [*OB*]

1	**u$_4$-ri-a u$_4$-sù-rá-ri-a**	On that day, on that distant day,
2	**ĝi$_6$-ri-a ĝi$_6$-bad-rá-ri-a**	in that night, in that night long past,
3	**mu-ri-a mu-sù-rá-ri-a**	in that year, in that distant year,
4	**u$_4$-ul níĝ-ul-e pa è-a-ba**	in olden times, when eternal things were manifested,
5	**u$_4$-ul níĝ-ul-e mí-zi dug$_4$-ga-a-ba**	in olden times, when eternal things were taken care of,
6	**èš-kalam-ma-ka ninda šú-a-ba**	when in the shrines of the Land bread was tasted,
7	**^{im}šu-rin-na-kalam-ma-ka níĝ-tab ak-a-ba**	when in the ovens of the Land fire was blown,
8	**an ki-ta ba-da-bad-rá-a-ba**	when heaven had been separated from earth,
9	**ki an-ta ba-da-sur-ra-a-ba**	when earth had been delimited from heaven,
10	**mu-nam-lú-u$_{18}$-lu ba-an-ĝar-ra-a-ba**	(and) the name of mankind had been established on it;
11	**u$_4$ an-né an ba-an-de$_6$-a-ba**	at that time, when An had taken heaven,
12	**^{d}en-líl-le ki ba-an-de$_6$-a-ba**	Enlil had taken earth,
13	**^{d}ereš-ki-gal-la-ra kur-ra saĝ-rig$_7$-bi-šè im-ma-ab-rig$_7$-a-ba**	(and) when they had given the Netherworld to Ereškigal as a present;
14	**ba-u$_5$-a-ba ba-u$_5$-a-ba**	when he sailed away, when he sailed away,
15	**a-a kur-šè ba-u$_5$-a-ba**	when Father sailed away for the Netherworld,
16	**^{d}en-ki kur-šè ba-u$_5$-a-ba**	when Enki sailed away for the Netherworld:
17	**lugal-ra tur-tur ba-an-da-ri**	for the king small things were put on board,
18	**^{d}en-ki-ra gal-gal ba-an-da-ri**	for Enki large things were put on board;
19	**tur-tur-bi na$_4$-šu-kam**	the small things were like stones fitting in the hand,
20	**gal-gal-bi na$_4$-gi-gu$_4$-ud-da-kam**	the large things were like stones which make the reed dancing;
21	**úr-^{ĝiš}má-tur-re-^{d}en-ki-ga-ke$_4$**	they covered (l. 22) the bottom of Enki's boat
22	**níĝ-bún-na-du$_7$-àm ì-šú-šú**	fitting like a turtle;

lugal-ra a-^{ĝiš}má-saĝ-ĝá-ke$_4$	for the king the water at the stem
ur-bar-ra-gin$_7$ téš mu-na-gu$_7$-e	swallows up everything like a wolf;
^{d}en-ki-ra a-^{ĝiš}má-eger-ra-ke$_4$	for Enki the water at the stern
ur-maḫ-gin$_7$ saĝ ĝiš im-ra-ra	hammers like a lion.

The cosmogonic introduction of 'Gilgameš, Enkidu and the netherworld' tells a story that does not follow a logical time schedule [190]. Moreover it seems as if there are two different introductions that are told just one after another: the first introduction comprises the lines 1-10, the second one the lines 11-26 [191]. The first narrative, after having been set in primaeval times, begins by telling about the time when culture developed (4-7): everlasting things were manifested – one might add: 'for the first time' – and offerings were made (line 6: tasting of bread in shrines, thus by the gods). The story then switches to a moment in the history that must have happened earlier, as if to explain how culture could have come into being. Cultural development was possible after the separation of **an** and **ki**, which is told in a neutral 'passive' way (8-9); then the first introduction ends with the creation of mankind (line 10). Nothing is told about a cosmic intercourse or marriage of **an** and **ki**, let alone about the period in which both protagonists prepared themselves for it, as the third millennium texts did. On the other hand, this text mentions the separation of **an** and **ki**, though this event had also been told in earlier texts. In three texts from Abu Ṣalābīḫ [192], written in the so-called UD-GAL-NUN orthography, we read that Enlil separated heaven and earth. It is remarkable that in the third millennium texts, written in normal orthography that have been dealt with in this study, there is no mention at all of a separation of **an** and **ki** by Enlil [193].

From line 11 onwards the second 'introduction' starts: **u$_4$ ... -ba** [194]. The separation of **an** and **ki** is told again, but now in another formulation: An takes heaven, Enlil takes earth. Also the head of the netherworld, Ereškigal, is mentioned. Then there follows a further unknown story about Enki, who is sailing in the direction of the netherworld. This unique story is not easy to explain [195].

---

190 The content of these initial lines is repeated by Inanna (ll. 52 ff and 96 ff), but in an abridged version. See also Jacobsen's struggle with what he sees as the chronological incongruity and illogicality of the lines 1-10 (Jacobsen 1993, 122: 'a hodgepodge of snippets of introductions to other tales which the rhapsode threw together haphazardly as they came to mind.'). In my opinion, this incongruity is absent in the present translation.

191 Ferrara (2006, 58) proposes the same outline.

192 See ch 2.1.1b.

193 In ch 3, Excursus 1 'Enlil and Ninlil', the place of Enlil in the Sumerian pantheon and also Enlil's origin will be discussed in more detail.

194 According to Gadotti (2005, 47), the second part of the introduction starts at line 14. However, several arguments may plead against this division: 1. the content of the first ten lines forms a coherent unit; 2. the construction of line 11 starts with **u$_4$ ... -a-ba**, and 3. the reiteration of the separation of **an-ki** – now mentioned with the protagonists in this action. These arguments favour the division as proposed by the present author.

195 Kramer and Maier consider this episode as a struggle: ' "Enki's struggle with the *Kur*" is one of the unusual stories that depict Enki as a warrior.' (Kramer and Maier 1989, 82-83; the same idea had already been expressed by Kramer 1944, 76-96, and this was commented on by Jacobsen 1946, 143-148). An almost similar explanation is given by Clifford (1994, 24): 'The cosmogony

Jacobsen [196] has tried to explain why the episode with the descent of Enki to the netherworld (as he called it) was inserted: the turbulence accompanying Enki's ride in a boat on the Euphrates uprooted the **ḫalub**-tree, which then is the connection with the next story of Inanna and this tree. Pettinato renounced the idea of Jacobsen, who argued in favour of a descent by Enki to the netherworld [197], and he claimed that there is no textual evidence for such a descent [198]. On the other hand, according to Gadotti Enki is indeed travelling to the netherworld [199].

Katz wrote that Enki's travelling to the netherworld is part of the cosmological introduction to the text, but with unknown mythological context. Most likely a voyage by sea was described, which may have been influenced by the properties of Enki's divinity and his residence in the **abzu** [200].

That it was a turbulent passage is beyond doubt (lines 23-26). The episode is a reflection of the violence that Enki and his boat were faced with. It presumably refers to the danger of a descent into the netherworld, maybe it is a warning. The preceding episode with the stones, that were put on Enki's boat, might be a metaphor. I would like to propose two possible interpretations for this metaphor. Small and large stones were put on board of Enki's ship, resulting in its bottom being covered and giving it the appearance of a 'complete turtle' (lines 17-22). Turtles are protected against all kind of violence by means of their huge shield. Enki had to cross a turbulent sea and meets violence during this passage. The turtle-like covering of his boat may be a metaphor for the protection of Enki on his voyage. Another interpretation of this covering may be that the stones represent in a metaphorical way the roof of Enki's residence [201]. The **abzu**, the freshwater reservoir of Sumer, was located subterraneously, and as such it needed a robust partition from the upper world. The turtle shell protected, and at the same time hid, Enki's residence, 'a place the inside of which not a single god can observe' [202].

Gadotti, in search of a function for this story of Enki [203], argued that this episode may be a prelude to the forthcoming descent of Enkidu into the netherworld. It also may point to the difficulties inherent in a journey to the netherworld, even for a god. Moreover, it was Enki who was willing to intervene in favour of Enkidu's return from the netherworld (line 237).

---

explains the underworld as primordially violent and chaotic, a place that seizes Ereshkigal and fights Enki.'

196 Jacobsen 1993, 120. The **ḫalub**-tree was uprooted by **á-u$_{18}$-lu**: 1. the force of the south wind, or 2. the force of a demon. This second possible translation may be not so surprising: perhaps there is a connection between the 'Enki-myth' in the lines 14-26 and an incantation (Forerunners **udug-ḫul** ll. 299-357; *vide infra*).

197 Jacobsen 1993, 122.

198 Pettinato 2000, 865.

199 Gadotti 2005, 49-50.

200 Katz 2003, 40.

201 Green (2000, 1839, *ad* 'turtle') mentions that the symbol 'turtle' has been associated with Enki/Ea since prehistoric times.

202 'Enki and Ninmaḫ', line 13. The image of the coverage of Enki's boat with stones may have found a far echo in 'An account of the creation of the world by Marduk and the goddess Aruru' [82-5-22, 1048, CT 13 36: 17-18], where Marduk creates dry land by making a raft over the primaeval waters, on which raft he puts earth: 'Marduk wove a raft on the face of the waters, created earth and put it on the raft'.

203 Gadotti 2006, 57-58: ch. 4.6.

In search of a parallel or an explanation for Enki's journey, Pettinato studied the lines 299-310 of the 'Forerunners to Udug-ḫul' [204]. These lines comprise an incantation with an introduction in which Enki has 'turned his mind to the netherworld' (lines 299-303):

299	**[en-e an-gal-ta ki-dagal-šè] ĝéštu-ga-ni na[m]-g[ub]**	[The lord turned] his mind [from the great heaven to the broad Netherworld].
300	**[en-gal ^{d}en-ki-ke$_4$ an-gal-ta ki-dagal-šè] ⌜ĝéštu-ga-ni⌝ [nam-gub]**	[The great lord Enki turned] his mind [from the great heaven to the broad Netherworld].
301	**diĝ[ir-gal-e an-gal-ta ki-dagal-]-⌜šè⌝ ĝéšt[u-g]a-⌜ni nam-gub⌝**	[The great] god turned his mind [from the great heaven] to [the broad Netherworld],
302	**a[bul] ^{d}utu-šú-š[è] ĝéštu-⌜ga-ni⌝**	(he turned) his mind to the [great gate] of the west,
303	**ki-⌜nú da⌝-r[í diĝir-kam$^{?}$] ⌜urugal^{gal}⌝-ka-⌜šè⌝ ĝéštu-ga-ni**	(he turned) his mind to the eternal resting place [of the god?], the grave.

Geller pointed to the close resemblance between the opening lines of this incantation and the opening passage of 'Inanna's descent to the netherworld', and also to thematic similarities between these two compositions [205].

This incantation and 'Gilgameš, Enkidu and the netherworld' show also some parallels, sometimes even very similar expressions, when speaking about some ghosts who are present in the netherworld:

	UHF (Geller)		GEN version A (ETCSL 1.8.1.4)
345	**gù ⌜ba⌝-ra-an$^{!}$-na-ab-⌜dé-e$^{?}$⌝-(en$^{?}$)** you may not call to him	193	**kur-ra gù nam-mu-un-ĝá-ĝá-an** you should not shout in the netherworld
321	**l[ú] ⌜ur-e ba⌝-an-gu$_7$ ḫé-me-⌜en⌝** whether you are the one whom a dog devoured	287A	**/lú\ ur-mah-e gu$_7$-a [igi] /bí\-du$_8$-am$_3$** did you see him who was eaten by a lion?
323	**lú ⌜ùr-ta ba-šub⌝ ḫé-me-e[n]** whether you are the one who fell from the roof	287C	**lú ùr šub-ba /igi\ [bí]-du$_8$-am$_3$** did you see him who fell down from the roof?
318	**⌜lú⌝ ^{ĝiš}tukul-a ba-an-gaz ḫé-me-⌜en⌝** whether you are the man killed with a weapon	290	**lú mè-[a] šub-ba igi bí-in-[du$_8$-am$_3$]** did you see him who fell in battle?

204 Pettinato 2000, 865-869. 'Forerunners to Udug-ḫul' (UHF): Geller 1985. Transcription and translation are Geller's.

205 Geller 1985, 100-101, *ad* lines 299 and 300. Pettinato did not fully agree with Geller about the analogy between this incantation and 'Inanna's descent', because in the incantation it is only said that Enki 'turned his mind to the netherworld', and he did not really descend to it. Pettinato also did not accept Jacobsen's interpretation of the lines 14-26 of GEN as being a real descent by Enki into the netherworld (Pettinato 2000, 865).

In conclusion, the description of Enki's journey towards the netherworld in 'Gilgameš, Enkidu and the netherworld' as such is a unique story. 'Turning his mind to the netherworld' – as said in the incantation – is different from a descent to the netherworld. The present author – in agreement with Pettinato – has the opinion that Enki's sailing for the netherworld indeed is not intended to describe an actual descent by Enki into the netherworld. However, in the light of some parallel expressions in this particular incantation and in GEN that could be demonstrated, this journey of Enki fits very well into (the introduction of) the story of 'Gilgameš, Enkidu and the netherworld', especially in connection with Enkidu's later visit to the netherworld, his release from it, facilitated through Enki, and Gilgameš' questions about Enkidu's meeting with some ghosts in the netherworld. The story of the sailing journey is only a way of linking Enki with the netherworld and the forthcoming rescue of Enkidu out of it.

The cosmogonic information in this relatively long introduction, composed of two distinct introductions belonging to different stories, is limited. Besides the separation of **an** and **ki** – which implies that they must have been united once – and the division of the main cosmic realms – heaven, earth, the netherworld – among An, Enlil and Ereškigal, the creation of the gods is not mentioned at all. The coming into existence of mankind has been told in one neutral sentence: 'the name of mankind had been established on it (= earth)', without any further information about how and/or by whom.

2.1.8 *Enki and Ninmaḫ* [*OB*]

1	**u$_4$-ri-a-ta u$_4$ an-ki-bi-ta ba-an-[....]**	From that day, the day when heaven and earth [*were separated*$^{?}$],
2	**ĝi$_6$-ri-a-ta ĝi$_6$ an-ki-bi-ta x-[....]**	from that night, the night when heaven and earth [*were separated*$^{?}$],
3	**[mu-ri-a-t]a mu nam a[n-ki-bi-da$^{?}$] / ba-t[ar-ra-ba$^{?}$]**	[from that year], the year when the fate [of heaven and earth was determined$^{?}$],
4	**[diĝir-a]n-n[a]-ke$_{[4]}$-ne ba-tu-ud-da-a-ba**	after the gods of the heaven had been born,
5	**^{d}amalu nam-NIR-PA-šè ba-tuku-a-ba**	after the mother goddesses had been taken as spouses,
6	**^{d}amalu an-ki-a ba-ḫal-ḫal-la-a-ba**	after the mother goddesses had been distributed among heaven and earth,
7	**^{d}amalu [... ù$^{?}$]-ma-a-peš$_{11}$ ù-tu-da-a-ba**	after the mother goddesses [...] had become pregnant, had given birth,
8	**diĝir kurum$_6$-ma-bi [x...x]-bi-šè ba-ab-kéš-a**	after the gods ... because of ... of their food ration had set up an agreement,
9	**diĝir-šár-šár kíĝ-ĝá al-su$_8$-ge-eš [diĝ]ir-tur-tur dú-lum / im-íl-íl-e-ne**	numerous gods are assigned to the work, (but only) a smaller group of gods is bearing the toil;
10	**diĝir íd <im>-dun-d[u]n-[ù]-ne saḫar-bi ḫa-ra-li im-dub-dub-bu-ne**	the gods dig out canals; its sand they heap up: Ḫarali.

11	**diĝir im ì-hur-re-[ne] zi-bé inim àm-ma-ĝar-re-ne**	The gods scratch the clay! They complain of their life.
12	**u$_4$-ba ĝeštu$_2$-daĝal mud-diĝir-šár-šár-ĝál-ĝál**	At that time, the one with great wisdom, the creator among all the existing gods,
13	**^{d}en-ki-ke$_4$ engur-buru$_3$ a-sur-ra ki diĝir-na-me šà-bi u$_6$ nu-um-me**	Enki, in the deep Engur, the subterranean water – a place the inside of which not a single god can observe –,
14	**ki-nú-ni ì-nú ù-ku nu-um-zi-zi**	lay sleeping in his bed, and did not arise.
15	**diĝir ér-ra im-pà-pà-dè a-nir ĝál ì-ak im-me-ne**	The gods, continuously weeping, said: 'He caused the lament.'
16	**lú-<ù>-ku-ra ì-nú-a-ra ki-nú-bi nu-um-zi-zi**	To the one who is lying asleep – he did not arise from his bed – ,
17	**^{d}namma-ke$_4$ ama-palil ù-tud-diĝir-šár-šár-ra-ke$_4$-ne**	Namma, the primaeval mother, procreator of a multitude of gods,
18	**ér-ra-diĝir-re-e-ne dumu-ni-ir ba-ši-in-de$_6$**	has conveyed the weeping of the gods to her son:
19	**[ù]-mu-un ši-nú-ù-nam ù-mu-un ši-ku-ku-na-nam**	'Lord, while you lie there, while you are sleeping indeed, lord,
20	**[ ] /unu/$^!$ [nu-um]-zi-zi**	[*while you lie in your*] dwelling [*and do not*] arise [*from your bed*],
21	**dìm-me-er šu-dím-dím-ma-zu [x] gú$^?$-bi im-tu$_{11}$-tu$_{11}$-ne**	the gods, who have approached you with a complaint, [ ] tired themselves out.
22	**du$_5$-mu-mu ki-nú-zu zi-ga [ì-bí]-ma-al-la-zu-ta / na-áĝ-kù-zu ù-mu-e-kíĝ-ĝ[á]**	My son, arise from your bed! After you, with your wisdom, have let work your skill,
23	**kíĝ-sì-dìm-me-er-e-ne-ke$_4$ ù-mu-[e]-dím /** ***tir-ḫum*-bi ḫa-ba-tu-lu-n[e]**	(and) when you have created those who will take over the work of the gods, let them (= the gods) loose their basket.'
24	**^{d}en-ki-ke$_4$ inim-ama-na-^{d}namma-ke$_4$ ki-nú-na ba-ta-zi**	Now Enki, at the word of his mother Namma, arose from his bed.
25	**ḫal-an-kù niĝin$_2$ šà-kúš-ù-da-na TAR im-mi-ni-a[k]**	Roaming in Ḫalanku with a heavy heart, he has taken a decision.
26	**ĝeštu$_2$ ĝizzal$_x$ èn-tar [ ] nam-kù-zu mud-me-dím níĝ-nam-ma / se$_{12}$-en-se$_{12}$-šár im-ta-an-è**	The intelligent, understanding, investigative, [ ], wise one, creator of everything: he did emanate a birth goddess.
27	**^{d}en-ki-ke$_4$ da-né ba-ši-in-de$_6$ ĝeštu$_2$ ì-niĝin-e**	Enki brought her (= *the birth goddess*) at his side, (and) he directed the attention to her.

28	**^{d}en-ki-ke$_4$ mud-me-dím ní-te-a-na / šà-bi ĝeštu$_2$-ga$^!$ ù-mu-da$^!$-ni-de$_5$-ge**	When Enki had considered the character of the manifestations, which he created himself,
29	**ama-ni ^{d}namma-ra gù mu-un-na-dé-e**	he says to his mother Namma:
30	**ama-ĝu$_{10}$$^!$ mud mu-ĝar-ra-zu ì-ĝál-la-àm / zub-sìg-diĝir-re-e-ne kéš-da-ni**	'My mother, when the creature which you have suggested exists, fasten the carrier basket of the gods to him.
31	**šà-im-ugu-abzu-ka ù-mu-e-ni-in-šár**	When the birth goddesses (l. 32) have stirred for you in the clay on top of the **abzu**,
32	**se$_{12}$-en-se$_{12}$-šár im mu-e-gir$_8$-gir$_8$-re-ne / za-e me-dím ù-mu-e-ni-ĝál**	they will pinch off clay for you. When you have mould it into form,
33	**^{d}nin-maḫ-e an-ta-zu ḫé-ak-e**	let Ninmaḫ act as your companion.
34	**^{d}nin-imma$_3$ ^{d}šu-zi-an-na ^{d}nin-ma-da / ^{d}nin-šara$_2$ ^{d}nin-bara$_2$**	May Ninimma, Šuziana, Ninmada, Ninšara, Ninbara,
35	**^{d}nin-mug ^{d}mú-mú-du$_8$ ^{d}nin-⸢gùn⸣-na**	Ninmug, Mumudu and Ninguna
36	**tu-tu-a-zu ḫa-ra-gub-bu-ne**	assist you while you are giving birth.
37	**ama-ĝu$_{10}$ za-e nam-bi ù-mu-e-tar ^{d}nin-maḫ-e zub-sìg-bi ḫé-keše$_2$**	My mother, when you have determined their fate, let Ninmaḫ fasten their carrier baskets.'
38	**[ ] NI-dù nam-lú-[ulu$_3$][ ]**	[ ] … mankind [ ]
39	**[x] nam-lú-ulu$_3$ àm-[ ]**	[ ] mankind [ ]
40	**[pe]š$^?$ saĝ-e gù$^?$ àm-m[a- -sù$^?$]**	[pregn]ant$^?$; the first-born has screamed$^?$;
41	**[ ] dim$_4$ su-/unu/-RI^{su} dug-a à[m-]**	[ ] approached/was prudent; the afterbirth was [ ] in the jar;
42	**[ga-raš$^?$]^{sar}-e ĝiš-nu$_{11}$ mi-ni-in-íl nam-[maḫ$^?$-x]**	she made [the leek ?] raise his eyes; [ ]
43	**[ ] ság numun$_2$-e mi-ni-in-ri ù-tu na «bi» mu-de$_5$**	[ ] scattered; she has laid it down on the grass [206]; giving birth was consecrated.
44	**^{d}en-ki-ke$_4$ kíĝ ni$_{10}$-ni$_{10}$-da ∅**[207] **mi-ni-in-lá šà-bi ba-ḫúl**	Enki showed the work, while he completed it; their hearts rejoiced.

206 In the chapter 'Systems of religious and spiritual belief' *sub* 'Earth' [Encycl. Britan. 26 (1991) 532b-533a] we read: 'Although in polytheistic religions the earth is usually represented as a goddess and associated with the god of heaven as her spouse, only rare is there an elaborate or intensive cult of worship. (...) When the common people have displayed intensive attention to "mother earth" (*such as the practice of laying down newborn babies on the earth* [*it. JL*] and many other rites), this partially reflects older cults that have remained relatively free from warrior and nation-building peoples with their emphasis on war (as in western Sudan, pre-Aryan India, and the Indian agrarian ara of northern Mexico).' The second part of line 43 may refer to such a rite: laying down a newborn baby on the earth in order to honour Mother Earth. In my opinion the Sumerians were indeed a peaceful people.

Unless a new unbroken copy of the text of the first two lines can be found, we will never know what exactly happened with **an** and **ki** as described in these lines. From the texts that have been discussed already it may be inferred that in these lines the separation of **an** and **ki** is mentioned [208]. The suggestion made for a completion of these lines with a verb like **dím** 'to fashion, to create' is unlikely, because this would be unique [209]: until now no Sumerian or Akkadian text from the period before 1500 BCE has been discovered which describes 'the creation of heaven and earth'.

The next lines (4-7) tell in a general but short way about the birth, marriage and distribution of the gods. The story of the theogony itself most likely is not important in 'Enki and Ninmah̬'. Line 4 starts with the begetting of the gods of heaven (the Anunna?), but by whom [210]? The lines 5-7 make clear that new generations of gods appear, both in heaven and on earth. The introduction, telling that there are gods, leads to the first climax, a nadir for the gods who have to work for their own maintenance: the gods complain about their life. At the same time these circumstances form the prelude to the real climax of the first part of 'Enki and Ninmah̬': the creation of the first man. People have to take over the job of the gods. The ultimate reason for the creation of mankind (in 'Enki and Ninmah̬') or for their inspiriting [211] (in 'The Debate between Grain and Sheep') is that people are necessary in order to provide the means of sustenance for the life of the gods. People can do that by taking over the digging job of the gods ('Enki and Ninmah̬'), or by breeding cattle ('The Debate between Grain and Sheep').

The idea for the creation of substitutes for the gods, who can take over their jobs, originates from Namma (line 23), whereas Enki is responsible for the elaboration of the idea. The first thing Enki does is to create the **se$_{12}$-en-se$_{12}$-šár**, otherwise not known in the Sumerian literature. The literal meaning of this expression is proposed to be 'the one who makes both **en**s and a crowd come alive' [212]. Who is this **se$_{12}$-en-se$_{12}$-šár** ? Lambert assumed that **se$_{12}$-en-se$_{12}$-šár** = **šà-tùr** = *šassūru* 'womb; birth goddess', based on the bilingual version of line 32 [213]. However, this bilingual text, which dates from the first-millennium, sometimes differs from the Old Babylonian text. Moreover, there is no known lexical list which confirms the equation **se$_{12}$-en-se$_{12}$-šár** = **šà-tùr** = *šassūru* [214]. It may be that in the first millennium the old and unique expression **se$_{12}$-en-se$_{12}$-šár** was not understood any more. Because of the assignment of these **se$_{12}$-en-se$_{12}$-šár**, especially during the delivery, they function as the midwife.

In '*atra-h̬asīs*' [215] is related that man was created from clay mixed with the flesh and blood of a slaughtered god:

---

207 ∅ = erasure

208 In the edition of this text (no. 8) I have made a suggestion for the Akkadian verb in the broken NA text of line 2: the Dt-stem of *paṭāru* 'to be loosened, to fall apart' [CAD P, 286 ff].

209 Jacobsen 1987, 153; Klein 1997a; ETCSL 1.1.2.

210 The NA text deviates in this respect and speaks about the begetting of the gods of heaven and earth.

211 See ch. 2.1.6. In 'The Debate between Grain and Sheep' people already existed but they were uncivilized.

212 For a discussion of the meaning of **se$_{12}$-en-se$_{12}$-šár**: see Appendix Text Editions no. 8 *ad* line 26.

213 Lambert 1992a, 132.

214 CAD Š II 145: the lexical part at *šassūru* A only mentions **šà-tùr** = *šassūru*.

215 Translation of the present author. The content of this passage is discussed in more detail in ch. 4.5.2.

I:208 Let them slaughter one god, so that
I:209 the gods become freed (from their duties) [216] thereby [217].
I:210 Let Nintu (l. 211) with his flesh and blood
I:211 mix clay,
I:212 so that god and man are mixed
I:213 together in the clay.
I:214 Let us hear in future days a drum;
I:215 let there be the spirit from the god's flesh:
I:216 let it (= *the spirit*) make its characteristic recognizable [218] to the living, and
I:217 let the spirit be there in order not to forget this. [219]

In comparison with the story told in 'Enki and Ninmaḫ', there is in '*atra-ḫasīs*' a new element in the creation of man, viz. the slaughter of a god and the use of his flesh and blood, which should be mixed with clay, to create man. The flesh of the god contained *eṭemmu* 'spirit, ghost', and that spirit became present in man. The spirit had belonged to a god, thus in the newly created man a divine element was present. This created the basis for the continued existence of man, at least of his soul, in the netherworld after his death. His memory should be kept alive, as is expressed in line 217 of '*atra-ḫasīs*' [220]. That is a difference between the two stories: in '*atra-ḫasīs*' the immortality of the spirit/ghost/soul was explained and connected with the origin of life of man [221]. In 'Enki and Ninmaḫ' this aspect was not dealt with.

But in the discussion here the focus is on the violent action of slaughtering a god. Steinkeller studied early Semitic literature and third millennium seals with mythological motifs. He concluded that the motif of theomachy or the slaughter of a god, which is so 'conspicuous and prevalent' in the Semitic mythology, is completely unknown in the Sumerian mythology [222]. The difference observed between the creation story of man in 'Enki and Ninmaḫ' and that in '*atra-ḫasīs*' confirms the conclusion of Steinkeller.

The end of the first part of 'Enki and Ninmaḫ' (lines 38-44) is quite damaged. A tentative conclusion about the content of these lines is, that Namma's pregnancy and the birth of the first man was described (lines 38-40), and finally the usual birth rituals (lines 41-43) have been mentioned: putting the afterbirth in a jar, opening the baby's eyes, and laying the newborn baby on the earth.

---

216 Oshima (2012, 415, note 56) suggested the meaning: 'to be cleared (from their duties)' [CAD E, *elēlu*, Dt-stem)].

217 For the reading *i-na qé-⌜er⌝-bi* instead of *i-na ṭi-⌜i⌝-bi*: see George and Al-Rawi 1996, 187 *ad* 92.

218 CAD I-J 33-34 *ad* 6: *šūdû* 'to announce, to proclaim, to make recognizable, to mark'.

219 The present translation and interpretation differs from that given by Katz (2005, 58 and notes 12 and 13), who argues that Nintu is subject of line 216.

220 These elements will be discussed in more detail in ch. 4 *ad* Anthropogeny.

221 See also Katz 2005.

222 Steinkeller 1992, 246-247: 'I think that anyone familiar with Sumerian mythology will agree that the scene of general mayhem (...) somehow does not go together with the well-ordered system of divine relations projected by Sumerian religious sources. The same is true of seals (...) showing the mutilation and death of a mountain-god, for the slaughter of a god is likewise a motif that is foreign to Sumerian mythology.' Boehmer (1957-1971; 1965, 49-59) mentioned that struggling gods are scarcely to be found in the visual arts before and after the Akkad period.

In Appendix 1 to the edition of 'Enki and Ninmaḫ' short episodes from three greater stories are chosen to illustrate the problem raised of the gods who have to work for their maintenance and what their solution to this problem is. In Appendix 2 attention will be paid to a supposed 'other creation' of man, as some scholars believe.

Appendix 1: Gods who (have to) work

2.1.8.1a *Ninurta's exploits* [*Ur III*]

334	**u$_4$-bi-a a-silim ki-ta-du a-gàr-ra nu-um-dé**	On that day, no healthy water, welling up from the earth, did flow out over the arable land.
335	**ḫalba$_6$ du$_8$-du$_8$-du$_8$-ù u$_4$-zal-le-da-gin$_7$ kur-ra é-ri-a ba-ni-íb-íl-a**	When the ceaselessly accumulating ice, which looks like the dawn, had raised the wasteland in the mountains,
336	**diĝir-kalam-ma ba-súg-ge-eš-a**	when the gods of the Land had been assigned to a task,
337	**^{ĝiš}al ^{ĝiš}dusu-bi mu-un-lá-eš-a**	when they had taken charge of hoe and basket
338	**ur$_5$-ra-àm éš-gàr-bi ì-me-a**	– this indeed was their orderly duty –,
339	**kalam tar-bi-šè é-lú gù ba-an-dé**	there was talk of a community of men for the ploughing of the Land.
340	**^{íd}idigna nam-gu-la-ba an-šè u$_5$-bi nu-íl-e**	The Tigris, in its greatness, did not rise to its full height;
341	**kun-bi a-ab-ba-e li-[bí-lá-a] a-dug$_3$ nu-um-[ ]**	its mouth did not [reach] the sea, nor did it [carry$^?$] sweet water.
342	**nesaĝ$^?$ kar-re lú nu-gíd$^?$-e**	No one did transfer first-fruit offerings to the market-place.
343	**šà-ĝar ḫul-a níĝ nu-tu-ud**	The famine was hard, nothing was brought forth.
344	**íd-tur-tur-re šu-luḫ lú li-bí-in-ak saḫar nu-mu-da-an-zi-zi-i**	Nobody cleaned the little canals, the mud was not dredged up.
345	**gana$_2$-zid-da a nu-šub-bé ég ak nu-ĝál-la**	Water was not let down on fertile fields, the making of ditches did not exist.
346	**kur-kur-re ab-sín-na nu-gub-bu še bir-a i-im-ak**	For the lands there is nothing standing in furrows: grain was sown scattered.

2.1.8.1b *A hymn to Nibru and Išme-Dagan* [*OB*]

A 97	**^{d}en-líl**	Enlil
	^{d}nin$_x$ (EGI$_2$)-líl-bi	and Ninlil
	an-na	

igi im-mi-in-ĝál-eš	fastened their eyes upon the heaven (l. 99).
ki-a	On earth
zag mi-ni-in-ĝar-re-eš	they had established a sanctuary.
u$_4$-bi-a	On that day,
an-gal-	when in the great heaven
ki-daĝal-ba	and on the broad earth
šà-bi mu-un-è-a	their intention has come to light,
diĝir ^{d}a-nun-na-	the Anunna gods
an-ki-ke$_4$-ne	of heaven and earth
kíĝ íb-gi$_4$-gi$_4$-ne	gave orders.
^{ĝiš}**al**	Hoe
^{ĝiš}**dusu**	(and) basket,
níĝ iri ki ĝá-ĝá	things for founding cities,
šu-ba «na»	
mu-un-ĝál	were in their hands (l. 113).

2.1.8.1c *atra-ḫasīs* [*OB*]

From the Akkadian version of '*atra-ḫasīs*', two passages are discussed, that refer to the duty of the gods and their solution for it [223].

I:1	*i-nu-ma i-lu a-wi-lum*	When gods were 'man',
I:2	*ub-lu du-ul-la iz-bi-lu šu-up-ši-*[*i*]*k-ka*	they bore the corvée work, they carried the basket.
I:3	*šu-up-ši-ik i-li ra-bi-*[*m*]*a*	The basket of the gods was large, and
I:4	*du-ul-lu-um ka-bi-it ma-a-ad ša-ap-ša-qum*	the corvée work was heavy, extensive the hardship.
I:5	*ra-bu-tum ^{d}a-nun-na-ku si-bi-it-tam*	The great Anunnaku, their seven,
I:6	*du-ul-lam ú-ša-az-ba-lu ^{d}i-⌜gi-gi⌝*	let the Igigi bear the corvée work.

I:190	[*š*]*à-as-sú-ru* **lú-u$_{18}$-lu** *li-ib-ni-ma*	May the mother goddess create mankind, and
I:191	*šu-up-ši-ik ilim a-wi-lum li-iš-ši*	may man carry the basket of the gods.

In 'Ninurta's exploits' earth is depicted as a dry wasteland. The gods have a task: 'they have taken charge of hoe and basket', and this was their normal duty. It is not explicitly said, but of course this task was meant to provide food and drink for themselves. Then follows a rather mysterious sentence (l. 339): **kalam tar-bi-šè é-lú gù ba-an-dé**: 'there was talk of a community of men for the ploughing of the Land'. This sentence may be explained

---

223 The Akkadian text of these lines has been copied from Lambert and Millard 1999; 42, 56.

as follows: the gods are talking, probably complaining, about their task. The plan is proposed to create people in order to cultivate the Land (= Sumer). However, this idea is not further elaborated in this story. The text continues relating how miserable the circumstances of the Land are, and describing how Ninurta finally finds a solution with respect to water management [224].

The 'Hymn to Nibru and Išme-Dagan' mentions only in a very short and isolated episode that the gods were working with hoe and basket, but that they did not use these instruments for tilling the soil: hoe and basket are 'things for founding cities'.

'*atra-ḫasīs*' opens with one of the most discussed cuneiform lines: *i-nu-ma i-lu a-wi-lum*: 'when gods were 'man' ', they bore the corvée work, they bore the basket, in other words: it opens with the toil of the gods. In '*atra-ḫasīs*' the solution to free the gods from their heavy forced labour is the creation of mankind, indicated in line I:190, but fully worked out in a later episode. The creation of man as told in '*atra-ḫasīs*' differs substantially from the story in 'Enki and Ninmaḫ' (see above).

Appendix 2: Another 'creation' of man?

2.1.8.2 ***The Sumerian Flood Story*** [*late OB*] [225]

Obverse column i
[*About 36 lines missing*]

1'	**[...] im-ĝá-⌜ĝá⌝ [...]**	' [...] he/she/it places [...]
2'	**nam-lú-ulu$_3$-ĝu$_{10}$ ḫa-lam-ma-bi-a ga-ba-⌜ni]-[ib-gi$_4$-gi$_4$ $^?$]**	I want to reconsider the forsaking of my human people;
3'	**^{d}nin-tu-ra níĝ-dím-dím-ma-ĝu$_{10}$ sì-[sì-bi-a$^?$] / ga-ba-ni-ib-gi$_4$-[gi$_4$]**	on behalf of Nintu I want to reverse the degrading of my creatures;
4'	**ùĝ ki-ùr-bi-ta ga-ba-ni-ib-gur-ru-dè**	I will let the people turn from its habitat.
5'	**iri^{ki}-me-a-bi ḫé-em-mi-in-dù / ĝissu-bi ní ga-ba-ab-dúb-bu**	Let them (*the people*) build their cities (and) I will calm down their shadow.
6'	**é-me-a šeg$_{12}$-bi ki-kù-ga ḫé-em-mi-in-šub**	Let them put down the bricks of temples on holy places.
7'	**ki-eš-bar$^{!}$-a ki-kù-ga ḫé-em-mi-ni-ib-ri**	Let them establish places of divination on holy places,

224 The ultimate goal of this text is the glorification of Ninurta.

225 Although about 36 lines from the beginning of column *i* are missing, it is evident that at least this beginning of 'The Sumerian Flood Story' is different from the Akkadian introductions to the story of the flood, as told in '*atra-ḫasīs*' and 'Gilgameš, tablet XI'. But because of the incompleteness of 'The Sumerian Flood Story' it is not possible to make a full and justified comparison between that story and those of '*atra-ḫasīs*' and 'Gilgameš, tablet XI'.

8'	**kù a níĝ-izi-te-na si mi-ni-in-si-sá**	(and when) they have arranged there pure water that has quenched fire,
9'	**ĝarza me-maḫ šu mi-ni-ib-šu-du$_7$**	have completed the divine rules and exalted powers,
10'	**ki a im-ma-ab-dug$_4$ silim ga-mu-ni-in-ĝar**	(and) have irrigated the earth, (then) I shall establish well-being there.'
11'	**an ^{d}en-líl ^{d}en-ki ^{d}nin-ḫur-saĝ-ĝá-ke$_4$**	When An, Enlil, Enki (and) Ninḫursaĝa
12'	**saĝ-gíg-ga mu-un-dím-eš-a-ba**	had formed the black-headed people,
13'	**níĝ-gilim ki-ta «ki-ta» mu-lu-lu**	they made small animals crawl out of the earth,
14'	**máš anše níĝ-úr-limmu edin-na / me-te-aš bí-íb-ĝál**	they made goats, donkeys (and) all kinds of quadrupeds that are appropriate in the plain be there.

There has been some misunderstanding about the interpretation of the text of this obverse column *i*, first of all due to the word **ḫa-lam** (line 2'). As is discussed in my edition of this text [226], some scholars believe that already in this sentence the destruction of humankind was indicated. The second phrase that caused confusion was the text about the 'creation of the black-headed people' (ll. 11'-12'): how could these lines be explained, because people existed already?

As far as the remaining text of column *i* is understandable, it is not dealing at all with the creation of humankind. The text of line 2' is clear: people do already exist at the very moment that the remaining text of column *i* starts. The circumstances under which those people lived have to be inferred from the lines which follow. In the lines 2'-3' we read about 'forsaking my human people' and 'degrading of my creatures'. This means that human beings already existed, but were deteriorating for unknown reasons.

In the 'Sumerian Flood Story' line 3' we arrive at the moment that a god, presumably Enki, says that he wants to reverse the degrading of his creatures, and that he will do that on behalf of Nintu [227]. This may imply that the creation of mankind has not been a meaningless action [228]. It is the god's wish that from now on this people leave their way of life, which seems to be 'nomadic'. Their wandering can be inferred from Enki's promise: 'I will calm down their shadow', a very poetic way of saying that people are travelling around. In other words: he would give them a sedentary way of life. But therefore people have to do the following: building cities – another indication that people are not yet living sedentary – and temples, establishing places of divination, completing divine rules and exalted powers, and irrigating the earth. Thereafter the god will establish well-being. Although it is not explicitly stated, it is hardly possible to misunderstand these lines: it is the god's wish that the cultus will be installed and performed. Cultus means: people care

---

226 See the Appendix Text editions, 8.2: The Sumerian Flood Story.

227 The supposition that Enki is speaking in these lines is based on Enki's involvement in the creation of man in other stories ('Enki and Ninmaḫ'; '*atra-ḫasīs*').

228 As we have seen in some other stories, man was created to take over the tasks of the gods, who had to care themselves for their daily rations. See the Appendix Text editions: 8.1.a Ninurta's exploits; 8.1.c *atra-ḫasīs*; 10. KAR 4.

for the gods. Only then will there be prosperity for humankind. At first glance one might explain this conversion or promotion of mankind from uncivilized beings into a people living in cities as an aetiology of human civilization. The deeper background however is the self-interest of the gods, presented in a masked altruistic attitude.

The lines 11'-12' of 'The Sumerian Flood Story' mention the formation (**dím**) of the black-headed people, the Sumerians. This formation is not a 'creation *de novo*' of man in general, because people were already present [229]. It is a question of *re*-forming, improving, finishing off what had once been started, which results in civilized people worthy to be called Sumerians. Is it only to facilitate the life of this people that the gods later created all kinds of animals (the lines *i*: 13'-14') ? At this point in the story the text breaks off again for about 35 lines.

With respect to the initial presence of *uncivilized* people as described in the lines 2'-4' of 'The Sumerian Flood Story' there are a few parallels in other texts. In the first sentence of 'How grain came to Sumer' we read: '*Men used to eat grass with their mouths like sheep.*' [230] In 'The Debate between Grain and Sheep' the birth of the Anunna gods (ll. 1-2) is followed by an enumeration of who and what are absent (ll. 3-19). Then the presence of human beings who behave like animals is mentioned [231]. Yet, the Sumerian text in this passage is unambiguous in the matter of 'mankind': **nam-lú-ulu$_3$**. When appeared that the Anunna gods could not satisfy themselves with the products of Grain and Sheep, they decided to 'inspirit mankind'. This inspiriting of mankind may be translated as an amelioration of those beings from their uncivilized, animal-like status into the position of civilized humans. Only then would people be able to care for the gods.

In 'The Debate between Grain and Sheep' no reason is given for the initial presence or creation of those animal-like humans [232]. In 'The Sumerian Flood Story' the reason for their appearance may have been given in the lost sentences, because Enki regrets their deterioration.

2.1.9 *The Song of the Hoe* [*OB*]

1	**en-e níĝ-ul-e pa na-an-ga-àm-mi-in-è**	Then the lord has also let appear what was present in the bud!
2	**en nam-tar-ra-na šu nu-bal-e-dè**	The lord, who does not alter a fate decreed by him,
3	^{d}**en-líl numun-kalam-ma ki-ta è-dè**	Enlil, who makes the seed of the Land come out from the earth,

---

229 A real 'creation *de novo*' has been described in 'Enki and Ninmaḫ'.

230 ETCSL 1.7.6.

231 The 'Debate between Grain and Sheep' ll. 20-25: 'Mankind of those faraway days had not got to know eating bread, had not got to know a garment (or) something to dress! The people moved on all fours; like sheep they ate grass with their mouths, the water from the gardens they were drinking.'

232 This applies in general for animals: they are just there, without mentioning the reason of their presence (of course: they may function as food for the gods). 'Animal-like' is only a way of describing people who behave like barbarians, which is also a Sumerian denomination for nomads.

4	**an ki-ta bad-re$_{6}$-dè saĝ na-an-ga-àm-ma-an-sì**	has then set himself to separating heaven from earth,
5	**ki an-ta bad-re$_{6}$-dè saĝ na-an-ga-àm-ma-an-sì**	has then set himself to separating earth from heaven.
6	**uzu-è-a saĝ mú-mú-dè**	In order to let the first-born grow in Uzuea,
7	**dur-an-ki-ka bulug$_{2}$ nam-mi-in-lá**	he has then erected the axis at Duranki.
8	^{ĝiš}**al-e mu-un-ĝar u$_{4}$ al-è-e**	The hoe he used! And the sun rose.

The lines 9-17 have been omitted in this edition.

18	**uzu-è-a** ^{ĝiš}**al nam-mi-in-dù**	In Uzuea he indeed let the hoe enter;
19	**saĝ-nam-lú-ulu$_{3}$ ù-šub-ba mi-ni-in-ĝar**	he placed the first of humankind in a brick-mould.
20	^{d}**en-líl-šè kalam-ma-ni-šè ki mu-un-ši-in-dar-re**	In favour of his Land he (*the first one*) split open the earth towards Enlil.
21	**saĝ-gíg-ga-ni-šè igi-zi mu-un-ši-in-bar**	He (*Enlil*) looked with favour at his black-headed (*first*) one.
22	^{d}**a-nun-na mu-un-na-su$_{8}$-su$_{8}$-ge-eš**	The Anunna stepped forward to him (*Enlil*).
23	**šu-bi kiri$_{3}$-ba mu-un-ni$^{!}$-ĝál**	They greeted him respectfully.
24	^{d}**en-líl a-ra-zu-a mu-ni-in-ḫuĝ-e-ne**	They calmed Enlil with a supplication.
25	**ùĝ-saĝ-gíg-ga** ^{ĝiš}**al mu-un-da-ab-bé-ne**	They expressed before him their wish for a black-headed people.
26	**nin en ù-tu-da lugal ù-tu-da**	The lady who gave birth to the **en**, who gave birth to the king,
27	^{d}**nin-men-na-ke$_{4}$ tu-tu al-ĝá-ĝá**	Ninmena now establishes procreation.
28	**mas-su-an-ki-a en** ^{d}**nu-nam-nir-ra**	For the leader of heaven and earth, Lord Nunamnir,
29	**saĝ-zi saĝ-kal-la mu-ni-ib-še$_{21}$-še$_{21}$**	she (*Ninmena*) calls them, who are loyal and strong, by name for him;
30	**saĝ-bi gu-dili-àm mu-un-è-dè-a**	she, who forms a row of them for him,
31	**diĝir-re-e-ne-ra kurum$_{6}$-šè mu-un-dab$_{5}$-bé**	she makes them provide for the daily rations of the gods.
32	^{d}**en-ki-ke$_{4}$** ^{ĝiš}**al-a-ni zà-mí ba-an-du$_{11}$**	Enki praised his (*Enlil's*) hoe.
33	**ki-sikil** ^{d}**nisaba eš-bar-ra ba-an-gub**	The young lady Nisaba noted down the decisions.
34	^{ĝiš}**al-mul** ^{ĝiš}**al-kù-ba šu mu-un-ne-ĝál**	They (*the black-headed people*) had the shining and holy hoe in their hands.

In the introduction to 'The Song of the Hoe' [233] there is an important role reserved for Enlil. First of all, he is the only one who separates heaven and earth, as it is also related in some UD-GAL-NUN-texts [234]. This is in contrast with the lines 11-12 of 'Gilgameš, Enkidu and the netherworld', where An took the heaven and Enlil took the earth.

In line 6 the creation of mankind is announced, but no reason is given as to why this should happen. This creation is then carried out in the lines 19-20. This process is described as 'the placing of the first of mankind in a brick mould'. This means that the first man was fashioned with the aid of clay. This clay model was placed in a brick mould. The brick mould – in my opinion – functioned as a metaphor for the womb. Once this first human foetus had become mature, he broke through this clay model and his figurative womb, as the text says: 'he split open the earth' – a metaphor for the birth – (l. 20) and he appeared as a human being (l. 21). This is obviously another type of creation of mankind than that related in 'Enki and Ninmaḫ'. That Enlil only made a 'prototype' of a black-headed person, i.e. a Sumerian, is proven in the following lines in which the Anunna, after a greeting ceremony, ask Enlil for a whole people of black-headed ones. In a concise way the story tells that, probably on instigation of Enlil, Ninmena takes care of the procreation of mankind. Ninmena, in the list **an** = *ᵈa-nu-um* classed among the mother goddesses [235], literally means: 'Lady of the crown'. That is exactly what the description of this goddess says in line 26: 'The lady who gave birth to the **en**, who gave birth to the king'. At first this goddess seemed to be only responsible for the birth of the leading classes: the **en**s and the **lugal**s. But at the special request of Enlil and, via Enlil, of the Anunna, Ninmena takes care of producing 'loyal and strong individuals', reliable people of the working class who will be engaged in the daily care of the gods (lines 28-31). In line 31 the actual reason for the wish of the Anunna for a black-headed people is revealed. This reason seems always the same: human beings are necessary to take care of the gods, i.e. of their daily rations of food and drink.

With regard to the creation of man, its description in 'The Song of the Hoe' is a very brief one, in contrast to the one in the story of 'Enki and Ninmaḫ'. But of course, this is because this special text was mainly designed as a song of praise for the hoe, where the syllable ***al** has been used in as many ways as possible.

Although several scholars have some doubt about the seriousness of the text of 'The Song of the Hoe' – they consider the text to be mainly a play on the syllable ***al** [236] –, the mythical elements are known and have to be taken seriously: the separation of heaven and earth by Enlil, the creation of man who has to care for the gods. The way in which this creation took place is not attested elsewhere, but it is not influenced by the word play on the syllable ***al**.

---

233 The composition 'The Song of the Hoe' is one of the so-called Decad, a set of compositions belonging to the Sumerian school curriculum in the Old Babylonian period (Tinney 1999; Michalowski 2010).

234 IAS 113, 136, 203; see ch 2.1.1b and comments on the lines 8-9 of 'Gilgameš, Enkidu and the netherworld' in the Appendix Text editions, no. 7.

235 Litke 1998, 68, II-21. Ninmena as mother goddess will be discussed further in chapter 3 'God lists'.

236 See the edition of this text [Appendix Text editions, no. 9], note at the comments on line 8 for references.

2.1.10 *KAR 4* [*MB*]

Obverse

1	**u$_4$ an ki-ta tab gi-na bad-a-ta-eš-a-[ ]**	After heaven had been separated from earth, with which it was firmly joined,
2	**^{d}ama-^{d}inanna-ke$_4$-e-ne ba-se$_{12}$-se$_{12}$-e-dè**	so that the mother goddesses could live there (i.e. in heaven),
3	**u$_4$ ki ĝá-ĝá-e-dè ki dù-dù-a-ta**	after the earth had been built in order to be able to establish cult places there,
4	**u$_4$ ĝiš-ḫur-ḫur an-ki-a mu-un-gi-na-eš-a-ba**	when the designs in heaven and on earth had been fixed,
5	**e pa$_5$-re šu si-sá ĝá-ĝá-e-dè**	when the Tigris and the Euphrates had set out their banks (l. 6),
6	**^{íd}idigna ^{íd}buranun gú-ne-ne ĝar-eš-a-ba**	so that dikes could be built and ditches could be cut in good order (l. 5),
7	**an ^{d}en-líl ^{d}utu ^{d}en-ki / diĝir-gal-gal-e-ne**	An, Enlil, Utu and Enki, the great gods,
8	**^{d}a-nun-na diĝir-gal-gal-e-ne**	the Anunna, the great gods,
9	**bara$_2$-maḫ ní-te mú-a mu-un-durun$_x$-<eš> $^{!!}$/ ní-te-a-ni šu mi-ni-íb-gi$_4$-gi$_4$**	remained in an exalted shrine, grown with awe. He himself (i.e. Enlil) spoke there:
10	**u$_4$ ĝiš-ḫur-ḫur an-ki-a mu-un-gi-na-eš-a-ba**	'Now that the designs in heaven and on earth have been fixed,
11	**e pa$_5$ šu si-sá ĝá-ĝá-e-dè**	and the Tigris and the Euphrates have set out their banks (l. 12),
12	**^{íd}idigna ^{íd}buranun / gú-ne-ne ĝar-eš-a-ba**	so that dikes can be built and ditches can be cut in good order (l. 11):
13	**a-na-àm ḫé-en-bal-en-dè$^{!}$ (ZE$_2$)-en / a-na-àm ḫé-en-dím-en-dè$^{!}$ (ZE$_2$)-en**	what can we change? what can we create?
14	**^{d}a-nun-na diĝir-gal-gal-e-ne**	Anunna, great gods,
15	**a-na-àm ḫé-en-bal-en-dè$^{!}$ (ZE$_2$)-en / a-na-àm ḫé-en-dím-en-dè$^{!}$ (ZE$_2$)-en**	what can we change? what can we create? '
16	**diĝir-gal-gal-e-ne mu-un-su$_8^{!}$ (SUR)-ge$^{!}$ (RE)-eš-a / ^{d}a-nun-na diĝir nam-tar-re**	The great gods, who were standing there, the Anunna, the gods who decree the fate,
17	**min-na-ne-ne ^{d}en-líl-ra mu-un-na-ni-íb-gi$_4$-/gi$_4$**	two of them answered Enlil:
18	**uzu-mú-a^{ki} dur-an-ki-ke$_4$**	'In Uzumua, the bond of heaven and earth,
19	**^{d}NAĞAR ^{d}NAĞAR im-ma-an-šum-en-dè$^{!}$ (ZE$_2$)-en**	we shall slaughter the gods Alla and Illa,

20	**úš-úš-e-ne nam-lú-u$_{x}$-lu mú-mú-dè**	so that their blood makes mankind grow.
21	**á-ĝiš-ĝar-ra-diĝir-e-ne éš-gàr-bí ḫé-a**	Let the work assignment of the gods be its job (= *the job of mankind*):
22	**u$_{4}$-da-rí-šè e-sur / gi<-na-e>-dè**	to establish for ever the boundary ditches;
23	**^{ĝiš}al ^{gi}dusi šu-bí / ĝá-ĝá-e-dè**	to take in hand the hoe and the pannier;
24	**é-diĝir-gal-gal-e-ne / bara$_{2}$-maḫ-a túm-ma**	planning (l. 25) the houses of the great gods, befitting an exalted shrine,
25	**a-gàr-a-gàr-re ĝiš ḫur-ḫur-re**	(and) meadows,
26	**u$_{4}$-da-rí-šè e-sur / gi-na-e-dè**	to establish for ever the boundary ditches;
27	**e si sá-e-dè-«zé-en» / gi-na-e-dè**	to put in order and to consolidate the ditches;
28	**é-limmu-sud ú-ḫi-a / nun-nun-e-dè**	to increase all kinds of plants for the estates in the four corners;
29	**šeĝ$_{14}$-šeĝ$_{14}$ [x]**	... the rains ... ;

Reverse

1	**ki-ùr-sur gi-na-e-dè / gur$_{7}$ nam-mi-ni-íb-gur-gur-re**	to establish the boundaries of the dwelling grounds: then it (= *mankind*) will pile up heaps of grain.
2	*ḫi-e-pi* | *ḫi-e-pi*	*break*
3	*ḫi-e-pi* | *ḫi-e-pi*	*break*
4	*ḫi-e-pi* | *ḫi-e-pi*	*break*
5	**a-šà-^{d}a-nun-na-ke$_{4}$-e-ne dùg-dùg-e-dè**	to make the fields of the Anunna fertile;
6	**ḫé-ĝál kalam-ma nun-nun-e-dè**	to increase abundance in the Land.
7	**ezen-diĝir-e-ne šu-du$_{7}$-a**	The feast for the gods is completed,
8	**a-sed dé-dé-da**	when cold water is poured out,
9	**unu$_{2}$-gal-diĝir-e-ne bara$_{2}$-maḫ-a túm-ma**	and when the large abode of the gods is suitable for an exalted shrine.
10	**^{d}ul-le-ĝar-ra an-né-ĝar-ra / mu-ne-ne ì-pà-da**	Their names, Uleĝara and Aneĝara, are mentioned.'
11	**gu$_{4}$ udu máš-anše ku$_{6}$ mušen ne-ta-a / ḫé-ĝál kalam-ma nun-nun-e-dè**	In order to increase abundance in the Land by means of cattle, sheep, wild animals, fishes and birds,
12	**^{d}en-ul ^{d}nin-ul / ka-kù-ga-a-né zur-zur-re**	Enul and Ninul, while their pure mouth is honouring,

13	**^{d}a-ru-ru nam-nin-a túm-ma**	(and) Aruru, who is befitting the **nin**-ship,
14	**ĝiš-ḫur-gal-gal ní-ba mu-un-ḫur-ḫur-re-<ne>$^{!}$**	draw up their own great plans.
15	**gašam gašam lú$^{!}$-IM lú$^{!}$-IM**	The wise one after the wise one, ignorant person after ignorant person,
16	**še-gin$_7$ ní-bi ne ki-ta si$_{12}$-si$_{12}$ ki dím**	– like grain, that of itself becomes green from the earth and that adorns the earth,
17	**níĝ nu-kúr-ru mul-<an-na> da-ri-šè**	something that cannot be changed, just like an eternal star in the sky, –
18	**u$_4$-gi$_6$-na-ta ezen-diĝir-e-ne / šu-du$_7$-a**	to make day and night the feast for the gods perfect,
19	**ní-te-a-ni ĝiš-ḫur-gal-gal-la / mu-un-ḫur-ḫur-re**	he draws up himself great plans.
20	**an ^{d}en-líl / ^{d}en-ki ^{d}nin-maḫ / diĝir-gal-gal-e-ne**	Where [l. 70] An, Enlil, Enki and Ninmaḫ, the great gods,
21	**ki nam-lú-u$_x$-lu ba-ni-in-dím-eš / ^{d}nisaba ki-bé nam-en-na an-gub**	have created mankind, there is Nisaba honoured in her **en**-ship.
22	**ad-ḫal** *mu-du-ú mu-da-a lu-kal-lim* **al-til igi-kár gaba-ri libir-ra** / *šu^{II} ki-din-^{d}30* **^{lú}dub-sar-banda A** *su-ti-e* **^{lú}dub-sar-lugal**	Secret. The one who is competent should show it only to someone who is also competent. It has been completed and collated from an old copy. The hand of Kidin-Sîn, junior scribe, son of Sutû, the king's scribe.
23	[ ] A [ ]	

Just like the other texts from the second millennium, KAR 4 does not pay any attention to the time before **an** and **ki** had been separated, except in *Obverse* line 1 via the expression **tab gi-na**. Of course, what has been separated, must have been united before. But it is not inconceivable that the word **tab** in this respect has a deeper meaning. In ePSD we find **tab** 'to be parallel; companion, partner' (**tab** consists of two parallel horizontal wedges). The expression **tab gi-na** suggests that before the separation **an** and **ki** were partners; otherwise translated: 'were lying parallel'. This is a strong suggestion that there may have been a primaeval intercourse or marriage between **an** and **ki**.

The separation of **an** and **ki** seemed to be a *conditio sine qua non* for the mother goddesses to be able to live in heaven (line 2). This sentence is also somewhat cryptic, because what exactly these mother goddesses were doing is not related. From the lines 7-8 it may be inferred that the mother goddesses had given birth to the Anunna gods. These Anunna gods had prepared the earth so that cult places for them could be established; heaven and earth are definitively shaped, and even Euphrates and Tigris are present. As line 3 and line 5 make clear (by means of the verbal ***e-dè** forms), cult places were not yet present but had to be built, just like the dikes and ditches, but the necessary conditions therefore had been created by the gods. That all being done, Enlil asked the other gods: 'And what now? Where do we go from here?' (*obv.* ll. 10-15). Without wasting words some gods proposed to Enlil that two gods (Alla and Illa) should be slaughtered 'so that their blood makes mankind grow'. The killing of gods is a typical non-Sumerian motif, but a Semitic one. An outstanding textual example of this is found in '*atra-ḫasīs*' (I: 204-230), where one god is slaughtered, and Nintu mixed clay together with his flesh and blood, in order to create man [237].

Steinkeller has studied the possible correspondence between Semitic mythological texts and glyptic motifs of the third millennium [238]. His idea was [239], that the third-millennium narrative glyptics might illustrate the early Akkadian mythology. Steinkeller based this idea on the temporal and geographic distribution of the motifs on the seals, that correlate very well with the temporal and spatial perimeters of the political and cultural influence of the Akkadians. Steinkeller's conclusion was – and I fully agree with him – that the relations within the Sumerian pantheon were generally peaceful; the rare conflicts were scarcely if ever solved through violence. The motif of theomachy, which is so conspicuous and prevalent on Sargonic seals, is completely unknown in Sumerian mythology. Moreover, mythological narrative motifs are exceedingly rare in Sumerian art throughout its history, from the Uruk down to the Old Babylonian period [240].

In contrast to the situation in '*atra-ḫasīs*', where the god who had to be slaughtered was the leader of a revolt, in KAR 4 two gods are sacrificed – apparently arbitrarily –, without their having committed any crime; at least if so, we are not told. The reason for killing these two gods is that their blood is necessary in order to make mankind grow. Why two gods? Is it to create a man and a woman? Two names are mentioned in line *rev.* 10, but their gender is not indicated. The lines *obv.* 19-20 contain all the information of the KAR 4-text about the creation process of mankind. It has to be noted that strictly speaking the creation of mankind, as mentioned in l. 20, is only a plan; line 19 reads: 'we *shall* slaughter Alla and Illa…'. The contrast with e.g. the story as told in 'Enki and Ninmaḫ' could hardly be greater. Only now (*obv.* l. 21) do we learn why man had to be created: to take over the workload of the gods. Thereafter follows a large enumeration of the tasks that have to be fulfilled by man (*obv.* l. 22 - *rev.* l. 6). Finally man has to undertake the cults of the gods (*rev.* ll. 7-9).

About Uleĝara and Aneĝara there is no unanimity among the scholars, neither about the meaning of these names [241], nor about who these both are. Pettinato comments [242]: 'Die Annahme (…) daß Ullegarra und Annegarra die zwei ersten Menschen sein sollen, bleibt unbewiesen.' Indeed, no definitive conclusion can be drawn about Uleĝara and Aneĝara. If it was only the intention to indicate the creation of both sexes, it might have been done in other words. Now these creatures have specific names. Therefore my tentative suggestion is that Uleĝara and Aneĝara are the names of the prototypes of mankind, the first wife and man.

---

237 Lambert and Millard 1999, 56-59. Their translation of the lines 210-211 is rather strange: 'From his flesh and blood let Nintu mix clay'. The Akkadian *ina* may be translated as 'in' – in the sense of 'together with' –, resulting in a more meaningful sentence: 'Let Nintu, together with his (i.e. from the slaughtered god; JL) flesh and blood, mix clay.'

238 Steinkeller 1992.

239 This was also the idea of Frankfort and others; for literature: see the notes 6 and 7 of Steinkeller 1992, 245-246.

240 Steinkeller 1992, 246-247. The motif of killing a god is also clearly demonstrated in the text *enūma eliš* and in a text known as 'The theogony of Dunnu' (Lambert and Walcot 1965), also called 'The Harab myth' (Jacobsen 1984); both texts are of Akkadian origin.

241 See the edition of this text in the Appendix Text editions, no. 10.

242 Pettinato 1971, 81, *ad* 51-53.

If we now look at the stories about the *de novo* creation of mankind that are discussed here, it appears that the narrative in 'Enki and Ninmaḫ' and the very short report in 'The Song of the Hoe' – in both of which stories only clay has been used – are very different from the tales in '*atra-ḫasīs*' and KAR 4. In the last two texts the creation of man needed the blood of slaughtered gods, and thus we may conclude that these texts are most likely of Semitic origin. This kind of violence is completely absent in the former two texts, which are therefore to be considered as Sumerian. In some texts people are described who behave like animals, and these creatures are transformed later into 'civilized' people [243]. In these texts the creation of mankind is not the main theme.

2.2 *The main points of the texts*

Table 1 gives a schematic overview of the presence (+) or absence (–) of some items in the translated Sumerian texts which are relevant for the study of the Beginnings. The following subjects are selected in this scheme and discussed in more detail:
1. The expression **u$_4$-ri-a;**
2. The cosmos;
3. The gods;
4. Man;
5. Culture.

2.2.1 u$_4$-ri-a

Van Dijk was the first scholar to draw attention to the expression **u$_4$-ri-a** and its connection with compositions that have a mythological introduction. In his study *Le motif cosmique dans la pensée sumérienne* he devoted a chapter to the expression **u$_4$-ri-a** « in illo die » [244], and he introduced this chapter as follows: 'Les événements de la cosmogonie sumérienne se centrent autour de l'expression **u$_4$-ri-a** «ce jour-là».' Van Dijk [245] considers the phrase **u$_4$-ri-a** as closely connected with 'le motif cosmique', i.e. the marriage of **an** and **ki**. With reference to the introduction of 'Gilgameš, Enkidu and the netherworld (ll. 1-26) he concludes that there is a question of violence; and moreover: 'Cette violence est l'un des caractéristiques de la cosmogonie (...). Plus encore que le déluge, «ce jour-là» « **u$_4$-ri-a**», est le prototype de toute violence destructive.' [246]. In my opinion, this remark by van Dijk needs some nuancing. Indeed, the Barton cylinder tells about a gale that was blowing unceasingly, and that there were really flashes of lightning continuously. The other episode

---

243 See 'The Debate between Grain and Sheep', 'How grain came to Sumer'. With respect to 'The Sumerian Flood Story', no conclusion can be made about a possible *de novo* creation of man.

244 Van Dijk 1964, ch. 2: 16-34. For the theory of van Dijk about 'The beginning': see ch. 1.2.1. Comments on his theory are given in ch. 4.8.2.

245 Van Dijk 1964, 16. Dietrich (1995, 57), whose ideas with respect to the Mesopotamian cosmogony are based on those of van Dijk, even goes further: 'Bemerkungswert ist es jedoch, daß das **u$_4$-ri-a**-Motiv häufig dreigliedrig auftritt, indem auch eine Nacht (**gi$_6$**) und ein Jahr (**mu**) vor Zeiten einbezogen werden. Von daher legt es sich nahe, dort, wo die Dreiergruppe "Tag - Nacht - Jahr" auftritt, *eher an einem "Schöpfungszeitraum" als an einen "Schöpfungstag" zu denken.*' (*italics JL*).

246 Van Dijk 1964, 21. He continues: 'Le jour de la «terra parturiens» est le jour de violence par excellence. Ces deux aspects: naissance de la vie et violence reviennent dans les exemples ci-dessous (...)'. It seems as if according to van Dijk birth, and especially also labour pain, are connected with violence and hence with the expression **u$_4$-ri-a**.

	third millennium						*second millennium*				
	Early Dynastic				*Ur III - OB*			*Old Babylonian*			*MB*
	IAS 114	IAS 113 136/203	Ukg 15	Barton cylinder	NBC 11108	Tree / Reed	Grain / Sheep	GEN	Enki & Ninmaḫ	Song of the Hoe	KAR 4
u_4-ri-a			(+)	+				+	+		
Cosmos / **an-ki**											
an	+		+	+	+	+	+				
ki	+		+	+	+	+					
an x **ki** interaction	+		+ shout	+ shout	+ lie	+ inter-course					
an ↔ ki separation		+						+	+?	+	+
Gods											
díd-maḫ (with superscript d)				+							
Namma									+		
An								+			+
Enki - Ninki	+		–								
Enlil - Ninlil	+		–								
Mother goddess(es)									+	+	+
Ninḫursaĝa				+			(+)				
septuplets	+			+							
Anunna					–		birth			+	+
Gods of heaven					–				birth		
Enlil	+			(+)	(+)		+	+		+	+
Enki	+						+	+	+	+	+
Gods (who) work									+	+	+
Man											
absence / presence					–		1 animal-like 2 human	+			
creation with clay									+	+	
creation with god's blood											plan
Culture											
cultus					–		(–)	+			plan
agriculture / cattle breed-ing					–		origin				plan

Table 1: Survey of some important items of the translated Sumerian texts

of violence concerns the voyage of Enki for the netherworld, told in GEN lines 23-26. Enki's boat has been attacked by a turbulent primaeval sea, but this violence is not connected with cosmogony. There remains only one example – the Barton cylinder – concerning the so-called violence that should be characteristic of the cosmogony! Contrary to van Dijk I would like to say: the opposite is true. Most stories are suffused with a spirit of peace and quiet. The 'violence', labelled by van Dijk as 'caractéristiques de la cosmogonie', points to the cosmic circumstances of the beginning in only one text. Moreover, among the protagonists there is, at least in the Sumerian texts, no violence at all.

Krebernik points to some formal aspects of Sumerian literary texts, among them the opening phrase beginning with **u$_4$-ri-a** [247]. According to Black [248] this expression is a '... specialised opening formula found in Sumerian narrative poems from the earliest times: the introduction whose purpose is to set the action of the narrative in a past of distant antiquity, a technique familiar from many literatures.' Black also adds several examples, which show that the **u$_4$-ri-a** expression could be used with some nuances [249], which are also present in the texts of this study. This formal literary technique shows a great continuity, from Early Dynastic times onwards.

Vice versa: the expression **u$_4$-ri-a** in a literary text – occurring not always at the beginning of a text – does not necessarily introduce a cosmogonic event. Examples of instances where it introduces a cosmogonic event have been presented in this study. In other texts the expression is only used as a reference to 'former days' [250], e.g. in 'How grain came to Sumer', in 'Enmerkar and the Lord of Aratta' and in 'The instructions of Šuruppag'.

247 Krebernik 1998, 322-323 and note 818. My transcription of IAS 280 differs from that of Krebernik:

	My Transcription	*Transcription of Krebernik*
i:1	**u$_4$-ri$^{!}$-[ši$^{?}$] u$_4$-ri$^{!}$-⌜ši$^{?}$⌝**	**ud-[IGI?.]ḪU ud-⌜IGI$^{!?}$⌝.ḪU**
i:2	**na-nam**	**na-nam**
i:3	**ĝi$_6$-ri$^{!}$-ši ĝi$_6$-ri$^{!}$-ši**	**ĝi$_6$-IGI.ḪU ĝi$_6$-IGI.ḪU**
i:4	**na-nam**	**na-nam**
i:5	**mu-ri-ši [mu]-ri$^{!}$-[ši]**	**⌜mu⌝-IGI.RI$^{?}$ [m]u-[IGI.]ḪU**
i:6	**[na-nam]**	**[na-nam]**

Here **ši** is supposed to be written for **šè**.

248 Black 1992, 73.

249 Black 1992, 73-74, 93-95 (the examples): 'Although the formulae are in general similar, in some cases the distant antiquity referred to is a period before even the principal gods existed; otherwise it may be a period after the separation of heaven and earth when only the gods existed, or else a period in the earliest infancy of mankind; or (in *The Instructions of Šuruppag*) the immediately antediluvian period. Within the body of a poem, individual section boundaries can often be demarcated by adverbial constructions such as **u$_4$-bi-a** "then, at that time". In such cases **u$_4$-bi-a** has a structural, rather than purely adverbial function.'

250 Cosmogonic introductions are found in the 'Barton cylinder'; 'Enki and Ninmaḫ' [ETCSL 1.1.2], ll. 1-3; 'Enki's journey to Nibru' [ETCSL 1.1.4], l. 1; 'Gilgameš, Enkidu and the netherworld' [ETCSL 1.8.1.4], ll. 1-3. References to 'former days' are: 'How grain came to Sumer' [ETCSL 1.7.6], l. 2; 'Enmerkar and the Lord of Aratta' [ETCSL 1.8.2.3], l. 6; 'The rulers of Lagaš' [ETCSL 2.1.2], l. 65; 'The Debate between Grain and Sheep' [ETCSL 5.3.2], l. 20; 'The instructions of Šuruppag' [ETCSL 5.6.1], l. 1.

The present study shows that only three out of the texts studied which tell about 'the Beginning' start with **u$_4$-ri-a**: 'Barton cylinder'; 'Gilgameš, Enkidu and the netherworld'; and 'Enki and Ninmaḫ'.

With respect to Ukg 15, if it is accepted that the order in which the columns are presented [251] should be reversed, then Ukg 15 has also an introduction pointing to primaeval times, but with a unique formulation: **u$_4$-(da) im-ma** / **ul-[lí-a] im-ma**. This expression has not been found elsewhere in the Sumerian literature.

2.2.2 *The cosmos:* an-ki

The cosmos, the universe, is represented by the Sumerian compound **an-ki**. Under this heading it is indicated in the table in which texts **an** and **ki** are present, and in which relationship they both are described.

All the texts from the third millennium mention a relation between **an** and **ki**. Because the beginning of the text of IAS 114 has been broken, it is only possible to conclude that **an** and **ki** do something together. In the other texts this relation is expressed as shouting, lying together, or having intercourse. Of course, if there is report of a relationship between **an** and **ki** [252], this implies their presence. Some texts even describe, in more than one line, details about both **an** and **ki**, especially of their behaviour, such as their preparation for a cosmic marriage, but there is not one word about their separation. As 'The Debate between Tree and Reed' shares these details with the other third-millennium texts, my tentative suggestion is that this text also dates, at least originally, from the third millennium. In some pre-Sargonic texts that are written in the so-called UD-GAL-NUN script [253] the narrative tells that Enlil has separated heaven and earth, but this action is not found in Sumerian literary texts with normal orthography from the third millennium.

The reverse applies to the group of later texts, of the Old Babylonian period and later. They only mention the separation of **an** and **ki**, but what happened beforehand is not told.

2.2.3 *The gods*

In this section is indicated which gods were present or absent in the beginning. In some texts the mere absence or presence of e.g. the Anunna gods is mentioned; there are also texts which describe the role of one or another god in a creation process. One of the items – gods (who) work – is important because it is the starting-point for the idea of the creation of mankind.

Also with respect to the gods there is a difference between the third-millennium texts and the more recent ones from the Old Babylonian period, which describe 'the Beginning'. In the older text group IAS 114 has an exceptional and unique position. This text mentions that Enki and Ninki are Enlil's parents, which as far as I know has never been recorded in any other literary text [254].

---

251 See Sjöberg 2002, 230. For discussion about the order of the columns, see the edition of this text in the Appendix Text editions no. 2.

252 For the meaning of **an** – the god An or Heaven –: see the Excursus 'The animate *vs* inanimate class of **an** and **ki**' in the Appendix Text editions.

253 See ch. 2.1.1b.

254 This parentage will be discussed in more detail in the chapter 3 'God lists' of this study. Wang (2011, 238) considers An as father of Enlil, without giving any literature reference for this

Equally surprising is the passage that tells us that Enki has been born of **an** and **ki**. Although the spelling of the Enki in line 12' has no indication that this god is a different god from the Enki mentioned in line 11', Lambert translates without comment the Enki of line 12' as Enki(g), thus indicating that this is the Enki of Eridu [255]. Most probably this is true. It has been argued that ^{d}**am-an-ki** is the so-called *emesal*-spelling for Enki [256]. In my opinion, the name ^{d}**am-an-ki** has been derived from Enki's lineage, as offspring of **an** and **ki**. Therefore the name ^{d}**am-an-ki** is most likely not an *emesal*-spelling for Enki. Occasionally only one of Enki's parents is mentioned, e.g. An as his father, or Namma as his mother [257].

The rest of the third-millennium texts notes specifically names of gods who are absent, e.g. Enki-Ninki, Enlil-Ninlil, Anunna, the gods of Heaven. In NBC 11108 a god (Enlil) or

---

statement; I cannot corroborate this parentage, simply because – to my knowledge – there is no appropriate Sumerian text.

255 Lambert 1981, 84, *sub* 5). This point will also be discussed in the chapter 3 'God lists'.

256 E.g. Schretter (1990, 153, no. 49) gives ^{d}**am-an-ki** as the *emesal*-spelling for Enki. Kramer (1940, 73, *ad* line 18) wrote: 'The name ^{d}**am-an-ki** does not mean 'the wild ox of heaven and earth', as a synthetic and superficial analysis might suggest, but in all likelihood **am-an** is simply a variant pronunciation of **umun**, the eme-SAL form of **en** (...).'. First of all, Kramer gives no argument at all for his statement that 'The name ^{d}**am-an-ki** does not mean 'the wild ox of heaven and earth' '. Moreover, if /**aman**/ should be simply a variant pronunciation of /**umun**/, then this would most likely not be written in the very 'non-Sumerian' way as **am-an**, but analogously to /**u-mu-un**/ as /**a-ma-an**/. The spelling /**aman**/ is only found in this name for Enki; there is not one instance where the *emesal*-form of **en** has been written as /**aman**/; only /**umun**/ has been used in this respect [data via a search in ETCSL-corpus]. In 'A *šir-šag-ḫula* to Damgalnuna' [ETCSL 4.03.1], ll. 1-2, which does not use *emesal*-text, the names ^{d}**en-ki** and ^{d}**am-an-ki** are used in a parallel way. Other non-*emesal* texts that use the name ^{d}**am-an-ki** are the 'Nippur god lists': SLT 123 *ii*: 1 and SLT 124 *i*: 24, and the 'Isin god list' (Wilcke 1987, 94: A I 34, B I 5') (see also ch. 3 of this study). The *emesal* god lists probably date from the second half of the second millennium B.C.E. (Lambert 1957-1971, § 10, 477). In 'Inana's descent to the netherworld' [ETCSL 1.4.1] the following can be observed. In the lines 58-60 and 65, Inanna speaks to Ninšubur, her female servant, and Enki's name is written as ^{d}**am-an-ki**. But in line 212, when Ninšubur speaks, Enki's name is written as ^{d}**en-ki**, while other words in her speach are written in *emesal*. Also the opposite can be observed: in a 'neutral' narrative sentence (e.g. l. 209) the name of Eridu has been written in *emesal*: **uru$_2$-ze$_2$-eb^{ki}**. Therefore, in my opinion, this text does not prove that ^{d}**am-an-ki** should be an *emesal* spelling for Enki. Perhaps the variation /**aman**/ *vs.* /**umun**/ is a later development, while the 'normal' name ^{d}**am-an-ki**, as an epithet of Enki, was already in use. This could explain why ^{d}**am-an-ki** was included in the *emesal* vocabulary (MSL IV, I: 38). The possibility that ^{d}**am-an-ki** has been mistakenly included in the *emesal* vocabulary is not unlikely, because it contains more mistakes with respect to Enki: in the lines 2 and 3 of this vocabulary the primaeval pair Enki-Ninki have been equated with Enki (of Eridu) and his wife Damkina (Borger 1998, 18). My tentative conclusion is that ^{d}**am-an-ki** is not a specific *emesal* spelling for Enki. If ^{d}**am-an-ki** is an alternate (non-*emesal*) name or epithet for the 'Eridu' Enki, then the final k/g in the name of this god – often seen in genitive constructions like 'the X of Enki' – can easily be explained. At the transition of the god's name from ^{d}**am-an-ki** [= *^{d}**am-an-ki-ak**] to ^{d}**en-ki**, the final ***ak** of the first name was transferred to the second name: *^{d}**en-ki-ak** → *^{d}**en-ki-k**. Or, more simply: the name of Eridu's Enki means 'Lord <u>of</u> the earth', in Sumerian *^{d}**en-ki-ak** → *^{d}**en-ki-k** → ^{d}**en-ki**.

257 An as his father: 'Enki and the world order', [ETCSL 1.1.3], ll. 61-80; Namma as his mother: 'Enki and Ninmaḫ', [ETCSL 1.1.2], *a.o.* ll. 17-18.

goddess (Inanna) is mentioned, but only to refer to the absence of the cultus belonging to that god. The Barton cylinder presents active gods: Ninḫursaĝa has intercourse with a god who has been named most likely in the broken part of column *i*. Besides, there are **^{d}íd-maḫ** and 'a god of the river'.

Two OB texts report the birth of a group of gods: 'The Debate between Grain and Sheep' (birth of the Anunna), and 'Enki and Ninmaḫ' (birth of the gods of heaven). In the other texts of the OB group the gods are present, mostly in an active sense. A special subject in the more recent texts is the duty of the gods to work like men. This seems to provide a logical bridge to the next subject: the creation of man [258].

2.2.4 *Man*

The table shows in which texts man is mentioned and in which respect: his mere absence; his presence in animal-like form or as a civilized human being; his creation or the plan for it.

From the older text group, NBC 11108 mentions the absence of a particular group of people – priests and priestesses –, so that no cultic rites could be performed. In 'The Debate between Grain and Sheep' men are present, but originally they behave like animals. Only when the gods need their daily rations of food, they 'transform' the primitive animal-like man to someone who is able to work in agriculture and cattle breeding. The *de novo* creation of mankind, the idea of which was mentioned briefly in 'Ninurta's exploits' [259], is related in two texts of the more recent text group: 'Enki and Ninmaḫ' and 'The Song of the Hoe'. In both instances man is created with the aid of clay, without the blood and/or flesh of the slaughtered gods. In the most recent text edited in this study – KAR 4 – however, the gods only suggested the creation of man; for this they would use the blood of slaughtered gods.

The common goal for the creation or civilization of mankind is the same in all stories, and may be formulated as: man has to care for the gods, to take over the tasks originally carried out by the gods, and to provide the gods with food and drink.

2.2.5 *Culture*

In this last section several cultural aspects are brought together – the cultus, agriculture, cattle breeding – and the way in which these are mentioned in the texts.

NBC 11108 gives a rather detailed report about the absence of several cultic elements. In 'Gilgameš, Enkidu and the netherworld' the presence of activities that may be considered as belonging to a cultus (ll. 4-7) is mentioned. 'The Debate between Grain and Sheep' stands midway between: after an extensive summing up of what is absent in the realm of cultus and culture, the beginning of animal breeding is mentioned. That started from the moment that the people are urged to begin with the breeding – in this case of sheep – in order to fulfil the needs of the gods.

---

258 The text of 'Ninurta's exploits', in which 'there was talk of a community of men for the ploughing of the Land' (l. 339), might be composed in the Ur III-period (Cooper 1978, 10).

259 'Ninurta's exploits' [ETCSL 1.6.2], l. 339.

2.3 *Preliminary conclusions*

At the end of this chapter, in which the contents of the edited texts have been considered in more detail, I will examine the various stories with respect to the three main points of my study: cosmogony, theogony and anthropogeny. My special interest concerns questions such as:

- What do these texts tell us about these beginnings?
- Is there one story for each of the three beginnings, or are there more parallel narratives?
- Is it possible to identify a diachronic development in one or more of these beginnings?
- Is there a connection between the three beginnings?

In answering these questions – it must be repeated – we have to bear in mind the restrictions inherent in the fact that, especially for the third millennium, we have mostly only one unique tablet of a specific text at our disposal.

2.3.1 Cosmogony

2.3.1.1 Cosmogony – third millennium

All texts from the third millennium report the presence, in a distant past, of **an** and **ki**. Cosmogony starts with the presence of **an-ki**, but no indication is given of the origin of this unity: **an** and **ki** are just there. Sometimes the circumstances in which **an** and **ki** are present have been outlined. The descriptions of these situations vary from a neutral report of the absence of sun and moon, to tempestuous times with storm and gale. Those tempestuous times may be meant metaphorically, in the same way as the shouting of **an** and **ki** together. The interaction between **an** and **ki** is expressed in several ways: they are shouting to each other, which has been interpreted as a form a courting; they form a unity, are lying together. Also the ultimate act, the intercourse, is indicated several times. During the preparation for their intercourse or marriage **an** and **ki** are beautifying themselves. For **ki** this is described very extensively in 'The Debate between Tree and Reed'. In other words: the several phases of the relationship between **an** and **ki** are described. It may be surprising that the extant literary texts, with the exception of IAS 114, fail to make any mention of the results of the cosmic intercourse or marriage between **an** and **ki** [260]. In the section 'Theogony' it will be discussed that there may be indirect information about this topic.

The statement that An is looking to the netherworld (NBC 11108) is unique. No other third-millennium text used in my study mentions the presence or creation of the netherworld [261].

---

[260] The text of 'The Debate between Tree and Reed' has to be excluded from this discussion. Tree and Reed are the result of the marital union of **an** and **ki**, but this cosmological introduction has to be valued as a typical aetiological introduction, purely aimed at the appearance of the contestants Tree and Reed.

[261] The text DP 141 *rev. iii*: 2 [CDLI no. P220791] reads: **lú-^{d}ba-ú kur-ré la$\breve{h}_5$-$\breve{h}$a-me**: 'They are men of Bau who are carried away by the netherworld.' Bauer (1998, 486-487) has shown that **kur** in this text has the meaning 'netherworld' and not 'mountain' or 'foreign land'. In line 329 in 'Ninurta's exploits: a *šir-sud* (?) to Ninurta' [ETCSL 1.6.2] '**ur$_5$-ra-am$_3$ bar-bi irigal $\breve{h}$e$_2$-em**', **irigal** may refer to the netherworld: 'This, its body, shall be the netherworld.' The implication of this interpretation might be that Ninurta created the netherworld from the body of Asag (suggestion of Katz), although the late bilingual version has the Akkadian word *qabru* 'grave' as a translation of **irigal**. In my opinion, Ninurta transformed Asag into stone (the lines

Likewise the creation of **an** and **ki** is not described anywhere. It is obvious that the presence of the three or four (if the **abzu** is included) main domains of the cosmos has been accepted without any creation story.

2.3.1.2 *Cosmogony – second millennium*

All the texts from the second millennium show great differences from the texts from the millennium before. First of all, there is hardly any description of the primaeval circumstances as seen in the third-millennium texts. No attention is given to the absence or presence of light, darkness, sun, moon, gale, lightning… With respect to the cosmogony there is one striking detail, viz. the absolute absence of any description of the courtship between **an** and **ki**. Only in one instance, 'The Debate between Grain and Sheep', I argued that there may have been intercourse between **an** and **ki**, when it is accepted that in this case **ki** has been called **ḫur-saĝ-an-ki-bi-da** [262]. On the other hand, one aspect that was passed over in silence in the third millennium, at least in texts written in normal orthography, i.e. the separation of **an** and **ki** [263], is now described in at least three of the four second-millennium texts in this study [264]. We can only guess why the separation of **an** and **ki** has not been included in the third-millennium texts written in the normal orthography. One possibility might be that the UGN-orthographic tradition, written predominantly in Fara and Abu Ṣalābīḫ in the Early Dynastic period (as far as we now know), has prevented this aspect of **an** and **ki** being recorded in later texts, written in the normal script. This might imply that the mythology of this separation was not generally known in Sumer at that time [265]. Does it perhaps have any relation to the god who was held responsible for the separation of **an** and **ki**, viz. Enlil? The relevant texts were written in the northern part of Sumer. In all likelihood Enlil was not a Sumerian god of olden times; my hypothesis is that he was introduced into the Sumerian pantheon of northern Sumer during the last quarter of the 4th millennium BCE [266]. This may have been the cause of the unfamiliarity with Enlil's separation act in the South of Sumer.

327-328), and this stone material may represent mountains and as such the netherworld in future times.
An Ur III date of origin for 'Ninurta's exploits' (or: 'Lugal-e') might be possible, but this is not quite certain (Cooper 1978, 10; van Dijk 1983, 1). Manuscript H of Lugal-e [ISET 2, 23; Ni 4138] most likely dates from the Ur III period (Jagersma, personal communication).

262 See ch. 2.1.6 and the Appendix Text editions no. 6, comments at the lines 1-2.

263 The separation of **an** and **ki** is only mentioned in a few third-millennium texts written in the UGN-orthography. The examples are given in ch. 2.1.1b.

264 It is very likely that also in the broken part of the lines 1-2 of 'Enki and Ninmaḫ' the separation of **an** and **ki** has been mentioned.

265 According to Krebernik (1998, 298) there are no duplicates of UGN-texts which have been written in normal orthography (with one exception: a small list of personal names in UGN- and normal orthography from the Sargonic period [Krecher 1978b, 156; Westenholz 1975, pl. XIII no. 173; CDLI no. P020587]). According to Michalowski, Ur-Namma and Šulgi '(...) discarded almost all the existing narrative and poetic literature of the land, (...)' (2003, 195), and they '(...) threw out most of the older literary texts, which were almost exclusively mythological, (...)' (2007, 176). But in my opinion, the relative absence of these texts may also be caused by other calamities, e.g. the destruction of Sumerian texts by Sargonic kings. George (2005), in his article about the **é.dub.ba.a**, does not mention such destruction by e.g. the Ur III kings.

266 This aspect will be treated in more detail in the Excursus 1 'Enlil and Ninlil' in chapter 3 'God lists'.

2.3.2 *Theogony*

2.3.2.1 *Theogony – third millennium*

One of the oldest texts – IAS 114 – provides more information about theogony in primaeval times than any other third-millennium text. Immediately after mentioning '**an** and **ki** together' follows a statement about Enki and Ninki, the ancestors of Enlil, who brought forth a 'group of seven'; this group is not identified in more detail. A few lines further there is a report that Enki and Ninki brought forth Enlil. Here the lineage of Enlil is given in brief: Enlil is the youngest brother of that seven. The question is: from whom are Enki and Ninki descended? There is until now not one text that explicitly mentions their lineage. It is very surprisingly that in the same text, after the birth of Enlil, there follows the notice that **an** and **ki** produced Enki. Although there is no absolute proof that this Enki is the 'Enki of Eridu' and not the partner of Ninki, it is the order in which the several gods who are brought forth are mentioned, viz. Enlil - Enki - Nanna, that makes it plausible that the Enki brought forth by **an** and **ki** is Enki of Eridu [267].

The other texts discussed have no clear information about the birth of the first gods in primaeval times. Ukg 15 tells about **an** and **ki** during their courtship, after which is stated: 'Then Enki and Ninki did not yet exist.' The sheer mention of the absence of Enki and Ninki immediately after the presentation of **an** and **ki** who are 'shouting in their unity' – which has been explained as having a sexual connotation – suggests, though nothing more, that after the 'cosmic marriage' between **an** and **ki** Enki and Ninki will appear. There are other texts – incantations, a hymn [268] – in which **an-ki** and Enki-Ninki are mentioned after each other, but again without revealing the direct connection or the precise relation between both groups. My tentative conclusion about **an-ki** and Enki-Ninki is the following. Both groups are named after each other in these cases, but the fact that there is no direct mention of **an-ki**'s parenthood of Enki-Ninki, may be a strong suggestion that Enki and Ninki – Lord Earth and Lady Earth –, the pair that have brought forth septuplets and later Enlil, originated or developed from **ki**/earth alone before the separation of **an-ki** [269].

As I have argued earlier in this chapter, in the discussion of the Barton cylinder and IAS 174, it is in all likelihood **an** who has intercourse with **ki** (at the relevant places both texts are broken). In these texts **ki** may appear under several names: Ninḫursaĝa and Ningal, respectively. The result of this relationship is also septuplets. These seven are again not further identified, as was also the case in IAS 114. Thus there are two sets of septuplets, each with different parents: Enki-Ninki and **an-ki**, respectively.

We have come across a similar unclear situation in NBC 11108. After the description of the primaeval circumstances and the mention that An/heaven was lying with Ki/earth, but that An had not yet taken Ki as his wife, we read that the Anunna gods had not yet arrived. The suggestion here is that, once the cosmic marriage between **an** and **ki** was celebrated, the result would be the appearance of the Anunna gods.

---

267 The god Enki is also present in the next column of IAS 114, dealing with extispicy.

268 Biggs 1974, 48, ll. 65-69; Krebernik 1984, 102, no. 20: II 2-3; Conti 1997, 266, no. 8 (YOS XI, 83), ll. 7 and 10.

269 It might be that the origin of Enki and Ninki was mentioned before line *i*:1' of IAS 114, because an unknown number of lines are broken. I do not know of any text that mentions the origin of Enki and Ninki. I shall return to the couple Enki-Ninki and these seven in the chapter 3 'God lists' and in chapter 4.2.1.

After the study of the god lists, these two 'groups of seven' will be discussed and analysed further in chapters 4.2.1.2 and 4.2.1.3.

2.3.2.2 *Theogony – second millennium*

If my interpretation that **ḫur-saĝ-an-ki-bi-da** in 'The Debate between Grain and Sheep' represents **ki** is correct, then we learn from this text that **an** and **ki** engender the Anunna gods. Another text from this millennium, 'Enki and Ninmaḫ', only mentions that gods of heaven have been born. This announcement follows immediately after something that has been related about **an** and **ki**, probably their separation. This might imply that these 'gods of heaven' were brought forth by **an** and **ki**. It is very unlikely that these 'gods of heaven' are Enki and Ninki, because they play in the pantheon only a marginal role. It is more likely that these 'gods of heaven' are identical with the Anunna gods, 'the great gods' as is their usual epithet (see ch. 4.2.1.3).

The following information in 'Enki and Ninmaḫ' about mother goddesses is so general, that we cannot make further conclusions but: the gods multiply.

We have to conclude that the second millennium texts give hardly any information with respect to theogony in the very beginning.

2.3.3 *Anthropogeny*

2.3.3.1 *Anthropogeny – third millennium*

For the third-millennium texts I can be very brief with respect to anthropogeny: this subject has not been dealt with in the texts under discussion here. Only NBC 11108 mentions the absence of priests and priestesses, but the quintessence of these remarks is that there is no cult for any god. The absence of any culture and cultivation is emphasized: man does not play a part at all.

In 'Ninurta's exploits' (l. 339) the subject 'humankind' has been indicated briefly: 'There was talk of a community of men for the ploughing of the Land'.

2.3.3.2 *Anthropogeny – second millennium*

There is a dramatic change in the second-millennium texts: mankind is no longer absent. The quality of his presence varies. In some texts when men first appeared they were like animals ('The Debate between Grain and Sheep'; 'The Sumerian Flood Story'); their condition was like that of barbarians, or in other words: non-Sumerians. At a certain point, when the gods needed them for their sustenance, they were 'transformed' into civilized humans. From these stories we have no information about the creation of humankind itself. In any case these texts cannot be considered as a *de novo* creation of man [270].

That *de novo* creation was described in 'Enki and Ninmaḫ' from the very beginning, when clay on top of the **abzu** had to be mixed. Next this clay had to be moulded into a human form. The exact 'bringing to life' of this clay model is rather cryptically described. The text of lines 32-36 may be interpreted as follows: Namma, the mother of Enki, becomes pregnant with this 'model'. Ninmaḫ will assist her, though in which way we are

270 In 'The Sumerian Flood Story' that creation may have been described in the broken part of column *i*.

not told [271]. Besides there are eight birth-goddesses. The **se$_{12}$-en-se$_{12}$-šár** may represent the life spirit of the newly formed man, for the name may be translated as: 'he/she who makes both **en**s and a crowd come alive'. But this description most likely can be valid only for the prototypes of mankind. The rest of the text does not give any indication how or when normal human reproduction is put into operation.

The basic element clay was also used by Enlil (in 'The Song of the Hoe') to create the first human. He put clay into a brick-mould – which may be understood as a metaphor for the human womb – , from which the first black-headed man appeared [272]. Thereafter the normal human reproduction was set in motion by the goddess Ninmena.

How different is the plan that two gods propose to Enlil in KAR 4 in response to the question of Enlil: 'What can we change? What can we create?' Their answer was: 'We shall slaughter the gods Alla and Illa, so that their blood makes mankind grow.' How mankind should come into existence at all is not mentioned.

The process of anthropogeny has also been described in the Akkadian texts *atra-ḫasīs* and *enūma eliš*. The approach of these last two stories and of KAR 4 is fundamentally different from the former Sumerian ones. In the general discussion (ch. 4) it will be argued that the peaceful creation with clay without the blood of the slaughtered gods is a Sumerian idea; the other, more violent one is the Akkadian way of thinking about the creation process of mankind [273].

2.3.4 *Summary*

In the third millennium ample attention was given to **an** and **ki**, the interaction of both, and indirectly also to the generation of some primaeval gods [274]. The absence of cults and culture is often described. Man, his creation or presence, is not mentioned. After the Ur III-period a dramatic change in all these subjects can be observed. With respect to **an** and **ki**: their intercourse or the cosmic marriage is absent in these stories. Only their separation seems worth mentioning. The primaeval gods Enki and Ninki are passed over. Instead, the birth of the Anunna, the great gods of heaven, is now announced. A special topic is the *de novo* creation of man, and his indispensability for the well-being of the gods.

2.4 *'Sitz im Leben' of the texts*

In his review of Pettinato's 'altorientalisches Menschenbild' [275], Kümmel points to a shortcoming of this study: Pettinato does not make any difference between the various texts

---

271 In ch. 4.4.3 I will try to explain the 'companion-ship' of Ninmaḫ and the cooperation of Namma and Ninmaḫ.

272 For a discussion of the used Sumerian terms in this creation story: see ch. 4.3.3.

273 This general discussion is included in ch. 4, together with some Akkadian texts with respect to 'Beginnings'.

274 When J.G. Westenholz (2010, 304 and 308) used the terms 'mitosis' (when she described the separation of heaven and earth) and 'daughter cells' (when she mentioned the generation of **nin**-gods from **ki**/earth) – terms which are derived from the field of Cellular Biology –, she gave a biased, incorrect picture of the cosmogonic and theogonic processes as they have been described in the Sumerian texts discussed in this study.

275 Kümmel 1973-1974, 26-27: 'Hier wird ein völlig unreflektierter Mythos-Begriff unterschiedslos angewandt auf die Berichte von der Menschenschöpfung in so verschiedenen Texten wie 'Enki und Ninmaḫ', der sumerischen Sintfluterzählung, theologisch-lehrhaften Dichtungen (Lobpreis der Hacke, **u$_8$** und **ašnan**), dem Epos von Atraḫasīs und gar auf einen Text esoterischer

with respect to genre and as a consequence of this difference to their possible distinct 'Sitz im Leben'.

The same objection might be raised to the present study. Texts of different genres and periods have been used to analyse Sumerian thoughts about origins. Some of the texts may be attributed to a well-known genre, such as the Debate texts. Others, especially the texts from the third millennium BCE, are more difficult to classify, e.g. IAS 114 (omen?), NBC 11108 (incantation? [276]); for Ukg 15 the extant lines are even too few to guess at the genre. A certain tendency with respect to the main topics of the beginnings – cosmogony, theogony and anthropogeny – can be deduced, notwithstanding the difference in genre between the used texts (see Table 1). Let us take one of the main themes of our texts, e.g. the cosmic 'marriage' between **an** and **ki** and the preparation of both protagonists for this. Both mythemes have been phrased differently in the respective texts [277]. Certain details of a mytheme may have been elaborated in different ways, depending on the kind of text in which it has been incorporated, but that does not change the intrinsic meaning of that mytheme. Another theme, the *de novo* creation of man, is told in very different ways in 'Enki and Ninmaḫ' and in 'The Song of the Hoe'. These two texts are composed from different starting-points. In 'Enki and Ninmaḫ' there are two stories. The first one – the *de novo* creation of man – may be seen as an introduction to the second story, viz. the contest between Enki and Ninmaḫ with respect to the fate of the disabled man. Thus 'Enki and Ninmaḫ' has a double function: it describes the origin of normal people, and thereafter it becomes an aetiology for human malformations. 'The Song of the Hoe' has a completely different starting-point. According to several scholars, this text has to be seen as 'Kulturmythos' belonging to the Edubba [278]. This does not alter the fact that in both mythological stories the basic mytheme – the *de novo* creation of man with the aid of clay – is identical; only the elaboration of this mytheme is different.

In this study I have analysed the mythemes in texts that are related to Mesopotamian 'Beginnings'. Whether the 'Sitz im Leben' of a text could be determined or not, and how different the use or the intention of a text might have been, I assume that it did not influence the primary meaning of the mythemes; only their phrasing may vary.

***

Geheimlehre wie KAR 4. Dabei fehlt jeglicher Versuch, die innere Gesetzmäßigkeit literarischer Gattungen und ihre literarisch-historische Entwicklung, den "Sitz im Leben" des jeweiligen Schöpfungsberichts im literarischen Kontext, zu bestimmen.'

276 The emphasis on the absence of certain ritual aspects may suggest that this text is an introduction to an incantation.

277 Compare e.g. the embellishment of **ki** in Ukg 15 and in 'The Debate between Tree and Reed'.

278 Farber 1999a, 369.

Excursus

an x ki: the cosmic 'marriage' as model or motif for a ritual?

Undoubtedly one of the oldest observations in the Mesopotamian society is that plants start to grow after it has rained. In due time this may have led to the idea that the origin of everything that grows up from the earth was the result of a 'marriage' [279] between **an**/heaven and **ki**/earth. The rain, An's semen, fertilizes the earth. This is a metaphoric way of saying that earth generates several products after it has rained. Some texts concerning the Beginning also report the birth of gods as the result of the cosmic 'marriage'. Intercourse between man and wife too produces offspring. The analogy between mythological cosmic marriage and human marriage is evident. My question is: Is there a relation between the myth of the cosmic 'marriage' of **an** and **ki**, as described from earliest times, and what is known as *hierós gámos* – sacred marriage [280] – in the Mesopotamian texts about religion and religious rituals? Was the cosmic 'marriage' between **an** and **ki** enacted as a ritual? The relationship between myth and ritual in general has been described by several scholars [281]. With respect to a possible relation between the myth of the cosmic 'marriage' and the ritual of the *hierós gámos*, several aspects of these items will be discussed first [282].

1. Cooper wrote about sacred marriage that there are no archaeological artefacts bearing inscriptions or representations to support any such relationship [283]. In the texts of the Barton cylinder and of IAS 174, **ki**/earth is talking to a serpent after her marriage to **an**. As has been mentioned earlier [284], there are several stamp impressions dating from the beginning of the third millennium BCE in which a coupling pair is accompanied by a serpent. Impressions with such scenes have not been found in later periods. It might be that the scenes on these stamp seals refer to a ritual lovemaking marriage, or even to the **an-ki** marriage. Representations of a ritual marriage may also be found on some cylinders of the Early Dynastic time [285].

With respect to the copulating couple, von Wickede has a more explicit meaning [286]: erotic scenes on stamps were probably introduced at the end of the ʿUbaid period and in this

---

279 The use of the word 'marriage' for the copulation of **an** and **ki** is based on the text NBC 11108 lines 5-6.

280 For the texts concerning the sacred marriage and its rite: Sefati 1998. Sallaberger (1999, 155-156) is of the opinion that the usual interpretation of the *hierós gámos* – the sexual union of the protagonists during a wedding party – is only a modern concept of which no attestation exists. With respect to the study of Lapinkivi (2004) about the sacred marriage: see note 124.

281 Rappaport 1999, ch. 4.9, 7.7, 12.8; Doty 2000, 39-42; 78-79.

282 The points to be used for answering my question 'Is there a relation between the myth of the cosmic marriage of **an** and **ki** and the ritual of the sacred marriage?' are numbered 1-11.

283 Cooper 1972-1975, 259.

284 Ch. 2.1.3 of this study.

285 Frankfort 1955, nos. 340, 559, 796.

286 Von Wickede 1990, 260: 'Siegelthemen mit erotischen Szenen (*hieros gamos*) werden vermutlich gegen Ende der 'Obēd-Zeit eingeführt und in der Gaura-Periode sowie in der späten Uruk-Zeit fortgeführt.'

respect he has no doubt about the reproduction of a *hierós gámos*. Amiet [287] is more careful and does not corroborate the hypothesis that the sacred marriage would have been depicted.

2. Renger suggested that the **EN** of Uruk might already have participated in the ritual of the sacred marriage in ancient times [288], but he admitted that there is no textual evidence for the ritual performance of the primaeval cosmic marriage [289].

Several scholars have tried to interpret the scenes on the Uruk vase [290]. There is consensus in the literature about the female figure in the upper register: she is or represents the goddess Inanna. The figure behind the nude servant who is approaching her may be the **EN** of Uruk [291]. The two figures on a pedestal on the back of the rams [292] – placed behind two **MUŠ$_3$** signs as gateposts – are dressed similarly; they may represent the **EN** of Uruk (the left figure) and the **EN** of Inanna, the one with a **MUŠ$_3$** sign behind his back. Both functions, a secular and a ritual one, were executed by one person in archaic Uruk [293]. These two figures may represent votive statues [294].

There is no *communis opinio* about a possible relation between these depicted scenes and a sacred marriage ritual [295], as the following synopsis of the relevant literature may show.

3. In a critical article about 'sacred marriage' in ancient Mesopotamia, Sweet writes: 'A fair amount of "eisegesis" is evident in this interpretation of the reliefs (i.e. that the Uruk vase

---

[287] Amiet 1980, 134: 'A la vérité, nos documents ne permettent guère de vérifier cette hypothèse.'

[288] Renger 1972-1975, 258, § 22.

[289] Renger 1972-1975, 255, § 13c.

[290] Heinrich 1936, 15-17; Tf. 2, 3, 38; Strommenger 1962, figs. 19-22.

[291] A very similar composition of the male and female protagonists is found on the Blau plaque; the **EN** is bearing here some kind of mace, a symbol of his prominent position (Gelb *et al.* 1989, pl. 12; Gelb *et al.* 1991, 39-43). Schmandt-Besserat (1993, 212, fig. 11) reproduced a seal impression from Uruk, in which the **EN** may perform a fertility ritual, and in which both protagonists are presented opposite to each other. Boese (2010) discussed the period of origin of these plaqes or monuments – Jemdet Nasr or ED I – without a definitive conclusion.

[292] There are other examples of such pedestals on the back of an animal, with or without a figure on it; see e.g. Szarzyńska 2000, 72, the figs. 6 and 7.

[293] Steinkeller 1999.

[294] Hockmann (2008) supposed that some symbols in the upper register on the vase may be identified as semi-pictographic signs, and that these characters could be linked to and identified as toponyms. He proposed that the Uruk vase depicts the actual exchange of goods within a kind of archaic Ur III-like bala-system. With respect to the two figures inside the temple: he reads the pedestal at the left figure (the one with the **EN** sign) as **KID$_a$** and nearly the same base or plinth by the right figure (with the **MUŠ$_3$** sign behind him) as **AB** or **UNUG**. The sign combinations **EN-KID$_a$** and **MUŠ$_3$-AB/UNUG** should represent the cities Nibru and Zabalam, respectively. Somewhere between two big vases with fruit and/or vegetables he discovered the city Kutha, represented by a bull's head. But the similarity with the archaic city seals according to Matthews (1993, 37) is inadequate. Moreover: he did not explain the two male figures on the back of the rams. And why are only three cities, two very prominently represented and one hidden between vessels, in this 'bala-system'? Individual, unconnected motifs – indicated as infill by modern scholars because of a *horror vacui* – are not unusual in the plastic art of the archaic Uruk period. This could very well explain the bull's head between the vessels.

[295] References given by Cooper 1972, 259-260. Jacobsen 1975, 65; Amiet 1995, 491-492.

depicts the rite of the sacred marriage; *JL*). A more cautious interpreter will find in them nothing that requires them to be read as a wedding scene.' [296]

4. Van Dijk has no doubts that the offering of the **mí-ús-sa** as part of the sacred marriage ritual is depicted on the cult vase of Uruk [297]. A few years later he wrote, with reference to the cosmogonic introduction on the Barton cylinder [298], that the primaeval cosmic marriage of **an** and **ki** might have supplied the mythological motif for the sacred marriage ritual. A supporting circumstance for this hypothesis, according to van Dijk, is the fact that originally the New Year started in October, because that was the period in which the first rain fertilized the Mesopotamian earth.

5. The detailed description of this Uruk vase by van Buren is very clear and useful [299]. Some points in her description deserve attention:
- the Uruk vase was a cult vessel and must have been one of a pair;
- in the zone with plant motifs, ears of corn (male symbol) alternate with palm trees (female symbol); below this zone wavy lines indicate water;
- in the zone with animals, the ram and ewe are placed alternately;
- behind the entrance of the temple, symbolized by two **MUŠ$_3$** signs as gateposts, there are two rams bearing on their back architectonic supports on each of which stands a male figure: one with an object in the form of the sign **EN**, the other holding his hands in a ritual gesture and with one **MUŠ$_3$** sign behind him.

Van Buren concludes that the constant representation of objects in pairs suggests that the offerings were brought to Inanna and her consort, and that the decoration scheme brings out the pairing conception [300].

6. According to Schmandt-Besserat [301], the coalescence of several roles of the **EN** is well illustrated on the Uruk vase: 'Here the images evoke 1) the meeting of the En with Inanna for the sacred marriage; 2) altars prepared to perform the rituals; 3) the presentation of offering; and 4) the En's cosmic powers over the animal and plant kingdoms.' She also referred to some seal impressions in which the **EN** in the presence of a goddess or a female figure who represents her is depicted [302]. Her comment on this scene is, that it usually has been interpreted as probably depicting a special ceremony, perhaps a fertility rite performed annually in the spring, anticipating the later sacred marriage of Inanna. But if so, then the notion of bounty was emphasized rather than the sexual aspect of the ceremony, she said.

7. There may be also textual indications for the connection between the **EN** and a fertility ritual. Charvát supposed that from the Uruk-period up to the Fara-period inclusive, the **EN** was involved in 'fertility-generating ceremonies', most probably referred to by the sign

---

296 Sweet 1994, 91.

297 Van Dijk 1967a, 240. Jacobsen (1975, 65; 1976, 24-26) had also the opinion that the Uruk vase depicted the rite of the sacred marriage.

298 Van Dijk 1971, 456. For the text of the Barton cylinder, see ch. 2.1.3a.

299 Van Buren 1939-1940, 33-36.

300 Van Buren 1939-1940, 36.

301 Schmandt-Besserat 1993, 217.

302 Schmandt-Besserat 1993, 217. 'This scene' refers to a seal impression like the one depicted in fig 11, p. 212 (= Amiet 1980, pl. 45: 647. In Amiet's plate 45 are more analogous pictures of seal impressions, representing 'le thème de la rencontre').

NA$_2$ ('to lie down [of people]'), in which ceremony the **EN** in all likelihood was accompanied by the **NIN** [303]. His interpretation of the sign **EN**, a combination of several components, is: 'the one who infuses fertility into tilled land' [304].

8. Frymer-Kensky wrote about 'Agricultural Fertility and the Sacred Marriage' [305]. She assumed that the sexual union during the sacred marriage ritual was intended for the renewal of agricultural forces, for the promotion of the fertility of the land. She presented evidence for an ancient date of such a sacred marriage ritual: the Uruk vase, some inscriptions referring to bridal gifts brought by Ninĝursu for Ba'u, and texts in which Sumerian kings – since Eannatum of pre-Sargonic Lagaš – claimed the title 'beloved spouse of Inanna'.

9. Steinkeller showed that there is 'evidence that the custom of providing goddesses with human consorts did indeed exist in early Sumer' [306]. This evidence was supplied by an inscription of Ur-Nanše that commemorated the selection of a **dam-**$^{\mathbf{d}}$**nanše**, the spouse of Nanše; this 'spouse' was a priestly official. According to Steinkeller this 'proves the existence, in ED times, of male priests functioning as goddesses' consorts.'

10. Sefati has suggested that several of the 'Dumuzi-Inanna love songs' may be related to the ceremony of the sacred marriage [307]. There are descriptions of the preparation of a nuptial bed, the adornment of the woman, and metaphors for sexual intercourse [308]. Analogous descriptions of verdant and fragrant places for **an** and **ki** to lie down, and the adornment of **an** and **ki** are found in several texts, discussed in this study [309], in which the *cosmic* 'marriage' between **an** and **ki** is described.

11. After I had finished my PhD-study, an article of MacCaffrey was published about the Sumerian sacred marriage [310]. She offered a new and cogent view on this 'ritual'.

McCaffrey said that contemporary attestations speak about the concept of a marital union between the ruler and a goddess, but a description of the actual realization of such a ceremony is lacking. In her search for a visual representation of this sacred marriage – which might be only a symbolic ceremony – she investigated pictures of palm vase libations, a ritual that is not mentioned in any text. During this ritual a ruler is pouring a liquid on a plant (a palm) in a vase, a scene mostly positioned in front of a cult image of a god. The image of the palm 'poses a contradiction: the sex of the stylized palm is feminine but its form is masculine.' [311] Because of this double identity, the palm may symbolize the goddess Inanna. The libation ritual was explained as some kind of fertility ritual because of

---

303 Charvát 1998, 41-70.
304 Charvát 1998, 43.
305 Frymer-Kensky 1992, 50-57.
306 Steinkeller 1999. Citations on pp. 118 and 120, respectively.
307 Sefati 1998, ch. II.2, 38-44.
308 Sefati 1998, 89-99; 247-256 [DI-T]. For (the preparation of) the embellished bed, see also 'Enmerkar and Ensuhkešdanna' (Berlin 1979), 40, ll. 29-30, and 44, ll. 80-81; 'Gudea cyl.' B xvii:1-3 (Römer 2010, 37); the 'Temple Hymns', l. 210 (Sjöberg and Bergmann 1969, 29). For metaphors with sexual connotation, see also Lambert 1987a, 31-34.
309 NBC 11108 (ch. 2.1.4), 'The Debate between Tree and Reed' (ch. 2.1.5).
310 McCaffrey 2013.
311 McCaffrey 2013, 231.

the importance of water for the welfare of Sumer. This symbolic ceremony might be what modern scholars have called a sacred marriage.

The vessel used in this libation ceremony has a typical biconical shape, also termed pedestal. The most famous example of this type is the Uruk vase. McCaffrey explains the pictures on the Uruk vase as representations of the two seasonal *akitu*-festivals: one in autumn (sowing-festival) and one in spring (harvest-festival).

Let me at the end of this Excursus try to answer the question that was posed at its beginning: Is there a relation between the *myth* of the cosmic 'marriage' of **an** and **ki** and the *ritual* of the *hierós gámos*, the sacred marriage? Admittedly there is no direct proof for a positive answer. Despite all the arguments given by several scholars: there is – to the best of my knowledge – no textual evidence that links the myth of the cosmic 'marriage' of **an** and **ki** with the (supposed) ritual of the *hierós gámos*.

***

Chapter 3

God lists
or
Lists of divine names [312]

'Expedit esse deos, et, ut expedit, esse putemus.'
[Ovidius, *Ars Amatoria*, I: 637]

Lists of words are inherent in the ancient Mesopotamian culture. These lists appear among the earliest cuneiform tablets and they continue to be written as long as cuneiform writing was in use [313]. These lists consist of enumerations and classification of natural and cultural entities. Among these compilations are lists of divine names, which appear already in the Fara period, culminating in a final, canonical list: **an** = *anum*, dating from the second half of the second millennium BCE [314].

3.1 *God lists in relation to the Sumerian cosmogony and theogony*

In his discussion of the Akkadian myth *enūma eliš*, Jacobsen writes that this story has a short introductory section, '(...) dealing with the origins of the basic powers in the universe. (...) The story told here is known to us (...) in a more original and complete form in the great Mesopotamian list of gods called An = *Anum* and can be supplemented from older Sumerian myths.' [315] The basis for analysing and understanding lists of divine names, in my opinion, was laid by Jacobsen in 1946 [316]. At that time he wrote that to the Mesopotamian the world appeared neither inanimate nor empty, but abounding in life. Inanimate substances had personality and a will of their own [317]:

---

312 In the Appendix 'God lists', the relevant parts of the several god lists that are discussed in this study are noted.

313 Civil 2000a, 2305, 2311.

314 Lambert 1957-1971, 473-479; Rubio 2011, 97-101 and *passim*.

315 Jacobsen 1976, 168-169.

316 Jacobsen 1977a (reprint of the edition of 1946), chapter V: The cosmos as a state, 125-184.

317 Jacobsen 1977a, 131. Jacobsen's ideas bear a strong similarity to those as described by Cassirer in 1925. Cassirer published in 1925 *Sprache und Mythos: ein Beitrag zum Problem der Götternamen*. An English translation – *Language and Myth* – appeared for the first time in 1946. Cassirer referred in this study to a work by Codrington – *The Melanesians: Studies in their Anthropology and Folk-Lore* [1891] – as follows: 'Codrington shows the root of all Melanesian religion to be the concept of a "supernatural power", which permeates all things and events, and may be present now in objects, now in persons, yet is never bound exclusively to any single and individual subject or object as its host, but may be transmitted from place to place, from thing to thing, from person to person. In this light, the whole existence of things and the activity of mankind seem to be embedded, so to speak, in a mythical "field of force", an atmosphere of potency which permeates everything, and which may appear in concentrated form in certain extraordinary objects, removed from the realm of everyday affairs, or in specially endowed persons, such as distinguished warriors, priests, or magicians.' The core of this worldview is a power in general; in its positive aspect it is called 'mana', in its negative aspect it is the power

> 'So had any phenomenon in the Mesopotamian world whenever it was approached in a spirit other than that of humdrum, practical, everyday pursuits: in magic, in religion, in speculative thought. In such a world it obviously gives better sense than it does in our world to speak of the relations between phenomena of nature as social relations, of the order in which they function as an order of wills, as a state.'

To be able to understand nature and the many phenomena around him, the Mesopotamian had to understand the powers in these phenomena, had to know their characters and the arrangement of these powers. These powers evolved into gods, who finally were represented as anthropomorphic beings. All this was translated in social terms. Finally this culminated in the structuring of the cosmos and of the cosmic institutions in a way analogous to the structure of a state. The state structure of the universe was considered as a kind of axiom: Mesopotamian thought of the third millennium took no particular interest in its philosophical basis. Questions about the reasons for and the origin of that structure were out of order [318].

According to this line of reasoning Jacobsen arrives at the highest position for An, the god of the heaven, because of the dominant part heaven plays in every respect. Reflections of this view can be found in Mesopotamian myths. Jacobsen does not yet mention explicitly in this essay the god lists as supplementary sources for the information which the myths contain, but the seeds for their legitimation and justification have been planted in this chapter.

Kramer makes use of one of the god lists to answer a question which he asked in connection with the cosmogonic introduction of the myth 'Gilgameš, Enkidu and the netherworld' [319]. That question was: 'Were heaven and earth conceived as created, and if so, by whom?' Kramer found the answer in tablet TCL XV 10, which gives a list of Sumerian gods. Among them is the goddess Namma, described with the epithet 'the mother, who gave birth to heaven and earth'. This led Kramer to the conclusion that 'heaven and earth were therefore conceived by the Sumerians as the created products of the primeval sea'.

Van Dijk picked up the threads of Jacobsen's argument [320]. According to van Dijk the god lists, composed in the most conservative milieu, are some of the main sources of our knowledge about the Sumerian religion. The basis of the Sumerian pantheon is pluralism: external pluralism because of the influences of ethnic changes, and internal pluralism caused by the individual religion and by the particular pantheon of the various

---

of 'taboo'. (Cassirer 1953, 62-63; by courtesy of Katz for her reference to this study of Cassirer).

318 Jacobsen 1977a, 151. His ideas are adopted by Bottéro and Kramer 1993, especially 56-78.

319 Kramer 1972, 39 and note 41; he mentioned the tablet TRS 10, which is another name for TCL XV 10.

320 Van Dijk 1964, 2: 'Les listes de dieux, l'une des sources principales de notre connaissance de la religion sumérienne, composées certainement dans le milieu le plus conservateur, sont un reflet fidèle du monde politique et ethnique à l'intérieur et à l'extérieur du pays de Sumer. Le panthéon sumérien, tel que nous le connaissons par ces listes, n'est pas une unité en soi, construite sur une idée fondamentale avec une hiérarchie interne, dont il suffirait de trouver l'idée maîtresse pour en déduire la logique d'un système. Il n'est que trop évident qu'un pluralisme forme la base du panthéon sumérien; pluralisme externe, dû aux influences des changements ethniques. Pluralisme interne, ressortissant à la religion individuelle et au panthéon particulier des diverses cités.'

cities. A tendency from pluralism towards unity, says van Dijk [321], can be clearly perceived in the god lists dating from the second millennium BCE with their explanatory sections, although this in itself was no innovation. The unilingual Sumerian lists [322] mention the gods for the greater part according to a theological system, i.e. not according to origin or local pantheon, but according to shared qualities and character. In that way the mother goddesses are placed together in lists SLT 122-124.

The ideas of Jacobsen and of van Dijk were adopted by Lambert [323], who writes that the mythological thinking, which itself dated from prehistory, had been based on a thorough observation of the universe, in spite of the fact that the resulting texts showed fictitious elements. The personification of the forces of nature into anthropomorphic gods led away, suggests Lambert, from nature into theological fantasy. The development of cities and of municipalities reinforced this last aspect. Much attention arose for the arrangement of the gods. Moreover, the search for the ultimate origin of everything, occurring in some myths, had been transferred to the god lists. These lists are not just *ad hoc* creations of individual scribes. The older lists from the Early Dynastic period exhibit some variation, but yet they are traditional texts – found in various places – which display several principles or combinations of these, according to which the names are arranged.

The primary goal of a god list was to enumerate the gods, making use of various principles: 1. a theological/hierarchical principle, arranging the gods according to their position in the pantheon; 2. a lexical principle, whereby the arrangement of the gods was based on their name; 3. a genealogical principle, grouping the gods according to ancestry; and 4. a geographical principle, in which a city was the unifying factor for grouping gods together.

Van Dijk was the first to recognize that the god lists are a source for the ideas of the Sumerians about cosmic history [324]. Joining van Dijk, Wiggermann wrote [325]:

> 'The general outline of cosmic history was written down only when it needed restructuring, in *Enūma Eliš*. Earlier sources give fragmentary information, out of context or moulded to suit special purposes. (...) Foremost among the earlier sources are the god lists. The earliest god lists, those from Abu Salabikh and Fara, are generally organized after philological principles, not, like later, after theological ones. (...) When a group of items appears together in different lists, it must be concluded that the items were grouped independent of the purpose of the lists in which they occur.'

Wiggermann also gives examples of groups of gods who appear in contexts other than the god lists, e.g. where sacrifices are being made [326], in rituals, litanies and incantations [327].

---

321 Van Dijk 1964, 2-4.

322 SLT 122-124; TCL XV 10.

323 Lambert 1957-1971; Lambert 1975a, 49-51; Lambert 1990. It is remarkable that Lambert in not one of this publications refers to Jacobsen 1946 (reprint 1977), while he is clearly indebted to Jacobsen and his ideas which are articulated there. But as can be inferred from e.g. the study of Cassirer, these ideas were even older and have their origin in anthropological studies from the end of the 19th and the beginning of the 20th century.

324 Van Dijk 1964, 1-6.

325 Wiggermann 1992, 280.

326 Cavigneaux and Al-Rawi 2000, 23, ll. *obv.* 9-22 (= 'The Death of Gilgameš, Another version from Nibru', ll. 9-22, ETCSL 1.8.1.3). Katz (2003, 366 and 402) gives a detailed description of

The invocation of ancestors at rituals, e.g. the ancestor cult, is a well-known theme. Van der Toorn gives a clear example of the veneration of five generations where '(...) the food offerings of the ancestor cult are attendant only to the recitation of the name.'[328] Spronk writes: 'The offerings to non-royal dead usually included only one to five former generations, the offerings to the royal dead the entire dynasty.'[329] Lambert remarks in this connection: 'In principle a list of ancestors can equally well begin with the first known figure and work downwards both on the tablet and chronologically, or it can begin with the living person and list his ancestors in reverse order chronologically.'[330] Schmidt has collected data about the kings of Ebla and the rituals with respect to deceased persons. The names of deceased kings 'were proclaimed in retrograde order down to that of the immediate predecessor of the living king.' [331]

The order of the names of ancestors will be an aspect of the analysis of several god lists.

3.2 *God lists*

3.2.1 *The Fara god lists (ED III)*

Several tablets from the ED III-time, containing lists of various content, have been found in Fara, formerly Šuruppak. Among these are tablets with lists of god names. These belong to the oldest tablets of this kind known at this moment [332]. Originally Deimel published these texts as 'Schultexte aus Fara' [333]. The most important ones, with lists of gods, are SF 1, SF 5 and 6, and SF 23 and 24. Krebernik has re-edited SF 1, and SF 5 and 6 [334]. The last two lists, with a relatively small number of god names, appear to contain **'ku$_6$-kú'** gods: fish-eating gods. These lists will not be discussed any further in this study.

---

the list of gods mentioned in 'The Death of Gilgameš': the main netherworld gods (to whom offerings are made) are listed in the lines 9-13; the lines 14-22 contain just divine names, among them some 'ancestors' of Enlil in the lines 14-18.

327 The examples given by Wiggermann are from a later period than the Old Babylonian time.

328 Van der Toorn 1996, 52-55 (quotation: p. 52). Van der Toorn (1996, 159 and note 34) mentions that the ancestors of Enlil receive offerings. He refers e.g. to Cooper 1983, line 207. In 'The cursing of Agade' (Cooper 1983, 60 [ETCSL 2.1.5], the lines 207-208 belong together. The lines read: 207. **ír-bi ír-ama-a-a-^{d}en-líl-lá-ke$_4$** 208. **du$_6$-kù su-zi gùr-ru du$_{10}$-kù-^{d}en-líl-lá-ke$_4$ i-im-ĝá-ĝá-ne** 'their lament is (like) the lament for the ancestors of Enlil which they perform at the awe-inspiring holy mound, at the knee of Enlil'. In other words: this seems to be a mourning ritual *for* [genitivus objectivus] the ancestors of Enlil. In the translation of Cooper we read: 'Their laments were (like) laments which Enlil's ancestors perform (...)' and that of ETCSL: 'Their laments were as if Enlil's ancestors were performing a lament (...)'.

329 Spronk 1986, 108.

330 Lambert 1968, 2.

331 Schmidt 1996, 20.

332 Lambert (1957-1971, 473) mentions that 'Duplicates of the Fara lists generally have been found at Uruk, Ur and Tell Ṣalābiḫ, the first being earlier than the Fara lists, the latter two roughly contemporary. Thus these early lists were not simply ad hoc creations of individual scribes, but were traditional texts, with variants of course, handed down in several (and probably most) cities. There is no reason to suppose that the god lists differ in this respect, and the suggestion of Weidner [AfO 2 (1924-25, 3] that the Fara god lists are based on local cults in particular lacks any foundation.'

333 Deimel 1923.

334 Krebernik 1986.

The Fara list SF 1

The collation of tablet VAT 12760 (= SF 1) by Krebernik [335] resulted in a slightly different reconstruction of the tablet from the one published by Mander [336] in the same year. For my study it is important which gods are mentioned and in which order. Some prominent details of this list are the following.

The structure of the list proves to be a combination of hierarchical order – being a theological arrangement with An and Enlil at the beginning, followed by some other major deities – and a succession of groups of names containing the same element, e.g. **nin**, **lamma**. These last groups form a lexical arrangement. Within the **nin**-group several small groups, whose interrelationship is known from other sources, can be distinguished. On the other hand, according to Selz [337], many names of gods in god lists of the Fara period are not yet understandable. Moreover these god lists do not seem to have a uniform order of the god names.

The list [338] begins with a group of gods who, at least at later times, were regarded as major gods: An, ^{d}**en-E**$_2$ = Enlil [339], Inanna, Enki, Nanna, Utu, followed by ^{d}**men**, ^{d}**men-bar**, Nisaba, Nanibgal. Column *iii*: 1 reads ^{d}**nin-E**$_2$, followed by ^{d}**nin-sig**$_4$**-tu**, whereas column *vi*: 27 has ^{d}**nin-KID** = Ninlil [340]. Therefore it is unlikely that in *iii*: 1 ^{d}**nin-E**$_2$ has to be read as Ninlil, analogously to the reading of ^{d}**en-E**$_2$ (*i*: 2) as Enlil, because the name Ninlil in Fara was written as ^{d}**nin-KID**.

Some other gods who are of interest for the present study are:

* Namma (*viii*: 10 [341]), who is mentioned in an isolated position between names that are not likely to have any relation to the Namma who represents the primaeval waters;

* the small group in column *vi*: 25-28 [342]:

---

335 Krebernik 1986, 163. Krebernik also gives a general analysis of this list; his list is the starting-point for this study. CDLI no. P010566 shows a photograph of VAT 12760.

336 Mander 1986, 77-89.

337 Selz 1992, 197: 'Noch heute sind eine Vielzahl der in den Fāra-zeitlichen Götterlisten verzeichneten Götternamen nicht (sicher) deutbar. Zudem folgen diese anscheinend keinen einheitlichen Ordnungsprinzipien.' Lambert (1957-1971, 474) wrote: 'In general the Fara god lists contain so many obscure and otherwise unknown deities that little more can be said on their order (or lack of order).'

338 Indications of columns and numbers are according to the results of Krebernik 1986.

339 For a discussion about ^{d}**en-E**$_2$: see this chapter, Excursus 1 'Enlil - Ninlil'.

340 For a discussion about ^{d}**nin-KID**: see this chapter, Excursus1 'Enlil - Ninlil'.

341 After collation by Krebernik (1986), Namma's place has changed from *x*: 1 (Deimel 1923) to *viii*: 10. Namma is mentioned between ^{d}**PA-UNUG and** ^{d}**UM-ḪUR-ḪU**.

342 Column *vi*: 29 mentions a god(dess), not belonging to the group of ancestors. Krebernik reads ^{d}**NIN EN ŠE**$_3$ **GI KI** as follows: ^{d}**nin-GI.EN.KI.ŠE**$_3$. Especially the value /**šè**/ seems to be strange in this name. But the sign may also have the value **ĝir**$_{15}$, and the name may be read as ^{d}**nin-ki-en-**^{gi}**ĝir**$_{15}$ (with **gi** as gloss) 'Lord/Lady of the land of Sumer'. (For the spelling **ki-en-/gi̱r/**: see Wilcke 1974, 202-232 *passim*).
Another interpretation may be possible for the name in column *viii*:1 **AN TUM MA** *ìl-ma-tum*. This would be the only Akkadian name in a list which Krebernik supposed to contain 560 names (Krebernik 1986, 163). The name *ìl-ma-tum* is also absent in the Semitic pantheon described by Roberts 1972. Another possible reading is a Sumerian one: ^{d}**tum-ma** 'the divine cross-beam of the land'.

	transcription of Krebernik	*my transcription*
25	^{d}**nin-LAK 777.DU$_6$**	^{d}**nin-utua** [343]
26	^{d}**nin-ki**	^{d}**nin-ki**
27	^{d}**nin-KID**	^{d}**nin-líl**
28	^{d}**nin-BULUG$_3$**	^{d}**nin-bulug$_3$**

Three gods of this group, ^{d}**nin-utua**, ^{d}**nin-ki** and ^{d}**nin-bulug$_3$** are in other, later lists attested as belonging to the ancestor group of Enlil [344].

* Other gods that in the list **an** = *anum* appear as ancestors of Enlil may be equated with some gods who are spread over this Fara list:

	Fara			**an** = *anum* [345]
iii: 7	^{d}**nin-DU**	→	I: 105	^{d}**nin-DU**
iv: 19	^{d}**nin-á**	(?) →	I: 107	^{d}**nin-da**
vi: 12	^{d}**nin-du$_6$**	(?) →	I: 135	^{d}**nin-du$_6$-kù-ga**
vi: 13	^{d}**nin-ul-a**	(?) →	I: 99	^{d}**nin-ul**
x: 2	^{d}**nun-nun**	(?) →	I: 119	^{d}**nin-nun**

Some important remarks by Krebernik [346] have to be mentioned here:

1. SF 1 has a high number of names composed with ^{d}**nin-** (> 40%), whereas the ^{d}**en-**names are relatively few (4 - 5%). To this can be added that, as far as now can be analysed, there are at most three names which form possible ^{d}**en-X** - ^{d}**nin-X** couples [347].

2. With respect to content, i.e. the names formed mainly with **nin-**, one might conclude that this reflects a prominent role for the feminine element in the Sumerian pantheon in this

---

343 LAK 777 = **utua** (Borger 2010, 125, no. 449; Krebernik 1998, 275; ePSD *sub* **utuwa**). In my opinion **du$_6$** is a gloss at LAK 777. In the later god lists (TCL XV 10 and **an** = *anum*) this god has most likely been replaced by ^{d}**en/nin-amaš**. Wiggermann's (1992, 285) supposition, that it is highly probable that the ^{d}**en/nin-du$_6$-LAK 777** of the Abu Ṣalābīḫ list and of the two Fara god lists correspond with the ^{d}**en/nin-du$_6$-kù-ga** of the later lists, is very unlikely.

344 For the 'final' composition of this group I will use the list **an** = *anum* (Litke 1998). The position of ^{d}**nin-KID** = Ninlil, included as a member of the 'ancestor group of Enlil', is remarkable – apart from a scribal error – insofar as this goddess is supposed to be the spouse of Enlil (see also in this chapter the Excursus 1 'Enlil - Ninlil' for a possible earlier and other role of ^{d}**nin-KID**).

345 For **an** = *anum*, see Litke 1998.

346 Krebernik 1986, 165-166.

347 These possible counterparts are: ^{d}**en-E$_2$** (*i*:2) and ^{d}**nin-E$_2$** (*iii*:1), but dubious because ^{d}**nin-KID** (*vi*:27) is supposed to be the spouse of ^{d}**en-E$_2$** = Enlil; ^{d}**en-dag-ga** (*xiv*:1') and ^{d}**nin-dag** (*vi*:23); ^{d}**en-KA.SAR** (*xiv*:2') and ^{d}**nin-KA.SAR** (*v*:16'). The ^{d}**en-ki** of *i*:4 is no counterpart of ^{d}**nin-ki** (*vi*:26), because the first one is the Enki of Eridu, and the second one belongs to the 'ancestors' of Enlil.

period [348], and – we might add – therefore in the early third millennium Mesopotamian society. But at the same time we have to admit that names with **^{d}nin-** do not by definition only belong to female gods; for example, consider the names of the male gods Ninĝirsu and Ninurta.

3. A possible confirmation of the prominent female status may be that in the list SF 1 Inanna has been named in third position, immediately after An and Enlil, but before Enki, Nanna and Utu [349].

4. There may be another explanation for the high ranking of Inanna. The frequency of the element Uruk/Kulaba in this list may indicate that the tradition of this list originates from Uruk. Alternatively it may be supposed that Uruk had a dominant position at the time that this list was composed [350]. This idea may be confirmed by the positions, besides that of Inanna, of **^{d}nin-UNUG** (*i*:11), **^{d}nin-girim$_{x}$** (*i*:12), **^{d}nin-sún-LAMMA** (*i*:15), and the presence of **^{d}lugal-bàn-da** (*vii*:15) and **^{d}bìl-ga-mes** (*xiii*:7').

The Fara lists SF 23 and 24 [351]

Tablet VAT 12573 (= SF 23) is a tablet, the reverse of which is blank, and which contains lexical lists of geographical names [352] and divine names; these last ones comprise only less than 10% of the list SF 1. The god names are all without **diĝir** determinative [353]. The list of divine names does not start with An, who is missing in this list – just as Enki (of Eridu) –

---

348 See Steinkeller 1999, 113. This aspect will be discussed in this chapter, Excursus 1 'Enlil and Ninlil'.

349 In this respect Krebernik refers also to one of the incantations from Fara, in which the order is female - male (Krebernik 1984, 36 *ad* c; comments on 44-46).

350 Krebernik 1986, 166.

351 Deimel 1923; Mander 1986, 108-110; CDLI no. P010600 with transcription of Veldhuis.

352 For a parallel of the names in the first column: see Matthews 1993, 38, table 3.

353 In column *vi*: 17-18 the text reads: 17. **AN-MUŠ$_3$** 18. **AN-NAGA**. In column *i*: 6-7 two cities are mentioned: 6. **MUŠ$_3$-AB** 7. **AN-NAGA**, Zabalam and Ereš, respectively. This is a strong indication for the identification of the following two goddesses in *vi*: 17 and 18: Inanna (of Zabalam) and Nanibgal, respectively. For both goddesses the sign **AN** is not meant as a **diĝir** determinative; for Inanna it functions – I suppose – as a phonetic complement; for Nanibgal it is an integral part of the name. Some scholars presented a different transcription: **AN.inanna** and **AN.nisaba** (Mander 1986, 109, no. 21-22); **AN.MUŠ$_3$** and **nanibgal** (Veldhuis, CDLI no. P010600, *vi*:17-18). Van Dijk (1957-1971, 536; 1964, fig. 1) distinguished between heavenly and chthonic gods: this dichotomy results sometimes in two representations for the same god, a celestial and a chthonic manifestation, e.g. **^{d}an-inanna** and **^{d}inanna**. Selz (2008, 22) has the same opinion: 'A similar astral interpretation is suggested here for writings of deities such as **AN-^{d}NISSABA**, **AN-^{d}MAR.TU**, and **AN-^{d}INANA**. Such additional markings became possible or even necessary as soon as spreading use of the divine classifier AN overshadowed its reference to the celestial bodies.' Charpin (1986, 404; 1994), on the other hand, does not assign an astral value to a deity when the god name has been spelled with two **AN**-signs; in the case of **AN-^{d}Inanna** he argues for one female deity, i.e. Inanna of Uruk. Contra Charpin: Wilcke (1997, 414-415), who does not exclude the possibility of interpreting An and Inanna in the example of **AN-^{d}Inanna**. But already in 1979 Stol has demonstrated that e.g. 'the gods **^{d}MAR.TU** and **AN.AN.MAR.TU** are identical' (Stol 1979b, 178).

but with **en-ki** and **nin-ki** (*v*:17-18), immediately followed by **en-E$_2$** and **nin-KID** (= Enlil and Ninlil) [354]. This corresponds to the text IAS 114 *i*: 9'-11':

9'	^{UD}**GAL-NUN**	^{d}**en-líl**	Enlil
10'	**AMA-a**	**tud-a**	has been brought forth
11'	^{d}**GAL-UNUG** ^{ud}**nin-ki**	^{d}**en-ki** ^{d}**nin-ki**	by Enki and Ninki.

The lists SF 23-24 and the text IAS 114 make clear that Enlil is seen as the offspring of **ki** 'Earth'. In the same text (IAS 114 *i*: 3'-4') is mentioned that 'Enki and Ninki have brought forth the seven'.

In the present list, after Enki-Ninki and Enlil-Ninlil, follow five **en-** and **nin-** couples, who can be recognized as ancestors of Enlil in later lists [355]. This is in contrast to the list SF 1, where only a few of these ancestors – and moreover only the **nin-** partners – are named. The two names in SF 23 – **tùr** and **ĝiri$_3$** (*vi*:11-12) – following after the ancestor gods may be gods of animals, domesticated and wild animals, respectively, if **ĝiri$_3$** represents ^{d}Šakan [356]. After these 'gods of animals' **NI-NA** is mentioned, which might be indicating the goddess Nanše [357]; if so, then it is tempting to suppose that she is placed here in her aspect of goddess of fishes and birds [358]. Then follow two other names which have not been explicated until now: **LA-DIM** and **SAL-KID**. For **LA-DIM** we can only speculate [359]. The name **SAL-KID** probably is a scribal error for **NIN-KID** [360]; this again is

---

354 Deimel (1923, 23*) supposed that the list of divine names started at *v*:15-16 with ^{d}**BAD$_3$** and ^{d}**gi-bi(l)-[bád$^{??}$]**. The names *v*:14-16 are transcribed by Veldhuis (CDLI no. P010600) as: 14. **durum (EZENxAŠ)^{rum}**, 15. **an-durum (EZENxAŠ)** (but in my opinion the transcription should be: **an-durum (EZENxAŠ)^{rum}-X**, 16. ^{d}**gi⌜bil$_6$$^{?}$⌝ (⌜NE$^{?}$⌝-GI)-⌜duru⌝m$^{?}$** (also changed, JL). **durum** can be found as a city [between Uruk and Larsa] in several Old Akkadian administrative texts in F. Rasheed, *The Ancient Inscriptions in Himrin Area,* 1981 (CDLI: P212472, P212474, P212480, P212497, P212500).

355 For **en/nin-UḪ (=ḪI x NUN)** I would like to read **giriš$_x$** (according to van Dijk 1964, 7) and contrary to Veldhuis (CDLI no. P010600, *vi*: 1-2) who reads **giriš (= ḪI x ŠE)**.

356 Lambert 1986a. That **tùr** might represent the (mother) goddess Nintur seems less likely, because her name – as far as the present author knows – has never been written with this sign.

357 **NINA** (traditional transliteration), or Niĝin (**niĝin$_6$**, **nanše**, **sirara**, usually written as **ABxḪA**) [Borger 2010, 92 no. 236]**)** is the cult place of Nanše (Selz 1995a, 182-183, *ad* I.3 and I.4).

358 Veldhuis 2004, 24-25, ch. 2.2.2.

359 With respect to **LA-DIM** the following can be observed:
1. In SF 23-24 **NI-NA** (= Nanše ?) is immediately followed by **LA-DIM**;
2. In SF 1 Gatumdug (*ix*: 14) is mentioned shortly after Nanše (*ix*: 11);
3. The same can be observed in the AṢ god list: Nanše 26, Gatumdug 28;
4. In the AṢ **zà-mì** hymns (Biggs 1974, 49), the hymn for Gatumdug (ll. 108-109) shows that this goddess is already connected with Lagaš in the Fara period; the hymn for Gatumdug is followed by the one for Nanše (ll. 110-116);
5. As Selz (1995a, *sub* $^{(d)}$**gá-tùm-du$_{10}$**, 134-136, esp. 8 and 10) has pointed out: in NINA, city of Nanše, Gatumdug received, together with Inanna, offerings.
The conclusion is that Gatumdug was an important goddess in the Old Sumerian period in Lagaš (= **ŠIR-BUR-LA^{ki}**). My tentative suggestion is that **LA-DIM** may be an old spelling, used by Fara scribes, for Lagaš, or that **DIM** may be a scribal error for the sign **ŠIR**. If this is correct, than **LA-DIM/(ŠIR?)** may represent the goddess Gatumdug.

360 In fact **NIN-KID** is also mentioned in column *iii*:1, but it is not clear in which connection.

the goddess Ninlil. If this supposition is correct, then it immediately raises the question of why this goddess is mentioned twice. An explanation might be that the group of primaeval gods, starting with the pair Enki-Ninki, was seen as pure predecessors and as a coherent group, which comprises only the first stages of later gods; some of them have not survived in the pantheon, or have assimilated with major gods. The position of Ninlil in the predecessor group, among gods that indeed have not survived under their original names in the final pantheon, was not that prominent position which Ninlil and her husband Enlil occupied in the Sumerian pantheon. Therefore she has been mentioned again, now perhaps in her original position as Ninkid, the goddess of Tummal [361].

The rest of the list is made up of gods with the **nin**-prefix. Because of the defective state of this list it is difficult to decide according to which principle this list has been composed.

3.2.2 *The Abū Ṣalābīḫ (AṢ) god list* (ED III) [362]

The reconstructed list with god names from Abū Ṣalābīḫ [363] shows that this list contains only god names. Mander has analysed it in more detail, and concluded that the constitutive principles of this list – being 'acrographic, acrophonic and thematic' – alternated with each other [364].

For my study two groups of gods are important:
1. the gods mentioned at the beginning of the list;
2. the gods mentioned in nos. 273-288.

Ad 1. The first group of gods (nos. 1-9) – with a hypothetical but very plausible reconstruction of the first five places – represent the city gods of Sumerian towns important at that time, members of the **ki-en-gi** league: Uruk, Nippur, Eridu, Ur, Uruk and/or Zabalam, Lagaš, Adab [365]. The order of the gods at the beginning of the Abu Ṣalābīḫ and Fara god lists and that of the **zà-mì**-hymns show no conformity [366].

Ad 2. This group is comparable with the one in SF 23 *v*:17-*vi*:10, though in the AṢ-list one group has been added: ^{d}**en-an** - ^{d}**nin-an** [367]. The order of the god names in both lists is the

---

361 See the Excursus 1 'Enlil and Ninlil' at the end of this chapter.

362 Alberti 1985; Mander 1986; Krebernik 1986, 191 note 26.

363 Reconstructed from IAS 82-90 (Biggs 1974).
Wang (2011, 84), without any explication or motivation, starts IAS 82 with [Enki], [Ninki] instead of as usual with [An], [Enlil]; he also does not indicate the lacunas in the names of the next god names. Moreover, about ^{d}nin-KID he writes (Wang 2011, 86 n. 246): 'The only "theological" way to read ^{d}Nin-KID in IAS 082 Obv. Col. 1, line 3 is to regard it as in pair with ^{d}En-ki, between the pairs of ^{d}Enki-^{d}Ninki and ^{d}Nanna-^{d}Inanna while ^{d}Inanna and ^{d}Nin-gír-su formed another pair.' Wang omitted to explain what that 'theological' nature of these unique pairs should be.

364 Mander 1986, 133-134. The numbers which refer to the gods are the numbers in the study of Mander 1986, 24-32.

365 Krebernik 1998, 242, 312 *ad* 4.

366 Mander 1986, 40.

367 Besides in the 'ancestor'-group of Enlil, the AṢ-list contains two other ^{d}**en-ki**'s (no. 4 = Enki of Eridu, and no. 229,) and another ^{d}**nin-ki** (no. 415) (Mander 1986, 24, 28, 31).

same; the difference is that the **diĝir** determinative, that is absent in the Fara list SF 23, is present in the AṢ-list [368].

The repetition of the pair **ᵈen-E₂ - ᵈnin-KID** in this group – already mentioned, in all probability, in AṢ-list *i*: 2-3 – shows that the scribe has treated these gods as one coherent group. This group is also the only one in which some kind of regularization tendency becomes visible: the gods are organized in **en**- and **nin**-pairs. Just as in SF 23, the pair **ᵈen-E₂ - ᵈnin-KID** deviates from the other pairs in this group because of the difference in the second part of their name; all other pairs in this group share the same name in that part.

When we assume that the restoration of the beginning of the list and especially of the third name – **[ᵈnin-K]ID** – is correct [369], then it is surprising to find the name of Ninlil in this prominent position after, presumably, An and Enlil. Why is the goddess, who is in later texts known as the spouse of Enlil, mentioned here? She is even mentioned before Enki, Nanna and Inanna, gods who are not accompanied by their partners. A tentative conclusion is that **ᵈnin-KID** in this list does not (yet) represent the spouse of Enlil, but that she has this high ranking because of her own merits [370].

At the end of this discussion of the Fara and Abu Ṣalābīḫ god lists it is worthwhile summarizing some of Selz' conclusions [371]. These are as follows. In the Fara god list SF 1

---

368 Wiggermann (1992, 285) wrote: 'If the **ᵈEn/Nin-du₆-LAK777** of the Abu Salabikh god list and the two Fara sources corresponds with the **ᵈEn/Nin-du₆-kù-ga** of the later lists – highly probable since all ED items recur in the later lists – then the Abu Salabikh list agrees with the forerunner of *An-Anum*.' I tend to disagree for several reasons: 1. **ᵈEn/Nin-du₆-LAK777** = **ᵈen/nin-utua**, and thus they do not correspond to **ᵈen/nin-du₆-kù-ga** (see this study *sub* ch. 3.2.1 The Fara list SF 1); 2. not all ED items recur in later lists: **ᵈen/nin-gukkal** and **ᵈen/nin-á** do not return; 3. the Abu Ṣalābīḫ list does not correspond to the Genouillac list (= TCL XV 10 = forerunner of **an** = *anum*), for the last list contains 16 pairs of 'ancestors' of Enlil, the AṢ list only 7 pairs. Moreover the order of the shared pairs in both lists is different.

369 Biggs 1974, 83, *ad* no. 82. The reading **KID** is very probable.

370 See also the Excursus 1 'Enlil and Ninlil' at the end of this chapter.

371 Selz 1992, 199-200: '1) Bei Deimel Schultexte 1 überwiegt der von M. Krebernik * beobachtete Götterkreis von Uruk, aber auch Überlieferungen, die mit Eridu oder Ur zu verbinden sind, sind stark vertreten. Regional gesprochen scheinen in Deimel Schultexte 1 südmesopotamische Kulttraditionen zu überwiegen. Möglicherweise wird hier ein frühes religionsgeschichtliches Stratum fassbar. 2) Deimel Schultexte 1, die Götterliste aus Abū Ṣalābīkh und die Zami-Hymnen kennen in ähnlicher Weise eine gesamtsumerische Kulttradition. Der Norden dieses Gebietes wird dabei etwa von Sippar, Kiš und Dēr begrenzt. Die Akzentuierung innerhalb der einzelnen Traditionen ist verschieden. 3) In den Götterlisten haben semitische Gottheiten bereits Aufnahme gefunden. Die Zami-Hymnen kennen sie nicht, wenn man von den vielleicht semitischen Namen Zababa (Kiš) und Ištarān (Dēr) absieht. 4) Die Abū Ṣalābīkh-Götterliste inkorporiert elamische Gottheiten, wenn auch unter sumerischen Namen (Lugalaratta, Lugalelam, Ninšušinak). 5) Die Zami-Hymnen, im Gegensatz zu den Götterlisten, unterliegen der theologisch-politischen Absicht, Enlil als Oberhaupt eines gemeinsumerischen Pantheons zu etablieren **. Die bescheidene Rolle, die An in diesem Zusammenhang zukommt (Zami 41-43), ist dafür ebenso ein Indiz wie die Nennung Enkis nur unter seinem Namen Nudimmud (Zami 30-32). 6) Die Opferurkunden aus Fāra *** lassen lokale Kulttraditionen erkennen. Diese stehen wohl unter dem Einfluss des nahegelegenen Nippur.'
[* Krebernik 1986, 166. ** Auf völlig anderem Wege, nämlich über die Lesung und Deutung des Gottesnamens **ᵈen-líl** nach Quellen der vorsargonischen Zeit kommt P. Steinkeller [1999, 114, note 36; *JL*] zur Annahme eines alten semitischen Ursprungs dieses Gottesnamens. (...)

the gods of the Uruk pantheon are prevalent, but also the traditions from Eridu and Ur are present. In general: this list seems to represent predominantly the cult tradition of South Mesopotamia. The list SF 1 from Fara, the Abu Ṣalābīḫ god lists and the **zà-mì** hymns show a collective Sumerian cult tradition, with a different emphasis within the distinctive traditions. In these god lists Semitic gods are present, but not in the **zà-mì** hymns (except perhaps Zababa and Ištaran). The Abu Ṣalābīḫ god lists also contain Elamite gods (with Sumerian names). The aim of the **zà-mì** hymns, but not of the god lists, was to show Enlil as the head of the Sumerian pantheon.

3.2.3 *The Weidner god list* (*Ur III*)

Lambert and Richter present a survey of the texts with respect to the so-called Weidner god list [372]. Weidner was the first to draw attention to this special kind of god list [373]. The first beginnings of it could be traced back to the Ur III-period. The list was continued, with some extensions, until the late Babylonian period. The list shows a certain system, somewhat comparable with that of the list **an** = *anum*, e.g. the mention of the groups of the principal gods, although some of these gods appear in two separate places. According to Weidner, this god list has undeniably a scribal school character. The ancestor groups of both An and Enlil, present in other god lists, are absent in the Weidner god list. Lambert concludes the section of the Weidner list in the Reallexikon: 'At least at the beginning the arrangement of the list is apparently theological. In many places, however, it is difficult to discern the principles of arrangement, if there are any, and it is uncertain if there is even one case of lexical arrangement. Either our knowledge of the deities is inadequate to grasp the basis of the arrangement, or, more probably, various short lists have been compiled without any attempt at integrating them.'

3.2.4 *The Isin god list* (*OB*)

Wilcke published some lexical texts from excavations at Isin [374]. He points to the independence of the Isin lexical tradition which is impressively shown in the Isin god list. There exists no parallel of it outside Isin. The list has been composed according to a theological and a lexical order. This last one can be shown by the group of names beginning with **ᵈnin-**. The theological principle has been combined with local patriotism, so Wilcke.

Trifft dies zu, so ist die hier implizierte These, Enlil sei erst in frühgeschichtlicher Zeit an die Spitze des Pantheon befördert worden, geradezu notwendig. *** For references to these texts: Selz 1992, 200, note 50.]

372 Lambert 1957-1971, 474, § 2; Richter 2004, 18-20. Also by courtesy of van Soldt, who has given me an as yet unpublished copy of his score of all texts of the Weidner god list.

373 Weidner 1924-1925.

374 Wilcke 1987, 93-104; citation: 93, 98: 'Besonders eindrucksvoll zeigt sich die Selbständigkeit der Isinüberlieferung lexikalischer Texte bei der *Götterliste* (Beginn: *An*, *ᵈUraš*), zu der sich keine Parallelen außerhalb von Isin finden. Diese Liste umfaßte ca. 331 Zeilen und nahezu ebensoviele Götternamen, die aber nicht alle erhalten sind. (...) Die Liste folgt einerseits 'theologischen' Gesichtspunkte. Dies zeigt sich besonders am Anfang in der Folge An / Uraš, Enlil (+ weiterer Name) / Ninlil, 'Väter und Mutter Enlils' (erstaunlicherweise nach Enlil!) (...). Das theologische Prinzip verbindet sich mit dem Lokalpatriotismus (...). Andererseits zeigt die große Gruppe der mit ᵈNin- beginnenden Namen das lexikalische Prinzip der Ordnung nach Leitwörtern oder -zeichen.'

The god list begins with An, Uraš, Enlil/Nunamnir, Ninlil, followed by the ancestors of Enlil [375]: Enki-Ninki up to and including Enmena-Ninmena, 8 pairs in total. Enmena and Ninmena – otherwise not known as belonging to Enlil's ancestors – seem to be included here in the ancestor group, the only one that – besides Enlil/Ninlil – consists of **en/nin**-pairs. Ninmena is also included in the following group, that of the mother goddesses, consisting of Diĝirmaḫ, Aruru, Nintu, Ninmena, Ninḫursaĝa and Nin-X. A few lines further on several names of Enki follow; one of his names here is **ᵈam-an-ki**. Schretter claims that **ᵈam-an-ki** is the *emesal* name for Enki [376]. But this Isin-list contains – except possibly in the broken sections – no *emesal* names at all.

3.2.5 *The Nippur god list* (OB) [377]

Peterson gives a survey of the god lists, all from Nippur in the Old Babylonian period, that are preserved in the University Museum, Philadelphia [378]. There is one list, known as 'the Nippur god list', that is so far unattested from any other period or place, and the only edition until now has been produced by Jean [379]. Lambert writes [380]: 'In style this [list] is a simple string of names, arranged largely on theological principles, though toward the end some lexical grouping seems to occur.' The text, a school product, differs from the other god lists with respect to its structure; but on the other hand, it is also difficult to label it as a specific 'Nippur' composition, because of the shortness of the Enlil-section and the relatively low place (no. 34) in the list for the city god of Nippur at that time, viz. Ninurta.

The structure of the part of this list that is of interest for my study can be summarized as follows: the lines 1-3 contain An, followed by Antum and Uraš [381]; 4-5 Enlil - Nunamnir; 6. Ninlil; 7 Šulpa'e; 8. Ninḫursaĝa; then follow (9-16) other names of mother goddesses, some of which are epithets of Ninḫursaĝa (e.g. Nindiĝirene). Uraš is present a second time: at no. 37 of the composite text [382].

Antum and/or Uraš are absent in those god lists dating from the time before the Old Babylonian period. Antum is the Akkadian feminine form of An. Uraš is considered to be the wife of An in some Old Babylonian texts [383]; An and Uraš are the parents of Enki [384].

---

375 It has to be emphasized that the characterization of a group of gods as 'ancestors of Enlil' is not included in this list, but can be derived from the later god list **an** = *anum* (Litke 1998). The names of three pairs of 'ancestors' are not legible anymore.

376 Schretter 1990, 153, no. 49. In ch. 2.2.3 I have argued that **ᵈam-an-ki** is not an *emesal* name for Enki. See also the Excursus 2 'Enki' at the end of this chapter.

377 Because van Dijk interpreted the Nippur list as a genealogy (ch. 1.2.1.1), some of his conclusions cannot be corroborated by my study: 1. Enlil is not the son of An and Antum/Ki/Uraš; 2. Ninmaḫ and the next goddesses are indeed mother goddesses, but not spouses of Enlil; 3. Namma has not been passed over: she is present in this list (SLT 122, *iii*: 22), but indeed not at the beginning of this list, which does not represent a cosmogony or theogony.

378 Peterson 2009. The composite text of the Nippur god list is on the pages 14-16.

379 Jean 1931.

380 Lambert 1957-1971, 474, § 3.

381 Richter (2004, 17) considers these names (An, Antum, Uraš) as 'Vorfahren Enlils'. According to van Dijk (1964, 8) Antum (= **ki**) and Uraš are the names of spouses of An, and Enlil is their son; the equation Antum = **ki** has not been indicated in the Nippur list.

382 Peterson 2009, 14.

383 In 'A hymn to Ninisina' [Sjöberg 1982, 64, ll. 1-2; ETCSL 4.22.4, ll. 1-2] it is told that Ninisina is the daughter of An and she was borne by Uraš. Šu-ᵈSuen has the same parents (Kramer 1989, 304, ll. 4-5 ['A hymn for Šu-Suen', ETCSL 2.4.4a, ll. 4-5], as does Martu (Sjöberg 1977, 6-8 *ad*

The presence of the epithet ^{d}**am-an-ki** of Enki in this Nippur list [385], a non-*emesal* text, is another argument against the opinion that ^{d}**am-an-ki** is the *emesal* form of the name Enki. Absent from this list are the groups or members thereof, who were later known as ancestors of An and Enlil, respectively.

Besides the standard OB Nippur god lists, there are several non-standard lists. One of them – UM 29-15-90 – shows part of the ancestor group of Enlil [386].

3.2.6 *The Mari god list (OB)*

Lambert has published a unique list with god names, found at Mari [387]. It proved to be an exercise tablet with poor-quality signs and a strange order for the columns. Moreover, the list of the gods, of a theological rather than a lexical nature, is far from complete. According to Lambert, 'The presentation of the gods gives no hint as to the town in which it (i.e. this list, *JL*) might have been compiled.' '(...), this is a typical Old Babylonian god list from southern Mesopotamia.' [388]

The opening section concerns Enlil and Ninlil, followed by eight pairs of gods who are indicated as 'ancestors' of Enlil. Immediately after these 'ancestors' ^{d}**en-me-šár-ra** and ^{d}**nin-me-šár-ra** are mentioned, a pair that in TCL XV 10 and in **an** = *anum* is included in the ancestry of Enlil, but here this pair appears outside that group [389].

2; ['A hymn to Martu', ETCSL 4.12.2, ll. 3-4]). In other Old Babylonian texts, An and Uraš are mentioned together, but not explicitly as husband and wife (Falkenstein and von Soden 1953, 86 ['A *tigi* to Bau for Gudea', ETCSL 2.3.2, ll. 36, 39]; Sjöberg 1960, 65, l. 4 [An excerpt from a hymn to Nanna, ETCSL 4.13.c, l. 4]; 'An *adab* to Ninisina', ETCSL 4.22.5, ll. 1-3; no written publication available). There is only one text from the ED IIIa period [WF 108 = P011066, *obv. ii*:2] that mentioned ^{d}**uraš**, in connection with an offering.

384 'A *balbale* to Enki for Išme-Dagan': 'Nudimmud (...) born of Uraš (...) first son of holy An' (Green 1975, 66, ll. 10-11, 13 [with transcription ^{d}**uraš-e** in l. 11]; ETCSL 2.5.4.05, ll. 10-11, 13, with transcription **an uraš-e** in l. 11, and consequently the translation: 'Nudimmud (...) borne by An and Uraš (...) first-born child of holy An'. The interpretation 'borne by An and Uraš' is possible, and is in conformity with IAS 114, *i*: 12'-13'. On the other hand, in the texts of the late third millennium and OB period, Uraš has always the **diĝir**-determinative, with one exception: in 'The blessing of Nisaba by Enki' text B (a Lagaš-text) l. 2 Uraš is mentioned without the **diĝir**-determinative (Hallo 1970b, 124; ETCSL 4.16.1 ['A hymn to Nisaba'], the Lagaš version, l. 3). Therefore I prefer the transcription ^{d}**uraš-e** in l. 11 of the *balbale*; but with both transcriptions the final result of the lines 10-11 and 13 is that An and Uraš are the parents of Enki.

385 SLT 123 *ii*:1; SLT 124 *i*:24; composite text no. 24. See also the discussion about this name in Chapter 2.2.3.

386 Peterson 2009, 83-84.

387 Lambert 1985a. Edzard (1967a) and Lambert (1985c) have compiled lists of god names with the aid of texts in which deliveries to deities are mentioned. Durand (2008, 196-282) discussed the pantheon of Mari, but without any useful point of departure for this study of 'Beginnings'.

388 Lambert 1985a; 182 (first quotation); 188 (second quotation).

389 The pair Enmešarra-Ninmešarra – 'Lord/Lady of numerous **me**' – is only known from the god lists. Although every god has its own **me**, it is usual to consider the god Enki (of Eridu) as the god who possesses all the **me** (Farber 1987-1990, 610a). My supposition is that the pair Enmešarra-Ninmešarra in this Mari list belongs to the 'ancestor' group of Enlil, but by accident they have been placed after the 'concluding' lines 19-20 ^{d}**en**/^{d}**nin ama a-a** ^{d}**en-líl-lá**; as Lambert (1985a, 181) noticed: 'This arrangement of the tablet is clear proof that it is an exercise.'

Next some of the most important gods of the pantheon (Enki, Suen, Ninḫursaĝa,...) are cited, but there is no trace of An, at least no recognizable mention.

In his conclusion Lambert writes [390]: 'To sum up, this is a typical Old Babylonian god list from southern Mesopotamia. (...) Enlil's ancestry at the beginning is no doubt in origin from a distinct source. Though a similar ancestry appears already in the Ṣalābīkh list, its position is far from Enlil and Ninlil at the beginning.' Lambert does not explain clearly why Enlil's ancestry should have an origin outside South Mesopotamia. We have no other third millennium god lists, in which a group such as 'Enlil's ancestors' is present, other than those from Fara and Abu Ṣalābīḫ. In the Weidner god list, of which the first beginnings could be traced back until the Ur III-period, this ancestor group is absent. We have no actual indication as to the place of origin of Enlil's ancestor group [391].

3.2.7 *The list TCL XV 10 (= AO 5376) or 'Genouillac list'* (OB) [392]

A summary of what may be learned about this unique list from past publications has been given by Lambert and Richter [393]. The general conclusion is that this list is a forerunner of the list **an** = *anum*, but TCL XV 10 does not yet have an explanatory column. For the interpretation of the function of gods, the list **an** = *anum* and some mythological texts are helpful and necessary. TCL XV 10 is 'a simple string of names, arranged theologically' (Lambert). According to Richter, the list must originate from South Mesopotamia, as crucial gods from the North, the Diyala region and Syria are absent [394]. The list opens with 16 pairs **^{d}en-X - ^{d}nin-X** gods, in which we recognize the later ancestors of Enlil in **an** = *anum* [395]. Thereafter follow An, his theogony and family, though in a very abbreviated form compared to the corresponding section in **an** = *anum*. The next god is Enlil, accompanied by the major members of his court, but now completely separated from his 'ancestors'. Distinct groups are formed by Enki and his court, and the mother goddesses [396].

Enmešarra and Ninmešarra are also attested separately from each other. Sallaberger (1993, 103) classified Enmešarra, with respect to the Nippur cultus of the Ur III-period, as a minor deity, while Ninmešarra has not been mentioned. For a detailed discussion about Enmešarra: see ch. 3.3.5.1. In one text Ninmešarra appeared to be an epithet of Inanna (Zgoll 1997b; ETCSL 4.07.2); in another text, it was said about Ninlil: **^{d}nin-líl me-šár-ra-daĝal si-a kur-gal-da zag ša$_4$** 'Ninlil, comprehensively replete with numerous **me**, equal to the Great Mountain' ('An *adab* to Ninlil' ,Wilcke 1973, 7, l. 1 [ETCSL 4.24.1, l. 1]).

390 Lambert 1985a, 188.

391 If such an 'ancestor group' is present at all, then there is no mutual coherence between the various members of this ancestry list, e.g. no family order, and also: there is no conformity between the various lists, neither in length nor in order of the names.

392 De Genouillac 1930, pl. XXV-XXXI.

393 Lambert 1957-1971, 475, § 5; Richter 2004, 13-16.

394 Richter 2004, 14. TCL XV 10 originates from a clandestine excavation; the provenance is unclear.

395 Edzard (2004a, 583) considers the 16 pairs of the Enki-Ninki group as ancestors of An, although he also says that the Genouillac list is the forerunner of **an** = *anum*; the latter god list clearly indicates the Enki-Ninki group as ancestors of Enlil.

396 Some comments on van Dijk's conclusions with respect to TCL XV 10 (ch. 1.2.1.3) are:
1. It is evident that the conclusions of van Dijk with respect to the god list TCL XV 10 are based on the supposition that the list of Enki-Ninki gods – up to and including Enmešarra / Ninmešarra – are the 'ancestors' of An. However, these gods never existed before **an-ki**, but on the contrary they originate from **ki**. Moreover, there is no text in which heaven and earth were separated before any gods were born.

3.2.7.1 *The Enki-Ninki group*

The Enki-Ninki group is placed at the beginning of this list, before the gods who are known to belong to the Sumerian pantheon [397]. In the Fara lists SF 23-24 the Enki-Ninki group consists of seven pairs **en-/nin-X** gods, including Enki and Ninki and, as second group, **en-E$_2$** and **nin-KID**. The 16 ^{d}**en-/**^{d}**nin** pairs of the Enki-Ninki group in list TCL XV 10 share only four pairs with the corresponding group in the lists SF 23-24: **en-/nin-ki**; **en-/nin-giriš/**; **en-/nin-bulug$_3$**; **en-/nin-amaš** [398]. A striking difference, compared with SF 23-24, is the absence of the couple Enlil-Ninlil in the Enki-Ninki group of TCL XV 10. In SF 23-24 **en-E$_2$ - nin-KID** followed immediately after Enki-Ninki, and thereafter the other members of the Enki-Ninki group were mentioned. This 'embedded' rendering – at least as far as it concerns Enlil – corresponds to the text of IAS 114, in which is related that Enki and Ninki produced 'the seven' and also Enlil. In the order in which the gods in the Enki-Ninki group in TCL XV 10 are mentioned it is not possible to recognize any order of developmental stage or any connection between the successive members of this group.

3.2.7.2 *An – Enlil*

The Enki-Ninki group is followed by An and two gods – **an-šár-gal** and ^{d}**en-uru-ul-la** – who appear for the first time in a god list. Because the list itself lacks any information about these gods, these two names might be understood as epithets of An. But bearing in mind what Lambert has written about the mentioning of ancestors [399], these names could also refer to An's ancestors, the more so as ^{d}**ama-tu-an-ki**, 'the mother who has borne **an-ki**', follows after them [400].

Then follow ^{d}**uraš** and *Bēlet-ilī* [401]. It is very unlikely that here the god Uraš, a local deity of the northern Babylonian city Dilbat, who is mentioned in the Prologue to the Laws of Ḫammurapi, is meant [402]. It is more plausible that the goddess Uraš, as the wife of An, is referred to, considering also the epithet *Bēlet-ilī* 'Lady/Queen of the gods' in the next line. Uraš is very seldom found in texts from the third millennium [403], or in texts from the Old

---

2. The Enki-Ninki gods are called 'androgyne beings' by van Dijk. In the list TCL XV 10 the 'ancestor' group is composed of **en-** and **nin**-pairs, a male and a female god. There is no androgynous being present. Even **an-ki** is not androgynous: this unit is always a bipartite one.

397 Litke (1998, 21) remarks with respect to this group: 'In the section preceding the name of Anu, Ao (= TCL XV 10, *JL*) gives a long list of names (Ao: 1-30) which evidently is intended to represent the ancestors of Anu (...).' But there is nowhere any indication that the Enki-Ninki group constitutes the ancestors of An.

398 **en-/nin-amaš** in TCL XV 10 corresponds in all likelihood to **en-/nin-utua** in SF 23-24: **amaš** = **DAG-KISIM$_5$xLU-MAŠ** (sheepfold); **utua** = **DAG-KISIM$_5$xUŠ** (stud).

399 Lambert (1968, 2): 'In principle a list of ancestors can equally well begin with the first known figure and work downwards both on the tablet and chronologically, or it can begin with the living person and list his ancestors in reverse order chronologically.'

400 From a grammatical point of view, **an-ki** is not the object of the verb **tu** (the word order is incorrect). I suppose that **ama-tu** is a fossilized idiom, meaning 'natural mother'; then **an-ki** represents a genitive. See also PSD A III, 198a *ad* 1.4.1, and 207b. Cohen (1989, 80 and 83: l.8) translates **ama-tu-ud-da** as 'bearing mother'.

401 *Bēlet-ilī* is the Akkadian form of ^{d}**NIN**-*ì-li*.

402 Codex Ḫammurapi: *iii:* 18-22. Kienast 1985, 112, § 12.

403 Text WF 108, *obv. ii*:2 (P011066), ED IIIa, Fara. It seems to be an offering to Uraš. A second mention of Uraš is found in Temple Hymn no. 33, l. 421 (Sjöberg & Bergmann 1969, 41). It

Babylonian period. In some of these texts it becomes clear that An and Uraš form a couple: e.g. Ninisina is the daughter of An and Uraš [404], and Martu and Šu-Suen are sons of An and Uraš [405]. In other texts there is only a suggestion that An and Uraš are husband and wife [406]. **uraš** meaning 'earth' is found from the Old Babylonian period onwards [407].

3.2.7.3 *Namma*

Namma and her epithet **^{d}ama-tu-an-ki** will be discussed with regard to the list **an** = *anum*, and in ch. 3.3.2.

3.2.8 an = *anum* *(MB)* [408]

Because of the strong correspondence between **an** = *anum* and TCL XV 10, and because of the explanatory texts in the **an** = *anum* list, **an** = *anum* can be reliably used to analyse the Old Babylonian god lists, especially TCL XV 10. This in spite of the relatively late copy (end of the second millennium BCE) of **an** = *anum*.

In the introduction to his book, Litke gives a survey of the development of the god lists in Mesopotamia. Although there are no god lists known from the time between the ED III- and the Ur III-period, Litke thinks that it is justifiable to suppose that the composition of such lists was not an unknown practice in this interval. The indirect evidence therefore is the existence of several traditions of such lists in the Isin-Larsa and Old Babylonian periods. In this respect Litke remarks [409]:

> 'At the present time, the relationship between these various traditions (i.e. those of the Isin-Larsa time; *JL*) seems quite tenuous and difficult to analyze; nor does it seem possible to discover any demonstrable ties between them and the older Fara god-lists. The four lists doubtless reflect the local cult traditions in as many religious centers – possibly representing the ranking of the deities and sub-deities as worshiped in the respective religious centers. In this respect, since the various leading gods probably represent older patron deities of distinct political units, the ranging of the pantheon in these lists may ultimately reflect the play and counterplay of polit-

---

states that Ištaran is the son of Uraš; in this case it concerns the goddess Uraš (Wiggermann1997, 42 *ad f. Ištaran*).

404 Römer 1969b, 282, with references; 'A *šir-gida* to Ninisina' [ETCSL 4.22.1], ll. 84-89; 'A hymn to Ninisina', Sjöberg 1982, 64, ll. 1-2 [ETCSL 4.22.4 -ll. 1-2].

405 Sjöberg 1977, UM 29-13-509, 6-8 *ad* 2 ['A hymn to Martu', ETCSL 4.12.2], ll. 3-4; Kramer 1989, 304, BM 100042, ['A hymn for Šu-Suen', ETCSL 2.4.4.a], ll. 4-5.

406 The expression that was used is: **mí-(zi)-dug$_4$-ga-an-^{d}uraš-(...)** 'treated kindly by An and Uraš...' (Falkenstein 1953, 86: 'A *tigi* to Bau for Gudea' [ETCSL 2.3.2], ll. 36, 39; Sjöberg 1960, 65, 'An excerpt from a hymn to Nanna' [ETCSL 4.13.c], l. 4). In 'An *adab* to Ninisina' [ETCSL 4.22.5 (no written publication available)], ll. 1-3, it is said about Ninisina that she was engendered by An ... and was brought up sitting on the holy knees of Uraš.

407 ePSD *sub* **uraš**; the lexical list Ea I, 338 b-c, reads: **ú-ra-áš IB** = *šá-mu-u er-ṣe-tum*.

408 Litke 1998. My comments on van Dijk's discussion of this god list (ch. 1.2.1.3) are: 1. Heaven and earth (**an-ki**) have never been androgynous (see e.g. IAS 113 *ii*: 5-10; IAS 114 *i*: 1'; IAS 136 *iii*: 1'-3'; IAS 203 *ii*: 3'-5'; Ukg 15 *ii*: 2; NBC 11108 *obv.* 5; 'The Debate between Tree and Reed' l. 6; GEN ll. 8-9; KAR 4 *obv.* 1); and 2. In the list **an** = *anum* van Dijk passes over the epithet of Namma, viz. **^{d}ama-ù-tu-an-ki**; this is *nota bene* the only god list in which this equation has been explicitly mentioned.

409 Litke 1998; first quotation: 2-3; second quotation: 6.

ical fortunes in early Babylonia even more than the independent development of theological concepts.'
'The series An : [d]A-nu-um can hardly be correctly classed as a lexical text for it makes no attempt to list Sumerian and Akkadian equivalents. It is, instead, an explanatory list that seeks to clarify the offices and relationships of the numerous members of the pantheon. Unlike most lexical texts, this series is not even Akkadian in its composition, but rather, Sumerian.'

Although the texts of **an** = *anum* currently available are written after the Old Babylonian period by Assyrian scribes, Litke considers these texts as directly or indirectly copied from Old Babylonian originals.

3.2.8.1 *The 'ancestors' of An*

The first group in **an** = *anum* that can be distinguished consists of 21 **en-ama-a-a-an-na-ke$_4$-ne**. Litke examines the possible meaning of this expression [410]:

> 'It has been generally assumed (...) that this group of deities represents the Väter-Mütter of Anum (although Tallquist [1938, 251] suggests the alternate possibility that the term involved 'ältere Entwicklungsstufen' of Anum). If this interpretation is valid, then the words **ama a-a** must be regarded as being in apposition to the word **en**; and the actual translation of the line would be, '21 exalted ones, the fathers and mothers of Anum.' Several other factors, however, complicate the picture. As is mentioned under line 3 above, these names (in a more abbreviated list) appear in Ao (i.e. TCL XV 10; *JL*) as additional names of Anum himself. In text a i 1 ff. (= K. 4338B [CT 24, 19], *JL*), the male names of each of these pairs are also equated with Anum. In fact, this older tradition is still preserved in the present series by the use of the dittos in lines 4-23, which would indicate that these pairs are still all identified with the one pair, Anum and Antum. This suggests that line 24 intends to sum up '21 exalted ones, father and mother pairs who belong to the circle of (and are identified with) Anum' (the summary mentions only Anum, but the dittos in lines 4-23 reveal that Antum is to be included as well).'

Some comments may be made with respect to this quotation:

1. The list **an** = *anum* is the first god list in which explicitly a group of gods has been indicated as 'ancestors' of An [411]. Two names we did already meet in TCL XV 10, viz. **an-šár-gal** and **[d]en-uru-ul-la**. In this list, however, any indication about who these gods are is lacking.

2. Litke's suggestion that the older tradition (= that of TCL XV 10; *JL)* is preserved by the use of dittos is without foundation. In order to substantiate this statement, a closer examination of the lines in question is necessary [412].

410 Litke 1998, 23-24.

411 For the discussion about 'ancestors' or 'older stages of development', as Tallqvist suggested, see ch. 4.2.3.2.

412 The following tablets CT 24, 1 and CT 24, 20 are the sources C and B respectively, which were used by Litke in reconstructing tablet I of **an** = *anum* (Litke 1998, 20).

CT 24, 1:

1 2 3 4 5	**an** **an** **an-ki** **an-uraš** **^{d}nin-uraš** *etc.*	*^{d}a-nu-um* *an-tum* *^{d}a-nu-um u an-tum* ‘ ‘ *etc.*
22	**21 en ama-a-a-**	**an-na-ke$_4$-ne**

CT 24, 20:

1 2 3	**an** **an : an-tum** **an-ki**	*^{d}a-nu-*[...] : **ki** [x...] *^{d}a-nu ù* [*x...*]
4	**an-uraš** *etc.*	**^{d}nin-[uraš]** *etc.*
14	**21 en ama-**	**a-a-an-na-ke$_4$**

The dittos [413] appear only on the tablet copied in CT 24,1. In general, the right column of this tablet contains the Akkadian counterpart of the Sumerian entry in the left column, or that column describes who or what the god in the left column represents, with the exception of line 2. The dittos indicate that the Akkadian name of the god is identical to its Sumerian name in the left column [414]. In CT 24, 20 the dittos are absent. The first three lines are ‘explicative’; in the lines 4-13 each line of both columns together mentions a pair of gods – the male god left and the female god right –. The conclusion seems to be that the ‘21 ancestors of An’ indeed only refer to An, but the dittos in CT 24,1 lines 4-23 are not meant to include Antum as well, as Litke supposes [415]. Antum is the wife of An [416], and this aspect is further developed in this list; starting from **an-ki**, the development of this original unity is continued, via several preliminary, intermediate entities – some of which contain **an** or **ki** – (20 in total), until the final phase, the god of heaven – An – has been reached. Since **an** is already present from the beginning – in a primaeval phase joined or united with **ki** – it is

---

413 For a detailed explanation of the use and interpretation of dittos – the sign **MIN** – and of the sign **šu** in these lists: Litke 1998, 6-15 (‘The inner structure of the series **an** = *^{d}a-nu-um*).

414 Because the dittos are absent in lines 1-3 of CT 24, 1 the Akkadian names are not *identical* with the Sumerian ones, but the Akkadian *equivalent* of the Sumerian name, at least in the lines 1 and 3. The first two lines of this list clearly demonstrate that the Sumerian pantheon (the left column) does not know a female partner of An like the one who is present in the Akkadian pantheon, viz. *Antum* (l. 2). J.G. Westenholz (2010, 320), however, interpreted the lines 1 and 2 as follows: ‘It is interesting to note that Sumerian An is considered bisexual and is equated with both Anu and Antu.’ In my opinion there is not any indication of the supposed bisexuality of An in any text whatever.

415 See the last quotation of Litke (1998, 24). In fact it might also be possible to speak about ‘developmental phases’ of **ki**, starting from **an-ki**, and further **^{d}ki-šár-gal**, and **^{d}ki-šár**.

416 Edzard 1965, 41-42.

more appropriate to speak about 21 'developmental phases' of An, instead of 'ancestors'[417]. If we accept that the 20 'ancestors' after **an-ki** form male - female pairs, and we suppose that **an-X** is male and **ki-X** is female, then the pair immediately following the line 3 **an-ki** ^{d}*a-nu-um u an-tum* is problematic.

That conspicuous pair is, according to the traditional transcription, ^{d}**uraš** and ^{d}**nin-uraš**. Usually **uraš** is supposed to be 'earth, (arable) land', a form or part of **ki**, which is generally accepted to be feminine [418]. The two consecutive lines in question are:

an-ki | ^{d}*a-nu (-um) ù an-tum*
AN-IB | ^{d}**nin-IB**.

The last pair is usually transcribed as: ^{d}**uraš** and ^{d}**nin-uraš**. But there is another transcription possible for the item **AN-IB**, viz. **an-uraš**. This unit represents a complex between on the one hand **an** and on the other hand a developmental phase of **ki**, viz. **uraš**.

In the excerpt tablet of **an** = *anum* [419], the lines *i:* 2 and 3 read:

[**AN**] $^{\text{ú-ra-aš}}$ **IB** | **AN** " *šá iš-pik ik-ri-bi* [420] An-uraš *of the storage of the votive offerings*
[**AN**] **šár-gal** | **AN** " *šá kiš-šat* **an-ki** An-šargal *of the whole of heaven and earth.*

In my opinion, each line of the right part of this column means: it concerns **an**, with the specification given in the left column now represented by the dittos, followed by some explanation in the Akkadian language. The entries are different 'manifestations' or 'aspects' of An [421].

From the point **an-ki** in the list **an** = *anum* onwards, this initial entity has developed into three separate **an-** and **ki-** entities. The three couples are: 1. **an-uraš** and ^{d}**nin-uraš**; 2. **an-šár-gal** and **ki-šár-gal**; 3. **an-šár** and **ki-šár**. This seems to justify the transcription of **AN-IB** as **an-uraš**. Transcribed in this way, the list of An's manifestations is not interrupted until ^{d}**en-šár** - ^{d}**nin-šár**. The tentative conclusion is that the transcription given by all authors, – ^{d}**uraš** and ^{d}**nin-uraš** – is nothing other than an artificial splitting up of

---

417 Lambert (1974, 15) also supposes that in the list TCL XV10 the gods between Enki and Ninki leading to Enlil are developmental stages.

418 See also the discussion about ^{d}**uraš** at § 3.2.7.2 *ad* An-Enlil. It is very unlikely that the male god Uraš, discussed there, is meant in this position in the list **an** = *anum*. Lambert (1980-1983a, 219b): 'A similar origin [*like Enlil and Ninlil from Enki and Ninki; JL*] for Anu and Antu is given in An = Anum I 4-23, where Uraš ('Earth') and Nin-uraš ('Lady Earth') head a nine-pair ancestry.' As will be made plausible, it is very unlikely that An/heaven should originate from 'earth' alone.

419 Tablet K. 4338B (CT 24, 19).

420 CAD I-J, 66a *sub ikribu*; AHw 397b *sub išpikū*.

421 Lambert (1975b, 196) too explained in this way the entries in the list **an** = *anu ša amēli*. But then, speaking about the identification of one god with a similar one with another name, he continues (p. 197): 'The name, and so the existence, was transferred to another owner. Still another kind of absorption can be quoted from An = *Anu ša amēli*. The last name of Anu given there is Uraš "earth". Elsewhere Uraš is very properly Anu's spouse: heaven and earth form a natural pair. So even a wife can be absorbed into her husband!' This explanation seems very unlikely: **an** and **ki** (or **uraš**) are never equated. The solution is to read **an-uraš** instead of ^{d}**uraš**. The line in question of **an** = *anu šá amēli* (Litke 1998, 229, l. 12) is clear on this point: **an-uraš** = *anu šá milki* 'An of the decision'; in other words: it concerns a certain aspect of An.

uraš [422], without any aspect of An, and suggesting a pair with procreative function. Moreover, **uraš** alone, being 'earth', cannot be a manifestation, ancestor or aspect of **an**, being heaven or An. Compare these **uraš**-entries with **ᵈuraš** in TCL XV 10: in this list **ᵈuraš** is considered to be the wife of An.

From the entry **an-šár-gal - ki-šár-gal** it might be concluded that the original entity **an-ki** has been separated into an **an**-part and a **ki**-part. On the basis of the different meanings of **ki** 'earth' and **uraš** 'arable land', – **uraš** being a more specialized form of **ki** – one might expect that the pair with **uraš** should follow after the pairs with **ki**, instead of preceding them.

The next pair, **ᵈen-šár - ᵈnin-šár**, is remarkable because a pair with the same names also belongs to the 'ancestor'-group of Enlil. It is hard to believe that this is a mistake. It may be that the authors of the list tried to indicate that An and Enlil shared at least some aspects of their power via a shared 'ancestor'-pair.

The 'ancestor' group of An continues, after the pair **ᵈen-šár - ᵈnin-šár**, with a heterogeneous list of pairs that are unknown in the earlier god lists: *ᵈdu-rí - ᵈda-rí; ᵈlàḫ-ma* - **ᵈla-ḫa-ma**; [*ᵈé-kur - ᵈgá-ra* [423]] ; *ᵈa-la-la - ᵈbe-li-li*. Lambert writes in this respect [424]:

> 'The first pair 'Ever and Ever' (i.e. *ᵈdu-rí - ᵈda-rí*; *JL* [425]) are a Babylonian phrase which is not grammatically masculine and feminine, but it is to be so construed for the purpose of this list. The second pair (i.e. *ᵈlàḫ-ma* - **ᵈla-ḫa-ma**; *JL*) are equally artificial, since they are respectively the Babylonian and the Sumerian forms of a word indicating some kind of sea-monster. However, Alala and Belili are correctly male and female in mythology.'

Alala is the predecessor of An in the Hittite *Kumarbi Myth* [426]. In a more recent paper [427], Lambert wrote that there is a strong evidence that the two *laḫmus* had a cosmic function to hold heaven up. In An's ancestry 'eternal Time is separated from Heaven and Earth by Laḫmu and Laḫamu. These latter were not muddy, but were pillars of the universe.'

If there were some doubt whether the name **ᵈen-uru-ul-la** signified an ancestor or an epithet of An in the list TCL XV 10, we now have to conclude that **ᵈen-uru-ul-la** and **ᵈnin-uru-ul-la** belong to the 'ancestors' of An [428].

---

422 Kienast (1985, 112, § 13): '(...) Nin-uraš 'Herrin Erde' in der Götterliste An = *Anum* ist vermutlich eine ad-hoc-Bildung.'

423 This pair is not present in all the sources.

424 Lambert 1975a, 53.

425 In two Middle Babylonian tablets from Nuzi the god *ᵈza-ar-wa-an* was mentioned (SMN 491, l. 10; SMN 690, l. 18; [Pfeiffer and Speiser 1935-1936]). This god might be identified with the Iranian supreme god Zurvān, who connoted 'Time' (Brandon 1963, 197). Acquaintance with this god in Mesopotamia may have caused the presence of Duri and Dari 'Ever and Ever' in the list of An's ancestors in **an** = *anum*. Other evidence that the Indo-Iranian deities in Mesopotamia were known can be found in two treaties between Suppiluliuma of Ḫatti and Šattiwaza of Mitanni: some Indic deities (Mitra, Varuna, Indra, the Nasatya gods) besides e.g. An, Enlil, Ninlil are mentioned (Beckman 1996, 43 [6A §14], and 49 [6B §11]).

426 Güterbock 1980-1983; Hoffner 1990, 40-43; Römer 1994.

427 Lambert 1985b, 199 (in response to Wiggermann's 'Exit *Talim*!' [Wiggermann 1981-1982]).

428 Several Mesopotamian cities bear the epithet **uru-ul**: among them Nibru, Tummal, Kullab, Unug, Ereš, Kisiga, Ur and Zimbir (Römer 1988, 57 *ad* 57).

The following questions can be raised: what is the origin of this ancestor group, and why has it been formulated so explicitly here? As we shall see, in the list **an** = *anum* there is also an 'ancestor group' of Enlil. Parents or ancestors of Enlil were already known from the Early Dynastic times onwards. In due time this group was steadily enlarged, resulting in 16 pairs in the list TCL XV 10. A simple explanation for the 21 ancestors of An in **an** = *anum* may be that the authors of **an** = *anum* held the view, that it was disrespectful to An, who was still the supreme god of the pantheon, although he was a *deus otiosus* [429], if they did not supply him with a respectable ancestry, while Enlil was endowed with it. But there is a striking difference between both groups of ancestors. The 'Enlil group' is always composed of **^{d}en- ^{d}nin** pairs. On the other hand, the 'An group' contains only two **^{d}en- ^{d}nin** pairs; surprisingly enough one of those pairs – **^{d}en-šár** and **^{d}nin-šár** – is also included in the **enki-ninki** group, both in **an** = *anum* and in TCL XV 10, thus Enšar-Ninšar also belong to the 'ancestors' of Enlil.

3.2.8.2 *Bēlet-ilī*

Immediately after the 'ancestor'-group of An follows the name *Bēlet-ilī* (**^{d}NIN**-*ì-li*), and the explication: the wife of An. The name is more an epithet, 'Lady/Queen of the gods', than a real proper name. This designation makes it possible to identify her with **^{d}uraš** in the list TCL XV 10.

3.2.8.3 *Namma*

After the spouse of An, Namma is mentioned, with the first designation as 'the mother of Enki'. But for the cosmogony the second entry is more important: **^{d}ama-ù-tu-an-ki** 'the mother who has borne **an-ki**' [430].

3.2.8.4 *The 'ancestors' of Enlil*

From line I:96 onwards [431], a group starts, which at the end is indicated as 42-**àm en ama-a-a-^{d}en-líl-lá-ke$_{4}$** (in line I:138). A comparable, but smaller, group (16 pairs) opens the list TCL XV 10. The number and order of these 'ancestors' are different in both lists; the only fixed point is the pair that opens the line of 'ancestors', viz. **^{d}en-ki** and **^{d}nin-ki**. In IAS 114 *i*: 3'-4' it is stated that Enki and Ninki have brought forth 'the seven', followed in *i*: 9'-11' by: Enlil has been created by Enki and Ninki. If we take this text literally, then we must conclude that Enlil is the younger brother of the seven. But if the text is interpreted more broadly, one could say that ultimately Enki and Ninki are the 'parents' or better: the primary source of Enlil, but that between Enki-Ninki and Enlil there are several other 'phases', viz. the seven [432]. The list **an** = *anum* was written more than a millennium later than IAS 114 and by then these seven had developed into a group of 21 pairs of 'ancestors'.

---

429 Being a *deus otiosus*, 'idle god', is a characteristic that sky gods share in general (Eliade 1958, § 14, 46-50).

430 Van Dijk (1964, 58) called Namma 'la Terre-Mère, dans la théologie d'Eridu'.

431 Line indication according to Litke.

432 The term 'phase' is used deliberately, to avoid the consequence that, if each following couple is considered to be 'son and daughter' of the previous couple, this should imply incest. But it cannot be excluded that such behaviour may have been acceptable for the divine world (see also the end of ch. 3.3.5).

Lambert remarks with respect to the intervening pairs in the 'ancestor' groups [433]:

> 'In all cases the first pair are the same, but the intervening pairs never agree completely between the various lists either in the number of pairs or in the distinctive element in each pair. These divergencies suggest that the intervening pairs were not in themselves important, but only serve to give remoteness to the first; and this conclusion is confirmed by the meanings of the names. In many cases the distinctive element in the pairs can be translated, but no cosmic sequence emerges when these meanings are taken together. (...) The rationale of this genealogy (i.e. Enki and Ninki being the parents of Enlil; *JL*) is clear. Enlil had a spouse Ninlil. It was desired to make them descendants of the prime element Earth, so this latter was made into a marital pair Enki and Ninki, which was coupled by intermediate pairs to Enlil and Ninlil.'

The present author disagrees with Lambert in some respects. First of all, this 'ancestor' list concerns only Enlil, not Enlil *and* Ninlil; the line I:138 '**42-àm en ama-a-a-^{d}en-líl-lá-ke$_4$**' is beyond all doubt in this respect. Further, why should it be '*desirable*' to make Enlil the descendant of the prime element Earth? Lambert does not explain this point, but I agree with his final conclusion: 'The only mythology here is that Enlil descended from Earth.'

Litke comments at line I:96, the starting-point of the Enki-Ninki 'ancestor' list, that it is evident that in the list TCL XV 10 the section beginning with Enki-Ninki was considered as the list of predecessors for both An and Enlil. This tradition, according to Litke, seems quite consistent with Babylonian theology, since Enlil was certainly regarded as An's son. As proof of it, Litke refers to a vase inscription of Lugalzagesi [434]. The mention of An as the 'beloved father of Enlil' has to be valued in its context, being a prayer to Enlil, as a respectful phrase; this has no theogonic implications. The cosmogonic introduction of the text IAS 114, included in this study [435], shows that Enki and Ninki are Enlil's parents, not those of An. The god lists SF 23-24 from Fara present Enlil and Ninlil after Enki-Ninki. Also in the Mari god list is the Enki-Ninki group regarded as 'ancestors' of Enlil alone.

3.3 *Cosmogonic and theogonic implications of the god lists*

I am fully aware of the risks inherent in 'backward reading' – the use of later text materials to explain older ones – especially in the study of religion, as was pointed out by Michalowski [436]. But the god list **an** = *anum* is the list *par excellence* to understand the relationships and the offices of the members of the Mesopotamian pantheon [437], because of the column with explications next to the god or goddess in question. Gods who play a role in the cosmogonic and theogonic history, as indicated or indirectly inferred from the list, can be traced in older texts, in which their position and function are not always clear at first glance, of course with some exceptions. Once the position of the gods with respect to cosmogony and theogony in the older texts and lists has been clarified, we can study anew

---

433 Lambert 1975a, 52.

434 Lugalzagesi 1, (CDLI no. P263413) *iii*: 14-16: **^{d}en-líl lugal kur-kur-ra-ke$_4$ an a-a-ki-áĝ-ni** 'Enlil, Lord of all lands, for his beloved father An'.

435 Ch. 2.1.1a.

436 Michalowski 1998, 239.

437 Litke 1998, 6.

the later texts and lists. In this way a sketch of the history of Mesopotamian religion – cosmogony and theogony being part of it – for the third millennium and the first half of the second millennium may be written.

Which parts of the Sumerian cosmogony and theogony can be written with the aid of the data deduced from the god lists, will be discussed in the next sections.

3.3.1 *Composition of the god lists: analysis of the order of several (groups of) gods*

As starting-point for an analysis of the 'historical' order of the single gods or of the groups of gods, the lists **an** = *anum* (AA) and TCL XV 10, which last one can be considered a forerunner of **an** = *anum* (FAA), serve the purpose very well. Both lists show the same groups, although these groups differ in the number of gods and in their order within the groups, e.g. the **ᵈen-ki - ᵈnin-ki**-group and the 'ancestor'-group of An.

The order **an** → **ᵈama-(ù)-tu-an-ki**, present both in AA and in FAA, is not logical from historical-cosmogenetic viewpoint: the one who is begotten has been mentioned before the creatrix. But as we shall see: god lists are never chronologically arranged. Lambert has given a solution to obtain the correct order of events [438]. For a chronologically correct reading of the list TCL XV 10, he supposed that the order in which the gods **an** → **ᵈama-(ù)-tu-an-ki** are mentioned has to be reversed:

> '(...) this is confirmed by the appearance of "Mother-who-gave-birth-to-Heaven-and-Earth" at the end, since this title suggests a primaeval being. As interpreted by the present writer, Nammu heads the list chronologically and the title just quoted describes her. (...) The second generation is Urash, another Sumerian word for "earth", and Bēlet-ilī, Babylonian for "Mistress-of-the-Gods", seems to be an epithet of Urash. The third generation is En-uru-ulla, Sumerian for "Lord-of-the-Primaeval-City", which illustrates the cosmological importance of cities in Sumero-Babylonian thought. The remaining names, Anshargal and An, Sumerian for "the whole heaven" and "heaven", are either two generations, father and son, or two names of Anu. Evidence can be quoted for both of these possibilities. In this case all the stages of the genealogy are cosmologically important, and the final result is that Anu descended from cosmic water.'

Wiggermann suggested another solution [439]:

> 'He (= Lambert, *JL*) considers the "Mother-who-gave-birth-to-Heaven-and-Earth" (5b) to be a primeval being, since this is what the name suggests. Since there are no essential differences of opinion on the grouping of items, we start with the groups, not with single items. On the basis of the meaning of 5b, Lambert reads the list "upward": 5-4-3-2. This makes 5b the ancestor of them all, and 3 the ancestors of Anu, the desired result. There is, however, another way of obtaining this result, at first sight less elegant, but not objectionable once the upward reading principle is accepted: applying this principle twice, that is on 2-3, and on the resulting list, giving 5-{(3-2)-4}.

438 Lambert 1975a, 53.

439 Wiggermann 1992, 290-291. The present table is made after the table on pag. 290.

Another point that needs attention is the place of 5 in the second solution. When upward reading is applied to a series 1-(3-2)-4-5, this results in either 1-5-{(3-2)-4}, or in 5-{1-(3-2)-4}.

Before we choose one of the alternatives, we must look at the order of items in *AA*. This text explicitly calls 3 the ancestors of An, and on this basis it is universally accepted that the order is reversed. Inside section 3, however, the reading is downward, since form (single god) and meaning (Heaven-Earth) of An-ki do not allow it to be derived from a pair (a more advanced type of procreation) denoting a much less general cosmic entity ("Earth").

What was valid for 5 (Amatuanki) in *FAA* is valid for her in *AA* too: her name shows that she precedes 2-4. In *AA*, however, simple upward reading is impossible: the wife of An would precede An, and ancestor 3d would precede ancestor 3a, the much more general single An-ki. Here too the alternative solution can be applied: the upward reading principle is applied first to 2-3, and then to the whole group, giving 5-{(3-2)-4-6a}.'

	FAA	*AA*
1	^{d}en-ki ^{d}nin-ki	
2	an	an
3 a		an-ki (*single god*)
b		^{d}uraš ^{d}nin-uraš
c	an-šár-gal	an-šár-gal ^{d}ki-šár-gal
d	^{d}en-uru-ul-la	^{d}en-uru-ul-la ^{d}nin-uru-ul-la
4 a	^{d}uraš	
b	^{d}*bēlet-ilī*	^{d}*bēlet-ilī* = dam-an-na-ke$_4$
5 a	^{d}namma	^{d}namma = ama-^{d}en-ki-ga-ke$_4$
b	^{d}ama-tu-an-ki	^{d}ama-ù-tu-an-ki = ^{d}namma
6 a		^{d}en-ki ^{d}nin-ki
b	^{d}en-líl	^{d}en-líl

Some comments may be made in connection with this method of analysis:

1. God lists are one-dimensional, in other words: items of equal importance will still be found arranged after each other.
2. The arrangement, the order of the gods, in these both lists point to a combination of theological order and – in some divisions – of a historical/cosmogenetical order. The lists are not intended to present a linear cosmogony or theogony.
3. The 'upward reading principle' as applied by Wiggermann is somewhat incomprehensible. For instance: his first treatment, 'applying this principle twice', i.e. on 2-3 and then on the resulting list, 'results in 5-{(3-2)-4}'.
   In my opinion the result of this exercise – upward reading twice – is as follows:
   a. starting-point: 2-3-4-5.
   b. first upward reading on 2-3 results in 3-2; halfway score is (3-2)-4-5.
   c. second upward reading 'on the resulting list', i.e. on the halfway score; the final result is: 5-4-(3-2). This is exact the result of Lambert's upward reading once!
4. The place of item 5 seems to be optional; upward reading of 1-(3-2)-4-5 results, according to Wiggermann, in *either* 1-5-{(3-2)-4} *or* in 5-{1-(3-2)-4}. This is not simple 'upward reading', but choosing what result seems to be the most plausible one.

Perhaps the upward reading exercise may be omitted. Thanks to the explications given in the list **an** = *anum* and the information obtained from the translated texts, simple logical reasoning brings a solution for a reliable interpretation and order of the lists in cosmo-genetic, theogenetic and theological respects.

3.3.2 *Namma* [440]

In **an** = *anum* we find Namma twice:
1. **ᵈnamma** = **ama-ᵈen-ki-ga-ke$_4$**
2. **ᵈama-ù-tu-an-ki** = **ᵈnamma** [441].

The second entry makes the position of Namma clear as <u>the</u> primaeval element, giving birth to **an-ki**. The name **ᵈnamma** and the epithet **ᵈama-tu-an-ki** appear, one after the other, in TCL XV 10 (*i*: 36-37). The sign used to spell Namma is **engur**; this value is attested from the ED III-period onwards. Namma is present in several other god lists:
- Fara SF 1 *viii*:10; in the same list there is a name composed with Namma [442];
- in the Abu Ṣalābīḫ-list Namma could not be found; there are several lacunae, in which the name may be present; but there are a few names composed with **engur** [443];
- in the Nippur list [444];
- in the Mari-list Namma is present, without relevant context.

In all these lists there is not any indication about the function of Namma. This has been done for the first time in list TCL XV 10, in which Namma, mentioned after An and his wife, is followed by the epithet **ᵈama-tu-an-ki** 'Mother who has borne **an-ki**', a position that has been confirmed in **an** = *anum*. Namma as mother of **an-ki** is unknown in this

---

440 See also Wiggermann 1998-2001a.

441 Litke 1998, I: 27-28.

442 Krebernik 1986,175-176. In SF 1, *viii*:22: **ᵈnamma-URU x DARA$_3$**. **DARA$_3$** is the sign for 'ibex', and the ibex is the symbolic animal of Enki (Wiggermann 1993-1997, 226b). Does this name mean: 'Namma of the town of the ibex', in other words: Namma of Eridu?

443 Mander 1986, 31, nos. 376 (**ᵈ[ ]x-ENGUR-[ ]**) and 395 (**[ᵈ]ZI.NA-ENGUR**). This last name may be transliterated as **ᵈ⁽ⁿᵃ⁻ᶻⁱ⁾ENGUR**, and may be another spelling for the name of Nanše, as can be found in **an** = *anum* III: 67-68 (Litke 1998, 124-125; the identification with Nanše seems justified, because **ᵈna-zi** is the spouse of Nindara (Selz 1995a, 215-217). In **an** = *anum* the other spelling for Nanše, *viz.* **ᵈABxḪA**, has not been preserved, if ever present). In the AṢ-list **ᵈABxḪA** (no. 26) most likely has been differentiated from **ᵈ⁽ⁿᵃ⁻ᶻⁱ⁾ENGUR** (no. 395), because of their remote distance in the list. In the list SF 1 only the name **ᵈABxḪA** is left (SF 1 *ix*: 11). In the OB list SLT 122, **ᵈna-zi** and **ᵈABxḪA** are placed after each other (*iii*: 13-14), as in TCL XV 10 (no. 293-294 ; de Genouillac 1930, pl. XXIX). In the AṢ **zà-mì** hymns (Biggs 1974, 49, ll. 110-116) and in the Temple Hymns (Sjöberg and Bergmann, 1969, 33, TH no. 22), Nanše's name is written as **ᵈABxḪA**. My tentative hypothesis is the following. At first, there was a distinction between the goddess who – in the marshes of South Mesopotamia – was associated with birds and fishes, represented by **ᵈ⁽ⁿᵃ⁻ᶻⁱ⁾ENGUR** (**engur** representing the underground waters), and the goddess of Nina/Nimin/Niĝin (Bauer 1998, 507-510), part of the district Lagaš, which goddess was written as **ᵈABxḪA**. In the course of time, the goddesses **ᵈABxḪA** and **ᵈ⁽ⁿᵃ⁻ᶻⁱ⁾ENGUR** were syncretized. (For Nanše in general: see Heimpel 1998-2001; Veldhuis 2004; Selz 1995a, 181-212).

444 SLT 122, *iii*:22; in SLT 124 Namma is perhaps present, but at the expected place there is a lacuna.

capacity in the third millennium [445]: there is no text (not yet at least) in which Namma has been described as the primaeval mother. This idea must have been developed somewhere shortly after the Akkad period or in the Ur III-period.

Namma is mentioned in the **zà-mì** hymns from Abu Ṣalābīḫ [446], but not in the Temple Hymns. Further, her name is found in only a few texts from the third millennium, of course besides its appearance in the name Ur-Namma [447]. Is it only coincidence that three out of the four results originate from Sumer proper, southern Mesopotamia? Most interesting with respect to my research is the inscription of Lugalkisalsi. Let us cite the whole text:

1	**ᵈnamma**	For Namma,
2	**dam-an-ra**	spouse of An,
3	**lugal-kisal-si**	has Lugalkisalsi,
4	**lugal-unug^{ki}-ga**	king of Uruk (and)
5	**lugal-urim^{ki}-ma**	king of Ur,
6	**é-ᵈnamma**	the temple of Namma
7	**mu-dù** $^{!}$	built.

This text is interesting for several reasons:
1. There was a temple built for Namma, presumably in Uruk, as early as in the ED III-period;
2. Namma is called: spouse of An. In this respect she has the same position as **ki** = Ninḫursaĝa = Ningal, the mother goddess. This last relationship – spouse of An – reminds us of another aspect of Namma, viz. that she is the mother of Enki [448]; this was

---

445 According to ePSD, **namma** is: '(cosmic) underground waters'.

446 Biggs 1974, 50, lines 140-141: **šu:EŠ-gi še gu ti** / **ᵈnamma zà-mì**: '**šu:EŠ** reed, barley and flax to live, Namma be praised' (a tentative translation of the present author; the uncertainty of the translation is due to the unknown order of the cuneiform signs). The **šu:EŠ** reed may be some special kind of reed. As Biggs already observed (1974, 55 ad lines 140-141): 'The occurrence of **šu EŠ gi** in [IAS] no. 21 [*iv*: 1] and its duplicate SF 23 iii 14 suggests that these signs belong together here as well.' Lines 140-141 may express the praise for Namma because she facilitates the growth of reed, grain and flax, which are necessary for the sustenance of man. There is no indication of Namma as primaeval origin.

447 A literature search (April 2012) yielded the word '**nammu**' in six texts; in five Old Akkadian texts **nammu** was part of the name Ur-Nammu. The word '**namma**' appeared in 647 texts, nearly all with the name Ur-Namma, but there are a few others: 1. **sanga-namma** (Biggs 1974, OIP 99, no. 48 *obv. iv'*: 6' [lexical text, ED IIIa; Abu Ṣalābīḫ; CDLI no. P225926]). 2. the inscription of Lugalkisalsi (Neumann 1981) from Uruk (?) ED IIIb; ll.1-2 **ᵈnamma dam-an-ra**; l. 6: **é-ᵈnamma**. 3. **ša ᵈnamma-x-x** in an Old Akkadian letter from Umma (unpublished; collection Neuman et. al.; CDLI no. P339021). 4. the personal name **lú-ᵈnamma** in an Ur III administrative text from Ĝirsu (P. Notizia, *Nisaba* 13 [2006], 030; CDLI no. P355945). This personal name is, besides Ur-Namma, the only one I have found.

448 The text of this inscription might imply that An is the father of Enki, although it is not stated. In IAS 114, *i*: 12'-13', **an** and **ki** have created Enki. The epithet **ᵈam-an-ki** for Enki also points to **an** and **ki** as his parents. In 'Enki and the world order' (Benito 1969, 89, l. 68) Enki is called **dumu-saĝ-an-na** 'first-born son of An'. See also ch. 2.2.3 of this study. While these data agree with respect to An as the father of Enki, they disagree about his mother.

mentioned for the first time in 'Enki and Ninmaḫ' [449], a text copy from the Old Babylonian period, and also later in **an** = *anum* I: 27.

From the above it may be concluded:
- During the third millennium, until the Ur III-time, Namma has a minor position in the pantheon, but also in the religious perception of the Mesopotamians;
- Originally Namma's name was associated with the underground waters in South Mesopotamia [450], and she represented especially the vegetation of the marshes; the fauna of the marshes in South Mesopotamia was represented by Nanše;
- Namma's position as 'primaeval waters' probably originated shortly after the Akkad period or in the Ur III-period [451]; then she becomes **ama-(ù)-tu-an-ki**, mother of **an-ki**. This unit **an-ki** is well-known from ED-times on [452].

3.3.3 *'Ancestors' of An*

It is not until the list **an** = *anum* that the 'ancestors' of An are mentioned as such. As we have seen, this ancestor group is heterogeneous, being a combination of Sumerian and Semitic names: it is impossible to distinguish in it a linear development from **an-ki** to An. The only clear point is that at the beginning of these 'ancestors' is **an-ki**. Another striking detail is that the number of these 'ancestors' is only 21, half of the number of the 'ancestors' of Enlil, while in the earlier list TCL XV 10 An was accompanied by only two names, which might have been epithets. As is obvious, the 'ancestor' list of An is a very late development, from the Old Babylonian period or even later. It seems to be an *ad hoc* or occasional solution of the post-OB scribes for the absence or the shortness of an analogous list for An in earlier times, the more so as Enlil has a long list of 'ancestors'. In earlier times there is no trace at all of An's ancestors. An did not need such a 'family tree': he was from the beginning known as the 'god of heaven' in the Sumerian religion.

In the Old Babylonian list TCL XV 10, An has two epithets or ancestors, a differentiation between both is not possible. Both gods, **an-šár-gal** and **^{d}en-uru-ul-la**, could not be traced back in literary texts of the third and second millennium up to and including the Old Babylonian period. The conclusion from the god lists with respect to the 'ancestors' of An must be as follows.

In the third millennium, if An is mentioned in a list he is present as the first god. However, there are some lists in which his name is absent. He appears without any further indication, epithet or 'ancestor'. An's ancestor list was developed during the second millennium, presumably to counterbalance the ancestor group of Enlil. In TCL XV 10 **an** is followed by two epithets or ancestors, and then by a form of **ki**, viz. **^{d}uraš** = ^{d}*belet-ilī*, An's wife. The ancestor group of An was named as such for the first time in **an** = *anum*; that of Enlil was so already in the OB Mari god list. Although An was the supreme god, he was in fact a *deus otiosus*: the actual power rested with Enlil. Or, to put it differently: the theologians/priests of Nippur were more influential than those of Uruk. One way to reflect this difference in power might have been the number of 'ancestors' of both gods: in **an** = *anum* An had only half of the number of Enlil's ancestors: 21 *vs* 42.

---

449 See the translation of this text (ch. 2.1.8; Appendix Text editions, no. 8).
450 Jacobsen 1946, 139-140, note 21.
451 This aspect will be discussed in ch. 4.
452 See the translations in this study.

The tentative conclusions with respect to the two units **an-ki** and **an-^{d}uraš** are:

1. **an-ki** is the primaeval unit *before* their separation, in which **an** and **ki** are partners [453];
2. **an-^{d}uraš** – An and his wife Uraš in TCL XV 10 – form a pair *after* the separation of **an-ki**.
3. The pair **an-^{d}uraš** has clearly to be differentiated from the 'ancestor' **an-uraš** (**uraš** without determinative) in **an** = *anum*; **an-uraš** only represents a genealogical aspect of **an** [454].

3.3.4 *Uraš*

Wiggermann recommended his method – to apply the upward reading principle twice – because of 'a number of advantages' over the method of Lambert, viz. upward reading of single items [455]. One of the advantages would be: 'Uraš can be taken as wife of An'. As pointed out in section 3.3.1, Wiggermann's method seems not to be free from arbitrariness. For during his 'upward reading', Uraš always arrives at the desired last place, which is impossible if one is reversing the series in a correct way.

^{d}uraš has been found in the Nippur-list [456], after **an** and *an-tum*. It has been shown that since the Old Babylonian period Uraš has been considered as the wife of An, examples of which are given in section 3.2.5. In list TCL XV 10, Uraš is followed by ^{d}*bēlet-ilī*, who appears to be the wife of An in **an** = *anum*. When these results and those from the texts are combined, it can be safely said – without any 'upward reading' – that Uraš in these lists represents also the wife of An. According to Jacobsen [457], **uraš** is another name for **ki**, 'probably an early loan from Akkadian', ' 'the tilth' '. In the development of the cosmos, the appearance of **uraš** after **ki** – **ki** being the primaeval form of 'earth' in the unit **an-ki** –, is understandable. The first development in this respect after **an-ki** is **an-uraš**. Due to the desire to form pairs, the authors of the list **an** = *anum* created a female counterpart: **^{d}nin-uraš**.

In this respect Lambert's idea about the 'ancestor' list of An in **an** = *anum* [458] has to be cited: 'The most important innovation [of **an** = *anum*; *JL*] is the placing of Urash and Ninurash at the head of the list: 'Earth' and 'Lady Earth'. This makes Anu the descendant of Earth, like Enlil, and it may well be the direct result of the ancestry of Enlil, since *An* = *Anum* contains both.' It may have become obvious from the discussion in the previous sections, that I do not share Lambert's view that 'An is a descendant of Earth'.

---

453 Texts: IAS 114, *i*: 12'-13'; Ukg 15, *ii*; Barton cylinder *i*: 12-14; NBC 11108 *obv.* 5-8; Debate Tree-Reed, 6.

454 See also ch. 3.2.8.1, The 'ancestors' of An.

455 Wiggermann 1992, 291; see also ch. 3.3.1.

456 SLT 122-124; see ch. 3.2.5.

457 Jacobsen 1976, 95, 249 note 86. Jacobsen considers **uraš** as a *purās*-form of *erēšu* [CAD E, 285b: *erēšu* B (e.g.) 'to cultivate or plant (a field)']; if so, then **uraš** may have a meaning like 'cultivated land'. In the bilingual text 'A Letter from X to the god Nanna' [ETCSL 3.3.22; CDLI no. P349116, exc. no. W 17259,w] we read in line 3: (...) **men-an-IB (=uraš)-a** / (...) *a-ge-e ša-me-e ù er-ṣe-tim* '(...) the crown of heaven and earth'. **uraš** was equated with *erṣetu* 'earth'. Black and Green mentioned (1998, 182b, *ad* Uraš [goddess]) that 'In ancient commentaries, the name Uraš is explained as meaning "earth" ...).' Theoretically, **uraš** may be derived from *(w)aršu(m)*: AHw 1471 'schmutzig'; CAD A II 309-310 'dirty, unclean'. But in the light of the comments made above, the derivation of **uraš** from *erēšu* seems the more obvious one.

458 Lambert 1975a, 54.

3.3.5 *Enlil and his 'ancestors'*

In **an** = *anum* there is a group of 21 pairs of gods clearly classified as 'ancestors' of Enlil. Some members of this group appear also in older lists, where their number and order are always different. Starting with the oldest lists, those from Fara and Abu Ṣalābīḫ, the following development can be shown.

The Fara list SF 1 has four successive names, and only **nin**-combinations, which may be attributed to the 'ancestor' group of Enlil. This group has **ᵈnin-ki** and **ᵈnin-KID** in second and third place, respectively. As is shown in ch. 3.2.1, there are some other gods who later possibly belong to Enlil's 'ancestor' group and who are scattered over the list. A tentative conclusion may be that, because of the strict hierarchical start of the list – An, Enlil, Inanna, Enki, Nanna, Utu – these later ancestor names are not yet thought of in combination with Enlil. It is also remarkable that **ᵈnin-KID** = Ninlil is mentioned within the small group which contained three later ancestors. But other Fara lists are well aware of the particular nature of a group gods and goddesses [459]: they list a group of seven pairs **ᵈen-** and **ᵈnin**-gods, starting with Enki-Ninki and Enlil - Ninlil in second position. This is fairly well in conformity with the text in IAS 114 [460], in which Enki and Ninki gave birth to the seven and to Enlil. The insertion of Ninlil, who is not known as the daughter of the primaeval pair Enki-Ninki, in the 'ancestor' group in the lists SF 23-24 may be seen as being because of the symmetry of this **ᵈen** - **ᵈnin**-group.

In the god lists from Abu Ṣalābīḫ exactly the same 'ancestor' group with the same order is present, but is now expanded by one pair, **ᵈen-an** - **ᵈnin-an**.

The conclusion may be that in the ED III-period a group of six or seven pairs of gods was already known as being associated with Enlil. I hesitate to say 'as an ancestor group', because 1. the position of Enlil is immediately after Enki and Ninki, while the other members of the group follow after Enlil; 2. text IAS 114 may lead to the conclusion that Enlil is just a brother of 'the seven', who were born first according this text [461]. Why then, in some ED III god lists, is Enlil mentioned immediately after Enki and Ninki, in that way suggesting that he was the first-born of this pair? A possible answer may be, that it is easier to give a 'historical' sketch of occurrences in a mythological text than in a one-dimensional list. IAS 114 is a very concise text, but the conclusion is not too far-fetched if we see in the lines 1'-16' a combination of cosmogony and theogony. There is a development from **an-ki** as a unity via the intermediary phase of Enki and Ninki and the seven to Enlil; these seven are not even mentioned by name! All attention is on Enlil. Therefore, to make clear that Enlil descends from Enki and Ninki, he is mentioned in the *lists* immediately after his parents. If his name had been placed at the end of the whole Enki-Ninki group, then one might conclude that the last pair of that group were his parents.

In the Old Babylonian time two exceptional lists appeared, at least with respect to the 'ancestor-group' of Enlil. In the Nippur-list (SLT 122-124) this group has been omitted. The Mari god list mentions this group, but the order is different: Enlil and Ninlil at the head, followed by Enki and Ninki and seven pairs of 'ancestors', while Enmešarra and Ninmešarra are placed after the conclusive lines **ᵈen-ama-a-a-ᵈen-líl-lá ᵈnin-ama-a-a-ᵈen-líl-lá**; thus they are excluded from this group, but included in 'Enlil's ancestors' in TCL

---

459 Fara lists SF 23-24.

460 See the edition of this text in the Appendix Text editions, no. 1.

461 From these data it also may be clear that the purpose of the list SF 1 must be different from the other texts, e.g. SF 23-24, the Abu Ṣalābīḫ god lists and IAS 114, which are nearly contemporary with SF 1.

XV 10 and **an** = *anum*. Another conclusion may be made from the Mari-list and **an** =*anum*: it is clearly stated that these ancestors are Enlil's ancestors, in other words: they are not involved in the genealogy of Ninlil.

From the OB period onwards a growing list of 'ancestors' is awarded to Enlil, in TCL XV 10 and, even more, in **an** = *anum*. These groups, although different in their number of pairs, show more coherence in their order. Enmešarra and Ninmešarra are in both lists the last pair of the group. Why this growing number of pairs? TCL XV 10 – provenance unclear – was written in the Old Babylonian period. Although a more specific date cannot be given, the list was composed nearly at the same time as when the Sumerian King List (SKL [462]) was copied. The SKL, a form of genealogical charter, was an important component in the ideology for the royal legitimation of the rulers of the Isin-dynasty [463]. Another example to demonstrate the importance of lineage at that time is the genealogy of Ḫammurapi with his long list of ancestors [464]. It may be that, by analogy with the lineages of the worldly rulers, the authors of TCL XV 10 felt the need to honour Enlil's position by providing him with a long genealogy. As I will try to demonstrate, Enlil is not a Sumerian god of olden times, but was probably 'created' at the end of the 4th millennium BCE in the northern part of Sumer, in Nippur [465]. The theologians of the Old Babylonian time must still have realized that, e.g. being reminded of it by the god lists, which seemed to have existed uninterruptedly from the ED times onwards. In the time after the Ur III-period, when people of Semitic origin had taken over the rule of Mesopotamia, the priests tried to consolidate the Sumerian origin for Enlil. Only in **an** = *anum* did the 'ancestor' list of Enlil reach a 'perfect' number, 21 pairs, being a multiple of the holy number 7. While in due course number and order of the 'ancestors' changed, the real parents of Enlil, viz. Enki and Ninki, were always the first of the 'ancestor' lists. The conclusion is inevitable: Enlil always has been considered to originate from a form of **ki**, earth.

With respect to the 'ancestor'-group of An: it is now clear that this is only a secondary development, originating from the need to provide An with a certain extent of similarity with Enlil with respect to genealogy. This development stems from the Old Babylonian period, in the beginning very hesitantly with two ancestors whose names even might be considered as epithets (TCL XV 10), ultimately resulting in the heterogeneous group of **an** = *anum*. This heterogeneity may be explained by the fact that An is (accepted as) the superior god, although a *deus otiosus*, in the various religious systems (Sumerian, Semitic) of Mesopotamia; the influence of these various religions can be traced thanks to this heterogeneity.

There is another reason not to take the 'ancestors' as real ancestors. According to Lambert [466], An and Enlil had 'somewhat artificial ancestries supplied to explain their origins'. If each pair gave rise to the next by bisexual reproduction, then 'one must accept the idea of brother and sister marriages, which were as taboo in ancient Mesopotamia as in most other parts and periods of the world.' Lambert continues: '(...) generally it (*incest; JL*) was avoided, and An = *Anum* and some smaller, related lists merge all the males in Anu's

---

462 ETCSL 2.1.1, a list starting with the kingship of Eridu; this list also mentions that, after the flood had swept over, the kingship was in Kiš.

463 Michalowski 1983; Wilcke 1989. According to Hallo (1983, 10): 'In its own terms, then, the Sumerian King List is not so much a king list as a city list.'

464 Finkelstein 1966.

465 See Excursus 1 'Enlil and Ninlil' at the end of this chapter.

466 Lambert 1975b, 197.

ancestry with him, and all the females with Antum, his spouse. This is evolution, not creation. Anu became himself by developing through a series of stages represented by his ancestor's names.' In general I agree with Lambert that it does not concern real ancestry, but instead of evolution the term 'several aspects' or 'deified phenomena' of the gods An and Enlil may be more appropriate. Evolution supposes a strict succession of stages, which are absent in the respective 'ancestor' lists of both An and Enlil.

3.3.5.1 *Enmešarra*

Some authors [467] refer to Enmešarra, sometimes even as 'the most famous among Enlil's ancestors' [468]. However, Enmešarra is not mentioned in any god list or other text from the ED period. His first presence can be shown in a few Ur III administrative texts, containing offering lists [469]. Sallaberger classified Enmešarra, whose cult was celebrated in Nippur and Tummal, among the 'minor gods' [470]. Enmešarra, absent in the OB god list from Nippur (SLT 122-124), does appear in the OB god list from Mari, together with Ninmešarra; both are listed after the group of ancestors of Enlil, in other words: excluded from them. Because of the order of gods in this Mari list [Enlil-ancestors-Enmešarra-Enki-Enlilbanda-Nudimmud], we may – at first glance – conclude that in this list Enmešarra and Ninmešarra have to be understood as epithets of Enlil and Ninlil, respectively, analogous to the epithets of Enki, viz. Enlilbanda and Nudimmud. In the lists in which Enmešarra has been included in Enlil's ancestors, he and Ninmešarra are always the last pair of the ancestor list [471].

Lambert mentions that the position of Enmešarra in OB Sumerian litanies [472] outside Enlil's ancestor group 'no doubt reflects a myth known also from allusions in Babylonian incantations (...[473]), that Enmešarra was the first god, father of the rest, and he was succeeded by Enlil.'[474] The myth to which Lambert referred might be the OB tale known as 'Enlil and Namzitarra' [475]; this tale contains an allusion that Enlil regained the Enlil-ship, that was taken away from him by his uncle Enmešarra, who was taken captive thereafter. All other evidence about Enmešarra –'the protagonist of a myth extremely difficult to

---

467 Lambert 1957-1971, 470; Civil 1974-1977, 66-67; Saggs 1978, 99-102. Wiggermann (1992, § 7, 287-289) devoted a section to Enmešarra; the references to this god given in that section are for the most part from times later than the Old Babylonian period.

468 Civil 1974-1977, 66.

469 From Puzriš-Dagan: T. Fish (1932), CST 146, obv. 2 [CDLI no. P107658]; T. Gomi *et al.* (1990) Hirose 134, obv. 2 [CDLI no. P109605]; W.W. Hallo, *HUCA* 29 (1958) 77 6, obv. 10 [CDLI no. P110438]; M. Cig et al. (1954), PDT 1, 545, rev. 1 [CDLI no. P125961]; and from Umma: H. de Genouillac (1922), TCL 5, 6053, rev. *i*: 19 [CDLI no. P13176].

470 Sallaberger 1993, 103, 140.

471 TCL XV 10 and **an** = *anum*.

472 In a short 'Enki-Ninki' list in 'The Death of Gilgameš', Enmešarra is included in this ancestor group (Kramer 1944, 10: version B, ll. 14-18; Cavigneaux and Al-Rawi 2000, 23, ll. 14-18; 'The Death of Gilgameš', Another version from Nibru, ll. 14-17, ETCSL 1.8.1.3).

473 The incantation texts to which Lambert refers are from the Neo-Assyrian period.

474 Lambert 1957-1971, 470.

475 Alster 2005b, 327-338; ETCSL 5.7.1. The earliest sources for 'Enlil and Namzitarra' are of the early OB period. It is not unthinkable that the story 'Enlil and Namzitarra' is related with the Sumerian story 'Ninurta and the turtle' [ETCSL 1.6.3], and with the Akkadian story 'Ninurta and the Anzu'. With respect to this last story, Civil 1974-1977, 67 wrote: 'KAR 307 (...) suggests an affirmative answer.' [KAR 307 is a Neo-Assyrian text]. For 'Ninurta and the Anzu': see Foster 2005, 555-561.

reconstruct' [476] – is from a period later than the OB period 'in which the theogonies were reformulated to give a central role to Marduk' [477]. Enmešarra also appears in the *utukkū lemnūtu* as a god who exorcizes demons; in the 'forerunners' to **udug-ḫul** his name is absent [478].

My hypothesis about Enmešarra is as follows. Enmešarra was 'created' by theologians in the Ur III-period [479]. In the beginning he was only a local minor god with a restricted cult area, centred around Nippur. At first glance his name Enmešarra – 'Lord of all **me**' – might seem to be an epithet of Enki (of Eridu), in whose possession the **me** are generally thought to be [480]. But Enki is not a minor god, even not in Nippur, thus the possibility of Enmešarra being an epithet of Enki is very unlikely. If Enmešarra in the Ur III-period was known as the god who had stolen the Enlil-ship from Enlil, then it is hardly to explain why he was worshipped in Nippur, the town of Enlil from, at least, ED times onwards. Perhaps the origin of Enmešarra might be found in the 'theological-political' situation in the Ur III period. Already at that time there might have become manifest a rivalry between the Nippur theologians/priests and those from Eridu – a rivalry that probably had been latently present since the introduction of Enlil as supreme god of Sumer. Or was it perhaps a reaction to the preceding Akkad period, in which Inanna/Ištar was the most prominent deity? The Nippur clergy had to make a 'statement' that from now on Nippur was again the seat of the supreme divine power: Enmešarra contra the Akkadian Ninmešarra = Inanna. It is not unlikely that both motives may have reinforced each other. But after all, Enmešarra's position in the pantheon never reflected his name. In later times, when the 'Enki-clergy' had become successful, the theologians had to find a 'solution' with respect to Enmešarra's position; finally they classified Enmešarra in the growing list of ancestors as the immediate ancestor of Enlil. These hypotheses still do not answer the question of how to interpret Enmešarra in the Mari god list: as a god independent from Enlil or as an epithet of Enlil.

The apparent discrepancy between Enmešarra = 'ancestor of Enlil' and Enmešarra = 'uncle of Enlil' may be solved if the Sumerian view of 'ancestor' has a broader meaning than our modern view. Civil formulated a solution by assuming that 'the Sumerian kinship system belongs to the so-called «Hawaiian» type' [481].

---

476 Civil 1974-1977, 66.

477 Civil 1974-1977, 66.

478 Geller 1985.

479 A similar idea was already expressed by Saggs (1978, 102) who supposed that '(...) Enmešarra had his origin in theological speculation rather than that he was an otiose deity of popular religion (...).'

480 Farber 1987-1990, who also mentions the most important myths in which Enki acts as 'master of the **me**'.

481 Civil 1974-1977, 66: 'The terms **šeš** and **nin**$_9$ (*brother and sister; JL*) are not limited to siblings but extend to the relatives of *ego*'s generation (hence the absence of a special term for «cousin» in Sumerian); similarly, there is only one term for the males, and one for the females of the first ascending generation, which are the same as the ones applied to the biological parents (note the use of **ugu** and **ù-tu** with **a-a** and **ama** to specify the «real» parents). The term **šeš-ad-da** is typically restricted to inheritance documents and similar cases where biological paternity needs to be explicitly excluded.'

3.3.6 *The Enki-Ninki gods* [482]

3.3.6.1 *Enki and Ninki*

Even in the oldest god lists – those of Fara and Abu Ṣalābīḫ – gods can be discerned who in the Middle Babylonian god list **an** = *anum* belonged to the 'ancestors' of Enlil. In the ED-time their number is restricted: three **nin**-gods only in SF 1; six **en/nin** pairs in SF 23-24, and seven **en/nin** pairs in the AṢ-list. As we have seen, this number has increased to 16 **en/nin** pairs in TCL XV 10 and 21 such pairs in **an** = *anum*. While number and order have changed, the first pair of this group of gods are always Enki-Ninki, with one exception: the Fara list SF 1 starts with **ᵈnin-utua**. In the ED-time, Enki and Ninki are followed by Enlil and Ninlil, in this way suggesting their direct descent from Enki and Ninki, a fact that – at least for Enlil – is confirmed in the text IAS 114, *i*: 9'-11'. But from whom or from what Enki and Ninki descended? Of course: the names of both gods suggest that they descend unambiguously and unilaterally from **ki**, earth; this must have happened at a very early moment in the cosmogony [483]. But as far as I know, there is not one text or list that relates the real lineage or the origin of Enki and Ninki.

An Old Babylonian incantation [484] contains an oath sworn on the lives of Enki and Ninki.

The oldest attestations – outside the god lists and IAS 114 – for **ᵈen-ki** and **ᵈnin-ki** are found in some incantations from Ebla [485], in which 'the roots of tamarisks are Enki and Ninki' – this confirms once more the chthonic character of Enki and Ninki. There is another text, on Ur-Nanše's diorite plaque, where is written: 'may Enki and Ninki care for the reed of the Engur' [486]. Here Enki and Ninki are written: **en-ki nun-ki** [487], without **diĝir**-determinative, and they are clearly distinguished from **ᵈen-ki** in Urn 49 *ii*:1, the Enki of Eridu. These examples clearly point to the chthonic character of Enki and Ninki. A more precise indication for this character, i.e. a connection of Enki and Ninki with the netherworld, is given by the **zà-mì**-hymn for Nergal, god of the netherworld, from Abu Ṣalābīḫ [488]:

---

482 Van Dijk called these gods 'dema-gods' (van Dijk 1957-1971, 535a, § 2a; *id.* 1964, 12 note 21; *id* 1971, ch. 4, 449-452). The definition given by Jensen (ref. in Cavigneaux-Krebernik 1998-2001, 446b) makes the classification of the Enki-Ninki gods as 'dema'-gods questionable, because the Enki-Ninki gods do not seem to meet the qualities of dema-gods.

483 The complete story of the cosmogony will be discussed in ch. 4.

484 Conti 1997, 266 (YOS XI, 83; CDLI no. P307498).

485 Krebernik 1984, 96-104 (incantations nos. 19 and 20); 228-229;
Incantation 19, I:4- II:2 **ᵍⁱˢšinig ĝiš-gi ĝiš-an / úr-pi ki-šè / ᵈen-ki ᵈnin-ki**: 'tamarisk, strongly rooted tree, tree of An / its roots downwards / (are) Enki (and) Ninki'.
Incantation 20, I:3 **ᵍⁱˢšinig úr-be ᵈen-ki ᵈnin-ki**: 'tamarisk, its roots (are) Enki (and) Ninki'.

486 Steible 1982, 111, Urn. 49 *ii*:8-9; the **gi** in l. *ii*:8 refers to l. *i*:2: **gi ᵍⁱˢgi-engur** 'reed of the Engur's canebrake'.

487 Alster (1970) has given a survey of the attestations in which **⁽ᵈ⁾nun-ki** is written instead of Ninki. The most interesting examples for this study are Ukg 15, *ii*:3 and Urn 49, *ii*:8; the greater number of the references given by Alster originate from the first millennium BCE.

488 Biggs 1974, 48, ll. 65-69. Transcription and translation are by the present author. Although 'the natural inclination is to equate **KIŠ.UNU** with Kutha', Katz (2003, 406-407) has clearly demonstrated that it is unlikely that in the ED period **KIŠ-UNUG**, indeed referring to the cult centre of Nergal, is identical with Kutha, and therefore she suggests that 'his cult was transferred from **KIŠ.UNU** to Kutha during the Old Akkadian period.'

65	**KIŠ-UNUG**	Residence of Nergal,
66	**dag-gal-an-ki**	great residence of Heaven and Earth,
67	**ĝissu-sig**	low-lying aegis
68	^{d}**en-ki** ^{d}**nin-ki**	for Enki and Ninki;
69	^{d}**KIŠ-UNUG zà-mì**	Nergal be praised.

3.3.6.2 *Other goddesses with the name Ninki*

Except for the just mentioned primaeval Ninki there are some other attestations of goddesses with the name ^{d}**nin-ki**:

1. There is, besides the 'primaeval' Ninki, yet another Ninki in the AṢ god list [489], but in such a context that I cannot give any indication about the identity of this goddess.
2. In the text of the 'Stele of the Vultures' [490] is said that, if Umma should break the oath, Ninki would bring up a snake from the earth (**ki-ta**) to bite into Umma's foot. The snake and **ki** may refer to the netherworld. Therefore it is not unlikely that this Ninki represented the goddess of the netherworld at that time [491].
3. The name Ninki is found in some personal names in administrative texts of the ED IIIa and Old Akkadian period, and there are many personal names in Ur III texts that are composed with Ninki. It seems not very likely that the name of the queen of the netherworld was a favourite part of a personal name: names composed with Ereškigal are unknown, to the best of my knowledge. With respect to the supposed special status of the gods belonging to the ancestor group of Enlil: see the next section. I could not find any personal name composed with a member of this ancestor group. Therefore the Ninki in the personal names may be another Ninki than the primaeval and the netherworld Ninki.

---

489 IAS 83, *rev. viii*: 4'; Mander 1986, 31 no. 415.

490 Sollberger 1956, 9-16 (Ean. 1), ll. *iii*: 2 - *v*: 41. This text was written in Early Dynastic Lagaš.

491 Selz (1995, 255 *sub* $^{(d)}$**nin-ki(-k)**. Selz made a distinction between the 'primaeval' Ninki (a name of which he said that it is 'nicht genitivisch zusammen-gesetzt') and the supposed netherworld goddess Ninki(-k). Katz (2003, 386): '(...) this clause antithetically parallels the previous one, in which Utu would throw a net from heaven (**an-ta**) on Umma. It implies that before Ereškigal ("lady/queen of the big earth") was first mentioned in the written sources, a female deity governed the netherworld: Ninki "lady/queen of the earth".' For the transliteration **NIN = ereš/eriš** in the name **NIN-ki-gal**, see Marchesi 2004, 186-189. There are a few early attesta-tions of the name Ereškigal:
1. An offering list of the ED IIIb period from Ĝirsu (Allotte de la Fuÿe (1912), DP 51, [CDLI no. P220701]);
2. A royal inscription of Lu'utu, ensi of Umma, from the Old Akkadian period (Frayne (1993) RIME 2.11.06.02, ex. 1 [CDLI no. P216745]). This is the only temple built for Ereškigal, known from texts of the third and second millennium BCE (George 1993). In this text, Ereškigal, for whom Lu'utu has built a temple, is named **nin-ki-utu-šu$_4$-ra**: 'lady of the place of the sunset'. This 'place of sunset' is a metaphor or euphemism for the netherworld or grave. For a recent interpretation of this text, see Artemov 2012, 21-22.
3. In the Temple Hymns – ascribed to Enḫeduanna, thus roughly about the same time as the inscription of Lu'utu –, the netherworld is said to belong to Ereškigal (Sjöberg & Bergmann 1969, 27, TH 14, l. 179).

From these few attestations may be concluded that there were several goddesses with the name Ninki:

1. Ninki of the primaeval pair Enki-Ninki;
2. A goddess of the netherworld, who possibly held this position before Ereškigal;
3. Most likely the Ninki of the personal names has to be differentiated from the first two goddesses.

3.3.6.3 *The Enki-Ninki gods – as a group or as an individual god or goddess* [492]

Jacobsen wrote about the Enki-Ninki group in TCL XV 10 [493]:

> 'Among these are the powers manifest in Earth viewed in their male and female aspects as **^dEn-ki**, "The earth lord", and **^dNin-ki**, "The earth lady", and the powers manifest in Heaven represented by **^dEn-an-na** and **^dNin-an-na**, "The sky lord" and "The sky lady". At the beginning of the genealogy, before everything else, stands the active principle of the world itself, its *modus operandi*, personified as **^dEn-me-šár-ra**, "Lord (**en**) *modus operandi* (**me**) of the universe (**šár-a(k)**)" and **^dNin-me-šár-ra**, "Lady *modus operandi* of the universe". And from them issued life: "Lord days of life" and "Lady days of life". These highly interesting speculations concerning world origins cannot well be ignored in a statement of Sumerian cosmogonic concepts.'

For one or another reason, Jacobsen obviously read the list, starting with Enmešarra, upwards, but he did not justify this decision.

More detailed analysis of the Enki-Ninki group, from the Fara lists up to and including **an** = *anum*, shows the following. The Fara list SF 1 (obv. *vi*: 25, 26, 28) represents three aspects: **utua** 'breeding ram; stud' , **ki** 'earth', and **bulug$_3$** 'growth'. In SF 23-24 there are: earth, butterfly, growth, breeding ram, fat-tailed sheep, and power. To this last group the Abu Ṣalābīẖ list adds: heaven. It is clear that the Fara lists are limited to the domain of the earth, where the Abu Ṣalābīẖ list extends it to heaven. The enlarged list TCL XV 10 shows a combination of various aspects, items that all together try to enclose valuable aspects for the life on earth [494]. The ultimate list, **an** = *anum*, is even more extensive. The aspects that can be distinguished in the list – but which are not grouped together – are: earth, heaven/sky, animals, abstract concepts (like: growth, holiness, lifetime), extispicy, and medicine. Below the significant elements of the names in the Enki-Ninki groups of the ED period and of **an** = *anum* – omitting the **^den-/^dnin**-parts – will be summarized, together with the translations.

ED period: SF 23 [495]

v:	17-18	**ki**	earth
	19-20	**en-E$_2$ - nin-kid**	Enlil - Ninlil
vi:	1-2	**giriš$_x$**	butterfly

[492] The complete list of the members of the Enki-Ninki group, together with the translation of the names, is added in the Appendix 'God lists'.

[493] Jacobsen 1946, 138-139. My translation of Enmešarra/Ninmešarra is: Lord/Lady of all **me**.

[494] Lambert (2008, 29) translated the names of Enlil's ancestors of the list TCL XV 10.

[495] Deimel 1923; Mander 1986, 109.

3-4	**bulug$_3$**	growth
5-6	**utua**	breeding ram; stud
7-8	**gukkal**	fat-tailed sheep
9-10	**á**	power

The AṢ-list adds [496]:

no.	287-288	**an**	heaven, sky

MB period: **an** = *anum* [497]

I	96-97	**ki**	earth
	98-99	**ul**	bud, fruit
	100-101	**mul**	star
	102-103	**udu**	sheep
	104-105	**gin**	firmness
	106-107	**da**	power [498]
	108-109	**bulug$_3$**	growth
	110-111	**ḫal**	secret, extispicy
	112-113	**ug**	lion
	114-115	**ga-raš** [499]	decision
	116-117	**šár**	all
	118-119	**nun**	prince/princess
	120-121	**kur**	mountain
	122-123	**amaš**	sheep-pen
	124-125	**kin-gal**	high official
	126-127	**kù-ĝál**	holiness
	128-129	**an-na**	heaven, sky
	130-131	**u$_4$-ti-la**	lifetime
	132-133	**da-šurim-ma**	power of dung [500]
	134-135	**du$_6$-kù-ga**	holy hill
	136-137	**me-šár-ra**	all **me**

---

496 Mander 1986, 29.

497 Litke 1998, 30-34. Some of Litke's transliterations have been converted into transcriptions: 102-103 **LU** into **udu**, and 104-105 **DU** into **gin**. Instead of **garaš$_3$** (**GA** + **KAS**) [nos 114-115] I read: **ga-raš**.

498 CAD L 152: **da** *le'û*, lexical part. Most likely **da** replaces the **á** in the ED lists (see also **da-šurim-ma** in this list).

499 **ga-raš** might have substituted the **giriš** of earlier god lists (Volk 1998-2001). For the meaning 'decision': see CAD P 530, *purussû* (lexical part). The transcription **ga-eš$_8$** – with the same meaning 'decision' – is also possible (ePSD *ad* **kaš**).

500 Medicinal use (?): see CAD K 28-29, *kabû* A, *sub* lexical part and *sub* b). Lambert (2008, 29) has a different translation: 'side of dung' (see the comment on **da** 'power' in this list in the footnote 499).

The butterfly (ED lists) may symbolize the beginning of life, because of its several metamorphoses (from larva to adult form); ram and sheep may represent the development of **ki** and of its habitability, as animals are a *conditio sine qua non*, both for gods and, later, for men [501]. The abstract and more general concepts 'growth' and 'power' can also easily be connected with the development of earth from its beginning. Perhaps the ancestor group of the AṢ list may have another symbolic meaning, because it starts with **ki** and ends with **an**, in this way separating as much as possible the components of the primaeval unit **an-ki** within this group. For this separation Enlil was responsible, as related in some AṢ texts [502].

The most extensive list of ancestors, in **an** = *anum*, shows, besides a few elements that are identical or analogous with those of the ED lists [503], new elements. The term **ul** is a concept pre-eminently fitting the beginning of development. The idea of power covers more items (**gin**, **ga-raš**, **kin-gal** [?], and perhaps **ug**); moreover the list contains more items that may be conceived as epithets (**šár**, **nun**, **kur**, **kin-gal** [?], **u$_4$-ti-la**, **du$_6$-kù-ga**, **me-šár-ra**). But most conspicuous and surprising are **ḫal** and **da-šurim-ma**; why are they included in the ancestor list of Enlil? For in the daily life of the Mesopotamians, extispicy and medicine belonged predominantly to the sphere of Enki/Ea and Asarluḫi [504].

The enumeration of the items which appear as god names in these lists do not show any logical order. Therefore the 'ancestor' lists are not meant as consecutive phases in the development from the primaeval unit **an-ki** into the earth that is being prepared for the arrival and the reception of men [505]. Only some aspects that have contributed to this development have been mentioned. On the other hand, equally conspicuous is the absence of e.g. grain, barley or date palm in Enlil's ancestor list, as these products are nearly indispensable for sustentation and also used for offerings [506]. A tentative hypothesis therefore is, that this can be explained by the origin of Enlil. Except for being – most likely – 'created' in North Sumer [507], Enlil may have been related predominantly to **kur** 'mountain': his well-known epithet is **kur-gal**; his famous Nippur temple is **é-kur**. This **kur** 'mountain' may be an environment appropriate for (some) animals [508], but not for the cultivation of cereals or date palms. Therefore those are absent in Enlil's 'ancestor' list.

Outside the god lists a group of gods, known as or belonging to the 'ancestors' of Enlil, is sometimes listed. One example is found in 'The Death of Gilgameš': several members of the group receive offerings [509]. On the other hand, the whole group – without any specification or listing of individual members –, may be mentioned as the plural **^{d}en-ki-(e)-ne ^{d}nin/nun-ki-(e)ne** [510]. Some texts in which the group as a whole or these plural forms are present are:

---

501 Another possibility may be that the mention of these animals points to the community, i.e. a pastoral people, whose supreme god was Enlil.

502 IAS: 113 *ii*: 5-10; 136 *iii*: 1'-3'; 203 *ii*: 3'5'. See ch. 2.1.1b.

503 ED lists: **bulug$_3$**, **utua**, **gukkal**; MB list: **bulug$_3$**, **udu**, **amaš**.

504 For my hypothesis about Enmešarra: see ch. 3.3.5.1.

505 In the words of Lambert (2008, 28): 'The result is clear: the names of these pairs, save for the first, are irrelevant for cosmology.'

506 See e.g. Sallaberger 1993, *passim*.

507 See Excursus 1 'Enlil and Ninlil' in this chapter.

508 E.g.: a butterfly is to be found in the hills rather than in the dry plain of Mesopotamia.

509 Kramer 1944, 10: version B, ll. 14-18 ('The Death of Gilgameš', Another version from Nibru, ll. 14-17, ETCSL 1.8.1.3). 'Several members', that means: with respect to the list in **an** = *anum*.

510 Alster 1970. According to Alster, these examples are often found in incantations, but most of the examples given by Alster, however, are from the Neo-Assyrian period.

1. an incantation from Nippur, Ur III period [511];
2. 'Ninurta's exploits' [512];
3. 'the curse of Agade' [513];
4. 'an *adab* to Nanna' [514];
5. a magic text from Tell Haddad, OB period [515].

The individual members of the 'ancestor' group rarely occur outside the group. Some examples are given. An Early Dynastic administrative text from Fara mentions that Ninkur receives offerings [516], or that an **išib**-priest is in Ninkur's service [517]. Further, during the third millennium the name Ninkur can be found as a theophoric element in personal names. Most of the other names of the members of the 'ancestor' group of Enlil could not be found, neither as gods receiving offerings nor as part of personal names. Exceptions are: **nin-ul**, **nin-nun** and **nin-an-na** in personal names; a few texts with offerings for Enmešarra in Ur III texts [518]; there is one text in which Enmul and Ninmul appear, together with Enki and Ninki [519].

The tentative conclusion is that the individual gods of the 'ancestor' group of the chthonic god Enlil are not 'real' gods like those who have been worshipped in the various cult centres in Mesopotamia. The 'ancestor' gods seem for the most part aspects or deified phenomena, signifying some essential elements necessary for the development of the earth and for making life possible on it. Perhaps that supposed 'productive' element of these deified phenomena is the reason that the 'ancestor' gods appear in ^{d}**en-**/^{d}**nin** pairs, a rather artificial affair, because such pairs are very unusual in the 'normal' Sumerian pantheon.

3.3.7 *Enki of Eridu*

In general, the god lists reveal no peculiarities about Enki, the god of Eridu, but there are a few exceptions.

---

511 Lutz, PBS I/2, 107, rev. 7: ^{d}**en-ki-ne** ^{d}**nin-ki-ne**.

512 Van Dijk 1983, 138, l. 661: ^{d}**en-ki** ^{d}**nin-ki-e-[ne] [saĝ nu-un]-da-ĝá-ĝá-ĝá** 'les dieux Enki - Ninki n'osent pas all[er contre] toi'. 'Ninurta's exploits', ETCSL 1.6.2, l. 658: 'the Enki and Ninki deities dare not resist (?) you'.

513 Cooper 1983, 60 ('The curse of Agade' [ETCSL 2.1.5]), ll. 207-208. The lines read: 207. **ír-bi ír-ama-a-a-**^{d}**en-líl-lá-ke**$_4$ 208. **du**$_6$**-kù su-zi gùr-ru du**$_{10}$**-kù-**^{d}**en-líl-lá-ke**$_4$ **i-im-ĝá-ĝá-ne** 'their lament is (like) the lament for the ancestors of Enlil which they perform at the awe-inspiring holy mound, at the knee of Enlil' ['lament *for...*': genitivus objectivus].

514 Sjöberg 1960, 35, Rs. ll. 7-9 ('An *adab* to Nanna', ETCSL 4.13.08, l. B7-9): 7. **èš-urim**$_2$^{ki} **nam-galam-ma-bi-ta** 8. **šag**$_4$**-bi-a** ^{d}**en-ki** ^{d}**nin-ki-ne nam si-si sá-bi-ta** 9. **me šu ši**$^{!}$**-im-du**$_7$**-du**$_7$**-ne diĝir-diĝir-a-nu-na àm-ma-su**$_8$**-su**$_8$**-ge<-eš>** 'Oh shrine Urim! In its interior the Enki and Ninki deities have, through their intricate craftmanship, perfected the **me** with their righteousness, (and) the Anunna gods stand there in service.'

515 Cavigneaux & Al-Rawi 1993, 178 ('Grande texte contre Namtar'), l. MA24: ^{d}**en-ki-ke**$_4$**-ne nun** ^{d}**en-ki-ne**; l. MB24 (duplicate): **en-ki-ni nun-ki-ni**.

516 Martin *et al.* 2001, 87: no. FTP 102 [CDLI no. P010018].

517 Jestin 1937, no. 1, rev. *vi*: 4' [CDLI no. P010700].

518 Enmešarra acquired a more important reputation after the Old Babylonian period; the texts concerning this are outside the scope of this study.

519 Kramer 1985, 120, l. 7 ('A *šir-namšub* to Utu', ETCSL 4.32.e, l. 7).

In list SF 1 and in the Abu Ṣalābīḫ list, Enki is present in a prominent position, i.e. after An, Enlil and Inanna, and after An, Enlil and Ninlil, respectively. This high ranking of Enki is the only information obtainable from the third-millennium god lists.

In the Old Babylonian god list from Isin, Enki has an interesting epithet: **ᵈam-an-ki** 'wild bull of **an-ki**'. This epithet suggests that the primaeval pair **an-ki** has brought forth Enki, a mytheme that is also known from the Early Dynastic text IAS 114 [520]. Whereas the Old Babylonian forerunner of **an** = *anum* – TCL XV 10 or the Genouillac list – gives no information about the origin of Enki, **an** = *anum* does. This list mentions **ᵈnamma** : **ama-ᵈen-ki-ga-ke$_4$** 'Namma, the mother of Enki' [521].

In the list **an** = *anum*, Namma functions as mother both of **an-ki** and of Enki. According to the Isin-list and **an** = *anum*, Enki should have two different mothers: **ki** (Isin-list) and Namma (**an** = *anum*). The Isin-list is not inconsistent with the cited Early Dynastic text. That means that until the Old Babylonian period there seemed to be consensus about Enki's origin. The ancestry of Enki changed, somewhere during or shortly after the Ur III period. It is only since the Ur III period or later that Namma has become the mother of **an-ki**; in other words: **an-ki** was no longer the primaeval origin of everything. This function was taken over by Namma. Likewise Namma, replacing **ki** as spouse of An, became the mother of Enki. Giving Enki a reliable [522] and above all older ancestry than Enlil, lent him a higher prestige than the former head of the pantheon, Enlil. The length of the ancestor list obviously does not contribute to this prestige, for Enki has only one ancestor pair. An additional disadvantage for Enlil might have been the unknown lineage of the parents themselves of Enlil: the primaeval pair Enki-Ninki [523].

Another point of development in the Mesopotamian pantheon in the course of the second millennium, was that Marduk, son of Enki, finally achieved the most powerful position in the pantheon [524]. Besides Ḫammurapi's politics, the high position of Enki – or in other words: the power of the Eridu-Babylon clergy – may have contributed to Marduk's rise.

---

520 Text IAS 114, *i*: 12'-13'; this text has been translated and discussed in ch. 2.1.1 and in the Appendix Text editions no.1.

521 Litke 1998, 24, I:27.

522 The use of the term 'reliable' has to be explained. By 'reliable' is tried to express an ancestry based on widely recognized gods; in the case of Enki these are Namma and An. On the contrary, there are the 'ancestors' of Enlil, originally Enki and Ninki – relatively unknown in the daily religious practice (no temples, no offerings) –, and later a growing list of so-called predecessors, most of them likewise not an object of daily veneration. Although this list became more and more impressive, it seems beyond doubt that the relative obscurity of Enlil's predecessors compared with those of Enki does not contribute to a high genealogical ranking.

523 There is no text available that mentions explicitly the origin of the pair Enki-Ninki.

524 The gradual rise to power of Marduk, first of all as the city god of Babylon, started with the assumption of power by the first Dynasty of Babylon, an Amorite dynasty, and the subsequent increase of power of the Babylon clergy, who were 'Enki-minded'. Marduk was mentioned as son of Enki in the so-called Codex Ḫammurapi, the lines 8-10: 8. (...) **ᵈamar-utu** 9. **dumu** *re-eš-ti-im* 10. *ša* **ᵈen-ki** '8. (..) Marduk, 9. first-born son 10. of Enki'.
For the history and genealogy of Marduk, the syncretism Marduk-Asalluḫi – and in that way: Marduk as son of Enki/Ea – and Marduk's rise in the Mesopotamian pantheon: see Sommerfeld 1987-1990, and Black and Green 1998, 128-129.

3.4 *Preliminary conclusions*

At the end of the analyses of the god lists we should evaluate the contribution of these god lists to our knowledge of the Sumerian ideas about the Beginnings: cosmogony and theogony. The questions I try to answer are:
- Do these lists tell us something about the Beginnings, and if so, what?
- Do the lists show coherence with respect to the Beginnings?
- Is it possible to identify a diachronic development in these Beginnings?

3.4.1 *Cosmogony*

Without the god list **an** = *anum* with its explanatory texts, it would have been hardly possible to draw any conclusions about the Sumerian cosmogony from the god lists as discussed in this chapter. The only, though important, information about cosmogony is the mention of Namma in **an** = *anum*, with the equation ^{d}**namma** = ^{d}**ama-ù-tu-an-ki**: Namma is the mother of **an-ki**. This knowledge stems from a Middle Babylonian source. With the aid of this information it is possible to search for Namma as the mother of **an-ki** in the other god lists.

In the third millennium Fara list SF 1, Namma has been situated in an unclear, even incomprehensible, context. Although the arrangement of the whole list is not clear, it is possible to discern – in some places – a group of gods who share a specific principle, such as a theological or a lexicographic one. Namma is preceded by a number of names composed with **UNUG** (with one exception: ^{d}**pisan$_3$**), then followed by two unknown gods, and thereafter by ^{d}**aš$_8$-gi$_4$** and two more gods whose names are composed with **Ḫ̮I x DIŠ**, the last one of which is again from the circle of Uruk (Kulaba) [525]. In the same list SF 1, at *viii*: 22, ^{d}**namma-URU x DARA$_3$** is mentioned. A tentative interpretation of **URU x DARA$_3$** is 'the ibex town'. As the ibex is the symbolic animal of Enki [526], the 'ibex town' may be synonymous with Eridu. If this interpretation is correct, this is the only reference of Namma to Eridu in a third millennium text [527].

In the Fara lists SF 23-24, Namma is not traceable. In the Abu Ṣalābīḫ list there are some incomplete names composed with **ENGUR**, but Namma herself is absent, unless her name was present in a lacuna.

---

525 Krebernik 1986, 175-176. The entries in column *viii* are (some of my readings differ from those of Krebernik): 5. ^{d}**ŠU.UŠ-unug**, 6. ^{d}**muḫaldim-zi-unug**, 7. ^{d}**endub$_x$-unug**, 8. ^{d}**pisan$_3$** (Krebernik: ^{d}**šen**; but LAK 713 is not **šen**, but **pisan$_3$**), 9. ^{d}**ugula-unug**, 10. ^{d}**namma**, 11. [^{d}U]**M.ḪUR^{mušen}**, 12. [^{d}**B**]**AR**, 13. ^{d}**aš$_8$-gi$_4$**, 14. ^{d}**ḪI x DIŠ** (**ḪI x DIŠ** is here perhaps the old notation for **TE**: ^{d}**TE** = ^{d}**múl** ? In any case, the transliteration of Krebernik as ^{d}**UTUL$_2$** is not in accordance with Borger's lexicon, in which **utul$_2$** = **ḪI x BAD** [Borger 2010, no. 640]), 15. and 16. are compositions with **kul-ab$_4$**.

526 Wiggermann 1993-1997, 226b.

527 Edzard (1965, 107, *ad* Nammu): 'Die Göttin, die wohl vor allem im Götterkreis von Eridu eine Rolle gespielt hat, (...).' Edzard does not give any reference for this statement. Green (1975, 84) wrote about Namma that a direct link between her and Eridu is provided by the epithet ^{d}**nammu ama urú-zé-ba-ke$_4$** 'Nammu, mother of Uruzeb' (Mullil balag **e-lum-gud-sún-e** Tablet I 32a). 'Uruzeb' is the *emesal* form of Eridu. For the OB version of this **balaĝ**: see Cohen 1988, 275, l. 25. Wiggermann (1998-2001, 137, § 2) mentions that 'N(amma) is an ancient goddess of Eridu, (...)', but just like Edzard without a clear reference.

In conclusion: third-millennium god lists provide no evidence at all that 'Namma is the mother of **an-ki**' [528].

What are the results for the second-millennium god lists? Namma is absent in the Weidner list, which was transmitted very conservatively since Ur III times up to and including the first millennium. The Nippur god list mentions Namma, but without any basis for drawing conclusions about the cosmogonic involvement of Namma. In list TCL XV 10, the name of Namma is followed by the entry **^{d}ama-tu-an-ki** 'mother who has borne **an-ki**'. The recognition of this entry as being an epithet belonging to Namma is only possible thanks to the explanatory text of the list **an** = *anum*.

With respect to cosmogony, it may be concluded that the idea that Namma was the primaeval mother who gave birth to **an-ki** came into being about the Ur III-period [529].

The primary goal of a god list was to enumerate gods, using various principles: theological, lexical, genealogical and geographical ones. Although the lists may be arranged in different ways, there is no inconsistency between them with respect to 'the cosmogonic beginning': in some second millennium lists Namma was introduced as the mother of **an-ki**.

3.4.2 *Theogony*

What has been said for the cosmogony, is true for the theogony: clear-cut information from the god lists about theogony in the Sumerian pantheon can only be obtained thanks to the explications in the list **an** = *anum*. Two important groups of gods with respect to theogony can be distinguished there: the 'ancestors' of An and the 'ancestors' of Enlil.

3.4.2.1 *The 'ancestors' of An*

In third-millennium god lists, An appears at the head of the lists, but in some lists An is absent. Not one of his 'ancestors' is present in these lists.

The list TCL XV 10 mentions two gods names after An – **an-šár-gal** and **^{d}en-uru-ul-la** – who in later times are recognizable as belonging to the 'ancestor' group of An. Had we not had that information from **an** = *anum*, both gods names could easily be understood as epithets of An. Nothing in the list TCL XV10 indicates that these both gods should be 'ancestors'. Finally, the group of 'ancestors' in **an** = *anum* has not to be interpreted as real ancestors, because of their heterogeneity and artificiality. One group – **^{d}en-šár - ^{d}nin-šár** – even belongs both to the 'ancestor' group of An and to that of Enlil. The introduction of An's 'ancestor' group seems to be a secondary development, triggered by analogy with the 'ancestor' group of Enlil.

As far as An is concerned: the god lists do not contribute new information about the origin of An; his development from the primaeval unit **an-ki** is well known from literary texts from the third millennium already.

---

528 In chapter 4 the results from god lists and mythological texts will be combined; the role and position of Namma will be discussed there at several points in more detail.

529 It is not known whether the name of the first Ur III king, Ur-Namma, who probably has come from Uruk, had any relation with the rise of the position of Namma as primaeval mother. Namma as a theophoric element could not be found in texts from the time before the Ur III period.

3.4.2.2 *The 'ancestors' of Enlil*

The tradition of classifying a group of gods as 'ancestors' of Enlil is older than that of the god list **an** = *anum*. Already in the Old Babylonian god list from Mari, there is a group of gods indicated as **^{d}en-ama-a-a-^{d}en-líl-lá ^{d}nin-ama-a-a-^{d}en-líl-lá**. This group is part of the 'ancestor' group of Enlil in **an** = *anum*. With the knowledge of the names of Enlil's 'ancestor' groups in the god lists of Mari and of **an** = *anum* – and also of an analogous group at the beginning of the list TCL XV 10, which was not indicated as **ama-a-a-^{d}en-líl-lá** – we return to the third millennium lists to look for 'ancestors' of Enlil.

The Fara list SF 1 shows only four of these names, beginning with **^{d}nin-**, of which **^{d}nin-ki** is not the first but the second one [530]. Their place in the list does not point to any relation with Enlil. But the contemporary lists SF 23-24 from Fara and the Abu Ṣalābīḫ list mention nearly the same group, starting with **^{d}en-ki - ^{d}nin-ki**, followed by **^{d}en-E$_2$ - ^{d}nin-KID** (Enlil - Ninlil), and then the other 'ancestors' [531]. Although it is difficult to believe that these gods are really ancestors of Enlil [532], this group is present from the first god lists onwards. The idea of an ancestor group of Enlil is continuously present in the third and second millennium, with the comment that: 1. this group is not present in every god list; 2. over the course of time this group has been enlarged, from 6 pairs in SF 23-24 to 21 pairs in **an** = *anum*, and 3. the order of the 'ancestors' is not fixed [533].

A remarkable difference between the lists of the third and those of the second millennium is the position of Enlil. In the third millennium lists Enlil and Ninlil follow immediately after the primaeval pair Enki - Ninki, except for SF 1. In the Mari list (OB period) the pair Enlil - Ninlil is the first one, followed by Enki - Ninki and the other members of the ancestor group. The list TCL XV 10 – also OB period – begins with the ancestor group of Enlil, without including Enlil – and eventually Ninlil –, and without any indication about this group. In an Old Babylonian lamentation, **e-lum gu$_4$-sún**, a number of gods were enumerated as ancestors of Enlil (but also of Ninlil), starting with Enki - Ninki, but with the exclusion of Enutila and Enmešarra [534].

The conclusion might be that the third millennium lists suggest that Enlil and Ninlil originate directly from Enki - Ninki. The position of the other members of the later 'ancestor' group in these lists is unclear: do they all originate from the primaeval pair Enki - Ninki, or does each pair represent a new generation? In accordance with the text IAS 114, the tentative conclusion is that these gods were considered as being brought forth by Enki and Ninki too, thus being of the same generation as Enlil. This concept has been changed in the second millennium lists. There is a growing number of 'ancestors' placed between Enki - Ninki and Enlil, the expression **ama-a-a-^{d}en-líl-lá** has been introduced, and there is no uncertainty about the origin of Ninlil as being a descendant of Enki and Ninki, because there is no summarizing line saying **ama-a-a-^{d}nin-líl-lá**. Was the idea that Enlil was only one of the numerous descendants of Enki and Ninki – and so had a number of contemporaries – incompatible with Enlil's position as supreme god? The tentative conclusion may be that in the ED period the Enki-Ninki group was conceived of as 'developmental stages' between the primaeval **ki**/earth and Enlil (see ch. 3.3.6.3). Perhaps

---

530 Krebernik 1986, 174, *vi*: 25-28.

531 SF 23, *v:* 17 - *vi*: 10; for the final text of the AṢ-list: see Mander 1986, 29, nos. 273-288.

532 See ch. 3.3.6.3.

533 For a translation of the several members of Enlil's ancestor group: see ch. 3.3.6.3, and the Appendix 'God lists'.

534 Cohen 1988, 280-281: e+174 - e+176 for the ancestors; e+178 for Enutila and Enmešarra.

the realization of genealogical lists, like the Sumerian King List [535], and the ancestor list of Ḫammurapi, was the stimulus for providing the supreme god Enlil also with a more extensive and 'real' ancestry.

3.5 *Summary*

To summarize what the god lists have contributed to our knowledge of the Sumerian ideas about cosmogony and theogony:

1. Namma was conceived of as the primaeval mother of **an-ki**. This idea, as far as we know, emerged after the Ur III epoch. But its conception may originate from the previous Akkad period, if the primaeval sea as the origin of **an-ki** was a Semitic idea. The Sumerians then may simply have borrowed it [536].
2. Enki and Ninki were from the very beginning the parents of Enlil.
3. The concept of the 'ancestors' of Enlil, – with a growing numbers of 'ancestors' over the course of time, but always with Enki and Ninki as the primaeval pair – was developed during or after the Ur III period.
4. Parallel to the development of the idea of Enlil's ancestors is that of An's ancestors, although on a smaller scale: fewer ancestors and also with a lesser coherence between the various stages of development; moreover, the final genealogy in **an** = *anum* combined Sumerian and Akkadian/Semitic names.

***

535 The OB-copies of the SKL [ETCSL 2.1.1] start with the kingship of Eridu. Steinkeller (2003) published an Ur III manuscript of the SKL which starts with the kingship of Kiš. Steinkeller wrote: 'There were, without any doubt, several distinct stages in its (*SKL; JL*) development: the original version dating perhaps as early as the Sargonic period, two or more recensions belonging roughly to Ur III times, and various OB redactions.' (Steinkeller 2003, 268).

536 See also ch. 4, where this hypothesis will be discussed in more detail.

Excursus 1

Enlil and Ninlil

In this chapter 3, there has been ample discussion about the ancestry of the god Enlil, but much less about his origin – Sumerian or Semitic –, and about his wife, the goddess Ninlil. The following points summarize some relevant data about these both gods; the writing and pronunciation of their names will be discussed. Eventually this will lead to my hypothesis about the origin and identity of Enlil: he is a Sumerian god, invented in Nippur during the last quarter of the 4th millennium BCE.

1 *Enlil*

1.1 *Enlil: Pre-Early Dynastic evidence*

Considering the way the names of Enlil and Ninlil were written, I start with the rendering of Enlil's town, Nippur, which was written with the god's name, on the oldest known city seal, originating from Jemdet Nasr. Matthews [537] shows a reconstruction – a composite drawing – of city seal impressions from the Jemdet Nasr-period. According to Matthews, Nippur/Nibru was represented as **EN-NUN**, but Steinkeller [538] doubted this identification, especially that of the sign **NUN**:

> 'The alleged occurrence of Nippur was sought by Matthews in a pair of signs which he analyzed as **EN.NUN**. While the identification of the first sign as **EN** is possible, the interpretation of the second sign as **NUN** is very questionable. However, even if the signs were indeed **EN.NUN**, there would still be no reason to suspect that Nippur is meant here (the underlying assumption is that **NUN** stands for **E**$_2$, as in the so-called **UD.GAL.NUN** orthography of the ED period, but there is no evidence that this orthography was known already in Uruk III times). Here it is important to note that Nippur's name is spelled correctly (**EN.E**$_2$) in the contemporaneous Uruk City List, line 2 * (...). The same writing also appears on at least one of the Ur sealings (MSVO 2 **, fig. 12, no. 1, probably also fig. 14, no. 16), as well as on a sealing from Uruk (ibid., fig. 10b). In this connection, note further that the sign-group **EN.NUN** (if, in fact, this is the correct reading of the signs in question) is otherwise very common in Uruk III texts.'
>
> [* *see e.g. Englund 1998, 93*]
> [** *= Matthews 1993*]

In my opinion the sign combination referred to may be read as **EN-E**$_2$, which would fit the town of Enlil, Nippur, very well.

537 Matthews 1993, 37. See also Michalowski 1993b; Nissen 1985, 228.
538 Steinkeller 2002, 254-255, note 29.

On a pre-ED city sealing from Uruk [539] 'the first two signs are almost certainly **EN** over **KID**, the representation of Nippur.' [540] With respect to the reading of the sign **KID** in this case, Steinkeller argues that this sign has to be read as **E_2** [541].

In the reconstructed archaic lexical city list from Uruk, Nippur – written as **EN-E_2** – appears as the second city [542].

Englund and Nissen mention in the city list also the city 'Ninlil' [543]. This 'Ninlil' has been written as **KID-NUN** or as **E_2-KID-NUN** [544]. It may be that this city 'Ninlil' was the cult place of **^{d}nin-KID**, that in later times was known as Tummal [545]. Steinkeller wrote [546]:

> 'It appears quite certain that the earliest Sumerian pantheon was dominated by female deities. As I would reconstruct the situation existing during the Uruk period, most of the city-states (or proto-city-states) had goddesses as their titulary divine owners [547]. Those goddesses controlled broadly all aspects of human and animal life, namely fertility, procreation, healing, and death. Included among them were the birth goddess Ninḫursag, Nintu, and Gatumdug; the grain goddesses Nisaba and Ninsud; the cattle goddess Ninsun; the fish and water-fowl goddess Nanshe; the goddess of sex drive Inanna; the healer Gula; and the death specialist Ereshkigal.'

To these goddesses I would like to add **^{d}nin-KID**, possibly in prehistoric times the goddess of Tummal. The original pronunciation of this name is not known, but it is not necessarily Ninlil [548].

---

539 Nöldeke and Heinrich (1932), pl. 15g [W11456; CDLI no. P285673].

540 Matthews 1993, 39.

541 Steinkeller 1995c, 700, *sub* nos. 129 and 142. As can be concluded from fig. 1 in a note from Englund (2011), he agreed with Steinkeller's comment with respect to the reading **E_2** instead of /**KID**/ of a certain sign.

542 Matthews 1993, 38, table 3; Englund 1998, fig. 26. Matthews (p. 39) refers to Englund (1988, 132, note 9) and Englund and Nissen (1993, 145) for the reading **KID_a** instead of **E_2**, but the reading **E_2** for this sign has been confirmed by Steinkeller (1995c). On a fragment of a geographical list from the Archaic period from Uruk (Green 1977, 293: W21126), the second entry is clearly written **EN-E_2**, representing Nippur.

543 Englund & Nissen 1993, 147 *sub* Cities 38.

544 On tablet W21208,18 O0206, and W23998,1 O0204, respectively. Steinkeller (1995c, 706 *sub* 413) has some doubts about the correct reading of **é+nun** = /**ninlil**/. The difficulty is caused by the subtle difference between the signs **E_2** and **KID** in the archaic texts.

545 Krebernik 1998-2001b, 457-458, § 3.4.2. See also: Sollberger 1962, and 'The history of the Tummal' ETCSL 2.1.3.

546 Steinkeller 1999, 113. Michalowski (2002) does not add any new perspective to the idea of the supposed dominance of goddesses in the early Sumerian pantheon. See also Sallaberger (2003-2006, § 14, 305-307) for a discussion of 'Frauenrollen und das Pantheon als Spiegel der Gesellschaft'.

547 A reflection of this female domination may be found in the majority of **^{d}nin**-names in the god lists of Fara and Abu Ṣalābīḫ, although **^{d}nin**- does not necessarily represent a female god.

548 If this was Ninlil, then we come across the same problems in explaining the meaning of the /**lil**/-part as those mentioned for Enlil's name.

1.2 *Enlil: Early Dynastic times and later*

From the ED-period until the Ur III-period, in Mesopotamia as well as in Ebla, Enlil's name was written as **^{d}en-E$_2$**, which might be translated as 'Lord of the Household' or 'Pater-familias' [549], whereas Ninlil's name nearly always appears as **^{d}nin-KID** [550]. Thereafter the writing of Enlil's name was harmonized with that of Ninlil: Enlil was written as **^{d}en-KID** = **^{d}en-líl**.

The difference in spelling of the second part of the names of Enlil and Ninlil is also found in, at least some, UD-GAL-NUN-texts [551]. In some other UD-GAL-NUN texts, the /**lil**/ part of Ninlil is written, not with **KID**, but like that of Enlil, with **NUN** [552].

1.3 *Enlil in Eblaitic texts*

Also in Ebla texts [553] the name of Enlil was written as **^{d}en-E$_2$**, with the equation *i-li-lu*. I use 'equation' and not 'translation'; in my opinion the Eblaitic *i-li-lu* is an Eblaitic/Semitic approximation of the Sumerian logogram **^{d}en-E$_2$**. According to Steinkeller and Michalowski, Enlil's name in the Ebla spelling *i-li-lu* suggests a possible etymology **il-ilī* 'god of (all) the gods' [554]. Moreover the expression *i-li-lu* **A-MU DIĜIR-DIĜIR-DIĜIR** is found in two Ebla incantations, thus *i-li-lu* is called 'father of (all) the gods' [555].

2 *Ninlil*

A survey of what is known about Ninlil is given by Krebernik [556]. In the ED-period Ninlil's town was written as **NIN-KID** or **KID-NUN** [557].

---

549 Steinkeller 1999, 114, note 36. For a summary of writings and interpretations of Enlil's name: Edzard 2003c. In an article about the archaic writing of the name of Enlil/Nippur, Steinkeller (2010) has given clear evidence for the writing in third-millennium texts of Enlil's name as **$^{(d)}$EN-E$_2$**. He also pointed to alternative interpretations of Wang, which he considered as very unlikely (Steinkeller 2010, 240 n. 11 [=Wang 2011, 41-59, 218], and 241 n. 21 [=Wang 2011, 77-78]).

550 Exceptions are two texts from Lagaš/Ĝirsu (ED IIIb): Biggs 1976, 31, obv. *ii*:3 [CDLI no. P 221801] and Cros (1910), 180 (= AO 4153, Ukg 15; CDLI no. P315470). In these texts Ninlil is written as **^{d}nin-E$_2$**. Lambert (2008, 27) considers the writing of 'slightly different signs for **LIL$_2$** following En- as compared with **LIL$_2$** following Nin- ' as a scribal whim (without further explanation), but why should scribes act in that way? Such a 'whim' is unknown for any divine couple.

551 Enlil: **^{UD}GAL-NUN** (IAS 114 *i*:9'); **^{UD}KIŠ-NUN** (IAS 114 *i*:15'); **^{d}GAL-NUN** (IAS 129 *x*:4'). Ninlil: **^{UD}nin-KID** (IAS 114 *i*:16'); **^{d}nin-KID** (IAS 129 *x*:6').

552 **^{UD}nin-NUN**: IAS 118, *ii*:1; IAS 163, *ii*:2.

553 Pettinato 1982, Tav. VIII, Testo 4 v X-10', X-11'; Edzard 1984, 28, no. 6, II.7 (photo of tablet: Taf. XLIV obv. *ii*:7).

554 Steinkeller 1999, 114, note 36; Michalowski 1998, 241-242. Vanstiphout (2009, 21, n. 38) endorses to the viewpoint of Michalowski.

555 Textual references: Michalowski 1998, 241, note 8. See also: Pomponio and Xella 1997, 170-171.

556 Krebernik 1998-2001b. Such-Gutiérrez (2003, 109) wrote about Ninlil: 'Die genaue Bedeutung von NinKID kann man noch nicht feststellen, trotzdem kann man vermuten, daß der Name mit ihrem Charakter als Fruchtbarkeits-Getreidegöttin zusammenhängt. In der Tat war sie die Hauptgetreidegöttin in Nippur.' His reference to Jacobsen (1989, 269) ['(...) die Auffassung Th. Jacobsens, nach welcher **lil$_x$** (**KID**) die reife junge Pflanze der Gerste bezeichnete'] is not convincing; the sign **KID** represents a reed mat (Steinkeller 2010, 240).

^{d}**nin-KID** is present in the god lists from Fara and Abū Ṣalābīḫ [558] and in the Temple Hymns [559], but not in the AṢ **zà-mì** hymns.

Temple Hymn no. 3 was dedicated to the temple of ^{d}**nin-KID** in Nibru/Nippur. These Temple Hymns were written and/or compiled, in all likelihood, by the daughter of Sargon of Agade, Enḫeduanna [560]. The text and translation of Temple Hymn no. 3 are given here [561]:

39	**tum-ma-al^{ki} me-nun-e gal pàd-da ní su-zi ri-a**	Tummal, pre-eminently favourite for the princely **me**, provided with awe and dread;
40	**temen šu-luḫ-sikil-zu abzu-a lá-a**	the foundation – your pure lustration – extending as far as the **abzu**;
41	**iri$^{?}$-ul ĝiš-gi gi-sumun gi-ḫenbur sag$_{12}$-ga**	primaeval city, reed land, embellished with old reed and reed stalks;
42	**šag$_{4}$-zu kur-ḫé-ĝál-la nam-ḫé-a dù-a**	your interior – a mountain of abundance – built in plenitude;
43	**itid-zag-mu-ezen-ĝál-la-za u$_{6}$-di tag-ga**	at your feast held in the month of the New Year touched by admiration;
44	**nin-gal-ki-ùr-ra ^{d}en-líl-da zag ša$_{4}$-a**	the great lady of the Ki-ur, equal to Enlil,
45	**nun-zu ama ^{d}nin-líl nita$_{3}$-dam-ki-áĝ-^{d}nu-nam-nir-ra-kam**	your princess, mother Ninlil, beloved wife of Nunamnir,
46	**é-tum-ma-al^{ki} mùš-za é bí-in-gub bara$_{2}$-za dúr bí-in-ĝar**	has, O Household Tummal, erected a house in your precinct (and) has taken a seat on your dais.
47	**8 é-^{d}nin-líl-nibru^{ki}-a**	8 lines: the house of Ninlil in Nibru.

In my opinion this Temple Hymn may be a clue to the understanding of the name ^{d}**nin-KID**. Tummal is described as a primaeval city [562], that has been beautified by reed lands, with both old and young reeds. Obviously, reed was an important product in Tummal, at least in ancient times [563]. The sign **KID**, with the sound value **kid**, is a logogram for 'reed mat'. This Temple Hymn is the only one with such close connection between reed and (part of) the name of the god or goddess [564]. The conclusion is nearly inevitable: ^{d}**nin-KID**, or

---

557 SF 23, *iii*:1 and IAS 21, *iii*:4, respectively.

558 SF 1 *vi*:27 (Krebernik 1986, 174); SF 23 *v*:20 (Deimel 1923); AṢ no. 3 and 276 (Mander 1986, 24, 29).

559 Sjöberg and Bergmann 1969, 19 TH no. 3 (The Temple Hymns, ETCSL 4.80.1, ll. 39-47).

560 Sjöberg and Bergmann 1969, 5 and note 2.

561 The transcription is based on the ETCSL edition 4.80.1, ll. 39-47.

562 I could not verify the sign **iri**$^{?}$ in l. 41. See for commentary on **iri**$^{?}$: Sjöberg and Bergmann 1969, 58 *ad* 41.

563 Oates (1960, 50) formulates this as follows: '(...); and when we remember that one of the principal exports of the Marsh Arab communities at the present day is reed matting, we cannot claim that the surplus wealth necessary for the purchase of imported pigments was not available before the introduction of agriculture.'

564 There are a few more lines in the Temple Hymns with **gi** (transcription and translation: ETCSL 4.80.1): line 101: **urim$_{2}$^{ki} ninda$_{2}$ gi-duru$_{5}$ sug$_{2}$-ga** 'O Urim, bull standing in the wet reeds', (TH 8 for the temple of Nanna in Ur); line 540: **munus-zid naĝa-kù-ga gi-dù-e tud-da** 'the true woman, the holy potash plant, born of the stylus reed', (TH 42 for the temple of Nisaba in

better now: **^{d}nin-kid**, is the goddess of reed matting Ninkid in ancient times before she became the spouse of Enlil [565].

3 *Hypothesis*

In this Excursus I would like to formulate a hypothesis about Enlil and Ninlil, their origin and the meaning of their names, without repeating all the former ideas, the literature of which has been summarized by Edzard [566].

My starting-point is the original writing of Enlil's name as **^{d}en-E$_2$**, a logogram, of which the literal sense is 'Master of the Household, Pater familias' [567]. As we have seen: in Ebla texts the name of Enlil was written as **^{d}en-E$_2$**, with the equation *i-li-lu*. In my opinion, this logogram is not a Sumerian name, but only an epithet, more a description of the god's capacity or function than a real proper name. It is unclear how the name, represented by this logogram, was pronounced. Because of the connections, or better: cultural transmissions, emanating from Kiš to Abū Ṣalābīḫ – both places not far from Nippur – and Ebla [568], it is not unlikely that due to the Semitic rendering *i-li-lu* the pronunciation of **^{d}en-E$_2$** became Illil or Ellil [569].

Enlil is probably of Sumerian origin. He may have been 'invented' in Nippur, at the time when the expansion of the 'Uruk world system' ended. The Nippur-Adab region appeared to have been more stable, both demographically and with respect to settlement longevity, than the Uruk area [570]. At the end of the 4th millennium the focal point was not Uruk anymore but Nippur, which became a more and more important trade centre, with a concomitant rise of the power of its clergy. The Semitic civilization extended its influence

---

Ereš); in line 297 (TH 24) **gi** is present in an unclear context. In these examples the close relation between 'reed' and (part of) the name of the god or goddess – as is found in TH 3 for Ninkid – is absent.

565 In any case, this hymn does not explain why the sign **KID** in **^{d}nin-KID** has to be pronounced **/lil/**. Krebernik (1998-2001b, 460a) says that Ninlil in the texts almost exclusively appears as spouse of Enlil, whose characteristics also determine those of Ninlil; Ninlil's individual features are rarely expressed.

566 Edzard 2003c.

567 Steinkeller 1999, 114, note 36.

568 For Kiš and the 'Kiš Civilization' see: Gelb 1981, 71-73; Charvát 1981; Charvát 2002, 167-170. From the beginning of the third millennium Kiš developed a dominant position in the area outside Sumer proper. Under its influence were some adjacent Sumerian sites, such as Abu Ṣalābīḫ. Kiš was for the greater part 'Semitic', as can be concluded from the personal names. Gelb (1981, 56): 'The literary and administrative texts from Abu Salabikh, situated close to the Semitic and the Sumerian areas, revealed to us unexpectedly a Semitic type of writing, language, and names, which link Abu Salabikh with the ancient kudurrus and votive inscriptions around Kish and with Ebla and Mari in the North, "literary" compositions, linking Abu Salabikh with Ebla, and month names and system of dating, which link Abu Salabikh with Ebla and Mari.' It might not be surprising if it should appear that the so-called UD-GAL-NUN orthography, mainly found in Abu Ṣalābīḫ, is not a cryptography (as supposed by Krecher 1992), but just a local way of writing Sumerian, influenced by the Semitic environment. Rubio, writing about UGN (2011, 103): ' (...) rather than a cryptography, all these apparently substitutional mechanisms may well reflect an early attempt to devise an alternative orthography for Sumerian texts.'

569 None of the attempts to explain the name **^{d}en-líl** on the basis of the various meanings of **líl** was convincing for the functional head of the Sumerian pantheon; for a summary: Edzard 2003c.

570 Ch. 1.1.1.

from Ebla to the north of Sumer, including Nippur [571]. The personal names of that time were predominantly composed with *ilum* and *ištar* [572]. In this social environment the (new) supreme god in the Nippurian pantheon was named *i-li-lu* 'the god of the gods', which name was paraphrased and written in Sumerian as **^{d}en-E$_2$**. Westenholz also noticed that Enlil was never equated with any Akkadian god [573]. Perhaps the reason is that this 'equation' was not possible, because Enlil was originally a Sumerian god who had no Semitic counterpart.

The 'invention' of Enlil in Nippur, the city with increasing influence, might also explain why Enlil had to descend from **ki**, represented as the gods Enki and Ninki. The clergy of Nippur could hardly approve of Uruk's An, the supreme god, as the father of Enlil. An came into being from the entity **an**, present in the primaeval unit **an-ki**. The only other entity in this primaeval unit was **ki**. From this **ki** the parents of Enlil, Enki and Ninki, developed. In this way is Enlil a chtonic god *par excellence*.

3.1 *The meaning of the name* ^{d}en-líl

Edzard [574] has summarized the literature about the writing and interpretation of the name Enlil. Edzard does not subscribe to the viewpoint of Steinkeller and Michalowski [575] about the origin of Enlil and the interpretation of his name, because of the – in his opinion – insufficient foundation of their thesis [576]. As a consequence he does not accept their hypothesis.

Both Steinkeller and Michalowski, reject the translation of **líl** as 'air' or 'wind' or the like, because **líl** was never used in Sumerian with this kind of meaning. Steinkeller [577]: '(...), from the very beginning of cuneiform script, the name of Enlil was written with a logogram, whose form was **EN.É**. Of course, this fact has important bearing on the etymology of Enlil's name and the history of Enlil's cult in archaic Mesopotamia. While this evidence does not preclude the possibility that Enlil is a Sumerian divine name, it emphatically excludes any chance of it being etymologically **en** + **líl**, "Lord-Wind", as thought by Jacobsen and other scholars before him.'

It is also my opinion that the **líl** in Enlil's name most likely has not the meaning 'wind, breeze; ghost' – the chtonic character of Enlil seems incompatible with such meanings –, but that it could be explained as an attempt to reconcile the Semitic epithet or name *il-ilī*, the supposed pronunciation of **^{d}en-E$_2$** as *Ellil or *Illil and the Sumerian writing **^{d}en-E$_2$** [578]. UGN-writing betrays a number of Semitic or northern characteristics,

---

571 Gibson 2007.

572 Westenholz 1988, 103.

573 Westenholz 1999, 84.

574 Edzard 2003c.

575 Steinkeller 1999, 114, note 36; Michalowski 1998, 241-242. Vanstiphout (2009, 21, n. 38) endorses to the viewpoint of Michalowski.

576 Edzard 2003c, 184: 'Wie dem auch sei, die *interpretatio sumerica* des Namens Enlil erscheint mir als die einzig zu rechtfertigende. Wir haben gesehen dass sich die Ebla-Schreibung (^{d})*I-li-l(u)* problemlos mit einer sumerischen Form *Yēn-lil verbinden lässt. Die These von Steinkeller und P. Michalowski, Illil/Ellil/Enlil sei ein aus dem semitischen Bereich stammender uralter Eindringling im sumerischen Pantheon, müsste mit kräftigeren Argumenten untermauert werden.'

577 Steinkeller 2010, 242.

578 The extant texts seem to suggest that the meaning of the UGN-sign **NUN** as **E$_2$** or **lil$_2$** is only applicable in the UGN-writing of Enlil's name. Based on Krecher's article, Woods writes that

with Sippar and Kiš as candidates for the centres of influence [579]. One of these characteristics is the prominent figuration of Enlil and Zababa in UGN literature: 'Possibly the former, but certainly the latter, is a northern import, perhaps Semitic in origin. (...) UGN texts regularly give supremacy within the pantheon to Enlil – **a UD.GAL.NUN** "father Enlil" – placing him above Enki and Nanna, while seeming to relegate An to a lesser position, in contrast to the big Fara god-list which begins with An.' [580] If this is correct, it may be assumed that the **NUN** in Enlil's name has not to be seen as another spelling for /**lil**/, but that it expresses exactly what the sign in 'normal' orthographic writing means, viz.: prince; foremost, best [581]. The Semitic spelling *il-ilī* for **ᵈen-E$_2$** has been rendered as adequately as possible by the UGN-spelling (**GAL-**) **NUN**, the addition **GAL** in order to emphasize that the god is indeed 'the greatest one'.

3.2 *Enlil as the head of the Sumerian pantheon*

The spelling of Enlil and of Nippur were already linked to each other in the pre-ED period. Enlil must have had a high position in that place. He was *pater familias* of the gods for those people who were belonging to the Kiš-civilization, but that does not mean that he was already in pre-ED times head of the Sumerian pantheon.

This god Enlil, sometime [582], has become the head of the Sumerian pantheon, with Nippur as his residence [583]. Charvát, sharing Kramer's view [584] about the building activities

---

'UGN and the standard orthography are essentially two adjoining facets of the same writing system' (Woods 2005, 27 and note 86).

579 Woods 2005, 30.

580 Woods 2005, 31. Woods also refers to Lambert 1981, 92-93.

581 In this respect I tend to disagree with the hypothesis of Michalowski (1998, 242): 'The usurpation of Enki's place by the new god (= Enlil; *JL*) may also be the source of the use of the **NUN** sign, a symbol widely associated with the older god, in the **UD.GAL.NUN** (= **dingir.en.líl**) writing of Enli's name, (...).' In my opinion, the sign **NUN** has been used independently in both cases (see also the Excursus 2 'Enki' in this chapter).

582 In the **zà-mì** hymns from Abu Ṣalābīḫ, Enlil seems to function as head of the pantheon, because he has awarded to the Anunna lots for their cult places (text: Biggs 1974, 46, lines 1-14). Since there is some debate about the translation of especially the lines 11-14 [Krebernik 1994], the present author will give his translation and comments:

11	**ᵈen-E$_2$ a-nun**	Enlil has (l. 12) to the Anunna
12	**ki mu-ĝar-ĝar**	domains allocated.
13	**diĝir-gal-gal**	The great gods
14	**zà-mì mu-dug$_4$**	he has praised.

Comments: line 14: The verbal form **dug$_4$** is singular; therefore it is Enlil who has praised the Anunna, and not the reverse. Lambert (1976, 430, note 1) has the same opinion.
See also Klein (2001, 295 and note 83): 'In unknown circumstances and for unknown reasons, the local god of Nippur, Enlil, became the head of the Sumerian pantheon no later than the ED II period.' Selz (1992, 190-191) has tried to establish – on the basis of a transcription of a personal name in WF 117, *i*:3 **ᵈen-líl-ukkin** – that Enlil had a leading function in the (divine?) assembly. However, the tablet WF 117 [= CDLI no. P011075] shows both in *i*:3 and in *ii*:1 the same sign **ukkin**, connected with two different gods, in personal names: **ᵈen-líl-ukkin** and **utu-ukkin**, respectively. If we accept that Utu has been written without **diĝir**-determinative, then it is hardly comprehensible that two gods should have a leading function in the divine assembly. In the same article, Selz (1992, 195) also suggested a prominent position for Enki due to his epithets **lugal-eridu^{ki}-ga** and **lugal-abzu**; but these epithets do not seem to be sufficient for a leading

of the Kiš-rulers, wrote [585]: 'At any rate, the "post-diluvial" Kiš-I dynasty sovereigns systematically developed (socially desirable?) sponsorship of religious institutions (...). This trend may well have started with (En)Mebaragesi, listed in the Nippur Tummal inscription as the builder of that temple, whose authentic text turned up in one of the Diyala region shrines (?), the Temple Oval, obviously the earliest of the oval sacred precincts (?) of ED III Mesopotamia (...).'

There is evidence that Enlil obtained his high position in the Sumerian pantheon only in a relatively late period, but certainly from ED time on. Englund [586] presented a list with non-numerical signs of greatest frequency in the administrative text corpus of Uruk IV-III periods. The fourth and fifth position are occupied by **AN** (An$^?$) and **NUN** (Enki$^?$), respectively [587]. Strikingly absent in this list is Enlil. Michalowski supposes that the use of the sign **NUN** in the UGN-writing for Enlil's name may be an indication for the usurpation by Enlil of Enki's place at the head of the Sumerian pantheon, for the sign **NUN** was associated with Enki [588]. Selz concluded, after a study of documents, votive inscriptions and god lists from the third millennium (Fara, Abu Ṣalābīḫ), that Enlil achieved primacy at that time at the expense of Inanna and Enki [589]. Sallaberger commented on this point that in his opinion it is not convincing to conclude that a cult has a relative meaning on a supraregional level based on its being widespread [590]. Westenholz inferred from lists with

---

position in the Sumerian pantheon. In all the god lists discussed from the third millennium, it is Enlil who is mentioned before Enki! Also the widespread cult places with **abzu** in their name are insufficient to attribute supremacy to Enki (Selz 1992, 195), because **abzu** also may be 'a ritual water container in a temple' (ePSD), something that is present in (nearly) every temple; in my opinion this only emphasizes Enki's association with the **abzu**. Selz's reference to the study of Green (Selz 1992, 195, note 31) with respect to the supposed prehistoric supremacy of Enki does not seem to be valid for the third millennium, because the references of Green date mainly from the second millennium and later. For my hypothesis about Enki: see Excursus 2 of this chapter.

583 Sallaberger (1997, 150-153) emphasizes that in the ED period Nippur has the first position only with respect to the cultus, because it is the residence of Enlil. In the political unity between Uruk, Adab, Nippur, Lagaš, Umma and Šuruppak, Nippur has the same position as the other cities. Such-Gutiérrez (2003, 34-35) mentioned a competition between Nippur and Uruk, but with evidence for this only in the EDIIIb period, viz. in the inscriptions of Lugal-kiĝine-dudu (see Frayne 2008, E1.14.14, 413-419). Lugal-kiĝine-dudu exercised the **en**-ship in Uruk and the **lugal**-ship in Ur (see also Edzard 1987-1990, 146, *sub* Lugalkinnedudu). In Frayne's text no. 1 it was Enlil who combined these functions for him (mentioned on numerous stone vessel fragments from Nippur); in text no. 2 Inanna did so (mentioned on two stone vessels from Nippur). Perhaps this may be explained, not so much as 'competition' between Enlil / Nippur and Inanna / Uruk, but as the real assumption of power by Enlil, i.e. the clergy of Nippur.

584 Kramer 1960, 277, note 25.

585 Charvát 2002, 214. ETCSL 2.1.3: 'The history of the Tummal'.

586 Englund 1998, 70. Englund gives an account of the texts used (65, note 123). The fact that 86% (5000 of 5820) of these texts originates from the district Eanna of Uruk may explain the absence of Enlil in this list.

587 The question marks are Englund's.

588 Michalowski 1998, 242. This argument seems not quite convincing, because the sign **NUN** in the UGN-orthography can have more meanings (Krebernik 1998, 301). See also the Excursus 2 'Enki' of this chapter.

589 Selz 1992, 199-203.

590 Sallaberger 1997, 149, note 6: 'Methodisch halte ich es jedoch nicht für überzeugend, aus einer weiten Verbreitung eines Kultes auf dessen relative Bedeutung im überregionalen System zu

functionaries involved with the cults for gods, that the absence of those functionaries for Enlil may be 'the theocratic evidence of the late arrival of Enlil', and she also pointed to the fact that: 'Although Enlil is the beloved high king of several of the southern rulers, he is not credited with giving them their kingship until the above inscription of Lugalkiginnedudu of Uruk.' [591]

The findings of Such-Gutiérrez [592], who studied the pantheon of Adab in the third millennium, are the following. Semitic influences could be shown by the presence of the names of Semitic gods in personal names – e.g. Dagan, Ea (besides the Sumerian name Enki), Erra, Ištar, Ilaba, Ištarān, Suen and Šamaš. Enki, and to a lesser extent Ea, and Enlil – in the onomastics the god with the most attestations – are a theophoric element as early as in the pre-Sargonic period. On the other hand, while offerings to Enki/Ea were performed from pre-Sargonic times onwards, the cult with respect to Enlil could not be proven before the Old Akkadian period. Such-Gutiérrez wrote that the mention of Enlil's cult besides that of An from the Sargonic period onwards may point to a growing influence of Enlil [593]. These results for the Adab pantheon confirm the above conclusion of Westenholz.

Enlil, supposed to be at first the most important god for the people of Nippur and the north of Sumer, viz. the people belonging to the Kiš-civilization, may have been promoted by the clergy of Nippur (in pre-ED times) and by the rulers of Kiš, (En)Mebaragesi being the first king mentioned to have built Enlil's temple in Nippur [594]. Due to the power of the Kiš dynasty and consequently that of the clergy of Nippur especially, Enlil may have 'conquered' the Sumerian pantheon. An, nominally the head of this pantheon, but in fact a *deus otiosus*, a peaceful non-violent god – so typical Sumerian, as Steinkeller assumed [595] – may then have been surpassed by a more powerful and active Enlil.

Steinkeller [596] wrote about the turning-point in history, when male gods became more important, or better: more mighty than the original city goddesses, and so became city gods in their place: 'It may be speculated that this growing masculinization of the Sumerian pantheon was partly an internal development, reflecting changes in the organization of the Sumerian society, and partly the result of a contact with the Akkadian population of northern Mesopotamia. Here it must be pointed out that, in contrast to the Sumerian pantheon, the Akkadian one was controlled by male deities, with goddesses generally

---

schließen.' But Sallaberger (1997, 149) agreed that: 'Die ältesten Götterlisten und ein Zyklus von Tempelhymnen zeigen, daß zu dieser Zeit (= *ED period; JL*) Enlil als bestimmender Gott an der Spitze des Pantheons steht.'

591 J.G. Westenholz 1992, 301; J.G. Westenholz 2000, 79 (for both citations).

592 Such-Gutiérrez 2005-2006.

593 Such-Gutiérrez 2005-2006, 14: 'Die Erwähnung des Kults von Enlil neben dem des An nicht vor der altakkadischen Periode scheint auf eine zunehmende Bedeutung Enlils ab der sargonischen Zeit in Adab hinzudeuten.'

594 Sollberger 1962, 42, ll. 1-1a; 'The history of the Tummal', ETCSL 2.1.3, ll. 1-2. According to Michalowski (2006, 162), this text is 'not a historical document, but quite simply a school concoction based primarily on the SKL (Sumerian King List; JL) and on the Nippur version of the lexical list Proto-Kagal.'

595 Steinkeller 1992, 246-247.

596 Steinkeller 1999, 114. Lambert (1987b, 128-129) wrote about the declining dominance of Mesopotamian goddesses: '(...) sexism does not seem to explain the decline in the number of goddesses as city patrons as between the third and second millennium. Rather the accidents of city decline are to blame.' But in my opinion this may not be the cause. After all, why should precisely the cities with a goddess as city god decline?

lacking any individual characteristics and functioning merely as reflections of their divine husbands.'

At this point ^{d}**nin-KID** lost her prominent position in Tummal, which was taken over by ^{d}**en-E$_2$**, who was 'Lord of the Pantheon' and city god of Nippur. Could it be that the story known as 'Enlil and Ninlil' [597], in which it is told that Enlil violated Ninlil, is a far echo of this change of power [598]? At the end of the third millennium even the spelling of the second part of the names of Enlil and Ninlil was uniform; the sign used was **KID**, with pronunciation /**lil**/, so that, regardless of the original pronunciation of the name of the goddess ^{d}**nin-KID** in earlier times, Ninlil now was 'but a female reflection of Enlil' [599], even in her name.

The other story, 'Enlil and Sud' [600], in which Sud, city goddess of Šuruppak [601], became the spouse of Enlil, after which her name was changed into Ninlil [602], might have its origin in a close relationship between Sud and Enlil in the ED period. Šuruppak was at that time an important city, a member of the **ki-en-gi** league [603]. The relationship between Sud and Enlil may be inferred from administrative texts of Šuruppak, in which Enlil is mentioned immediately after Sud or her sanctuary [604]. Perhaps Sud was – in Šuruppak – considered as the spouse of Enlil. After the destruction of Šuruppak [605] by fire at the end of the ED III-period, Sud was deprived of her city, and as a consequence she lost importance.

4 *Enlil's epithet*

In this discussion the fact remains that one of the epithets of Enlil is **kur-gal**, and that Enlil's temple in Nippur has been called **é-kur**. In other words: the element **kur** has been connected with Enlil. **kur** has several meanings: 1. mountain, 2. (foreign) land, 3. netherworld. As I will try to demonstrate (next paragraph): in all likelihood Enlil has never

---

[597] Behrens 1978 and ETCSL 1.2.1.

[598] Steinkeller (1992, 246-247) described the difference between the Sumerian pantheon, being relatively peaceful, and the Semitic pantheon, in which the gods used more violence.

[599] Steinkeller 1999, 114 note 36.

[600] Civil 1983 and ETCSL 1.2.2.

[601] Sud – ^{d}**SU-KUR-RU** –, the city goddess of Šuruppak, is mentioned in the Fara list SF 7 *i*:7-8 (Deimel 1923), but not in lists SF 1 and SF 23-24, or it might be in the lacunae. The god lists of Abu Ṣalābīḫ do not show the presence of Sud in the legible parts, but in the AṢ-**zà-mì** hymn Sud is praised (Biggs 1974, 51, the lines 180-181). Krebernik (1998-2001b, 455, §3.1.3) mentioned: 'In mehreren Textzeugen von 'Nannas Fahrt nach Nippur' (...) erscheint statt Sud als Stadtgöttin Šuruppaks Nin-irigala (^{d}**Nin-UNUG**). (...) Möglicherweise konnte Sud also auch mit Nin-irigala gleichgesetzt werden.' ^{d}**nin-UNUG** is present in the AṢ-hymns as goddess of Kulaba (ll. 15-18), and in the god lists of AṢ (no. 76) and Fara (SF 1 *i*:11; SF 23 *vi*:19), but here the goddess is presumably not Sud, but Ninirigal, goddess of Kulaba. Sud and ^{d}**nin-UNUG** appear together only in the AṢ **zà-mì** hymns, but – as far as the remainders of the respective tablets show – they are not mentioned together in the same god list. My tentative conclusion is that the ^{d}**nin-UNUG** in 'Nanna-Suen's journey to Nibru' (Ferrara 1973, 62, l. 233 [ETCSL 1.5.1] is not Sud, but that her mention may be due to a scribal error.

[602] This was already mentioned in advance (in l. 35) before the marriage between Enlil and Sud.

[603] Krebernik 1998, 242, and note 53: texts WF 92, WF 94.

[604] Martin *et al.* 2001, 25; 105 *sub* ^{d}**en-líl**; 112 *sub* ^{d}**sùd**. The position of Enlil immediately after Sud or her sanctuary may of course also be explained by the fact that, even at that time, Enlil was head of the Sumerian pantheon.

[605] Krebernik 1998, 242.

been a netherworld god. The **kur** 'mountain' is observable outside Sumer, in the north-east direction. **kur**, not being 'eine Gottesidee' [606], may be very well a symbol, to which Enlil and/or his power and authority were compared or even identified with. In that case the use of **kur** may be considered as a metaphor. **kur** 'mountain' may also function mythologically as a pillar upholding the heaven/sky and may therefore have become the symbol of the god Enlil, who once separated the heaven and the earth.

5 *Enlil as a god of the netherworld in prehistoric times?*

Katz supposed that in prehistoric times Enlil might have been the husband of Ereškigal, but due to changes in the Sumerian pantheon, in which Enlil obtained his supreme position, he could not be associated anymore with the netherworld [607]. This hypothesis is very difficult to prove. There is no ED text known in which Enlil's involvement with the netherworld is described. The oldest attestations show that even in the ED period Enlil (**^{d}en-E$_{2}$**) and Ninlil (**^{d}nin-KID**) formed a couple. Wiggermann wrote that before the Ur III-period southern and northern Mesopotamia had a different view on the rulers of the netherworld [608]. In South Mesopotamia Ereškigal was known as the queen of the netherworld from the Old Akkadian period onwards (inscription of Lu'utu; TH 14, line 179) [609]. Steinkeller supposed that at first the goddesses dominated the Sumerian pantheon [610]. As a consequence a male consort may have been less important. This may be the reason that Ereškigal's husband has not been mentioned in third-millennium texts. In the North, Nergal was acknowledged as the ruler of the netherworld [611]. The marriage between Ereškigal and Nergal brought these two views together, in the OB period at the latest [612].

Only in one, Old Babylonian, text – an incantation against evil spirits – are Enlil and Ereškigal mentioned together as parents, viz. of Namtar [613], but they are never recorded together as parents of Ninazu, a god of the netherworld [614].

---

606 Hruška 1971, 194.

607 Katz 2003, 419 note 127; 439-442.

608 Wiggermann 1998-2001b, 218b.

609 In the pre-Sargonic text DP 51 *ii*:6 - *iii*:5 (Ukg. year 2) Ereškigal receives offerings. In this text she is not indicated as queen/head of the netherworld. Katz (2003, 378, note 44) assumed that Ereškigal's function was related to the netherworld due to the association with Ninazu in this text. TH 14, l. 179: **enegir^{ki} a-pap-gal a-pap-ki-a-^{d}ereš-ki-gal-la-ka** 'Enegir, great libation pipe, libation pipe of the netherworld of Ereškigal' (Sjöberg and Bergmann 1969, 27; ETCSL 4.80.1).

610 Steinkeller 1999, 113.

611 Wiggermann 1998-2001b, 218b. In Temple Hymn 36, line 464, Nergal's epithet is **lugal-utu-šú-[x]** 'Lord of the sunset'; compare this with Ereškigal's epithet in Lu'utu's inscription **nin-ki-utu-šu$_{4}$** 'Lady of the place of the sunset' (ch. 3.3.6.2).

612 Wiggermann 1998-2001b, 219a.

613 Geller 1985, l. 360. But even here one may have doubts about the real paternity of Enlil. The line says: **nam-tar dumu-ki-áĝ-^{d}en-líl-lá ù-tu-da ^{d}ereš-ki-gal** 'Namtar, beloved son of Enlil, born of Ereškigal'. But remarkably: in l. 308 of these 'Forerunners', Ereškigal is the wife of Ninazu. In general, when in a text it is said: 'God A is the father / son of god B', this may be only an expression of respect or devotion, but not of real family relationship. To give only a few examples: in 'Enki's journey to Nippur', Enlil is mentioned as father of Enki (Al-Fouadi 1969, ll. 104-105); in 'Enki and the world order', Enlil is called the father of Nanše (Benito 1969; ll. 417-420). Neither statement is in conformity with the usual genealogy of these gods.

In my opinion there are no indications at all for an association of Enlil with the netherworld. In the myth 'Enlil and Ninlil' he *goes to* the netherworld [615]; but Inanna did likewise. Those mythemes alone do not designate these gods as gods of the netherworld. Enlil and Ninlil created the moon god Suen-Ašimbabbar, the chthonic gods Nergal-Meslamtaea and Ninazu, and finally Enbilulu, the inspector of the canals [616]. On the basis of this text, Ninlil has never been associated with the netherworld; so why then Enlil?

One interpretation of a passage in the text 'Gilgameš, Enkidu and the netherworld' is that the netherworld was given to Ereškigal by Enlil as a (wedding) gift [617]. Indeed, according to this text the netherworld was given to Ereškigal, but by both An and Enlil, and not by Enlil alone as a gift [618]. As an indication for a prehistoric relation between Enlil and Ereškigal this OB text seems to me not suitable.

My tentative conclusion is that the question: 'Was Enlil a god of the netherworld in *prehistoric* times?' cannot be answered with certainty, neither for the Semitic pantheon nor for the Sumerian one, but in my opinion this is highly improbable.

6 *Another idea about Enlil's origin*

About the origin of Enlil and in respect of Enli's epithet **kur-gal**, Piesl [619] has tried to show that **kur** always has remained a pre- or non-anthropomorphous stone totem. The cult of this **kur**-totem would date from a pre-Sumerian period, and would finally have been overcome by the Enlil-theology. Hruška, on the other hand, disputed Piesl's thesis [620]. He wrote that

---

Real parenthood is expressed by the verb **(ù)-tud**. With respect to Namtar: as a deity he seems not to be attested in third-millennium texts (Klein 1998-2001, 143a).

614 For literature: see Katz 2003, 439-442. Mostly Ereškigal is named as the mother of Ninazu, whereas the father remains unknown. In 'a *balbale* to Ninazu' (van Dijk 1960, 57-80), Ninazu has two fathers: Nanna (l.8) and Enlil (l. 18), whereas his mother is indicated as **nin-maḫ** 'exalted lady' (l. 19). In the Temple Hymn no. 34, ll. 444-445, it appears that Enlil and Ninlil are the parents of Ninazu; but this fact does not associate both parents with the netherworld. Another example in which the father and mother of a god are not known as a couple is e.g. Enki, when in OB texts his father is called An and his mother Namma; An and Namma are not known as a pair in the Sumerian literature.

615 In this passage (ll. 91-116) Enlil takes the form of **lú-i$_7$-kur-ra** 'the man of the Ikura (*river of the netherworld*)'. The mere fact of this transformation seems to me not sufficient for an 'association of Enlil with the netherworld' (Katz 2003, 440).

616 For the text 'Enlil and Ninlil', see Behrens 1978; ETCSL 1.2.1. For a recent interpretation of this text, see Zgoll 2011.

617 Shaffer 1963, 99, ll. 12-13.

618 See the translation and the accompanying comment on these lines in 'Gilgameš, Enkidu and the Netherworld' in the Appendix Text editions no. 7.

619 Piesl 1969, 116: 'So blieb **kur**, das sich als präanthropomorphes Numen, als Steinheiligtum, als Steintotem zu erkennen gab, im Grunde genommen immer prä- bzw. an- anthropomorph. Die **kur**-Totem-Verehrung gehört einer Schicht an, die zeitlich vor das Eintreffen der Sumerer anzusetzen ist; sie ist präsumerisch. Diese **kur**-Verehrung wurde überlagert durch die **^{d}en.lil$_2$**-Theologie.'

620 Hruška 1971, 193-194: 'Das numinose **kur** ist selbst in den aufgeführten Texten nie einem Gott unterstellt, differenziert nicht die Götter nach ihrer Funktionen, es ist nicht das Ordnungsprinzip. Hierin unterscheidet sich **kur** von den anderen numinosen Begriffen wie **me**, **ĝiš-ḫur** oder **ĝarza**. **kur** ist also nicht für ein göttliches Wesen konstitutiv. Jan van Dijk [OLZ 1967, Sp. 232 f.] hat bewiesen, daß man in Sumer unter Numina immer nur konkrete individualisierte Sachen und keine platonischen 'logoi' verstehen muß. Deshalb ist schon die erste, für die besprochene

kur never had been equated with a divine principle in the texts cited by Piesl. Also the existence of a Sumerian totemism cannot be proven.

***

Arbeit aber entscheidende These, **kur** sei eine Gottesidee (ein Totem), falsch. (...) Die wichtige religionsgeschichtliche Frage, ob es in einer frühen Hochkultur wie sumerischen einen primären Totemismus und dann eine Entwicklung zum Anthropomorphismus gab, ist mit den uns vorhandenen Quellen zur mesopotamischen Religion nicht zu beantworten. Auch Helga Piesl konnte sie nicht befriedigend klären.'

Excursus 2

Enki [621]

1 *The primaeval Enki*

In texts and god lists we have met the name **^{d}en-ki**. Two different gods are represented by this name, the first of them is the primaeval god **^{d}en-ki**, whose partner is **^{d}nin-ki** [622]. In the text IAS 114, Enki and Ninki – Lord and Lady Earth – are the parents of Enlil. Nowhere has the origin of Enki and Ninki been mentioned. There are two possibilities: a. The pair Enki-Ninki was the first, primaeval, pair of gods created by **an** and **ki**; and b. Enki and Ninki split off from **ki** alone. The aspect of Enki-Ninki is clearly limited: both have only the aspect of **ki** 'earth'. This Enki is encountered in the context of the primaeval stage of cosmogony and theogony, and he will not be further discussed in this Excursus.

2 *Enki 'of Eridu'*

The second **^{d}en-ki**, by far the most known and famous one and who has to be differentiated from the primaeval Enki, is the city god of Eridu. There are several interpretations of this name. The difficulty for some of the interpreters is the apparent discrepancy – at least in their view – between the name of this god **^{d}en-ki** 'Lord of the Earth', and the area of his authority and his abode: the **abzu**, the subterranean water. Jacobsen tried to reconcile both contradictory areas – earth and water – [623]: the power in the water was a creative one, akin to the powers in the earth, the fertile soil. On the other hand, Sollberger asserted [624]: 'In the name of En-ki, god of the (underground, sweet) waters, **-ki** cannot be "earth" (...); in view of the frequent ending **-g (...)** and of the well-attested rôle of the god as man's friend, I assume a translation "Lord Love", parallel to En-lil "Lord Breath" (and, perhaps, En-sun "Lord Wisdom").' On several occasions, Lambert emphasized that **ki** 'earth' and **ki(g)** in Enki's name are different nouns, and that Enki's name is in fact Enki(g) [625]. Alster disagreed with Lambert [626]: 'W.G. Lambert's conviction that **^{d}en.ki(-g)** does not contain the word **ki** "earth" is not necessarily true. **^{d}en.ki-ak** may have developed **^{d}en.ki(-g)** by dissimilation.' If Lambert is right, and **ki** 'earth' and the **ki(g)** in Enki's name represent

---

621 Espak's thesis (2010) discusses among others the god Enki in Sumerian mythology. For the topics relevant for the present study, Espak utilizes existing translations of Sumerian texts. I disagree with some of Espak's views: 1. The primordial pair Enki-Ninki has brought forth Enki (of Eridu) [pp. 96, 236]; 2. Men grew from earth like plants [pp. 75, 116]; 3. The Anunna gods are probably the offspring of Enki and Ninki [p. 174]; 4. He supposes that the first humans as they appear in KAR 4 (named Uleĝara and Aneĝara by the present author; see also my comments at line *Rev.* 10 at the edition of this text) are deified [p. 195].

622 These names have the form of appositions, not that of a genitive construction.

623 Jacobsen 1977a, 146.

624 Sollberger 1966, 141, *sub* 393 **ki(g)**.

625 Lambert 1976, 432; Lambert 1981, 85; Lambert 1989, 116a. In this last publication – a review of Gordon *et al.* (eds) *Eblaitica* – Lambert is contending very strongly: '(...) Enki, the Sumerian god, is rendered "Lord of the Earth". This is doubtful, first because ancient Babylonian scholars seem never to use or imply such a meaning, despite their obsession with the etymology of divine names; secondly, because the **-ki** is not "earth", since it ends with an amissible **-g**. It is properly **-kig**, of unknown meaning.'

626 Alster 1982, 6 note 1.

different nouns, then it is peculiar – but of course not impossible – that this noun and writing of **ki(g)** is only found in Enki's name.

In the next lines I will try to make a reasonable case for **ki** in the name of **^{d}en-ki** being really **ki** and not **ki.g**.

a. Texts in the CDLI- and ETCSL-catalogue showed predominantly the spelling **^{d}en-ki-<u>k</u>Vowel**; the spelling **^{d}en-ki-gVowel** is found only in a few cases [627]. The present author is fully aware of the fact that part of the Sumerian literature is not included in CDLI and ETCSL; but in spite of this fact, – and this search was done only to get a rough idea – these results may be extrapolated, and they appear to be in favour of the origin of the name * **^{d}en-ki-ak** → **^{d}en-ki** for Enki of Eridu.
b. In the so-called UD-GAL-NUN text IAS 114, the name 'Enki' is written in several ways: **^{d}GAL-ki** (*i*:3') and **^{d}GAL-UNUG** (*i*:11') for the primaeval Enki [628]; **^{d}GAL-UNUG** (*i*:12') for Enki of Eridu. **ki** = **UNUG** for the primaeval Enki, where **ki** is 'earth'; there is no discussion about this meaning in the literature. Therefore we may assume that **UNUG** = **ki** = 'earth' also in the case of line *i*:12', for Enki of Eridu. But we have to make a reservation in this respect: some UGN-signs represent more than one 'normal', orthographic sign, e.g. **GAL**, **LAGAB**, **RU** [629].

In conclusion: in my opinion the name 'Enki' of the city god of Eridu has been developed from * **^{d}en-ki-ak** → **^{d}en-ki** 'Lord of **ki**'.

3 *The parents of Enki*

The few Sumerian texts that mention the parents of Enki do not agree in this respect. In IAS 114 (ll. *i*: 12'-13') it is said that 'Enki has been brought forth by the mighty An and the luxuriant Earth'. One of the epithets of Enki is **^{d}am-an-ki** 'Wild Bull of **an-ki**', referring to this origin [630]. In 'Enki and the World Order' (l. 68) Enki says about himself: 'I am the first- born of An'. In 'Enki and Ninmaḫ' it is Namma who is the mother of Enki (ll. 17-18; 24; 29-30). While in all cases Enki's father is An, there is ambiguity about his mother: **ki**/earth and Namma, respectively [631].

4 *The meaning of the name Enki*

In the next sections the meaning of **ki** in the name of Enki of Eridu will be studied. If indeed **ki** in Enki's name should mean 'earth', and consequently his name has to be translated as 'Lord of the earth', then it is understandable that some authors could not agree with this meaning, apparently so incompatible with Enki's main domain **abzu**, the subterranean waters. Therefore I shall examine the possibility of another meaning of **ki** in

---

627 CDLI: **^{d}en-ki-ga** (2x), **-ge** (1x), **-ke$_4$** (14x), **-ka** (239x); ETCSL: **-ga** (0x), **-ge** (0x), **-ka-** (4x), **-ke$_4$-** (183x). On a tablet with incantations from the OB-time (A 7479), the first incantation shows **^{d}en-ki-ga-ke$_4$** (*i*:1), but incantation 4 has **^{d}en-ki-ke$_4$** (*iii*:2') [lit.: Farber and Farber 2003, 101-102]. In **an** = *anum* I:27 we find: **^{d}namma** = **ama-^{d}en-ki-ga-ke$_4$**.
Searches in CDLI and ETCSL were performed in January 2010.

628 The name of Ninki in IAS 114 has always been written as **nin-ki**, not as **nin-UNUG.**

629 Krebernik 1998, 299-302.

630 As I have argued in ch. 2.2.3, note 256, **^{d}am-an-ki** is not an *emesal* name for Enki.

631 This point will be discussed further in chapter 4, *sub* theogony.

order to attain a better understanding of Enki's name. Several diverse aspects, all related to the god Enki of Eridu, will be examined first.

5 *The relation of Enki and* NUN

An old representation of the name Eridu is **NUN**^{ki} [632]. **NUN** in its original meaning is palm, but in later texts it has been used only in the metaphorical sense: a) prince [633]; b) foremost, best. Legrain recognized the sign **NUN** in some archaic seal impressions from Ur, and he associated this sign with Eridu [634]. Matthews published a reconstruction of the archaic city seal impression (Uruk III-period) from Jemdet Nasr [635]. According to Matthews, the sign combination **EN-NUN** on the seal represented Nippur. Steinkeller had doubts about the interpretation of the second sign as **NUN**, because Nippur in the contemporaneous Uruk city list has been spelled as **EN-E$_2$** [636]. Steinkeller does not go as far as supposing that the sign combination **EN.NUN** represents Eridu. On the other hand, Eridu, in all likelihood, was included in the archaic lexical city list as **I.NUN** [637]. In addition, it would be strange if Eridu, one of the oldest cities in South Mesopotamia and moreover the city of one of the supreme and most important Sumerian gods – Enki –, should not be represented on those old city seals.

Englund presented a list with the non-numerical signs of greatest frequency in the administrative text corpus of Uruk IV-III periods [638]. The fourth and fifth position are occupied by **AN** (An$^?$) and **NUN** (Enki$^?$), respectively [639]. Strikingly absent in this list is Enlil.

The tablet W21671, dating to the Uruk III-period, might contain evidence of a cultic calendar in the south of Mesopotamia [640]. The distribution of commodities is supposed to be associated with cultic festivals, including **GIBIL NUN** 'new growth (festival) of Enki' and **SU NUN** ' ... (festival) of Enki' [641].

---

632 Unger 1938, 465, § 3.

633 *niššīku* 'prince' (CAD N II, 282-283) is an Akkadian name and epithet of Ea/Enki.

634 Legrain 1936, 12 *sub* Palm; 14 *sub* City of Eridu and **nun**. Szarzyńska (1987-1988, 9) holds the view that **nun** in archaic Uruk belongs to a series of archaic reed symbols; she did not connect this sign with Eridu, nor did she identify it as a reed cult symbol of Enki. A tentative conclusion is: '(...) the hypothesis that all reed symbols represented deities (at least during the archaic period and in particular regions) seems possible', but the sign **NUN** preserved its 'primarily abstractive meaning' (pp. 10-11). Michalowski (1993b, 122 note 9) concluded, based on this publication of Szarzyńska, that the sign **NUN** was used to write the name of Eridu and that it was also a reed cult symbol of the deity Enki.

635 Matthews 1993, 37. This reconstruction has been republished by Englund (1998, 93), who also borrowed the conclusion of Matthews with respect to the interpretation of **EN.NUN** = Nippur.

636 Steinkeller 2002, 254-255, note 29; see this chapter, Excursus 1 'Enlil and Ninlil' for quotation.

637 Matthews 1993, 38, *sub* Cities 15.

638 Englund 1998, 70. Englund gives an account of the texts used (65, note 123). The fact that 86% (5000 of 5820) of these texts originates from the district Eanna of Uruk may explain the absence of Enlil in this list.

639 The question marks are Englund's. The sign **MUS$_{3a}$** 'Inanna$^?$ ' is 'quite low in this list!'

640 Englund 1998, 127 and fig. 44. W21671 has CDLI no. P004434.

641 W21671 *i*:8 and *iii*:5 **GIBIL NUN**; *iv*:5' **SU NUN**. Englund adds: 'all translations highly speculative'.

The Fara tablet SF 7 (VAT 12761) shows the sign **NUN** after **ᵈen-E₂** and **ᵈMUŠ₃**, each sign in duplicate [642]. According to the order of the gods at the beginning of the Fara god list SF 1 (VAT 12760), the sign **NUN** in SF 7 most likely represents Enki [643].

In the Abū Ṣalābīḫ **zà-mì** hymn dedicated to Enki (ll. 30-32), in which Enki is called **ᵈen-nu-te-mud**, it is said:

30	**abzu ki-kur-gal**	abzu, area of the great mountain,
31	**nun men an-ki**	palm, crown of heaven and earth, [644]
32	**ᵈen-nu-te-mud zà-mì**	lord Nudimmud be praised.

With respect to the oldest texts, it may be suggested that Eridu and Enki, because of the sign **NUN**, were related to palms [645].

The old name of Enki, **ᵈNUN**, became part of the name of his wife: **ᵈdam-gal-nun-na**, Damgalnuna. This name is attested from the ED period onwards [646]. Her Sumerian name has been preserved: even in the god list **an** = *anum* [647], first of all she is mentioned under this name.

6 *Enki and the reed*

Another plant mentioned in relation with Enki is reed:

* The text Ukg 4 *vi*: 15-16 relates: 6. **gi-ᵈen-ki-ka-ka** 7. **lú ù-de₆** 'when someone was brought in the reed of Enki' [648].
* In Guaba – far south-east of Lagaš – there is a cult place dedicated to Enki, which is named **ki-ᵍⁱˢgi-gíd** 'place of the long reed' [649].

7 *Enki and the ibex*

Except plants, there is also an animal connected with Enki; the homeland of this animal – the ibex – cannot be found in a marshland like South Mesopotamia. Wiggermann: 'From

---

642 VAT 12761: *obv. i*: 1-6.

643 The **diĝir** determinative is missing at the sign **NUN**, as it seems to be also the case at the second entry **en-E₂**.

644 Of course, due to the uncertainty of the order of the signs in the respective cases, line 31 may be read as **men nun an-ki** 'princely crown of heaven and earth'.

645 The spelling **NUNᵏⁱ** for Eridu could be attested from ED-times up to and including the Old Babylonian period.

646 **ᵈdam-gal-nun**: Biggs 1974, OIP 99, 47: l. 45 (A.Ṣ. **zà-mì** hymn); Steible 1982, FAOS 5/2, AnHaf 7, l. 3 [CDLI no. P222732 (Tutub)]; idem, AnUr 7, l. 3 [CDLI no.P222853 (ʿUbaid)]; Burrows 1935, UET 2, supp. 14, obv. *iv*: 7 [CDLI no. P250339 (Ur)]. If we accept that **⁽ᵈ⁾dam-gal(-la)** also represents Damgalnuna as a theophoric element in personal names (Edzard 1965, 50), then I can add for the ED period: 1. Jestin 1937: TSS 130 *rev. i'*:6' [CDLI P010759]; TSS 210 *obv. ii*:2 [CDLI P010778]; 2. Deimel 1924: WF67 *rev. v*:13 [CDLI P011024]; WF68 *rev. ii*:8 [CDLI P011025]; WF69 *rev. ii*:7 [CDLI P011026]; WF77 *obv. ii*:4 [CDLI P011034] (all from Fara). In later periods Damgalnuna appears frequently as a theophoric element.

647 Litke 1998, 88, II 173.

648 Ukg 4: Sollberger 1956, 50-53, Ukg 4-5 [CDLI no. P222607]. Of course, the association with water can also easily be made, because reeds grow in marshland.

649 Bauer 1998, 474, 510; text BIN 08, 371, *obv. i*:5 [CDLI no. P221502].

the fifth millennium B.C. onwards the ibex is prominently present in Mesopotamian iconography, and it is known that at least in the second half of the third it was associated with Enki/Ea, (...).' [650]. Wiggermann continued with the description of the ibex motif, both in South Mesopotamia and in Susa. The ibex motif was combined with a human form in the fourth millennium. The ibex-god is the master of snakes and sometimes he wears a seal of authority. To cite Wiggermann again: 'Thus, in our view, the Susanian ibex-god constitutes a special development, an adaptation of a preexisting god to the requirements of the growing state. It is interesting to note that the Mesopotamian god Enki/Ea, the later master of the ibex, has, under his local name Ḫaja, a similar function: he owns a seal and is married to the grain goddess Nisaba (...).'[651] About this ibex-motif Michalowski commented [652]: '(...) the ibex – **darah** – is, in historical periods, associated with Enki/Ea. The ibex-god is the master of snakes and actually wears a seal of authority. Many interpretations of this motif come to mind, but one should not discount the possibility that this is indeed Enki, or whatever he may have been called at the time.'

Koch has established that it is remarkable that the ibex often has been depicted together with motifs of water or plants [653]. Hole supposed that 'It is possible that the wild goat refers not to the hunt but rather to powers of procreation and life-giving fresh water.'[654] These aspects fit well with the image of Enki.

8 *Ḫaja: another name for Enki?*

As the name Ḫaja has been mentioned in connection with Enki, a closer search to establish their possible relationship is justified.

In 1983 Civil wrote the following [655]: 'Originally, the name [= Ḫaja, *JL*] may have been a variant spelling of **é-a** and therefore identical with Enki, but Ḫaia and Ea are already considered two different deities in the early OB period.' Charpin, commenting on a 'Hymn to Ḫaia for Rīm-Sîn' [656], discussed the position of Ḫaja and his close relations to Enki.

---

650 Wiggermann 1995, 87-88.

651 Wiggermann 1995, 90. With respect to the 'similar function', Wiggermann remarked (pp. 88-90) that a Susanian ibex-god was wearing a round object on his breast, with a strong resemblance to a fourth-millennium double-sided stamp seal; on that stamp seal was an ibex-god on one side and a temple front on the other one. Wiggermann suggested that this representation of the ibex-god, wearing such a stamp seal, was a kind of self-reference which explains the meaning of the image: 'the ibex-god wears a seal with his own image, which defines him as the top figure in the redistributive hierarchy. In fact this is exactly what we find somewhat later in Mesopotamia, where the god is the head of the state, the temple his residence, and the human priest-king his first servant, the one who actually receives and redistributes.' (quotation pp. 89-90).

652 Michalowski 1998, 244.

653 Koch 2000, 586.

654 Hole 1992, 36.

655 Civil 1983, 44a.

656 Charpin 1986, 352-353: 'Haya ne semble pas avoir jamais été une divinité poliade; (...). Les relations de Haya avec le monde d'Eridu et le dieu Enki sont certainement historiquement complexes.' Furthermore: 'On voit donc les pouvoirs considérables dont Haya est doté, et en même temps sa position subordinée à Enki, qui lui a précisément délégué ces pouvoirs. La même ambiguïté existe dans les rapports entre Haya et Enlil. Haya est l'époux de Nunbaršegunu-Nissaba; leur fille Ninlil ayant épousé Enlil. (...) Mais en même temps, Haya est

Charpin remarked that Ḫaya probably never has been a city god, and that the relations between on the one hand Ḫaya and on the other hand Eridu and Enki have been complex, historically seen. Charpin says that in spite of Ḫaya's considerable power, he was subordinate to Enki, from whom he has received those powers. In any case, in the above-mentioned hymn Ḫaja and Enki are different gods, but they are strongly connected. Charpin does not explicitly exclude the possibility that in an earlier period Ḫaja and Enki might have been identical.

9 *Enki in Akkadian texts*

A search for the name **ḫa-ia₃** [657] shows the presence of this name – predominantly as a theophoric element in personal names – in two texts from the ED IIIb-time [658], two of the Old Akkadian-period and one of the Lagaš II-period. Roberts gives a survey of divine names in the old Semitic personal names [659]. He could attest the name Ea/ ˁAy(y)a even in a pre-Sargonic name; he also discusses the possible etymology of this name. If the name ˁAy(y)a were derived from a root, then Roberts suggested that it may be from the root *ḥyy* 'to live'.

Green ends the discussion of Ḫaja as follows [660]: 'Perhaps the names Haja and Ea (ˁAja) represent two renderings of the name of an ancient (pre-Sumerian?) deity adopted into both the Sumerian and the Akkadian pantheons, Haja persisting as the name of a separate but closely related deity after the form Ea became assimilated to Enki.'

10 *Enki in Eblaitic texts*

In a bilingual lexical text from Ebla [661], the Sumerian **ᵈen-ki** is equated with the Semitic *ˁà-u₉*. As was pointed out by Biggs and by Westenholz [662], in the middle of the third millennium BCE there was an 'unbroken linguistic continuum, a cluster of closely related dialects' in the area between Ebla and North-Mesopotamia, including Abu-Ṣalābīḫ and Kiš. Lambert agrees with this opinion [663]: 'The orthographic system of Ebla (...) is part of a third millennium Semitic writing system very wide-spread in Mesopotamia that the route by which it reached Ebla cannot yet be ascertained.' Pomponio and Xella give a survey of god names attested in Ebla texts. Discussing the Eblaitic lexical lists they write that the equation **ᵈen-ki** = *ˁà-u₉* in these lists indicates a rather clear semitic etymology, viz. of the stem **ḥyy*,

---

au service d'Enlil (...), dans son temple de Nippur.' Charpin's edition of this hymn [ETCSL 2.6.9.2] is on pp. 344-351.

657 **ᵈḫa-NI** was present in two Old Akkadian texts; **ᵈḫa-ia₃** gave 597 results in 472 texts (CDLI; sept. 2012), most of them dating from the Ur III-period. See also Galter 1983, 134; Weeden 2009, 90-103.

658 In the text SF 77 (= VAT 9128) – a lexical list – **AN ḪA NI** (**ᵈḫa-ia₃** ?) is mentioned (*obv. iv*:15). See for a discussion of this text: Weeden 2009, 92.

659 Roberts 1972, 19-20.

660 Green 1975, 75 and note 4.

661 Mander 2008, 19; text VE no. 803.

662 Biggs 1981, 132-133; Westenholz 1988, 101 (quotation).

663 Lambert 1992b, 51. Lambert shows (pp. 60-62) the great similarity between the text of ARET V 6 and that of OIP 99, IAS 326.

'to live' [/*ḥayyu*(*m*)/] [664]. Both authors refer in this respect to e.g. Lambert, but he is more careful in his statement [665]:

> 'The declined form of Ea here seems to be unique. It can be argued that **É** should always be read *ˁà* at Ebla, but that ignores the fact that the orthography of names may have been fixed elsewhere. From southern Mesopotamia there are late Akkad period writings *è-a* (J.J.M. Roberts, The Earliest Semitic Pantheon 19-20) and Ur III personal names *i-ti-ne-a* (Oppenheim, Eames 188) and *na-ra-me-a* (ITT II/1, p. 13, 686), which, by our current transliteration system, do not support Ḥaya. Explicit evidence for Ay(y)a comes mostly from the west (Hittite and Hurrian) and is of second millennium date (...). The whole question involves the phonetics of the vowels in question and cannot be settled without a much longer inquiry than is appropriate here.'

11 *The spelling of Ḫaja and Nanše*

The Sumerian name Ḫaja is written as **ᵈḪA-ia₃**, with the sign **ḪA**; in its original archaic form this sign clearly represents a fish, and **ḪA** is also the logogram for the Sumerian word 'fish'. The name Nanše – and also her city Nigin – is written as **AB x ḪA**. Thus also in the spelling of her name the animal 'fish' is represented. It is known that from oldest times (ˁUbaid period) there were fish offerings, e.g. in Eridu and Lagaš [666]. It is not unlikely that the name Ḫaja originally – in prehistoric times – referred to a 'fish god'.

12 *Hypothesis about the history of the (city) god of Eridu*

The debate about the beginning of the presence of the Sumerians in Mesopotamia has not yet ended [667]. The present author is not in the position to solve 'the Sumerian problem', but nevertheless I would like to give my hypothesis about the history of the god of Eridu, in later times known as Enki.

At a certain point in the prehistory, the Sumerians started to inhabit the south of Mesopotamia [668]. This point cannot be defined at this moment, but it is not impossible that their appearance there was before, or just at the beginning of the ˁUbaid period. Excavations in Eridu [669] showed a series of 17 building levels, exhibiting reconstruction or restoration of successive temples, which demonstrated the continuity of architectural tradition without remarkable breaks. Fish offerings can be confirmed from the earliest times. Oates argued in favour of continuity without foreign 'invasion' at any moment [670]. The earliest settlers in South Mesopotamia were marsh dwellers: marsh-like strata with indications of settlement have been found at Ur and Uruk. South Mesopotamia in pre-Uruk times is

---

664 Pomponio and Xella 1997, 168. See also: Tonietti 2003.

665 Lambert 1984, 399 *ad* 803.

666 Oates 1960, 50; Roux 1969, 137. In my opinion, both authors prematurely call the god to whom offerings were made in Eridu during the ˁUbaid period 'Enki'.

667 The theme of the 48th Rencontre Assyriologique Internationale in 2002 was 'Ethnicity in Ancient Mesopotamia' (van Soldt 2002). For the 'Sumerian Question', see especially the contributions of van Driel, Rubio, Steiner, and Wilcke.

668 See ch. 1.1 for a short survey of the Mesopotamian history.

669 Green 1975, 16; Safar *et al.* 1981, especially chapter 3.

670 Oates 1960, 46-47.

known as marshland, rich in fish, birds and reeds. In Eridu the excavations stopped at a sand level, but below that similar evidence of marshes and possible occupational debris might be found [671]. Although all this 'continuity' does not prove the presence of the Sumerians at a period as early as the beginning of the ʿUbaid period (or even earlier), the conviction of Englund, on the other hand, that the Sumerians were entering South Mesopotamia 'burdened with plano-convex bricks' – that means: after the Late Uruk and Jemdet Nasr period – has been convincingly disproved by van Driel [672], who showed that the use of this kind of brick 'cannot be regarded as something specifically Sumerian.' To cite Oates [673]:

> 'It is tempting to see some confirmation of the role which may have been played by the marsh-dwelling communities in the formation of al ʿUbaid-Sumerian culture, in the traditional temple offerings at Eridu and Lagash. The finds at Eridu, and at a considerably later period at Lagash, show that in these places the people dedicated to Enki his portion of their goods, not in the form of grain or meat, the basic form of wealth among farming communities, but in fish, the product of river, lagoon, and marsh – a tradition which is hardly likely to have been derived from a population principally dependent on the success of their fields and flocks.'

One element in this statement I would like to change: the name of Enki. At that early time in history it is quite uncertain that the god who was worshipped in Eridu was named Enki. In any case: the city itself served as a ceremonial centre for the region Ur/Eridu during the ʿUbaid and the Uruk period [674]. If Steinkeller's conviction is right that the earliest Sumerian pantheon was dominated by female deities [675], this female domination of the metaphysical world may be extrapolated to earlier times [676].

It is not impossible that Nanše was the original goddess to whom in the marshlands of South Mesopotamia fish offerings were made. The connection of Nanše with fish is clearly demonstrated by the spelling of her name: **AB x Ḫ̮A**: a fish (**Ḫ̮A**) inscribed in a dais or shrine (**AB, èš**). Finally the male gods became dominant. Nanše's place in South Mesopotamia was taken over by the god **Ḫ̮A-ia$_3$**, whose name referred to his quality as a fish god. The pronunciation of this name cannot be given with certainty because several sound values are possible for **Ḫ̮A** [677]. One of the places in which **Ḫ̮A-ia$_3$** was worshipped was Eridu; 'The architectural development of the Eridu temple during this period (*the ʿUbaid period; JL*) suggests that the town served as a ceremonial center for the area' [678].

I assume that the name Eridu, being an epithet [679], was not the original name for this place; instead I propose that this name was **NUN^{ki}**. While in the Uruk period there was a gradual increase in population, in the number of settlements, in settlement size and

---

671 Oates 1960, 47-50; Green 1975, 15.
672 Van Driel 2000, 494-495.
673 Oates 1960, 50.
674 Green 1975, 17, 19.
675 Steinkeller 1999, 113. See the citation in this chapter, Excursus 1 'Enlil and Ninlil'.
676 Charvát 2002, 37-39, 46-48.
677 Borger (2010, 218, no. 856) lists the following sound values for **Ḫ̮A**: **a$_7$**, **ḫ̮a**, **kua**, **ku$_6$**.
678 Green 1975, 17.
679 The name written as **eri-dug$_3$^{ki}** literally means 'good city'.

eventually in region differentiation [680], Eridu became – as one of the oldest settlements in the South – the place with the most famous reputation and high prestige, especially as regards religion. This may have been reflected in the Sumerian Temple Hymns, where Eridu is mentioned first [681], and in the later so-called 'Sumerian King List' of the OB period, which tells that Eridu was the first place to which the kingship descended [682]. As a consequence of this increased prestige, the god **ḪA-ia$_3$**, being the most important god, became titulary god of **NUN^{ki}**, and his name changed into **^{d}NUN**. Eridu, belonging to the circle of Ur-Uruk, became part of a larger district, the importance of which was growing. The god of the oldest town in the South became more than only a 'god of the fish': he represented also other aspects of fauna (the ibex, although not an indigenous animal in South Mesopotamia [683]), and flora (palm, reed). This god was gradually considered master of the whole region, and he obtained the epithet **^{d}en-ki**, 'Divine Lord of the Region' [684], under which name he became known in later times. In addition to this it may be possible that the god **ḪA-ia$_3$** as a typical *fish god* in South Mesopotamia had lost more and more importance. The reason for that may have been, that: 'At about the end of the Ubaid period or during the Early Uruk, the southern Mesopotamian plain changed from a swamp- and flood-ridden zone into dry land (...).' [685]. Therefore the original name **ḪA-ia$_3$**, a typical name for a 'fish god', became less appropriate. The supreme god of southern Mesopotamia became increasingly known under his new name Enki [686].

ḪA-ia$_3$, as an originally important god in South Mesopotamia, must have been known also in the region of Nippur – Ereš – Kiš – Abu Ṣalābīḫ. In this region the Semitic culture, predominantly present in the northern part of Mesopotamia, had its influence. The Semitic-speaking people, hearing the Sumerian name **ḪA-ia$_3$**, presumably interpreted this name from Semitic point of view. Perhaps there existed a Semitic god with comparable characteristics, whose name was derived from the Semitic root **ḥyy*. This may have resulted in a *Semitic* rendering of Enki's name, viz. **^{d}E$_2$**-*a* / ^{d}*ʾà-a* (*ʿà-u$_9$* in Ebla).

---

680 These data, based on the results of the surveys of Adams, are summarized by Pollock 2001, especially in the section 'Settlement Differentiation' (pp. 187-194).

681 Sjöberg and Bergmann 1969, 17-18. The Temple Hymns start with Enki/Eridu, suggesting a higher position for Enki than for Enlil who is mentioned in TH no. 2. Therefore these hymns – although they most probably are of Old Akkadian origin (with later editing) – might belong to another and perhaps older tradition than the Abu Ṣalābīḫ ED **zà-mì** hymns in which Nippur and Enlil are mentioned first (southern versus northern tradition?).

682 Jacobsen 1939, 70, ll. 1-2 (The Sumerian Kinglist: ETCSL 2.1.1, ll. 1-2). An Ur III manuscript of the SKL, the original version of which perhaps dates as early as the Sargonic period (Steinkeller 2003, 268), starts with the descent of the kingship from heaven to Kiš. If indeed the original king list was conceived in the Sargonic period, then it is understandable why this list starts with the Semitic-oriented kingdom of Kiš. The later version, Sumerian-oriented, gives the primacy of kingship to the Sumerian city of Eridu; the kingship was in Kiš only 'after the flood had swept over'.

683 Not in South Mesopotamia, but in all likelihood in the Susiana and the neighbouring Zagros mountains; the influence from southern Mesopotamia reached in this early period as far as the Iranian plateau (Roaf 1996, 63-66), thus the Sumerians may have been acquainted with the fauna there.

684 **ki** has several meanings, e.g. 'place, ground, earth, land, country, underworld' (ePSD).

685 Nissen 2001, 171.

686 Steinkeller 1999, 113: 'And then there was one dominant male figure. That was Enki, a personification of male reproductive power, the god of fresh water and creative intelligence. Enki undoubtedly was the original head of the pantheon.'

Another explanation for the writing ^{d}**E**$_{2}$-*a* may be possible. Therefore I would like to refer to what Jagersma wrote about phonology, and especially about the Akkadian and Sumerian phonemes */h/, /ḥ/ and /ḫ/* [687]. The word **é** 'house' originally contained the consonant /h/. For this noun an older form */haj/* can be reconstructed. In Ebla and Old Akkadian orthography, **é** could be used as a sound sign with the values /ha/ or /ḥa/. These values must come from the phonemic form of **é**, i.e. */haj/* 'house'. A standard phonetic development in the Akkadian language was */ḥaja/* > */eja/* [688]. From these considerations we tentatively may conclude, that the original name of the god of Eridu – ^{d}**ḪA-ia**$_{3}$ – may have had a pronunciation like /haja/ or /ḥaja/, and that therefore the name of this god was written as ^{d}**E**$_{2}$-*a* in the Akkadian spelling.

At any time, ^{d}**ḪA-ia**$_{3}$ and Enki have become different gods. A god named Ḫaja was known at least in Ereš as the husband of Nisaba [689]. In the above-mentioned OB 'Hymn to Ḫaia for Rīm-Sîn' [690], Ḫaja and Enki appear as two different, but somehow related gods [691]. It may be that in the end it was not recognized anymore that ^{d}**ḪA-ia**$_{3}$ and Enki once had been one and the same god.

In conclusion, my hypothesis is:

1. The original male god of Eridu in prehistoric times is ^{d}**ḪA-ia**$_{3}$.
2. As the influence of ^{d}**ḪA-ia**$_{3}$ extended from Eridu to the whole of Sumer he obtained the epithet ^{d}**en-ki** 'Divine Lord of the Region'.
3. The epithet Enki eventually became the common name of the god of Eridu.
4. In the Kiš-region the name ^{d}**ḪA-ia**$_{3}$ was not clearly understood, at least not in the original meaning it had in South Mesopotamia. The function of ^{d}**ḪA-ia**$_{3}$/Enki may have been connected with the Semitic root **ḥyy*. This resulted in the 'Semitic' spelling ^{d}**E**$_{2}$-*a* (^{d} *ʾà-a*) = Ea for ^{d}**ḪA-ia**$_{3}$ / Enki. While '^{d}**ḪA-ia**$_{3}$ of Eridu' became 'Enki of Eridu', the Semitic way of writing the name of this god remained ^{d}**E**$_{2}$-*a*, because the god himself did not change.
5. The alternative for 4: ^{d}**ḪA-ia**$_{3}$ had a Sumerian pronunciation like /haja/; this resulted in an Akkadian spelling ^{d}**E**$_{2}$-*a*.
6. There was also a god with the name Ḫaja, written by Sumerian scribes as ^{d}**ḫa-ia**$_{3}$. This Ḫaja became the husband of Nisaba in Ereš, and as a consequence Ḫaja and Enki/Ea became different gods.

***

---

687 Jagersma 2010, ch. 3.4.3 and 3.4.4.

688 The */a/* changed into an */e/* under influence of an adjacent laryngal (von Soden 1995, 13-14).

689 A separate article on the god Ḫaja (^{d}**ḫa-ia**$_{3}$) did not appear in the Reallexikon. Nisaba was city god of Ereš (Sjöberg & Bergmann 1969, TH 42, 48-49). Ḫaja and Nisaba (under her name Nunbaršegunu) appeared as the parents of Sud in the myth 'Enlil and Sud' (Civil 1983, 50, ll. 3-4 [ETCSL 1.2.2, ll. 3-4]). Further: Edzard 1965, 115; Michalowski 1998-2001, § 5, 576 (both Edzard and Michalowski: without literature references).

690 Charpin 1986, 344-347; ETCSL 2.6.9.2.

691 For this Ḫaja Weeden supposed a Semitic origin, the name being derived from the same root **ḥyy* as has been generally accepted for Ea (Weeden 2009). Galter (1981, 136) was very doubtful about a 'Wurzelverwandschaft der beiden Götternamen Ea und Haja'. According to Galter, there is no proof of a Semitic origin of Ḫaja; on the contrary: he is a Sumerian god. Weeden's answer to Galter's opinion is: 'Notwithstanding a Semitic origin, Ḫaya could still have become a Sumerian god, with an established Sumerian writing: ^{d}**ḫa-ià**.' (Weeden 2009, 98).

Excursus 3

Enlil versus Enki

While An was at home in Uruk from time immemorial, Enlil was in all likelihood introduced into the Sumerian pantheon [692] – first of all in the northern part of Sumer, probably already before the Early Dynastic time. An was the supreme deity of the Sumerian pantheon. According to the texts that are discussed in this study, the primaeval unit **an-ki** was considered the ultimate origin of the universe, of the gods and of everything at that time, the ED period [693]. The result of the theological thought was, that on the one hand the initially inanimate **an** evolved into a cosmic god, the (supreme) god of heaven: An. On the other hand, the inanimate **ki** developed the chthonic gods **ᵈen-ki** and **ᵈnin-ki**. At that time there is no text known which states that Enlil was brought forth by **an-ki** [694]. There is only one text that mentions the ancestry of Enlil: in the AṢ text IAS 114 *i*: 9'-11', it is explicitly described that **ᵈen-ki** and **ᵈnin-ki** are the parents of Enlil. This implies the ancestry of Enlil from **ki**/earth only. In one other text, An is called 'the beloved father' of Enlil [695]. This text of Lugalzagesi, the last Sumerian king of Uruk before the rise of the Akkad kingdom, seems to be only a respectful invocation of Enlil, in which at the same time An – whose cult place was Uruk – was honoured. The expression used, **a-a-ki-ága-ni**, is not meant to render a filiation. As far as known so far, there are no other texts indicating Enlil's origin.

The god lists showed, in due time, a growing series of 'ancestors' of Enlil. These gods have to be considered not as real ancestors, but as 'developmental stages' or 'spheres of concern' of Enlil. The question remains of why from the beginning Enlil had been provided with an 'ancestor' list, in contrast to An. It could be that this ancestor list had to clarify the background and the sphere of influences of the newcomer in the Sumerian pantheon. In the ED time, Enlil's ancestors were represented by a butterfly (the beginning of life?), animals like ram and sheep (important for the life of gods and men), and abstract ideas like growth and power. Most items have a 'chthonic' character, but with a striking absence of floral items. In my opinion this remarkable preference for animal items may point to the origin of Enlil, i.e. Enlil was originally the supreme god of cattle breeders.

The dominant position of Enlil has been demonstrated in several ways. This can be observed in the AṢ **zà-mì**-hymns: Enlil has assigned cult places to the Anunna gods [696]. In these hymns the Sumerian god of heaven An is just one of the gods, appearing after Ninirigal, Inanna and Enki [697]. From the perspective of the Nippurian theologians, it is understandable that Enlil has a prominent position. The fact that An has been mentioned

---

692 See ch. 3: Excursus 1 'Enlil and Ninlil'.

693 See also ch. 4.

694 See ch. 3: Excursus 1 'Enlil and Ninlil'.

695 Lugalzagesi 1, *iii:* 13-18 (CDLI no. P263413): 13. **šu-tur** 14. **ᵈen-líl** 15. **lugal-kur-kur-ra-ke$_4$** 16. **an a-a-ki-ága-ni** 17. **nam-šita-ĝu$_{10}$** 18. **ḫ̮é-na-bé** ... 'If Enlil, Lord of all lands, for his beloved father An utters a prayer for me ....' On this text Wang based the genealogical link between An and Enlil, An being the father of Enlil (Wang 2011, § 3.2.4.1; 237-238).

696 Biggs 1974, 46, ll. 1-14.

697 I have no plausible explanation for the order of the gods in the **zà-mì**-hymns. Biggs (1974, 45) wrote about the order of the gods: 'Deities whose cult cities are near each other are often in proximity in the hymn as well, but since deities from Uruk/Kullaba occur in three widely separated parts of the hymn collection, it is clear that the principle of organization was not mainly geographical.'

apparently somewhere in the sequence of these hymns might be explained by the following hypothesis. Although in the genuine Sumerian pantheon – i.e. the pantheon without Enlil – An is the supreme god, perhaps the weight of his position has more the character of *primus inter pares*. The gods and goddesses of the Sumerian cult places, each belonging to a different clan in the pre-urban, sedentary Sumer, share a more or less equal importance, and the hierarchical relation between them may be less significant. At least in historical times, An does not seem to have had that dominant position which Enlil has, but yet he is always considered as the supreme god, in view of his position as first in the god lists. Presumably he owes this position to the fact that he had developed from the primaeval unit **an-ki** into the god of heaven. Another indication for Enlil's supreme position is presented by some AṢ-texts (ED period) mentioning that Enlil was responsible for the separation of heaven and earth [698]. These texts may have been influenced by the environment – the north of Sumer, in the neighbourhood of the homeland of Enlil – where they were written. In 'The Song of the Hoe' the active role of Enlil in this separation is described again. In 'Gilgameš, Enkidu and the netherworld' the separation itself of heaven and earth has been described in a passive way [699], after which An took heaven and Enlil took earth. This text is explicit in saying that the spheres of influence have been divided between An and Enlil. An, the god of heaven, rather far away from daily life; Enlil, who possesses the earth, i.e. who has the actual supreme power in daily life on earth. Before the introduction of Enlil into the Sumerian pantheon, the situation in Sumer may be described as follows.

Initially Sumer consisted of small communities, each with its own goddess or god. The gods of the respective clans were represented each by their own emblem [700]. Several factors caused a gradual change in the Sumerian landscape: the growth of the population, increasing trade activities, changing climatological circumstances, resulting in more cities and less small settlements [701]. The respective gods were at home in the Sumerian communal pantheon. In due time in the prehistory a certain hierarchy developed, in the course of which the god of Eridu became the principal god of the genuine Sumerian pantheon after An [702]. As a consequence, he finally gained the name Enki, 'Divine Lord of the Region'. Inherent in the primacy of Enki, in those days of a growing coherency between the various, initially separate, communities in Sumer, were his control and his authority over all the principles that governed in the broadest sense the welfare of the Sumerian people. These principles are the **ME**, a hardly translatable concept, but one which can perhaps be summarized as 'divine powers' [703].

---

698 For details: see ch. 2.1.1b and Appendix Text editions, 7. GEN, note at ll. 8-9.

699 Although the result of the ED texts and that of GEN is the same, viz. the separation of heaven and earth, the position of Enlil in GEN with respect to An has been 'neutralized'; probably it was not necessary anymore to emphasize Enlil's position in the Sumerian pantheon, as was the case in the ED period when his introduction had taken place only shortly before. That in 'The Song of the Hoe' it is again Enlil who separated heaven and earth may be not a surprise, because Enlil is the protagonist in the first part of this text (ll. 1-42), the most duplicates of which originate from Nippur.

700 Szarzyńska 1996. For Egyptian equivalents: O'Connor 2011.

701 Maisels 1999, ch. 5; Nissen 1988, ch.'s 3 and 4; Nissen 1999, ch.'s 4 and 5; this study ch. 1.1.

702 See this chapter, Excursus 2 'Enki'.

703 Farber 1987-1990. For a better understanding of this section, a short discussion of the concepts **me** and **nam-tar** will be given. Farber's definition (1987-1990, 610a) is: '[**me**] bezeichnet die göttliche Idee, die allem Existenten immanent innewohnt.' These concepts could also be understood as as attributes or insignia, sometimes they were even visible. Farber: 'Zwar hat

Enki's control over the **ME** has been expressed e.g. in the myth called 'Enki and the World Order'. Vanstiphout [704] analysed this text in several ways, each time changing the emphasis in the question: 'Why did Enki organize the world?' Answering why just Enki had to do it, he argued that this text is very much centred on water [705]: 'water is the prime regulating principle; therefore Enki is the great regulator.' Moreover, Enki's character, 'his never ceasing love and care for mankind', is relevant, for 'what Enki is actually doing here is laying down the conditions for the good life on earth, which is taken to be identical to the (idealized) Sumerian way of public life.' Vanstiphout also claimed a didactic aim for this text: '(...), it is about what an ideal world would be and why this is so.'[706] To his analyses I would like to add the following.

Presumably the text had originally been composed shortly after the Old Akkadian period [707]. 'Enki and the world order' describes the situation of an earlier period full of prosperity. The text is clear evidence of Enki's concern with Sumer. The episode with Inanna is, in my opinion, crucial for a better understanding of this text. Therefore I start with the answer of Enki (ll. 424 ff.) to Inanna's complaints (ll. 387-423) that she has no function unlike other goddesses. Enki asks her: 'How can I disparage you, how can I enhance you?' In other words: you have already everything you need or you have the right to. Then Enki enumerates functions and properties – some typically feminine ones – he has given to her (ll. 428-444). All these properties belong to a peaceful Inanna. The next lines (ll. 445-448) are related to a warlike Inanna: the Inanna of Akkad. But even then, she has not lost her feminine character (ll. 449-450). After the Akkad period there should be – in the Sumerian perception – a return to peace; this is expressed in the lines 451-452 [708]. What

---

jeder Gott, jede Stadt, jedes Ding seine eigenen **me**, insgesamt jedoch werden sie im Besitze des Gottes Enki in der Tiefe des Abzu gedacht.' Klein (1997b) has made plausible that the **me** listed in the myth 'Inanna and Enki' refer 'to some form of emblems, i.e. two-dimensional representations of symbols, painted or engraved on banners or standards.' He supports his hypothesis with passages in Gudea, Cylinder B cols. 6-7, especially the phrase '**DN en-^{d}nin-ĝír-su-ra me-ni-da mu-na-da-dib-e**' 'He (= Gudea) lets DN go over to/pass before Lord Ninĝirsu with his symbol/emblem' (quotation: p. 215). See also Berlejung (1998, 20-25) for a detailed discussion of the concept **me**. In my opinion Rosengarten's treatment and French translation of **me** as 'Prescriptions' is not justified (Rosengarten 1977). Also her conclusion (p. 115) that there is no remarkable difference between 'les destins (**nam-tar**)' and 'les Prescriptions (**me**)' is debatable. The concept **nam-tar** means: fate, destiny; this may be changed by the gods, while **me** is an immanent principle.

704 Vanstiphout 1997e.

705 Vanstiphout 1997e, 130.

706 Vanstiphout 1997e, 132.

707 Bottéro and Kramer (1993, 181-182) observed several indications that this text originally may have been composed in the Ur III period, e.g. the city Ur was treated like Sumer's capital city (ll. 210-218), the Martu nomads appeared to be cattle-breeders without any hostile trait (ll. 131-133; 248-249), Elam and Marḫaši seem to be enemies (ll. 242-247). To these arguments I would like to add lines 445-448, which in my opinion describe the warlike, aggressive Inanna of the Akkad rulers, whose domination Sumer recently had experienced.

708 EWO 445-452 (ETCSL 1.1.3); transcription of ETCSL.

^{d}inana saĝ saḫar-re-eš ḫé-mu-e-dub saĝ numun-e-eš ḫé-mu-e-ĝar	Inanna, albeit that you have heaped up human heads like piles of dust, that you have sowed heads like seed;

may be inferred from the broken final part of the text is, that Enki emphasizes the peaceful and feminine properties of Inanna. The concern of Enki for the welfare of Sumer, as can be deduced from this text, may be indicated as the **ME** which are and have always been in the possession of Enki. This concern was present from the very beginning of the formation of what may be called **ki-en-gi** or Sumer.

At some unspecified time in history, but certainly before the ED period, the introduction of Enlil must have taken place [709]. From the moment that there are texts available, the positions of Enlil and Enki are already distinct. How were the respective spheres of influence divided among both gods? In an article about 'Magic in History', van Binsbergen and Wiggermann argued that as a religious counterpart of political centralisation in the third millennium, there emerged a centralistic idiom with Enlil as protagonist [710]:

> 'Enlil, and less prominently the other gods, govern by **NAMTAR** (Akk. *šīmtu*), that is by "allocating tasks", "determining the fates or destinies" of gods, man and the universe. An earlier layer of centralisation in the South focuses on Enki and his city Eridu. It would seem, however, that Eridu was a religious centre, the touchstone of tradition (**ME**), rather than a political centre.'

Both authors emphasize the contrast between **nam-tar** and **ME**: **nam-tar** is associated with governmental decisions, mostly made by Enlil, while **ME** 'evokes an impersonal and timeless order, the non-volitional state of equilibrium to which the universe and its constituent parts are subjected. (...) The **ME** are not created, but (...) they are rules of tradition. (...) As an impersonal cosmological principle **ME** would appear to stem from a religious repertoire predating the third millennium.' Van Binsbergen and Wiggermann argue further, that the idea of **ME** fits 'the loose association of small-scale village societies

---

446	**^{d}inana níĝ nu-gul-ù ḫé-mu-e-gul níĝ nu-sig$_{10}$-ge$_5$ ḫé-mu-e-sig$_{10}$**	Inanna, albeit that you have destroyed what should not be destroyed, that you have cast what should not be cast;
447	**šèm a-nir-ra-da túg ḫé-em-mi-si-ig**	even if you have removed the cover of the **šem**-drum of lamentations,
448	**ki-sikil ^{d}inana tigi a-da-ab é-ba ḫé-em-mi-gi$_4$**	maiden Inanna, and if you have shut up the **tigi**- and **adab**-instruments in their houses:
449	**lú u$_6$ di igi nu-kúš-ù-me-en**	you never will get tired of admirers looking at you;
450	**ki-sikil ^{d}inana pú sù-ra éš lá nu-zu-me-en**	maiden Inanna, you will not know how to hang a rope in a deep well.
451	**ì-ne-éš šag$_4$ gú-bi nam-gi$_4$ kalam ki-bi ḫé-em-gi$_4$**	Now, the heart has overflowed: let the Land become restored;
452	**šag$_4$ ^{d}en-líl-lá gú-bi nam-gi$_4$ kalam ki-bi ḫé-em-gi$_4$**	Enlil's heart has overflowed: let the Land become restored.

Comments:

445-448 For the translation of the verbal forms with the modal prefix **ḫé**, see Civil 2000b, esp. § 3.3. The lines 445-448 all express negative acts of Inanna.

446 The verb **sig$_{10}$** does not heave the meaning 'to create' (translation ETCSL), but inter alia 'to cast' (ePSD).

[709] See this chapter, Excursus 1 'Enlil and Ninlil'.

[710] Van Binsbergen & Wiggermann 1999, 20-21.

largely organised by kinship, while the obviously more hegemonic divine government exemplified by **NAMTAR** fits their reorganisation into cities and later a nation.'

nam tar 'to decree fate' [711] is not the exclusive right of Enlil; sometimes An, Enki, the Anunna or even other gods are associated with it. But Enlil has been mentioned most frequently in connection with it in the OB texts, just as with the expression 'the god whose word is irrefutable' or 'who never changes his utterances' [712]. To the best of my knowledge there is no text in which it is explicitly said that Enlil possessed the **ME**, in the way Enki does.

In conclusion: In prehistoric times Enki had – at some time – become the most important Sumerian god for Sumer and its well-being, and therefore the owner of the **ME**. His concern was primarily the prosperity of his land. Enki's involvement in magic may be traced back to this quality. Enlil, presumably introduced into the Sumerian pantheon at a time when Enki had already reached his position, has another field of activity: he has taken possession of the earth after the separation of **an-ki**. Enlil governs the world by **nam-tar**.

***

711 **nam-tar** has been equated with the Akkadian *šīmtu* (CAD Š III, 11-12).
dub-nam(-tar)-meš *ṭuppi šīmāti* is the 'tablet of destinies' [for references, some of which are of the post-OB period: CAD Š III, 13, *sub* 1 b) 1']. The several texts in which this tablet occurs always show the power to rule, inherent in this tablet, for the god who is in the possession of it. In the Sumerian OB text 'Ninurta and the turtle' (Alster 1971-1972; ETCSL 1.6.3; ll. 1-4) the Anzu bird had successively taken away from Enki: the **me**, the **ĝiš-ḫur** * ('plan') and the **dub-nam-tar-ra**, but all returned to the Abzu, to Enki. In the OB and SB Akkadian Anzu-myth (Foster 2005, 555-578; Hallo and Moran 1979) it is Enlil who was deprived of the tablet of destinies. In *enūma eliš* I-157 the tablet was given to Qingu.
* For **ĝiš-ḫur**: see Farber-Flügge 1973, 181-191.

712 In 'Enki's Journey to Nibru' [ETCSL 1.1.4], l. 68 this expression concerns Enki.
It is said of Enlil in: 'A Praise poem of Šulgi' [ETCSL 2.4.2.05], l. 40; in 'Enlil and Ninlil', [ETCSL 1.2.1], l. 152; in 'The Lament for Urim' [ETCSL 2.2.2], l. 168; in 'The Song of the Hoe' [ETCSL 5.5.4], l. 2.

Chapter 4

The development of Sumerian ideas about Beginnings

'Aussi bien, et les bons historiens en ont toujours conscience,
l'Histoire, comme la vie, ne connaît-elle que des développements,
et les commencements comme tels lui échappent:
il y a toujours quelque chose avant ...'
[J. Bottéro, *Mésopotamie. L'Écriture, la Raison et les Dieux*; 1987, 57]

Two kinds of texts were studied in the previous chapters in order to gain insight into the Sumerian view of cosmogony, theogony and anthropogeny. On the one hand there are the narrative texts, written in different periods. The oldest texts, from the Early Dynastic period, are IAS 114 (ED IIIa), Ukg 15 (ED IIIb) , and the Barton cylinder (ED IIIb-early Sargonic) with a possible parallel IAS 174. From the Ur III period there are relatively few texts recovered: NBC 11108, and presumably the 'debates' ('Tree-Reed', and 'Grain-Sheep'). From the Old Babylonian period I have discussed: 'Gilgameš, Enkidu and the netherworld'; 'Enki and Ninmaḫ'; and 'The Song of the Hoe'. Finally I included the text KAR 4 from the Middle Babylonian period in my discussion. From the same epochs a second kind of text, several god lists, are available.

Both kinds of text – lists and narratives – contain information about the Beginnings, though sometimes this information is available only in an indirect way. In the previous chapters both kinds of text were studied separately, and each one provided some information, without telling us the whole story of these Beginnings. Especially the cosmogonic and theogonic parts, almost without exception included at the beginning of a narrative text, contain gaps. These introductions never were intended to tell 'the whole story' of e.g. the cosmogony. Each introduction is different and characteristic, dependent upon and geared to the topic of the narrative itself.

The nature of the god lists, being enumerations of names without explanatory text – with the exception of the list **an** = *anum* – and arranged in theological, genealogical, geographical or lexicographical orders or combinations thereof, might lead us to suppose that these lists cannot add much to the story of the Beginnings as deduced from the narrative texts. But with careful application of an extrapolation of the data from the latest list – **an** = *anum* –, it is possible to derive a more complete story for these Beginnings, especially of cosmogony and theogony.

In this chapter I will try to reconstruct this story in its diachronic development, on the basis of the texts from the periods being studied: the Early Dynastic, the Ur III and the Old Babylonian period, with a small extension to the Middle Babylonian period.

4.1 *Cosmogony*

4.1.1 *The Early Dynastic period*

For this period we have four texts and several god lists at our disposal [713]. The god lists of the ED period are not at all informative with respect to the cosmogony. There is not any indication of the appearance of the unit **an-ki**.

Some texts have a cosmogonic introduction, focusing the attention on the principal subject of the text. The text IAS 114 is written in the UD-GAL-NUN orthography, and has been only partly translated. At least part of the text seems to be related to extispicy [714]. In the case of Ukg 15 the text comprises only a cosmogonic introduction; it is not known to which composition this introduction belongs, if such a composition exists at all [715]. The text of the Barton cylinder is difficult to understand, due to the damage to the text. As far as I can understand part of it: the cosmogonic introduction is followed by the preparation of a huge image of a god, absence of food and drink due to a taboo, the appearance of Ninurta, and – with great lacunas – finally a restoration of prosperity. Perhaps the end of the introduction (*ii*:13-16), telling about the supply of water, alludes to the forthcoming lack of food and drink. The picture that arises from these texts is as follows.

In the beginning there was the unit **an-ki** [716], the inanimate [717] unit of **an**/heaven and **ki**/earth. The origin of **an-ki**, how this unit has been produced or has come into being, is not known, i.e. it has not been indicated in the texts: in the ED period it is the ultimate origin, without being produced by something else [718]. This unit is active in a certain way, or

---

713 Texts: IAS 114, Ukg 15, Barton cylinder, IAS 174; god lists from Fara and Abū Ṣalābīḫ.

714 This may be concluded from the following lines, with UGN signs in capital and normal orthography in brackets: *iii*: 2' **me-te-ŠE$_3$(= ni $^?$) maš šu NUN(= mu)-gíd** 'he himself examined the goat's entrails'. *Comment*: by courtesy of Krispijn, who made the suggestion for the equation of the UGN sign **ŠE$_3$** with **ni** in the normal orthography. See also line *ii*: 3': **^{UD}GAL-UNUG (= ^{d}en-ki) me-te-ŠE$_3$ (= ni $^?$)** 'Enki himself'. *iii*: 5' **maš šu NUN(=mu)-gíd**; *iii*: 15' **maš ME(=šu) NUN(=mu)-gíd**; and *v*:13' **maš šu mu-gíd** 'he examined the goat's entrails'. The line *i*: 2', especially the word **ĝiškim** 'sign, omen' already points to the extispicies in the columns *iii* and *v*.

715 It might have been part of a composition used at a ritual for the opening of a new irrigation canal, but I could not find any such example in the Sumerian literature.

716 Selz (1990, 128-129) supposed that offerings for **AN-KI** are attested in RTC 7, *iii*: 1-2 and in ITT 1,1081, *i*: 10. Bauer (1998, 518) convincingly demonstrated that **AN-KI** has to be interpreted as **ki-AN**, an offering or cult place. The word order in the Fara period (RTC 7) had not yet been fixed; in later texts (e.g. ITT 1,1081; OA period) we find **ki-AN**. Also Such-Gutiérrez (2005-2006, 44) described texts with offerings for – in his transcription – **ki-an**. The present author prefers to write **ki-AN**; perhaps the transcription should be **ki-diĝir** 'divine place'. In an OB lexical text (SLT 18, *obv. iii*: 7'-10') we find **ninda** intended for **ki-sì-ga**, **ki-an-na**, **ki-^{d}utu**, and **ki-^{d}adad**, respectively, which seems to prove that these places **ki-X** are cult places.
For **ki-sì-ga**, see Lambert (1987c), who wrote that the meaning of **ki-sì-ga** could be 'silent/quiet place' where funerary offerings (*kispu*) were made.

717 For the differentiation between the inanimate and animate class of **an** and **ki**: see Excursus 'The animate *vs* inanimate class of **an** and **ki**' in the Appendix Text editions.

718 Westenholz' formulation (2010, 297) that 'An and Ki had their origin in an embryonic universe' in an ED IIIb inscription [= Ukg 15 *ii*: 1-5] suggests that there is an ultimate origin before **an-ki**. In my opinion **an-ki** *is* the embryonic universe. Further, Westenholz' remark (2010, 295): 'Their inseparable unit, An-Ki (...)' seems to be less appropriate, because ED texts describe that Enlil has separated this unit (see ch. 2.1.1b).

to put it differently: there is activity inside this unit. This activity, not mentioned in the earliest text IAS 114, has been described as 'shouting', which I have interpreted as 'making love', 'having intercourse'. During this process – from the inanimate unit **an-ki** up to the cosmic 'marriage' of **an** and **ki** –, there is a development of **an**. The initial inanimate **an** becomes the animate god An, the god of heaven. An is described as a youthful man and he is entitled **en**. In several texts is related that **ki**/earth was provided with water: the gift of water is present in wells, according to the oldest text (IAS 114 *i*: 5'-8'); and in a slightly later text (Ukg 15, *i*: 2-5) we read that An lowers the inlets of irrigation channels, and water is present in the holes. In these texts it is An who provides **ki**/earth with water; this may be seen as a metaphor for his ejaculation during the cosmic intercourse or 'marriage' between **an** and **ki**, so that **ki** appears 'in luxuriance', in other words: **ki** is fertilized. As a result of An's ejaculation, there originates a water reservoir in the earth, maybe an allusion to the **abzu** [719].

The text of IAS 174 and that of the Barton cylinder seem to be parallel texts. Although both texts are broken at crucial points when earth is chatting with a snake, we may presume – extrapolating from the importance of water for the inhabitants of Mesopotamia – that in this episode earth expresses her gratitude and happiness because of the presence of water. The presence of a snake makes the situation more complicated than in the preceding texts, where a methaphoric explanation for water – as An's seed – might be possible. From IAS 174 *ii:* 4'-5' it becomes clear that a snake is situated inside Ningal, who represents here **ki**/earth, and whose name may be an epithet for the mother goddess. For a snake inside the earth there are two possibilities: the snake is present in some hole in the earth or in the water, that is also part of the earth. This latter possibility seems the more probable one. In line IAS 174 *ii*: 8' the sign **a** 'water' is present, in a context in which Ningal and the snake are talking. In the Barton text (*ii*: 13), the **muš-ĝír**-snake is even

---

19 Compare this with the episode in 'Enki and the World Order' in which Enki filled the Euphrates and the Tigris with his seed [ETCSL 1.1.3, ll. 250-259]:

ki-bi-ta igi-ni ĝar-ra-[ta] **a-a ^{d}en-ki ^{id}buranun-na** **nam-mi-in-/íl\-a-ta** **gud-du$_7$-du$_7$-gin$_7$ ù-na mu-un-na-gub**	After he had turned his gaze from there, after Father Enki had lifted his eyes across the Euphrates, he stood up full of lust like a rampant bull,
ĝìš im-zi-zi dùb im-nir-/re **^{id}idigna a zal-le im-ma-/an\-[si]** **šilam ú-numun-na amaš ĝiri$_2$-tab-ba** **amar-bi gù di-/dam**	he lifted his penis, ejaculated (and) filled the Tigris with flowing water. He was like a wild cow mooing for its young in the wild grass, its scorpion-infested cow-pen.
^{id}/idigna\ gud-du$_7$-gin$_7$ á-na **mu-na-/ab\-[...]** **ĝìš im-zìg níĝ-mussa nam-de$_6$**	The Tigris ...... at his side like a rampant bull. By lifting his penis, he brought a bridal gift.
^{id}idigna am gal-gin$_7$ šag$_4$ im-ḫúl **ù-tud-ba mu-/ni\-[...]** **a nam-de$_6$ a zal-le na-nam kurun$_2$-bi** **na-dùg-ge**	The Tigris rejoiced in its heart like a great wild bull, when it was born .... It brought water, flowing water indeed: its wine will be sweet.

addressed as 'Exalted Divine River' [720]. In the discussion of the texts IAS 174 and the 'Barton cylinder' the suggestion was made that the **muš-ĝír**-snake might be a metaphor for a (great) river, with diverging branches like a delta [721].

The positive appreciation of snakes, as is evident from these Sumerian texts, may originate from Elam. Von der Osten-Sacken summarizes the significance of snakes appearing in several pictures on stamp and cylinder seals [722]. The pictures on third millennium seals show a relation between the snake and (sweet) water. At the same time there is an Iranian tradition which connected snakes with fertility. Van Dijk *et al.* mentioned the well-attested cult of snakes in Elam [723]. In addition, snakes also had negative connotations, judging by the ED incantations against snakes [724].

There is no mention of the **kur** in the texts from the ED period concerning the Beginning. Regarding the terms for the netherworld, Katz noted that it is not known which terms besides **kur** were in use before the Ur III-period [725].

It may be concluded that in the Sumerian concept about 'the Beginning', the realms of the universe – heaven and earth – and perhaps **abzu** are present from the very beginning. That the **kur** was absent in the respective texts does not mean that there were no ideas about the netherworld, but its origin was apparently not involved in the Sumerian thoughts about the cosmogony [726].

In some ED texts written in the UD-GAL-NUN orthography the separation of **an** and **ki** was described [727]. There are no third-millennium texts known at this moment, which were written in the normal orthography and which describe this separation. In all these UGN texts it is the god Enlil who separates **an** and **ki**. As I have discussed before [728], the UGN orthography of these texts was predominantly in use in the northern part of Sumer. It might be that this orthography was confined to this area and/or was not (completely) understood in the rest of Sumer. As a consequence this part of the cosmogony, viz. the

720 It is tempting to suggest that this 'water' or 'the divine river', perhaps – at least partly – subterraneously, may be related to the **abzu**, present from primaeval times. On the occasion of his translation of two texts relating to Nanše, Alster (2005a, 17) wrote: 'Abzu might, indeed, have denoted the waters of the marsh areas, as they were available for fishing and traffic by boat, basically without any cosmological connotations, (...). So maybe the common understanding of Abzu primarily as a cosmic entity in Sumerian texts is to be modified. The marshes and their fish and fowl, after all, were fascinating enough in themselves without cosmic mythology.' The present author agrees with Alster that the **abzu** primarily indicated the waters of the marsh areas in the South of Sumer. Since the marsh area was of vital importance for the inhabitants of Sumer, precisely this characteristic of the **abzu** may have secured it a place – albeit in veiled terms – in Sumerian cosmogonic texts.

721 Chapter 2.1.3.

722 Von der Osten-Sacken 1992, ch. 3.1.4.2 (pp. 72-80).

723 Van Dijk *et al.* 1985, 8 and note 23.

724 Krebernik1984, 20-24: Beschwörung 3.

725 Katz 2003, 58 note 141: 'The name Ereškigal indicates that **kigal** was an old appellation for the netherworld, but apart from her name, **kigal** is not attested in texts that can be traced to third-millennium sources.'

726 In chapter 2.1.3 the drawing on the tablet VAT 12772 from the Fara-period was described. The sign **kur**, present in its centre, may represent the mountains to the north-east of Mesopotamia. It cannot be excluded that even at that time **kur** also may have been used for the netherworld.

727 See ch. 2.1.1b. These texts are: IAS 113 ii:5-10; IAS 136 iii:1'-3'; IAS 203 ii:3'-5'.

728 Chapter 2 of this study.

separation act of Enlil, may not have been generally known. A second ground for the initial non-dissemination of this act of Enlil may be Enlil's origin. As I have tried to demonstrate in the Excursus 'Enlil and Ninlil': the origin of Enlil may be a north-Sumerian one; he may originally have been a Nippurian supreme deity. Connected with this origin may have been an initially small extent of acceptation of Enlil as head of the Sumerian pantheon – in any case not in the whole country –, and as a consequence the absence of the mention of his act of the separation of heaven and earth in the third millennium texts written in normal orthography [729]. It could also be that the separation of **an-ki** by a divine action specifically belonged to the Semitic cosmogony of olden times, and that it was not known in the genuine Sumerian cosmogony of the South. These possibilities – the relative obscurity of UGN texts, Enlil's origin in an environment that was influenced by the Semitic culture, and an unknown separation act of **an-ki** in genuine Sumerian mythology – may have caused the restriction of the mythological aspect of the separation of **an** and **ki** by Enlil to the northern part of Sumer, the border region with the Semitic-speaking people.

4.1.2 *The Ur III period*

The god lists from the Ur III period, just like those from the Early Dynastic period, provide no information about the cosmogony.

Tablets with literary texts surviving from the Ur III period are very scarce. A few relevant examples have been treated in this study: NBC 11108, and two debate texts [730].

The text of tablet NBC 11108 [731] starts immediately with 'the mighty An', as a person. It is clear from this first line that, besides **ki**/earth, the **kur** exists [732] from the very beginning: it has the attention of An. The **kur** is not further named in this short text, thus the mere mention in line 1 leaves us in uncertainty about what was meant by it. The two realms of the universe – heaven and earth – are placed in opposite 'positions': light and darkness. Lightening may be seen as a metaphor for activity, darkening for inactivity, situations that will be repeated later in the text. The dark earth, without any activity, reflects also the absence of man: there are no human beings to perform the rites for the gods. The core of

---

729 In the previous chapter some indications, based on cultic evidence, for the growing influence of Enlil only from the Old Akkadian period onwards were presented (ch. 3, Excursus 1 'Enlil and Ninlil', *sub* 3. Hypothesis, *ad* Enlil as the head of the Sumerian pantheon).

730 According to van Dijk (1953, 36) nearly all the debate texts originate in the Ur III period. For the texts remaining from the Old Babylonian period it is generally accepted that these were conceived at an earlier time. But as will be demonstrated, the debate texts have some aspects in common, that make them differ from the main stream of Old Babylonian texts, and which allow us to place the origin of these texts earlier in time.

731 Tablet NBC 11108 has the form and size used for small incantations, but its content does not match these. As the text predominantly relates about non-existing entities, without a conclusive ending and without a larger framework, it seems to be part of a greater entity, maybe the introduction to a ritual text, as the absence of rites is expressively and extensively mentioned in this 'introduction'.

732 The first meaning of **kur** is mountain. From Ur III times on, the term **kur** has been used among others for netherworld, but the other meanings – mountain and foreign land – cannot be excluded here (Katz 2003, § 2.3, pp. 102-112). Recently, Artemov wrote some critical notes on Katz' study about the **kur** and the netherworld geography in Sumerian tradition (Artemov 2012). For a possible pre-Sargonic attestation of the meaning **kur** 'netherworld', see the text DP 141 *rev. iii*: 2 [CDLI no. P220791] (discussed in footnote 261).

the text of NBC 11108 is formed by the lines 5-9 [733]. These lines describe several situations with respect to **an** and **ki**:
1. **an** and **ki** lay together, but this had not (yet) led to a marriage (ll. 5-6);
2. The text says (line 7): 'the day did not dawn, the night had spread over her (= **ki**/earth)'. In the given situation this can have only one meaning, viz. **ki**/earth was not yet prepared for the celebration of the marriage with **an**.
3. The same is described once again, in slightly different words: **an** is already dressed in his wedding apparel, but 'he still does not lie down' (ll. 8-9) [734]. The reason for the delay of the marriage is also given in an indirect way. If I translate line 9 more freely – 'on a verdant place he still could not lie down' –, then is immediately clear that the situation which has been described for **ki** in the previous lines has not changed: **ki** is still not ready for the marriage, for there is not yet an **ú-šim** 'a verdant place', a metaphor for an embellished nuptial bed.

In 'The Debate between Tree and Reed' [735] the intercourse between **an** and **ki** is explicitly mentioned (in the ED period there were only allusions to it). The preparations of **ki** for her 'marriage' [736] with **an** are extensively described in this debate. She embellishes her exterior with precious stones [737]. Also used is the same predicate of earth - **ú-šim** 'fragrant vegetation' – as in the previous text (NBC 11108 l. 9). Sefati mentions several examples of similar embellishment motifs in texts dealing with the ritual marriage [738]: the use of precious stones, a flowered bed, a bed covered with halfa-grass with cedar scent. According to Sefati, such embellishments apparently belonged to a tradition connected with the marriage ritual [739].

The second debate text in this study is 'The Debate between Grain and Sheep' [740]. Here is no explicit mention of intercourse between **an** and **ki**. However, this may be more or less implicitly described in the first two lines: 'When Heaven the mountain range of **an-ki** had let engender the Anunna gods, (...)' [741]. My interpretation of these lines is as follows: **an**/heaven has had intercourse with **ki**/earth, after which **ḫur-saĝ** – the mountain range –

733 The lines 2-4 say that there is no culture at all present on earth (l. 2), and that no rites are performed for Enlil (l. 3) and An (l. 4); the absence of rites is repeated, for Enlil (l. 10) and Inanna (l. 11). All these 'absences' are reflected by the darkness of the earth.

734 The verb used in these lines - **gíd** 'to be long' - differs from the one used in l. 5: **lug** 'to live/dwell in its natural habitat', said e.g. of animals in the pasture. **gíd** 'to be long' has been translated as 'to lie down (at full length)'. In my opinion the verb **lug** expresses a calm and peaceful gathering of **an** and **ki**, whereas **gíd** in this context has the connotation of having intercourse.

735 See ch. 2.1.5.

736 When 'marriage' (with ' ' marks) is written, then this means that the text does not use an expression for marriage explicitly, but there is at least intercourse.

737 For the appreciation of stones in different cultures, see Eliade 1958, 216-217.

738 Sefati 1998, 98-101. A specific example is the text 'A *kunĝar* to Inana (Dumuzi-Inana T)' [ETCSL 4.08.20].

739 Sefati 1998, 99.

740 See ch. 2.1.6.

741 For a discussion of this translation, which deviates from the other ones in the literature, see the edition of this text in the Appendix Text editions, no. 6. The common point in all the other translations is, that the first line of this text – **ḫur-saĝ-an-ki-bi-da-ke$_4$** – is considered merely as a location: 'upon the hill of heaven and earth'.

gave birth to the Anunna. I have argued that the **ḫur-saĝ** may represent the mother goddess Ninḫursaĝa. So far there is no fundamental difference between this text and that of IAS 114, in which **an** and **ki** have brought forth at least Enki [742]. Also in the tradition of the third-millennium texts – with the exception of some UD-GAL-NUN texts [743] –, there is no mention at all of a separation between **an** and **ki**. In this respect this debate text fits completely in with the third-millennium tradition.

In summary, the main cosmogonic themes in these texts dating back to the Ur III period are:

- NBC 11108: preparation of the marriage between **an** and **ki**;
- 'Tree-Reed': preparation and celebration of the 'marriage' between **an** and **ki**;
- 'Sheep-Grain': intercourse between **an** and **ki**, resulting in the birth of the Anunna gods.

In not any text from the Ur III period is the separation of **an** and **ki** mentioned.

4.1.3 *The Old Babylonian period*

It is generally accepted that some Old Babylonian Sumerian texts originated from the Ur III period, or perhaps even earlier. Yet there are significant differences between the texts which I gave the label 'third-millennium' and those with the label 'second- millennium' [744].

The texts from the second millennium have a different starting-point in the introduction from those of the third millennium. Some third-millennium texts describe the preparation of **an** and **ki** for their forthcoming cosmic 'marriage', and also the interaction between **an** and **ki**, being their intercourse, has its place in some of these texts. Moreover, the separation of **an** and **ki** had already been described in ED texts, albeit in the special group texts written in the UD-GAL-NUN orthography. The introductions of second-millennium texts leave out completely the courtship and/or marriage of **an** and **ki**. Instead, all texts only report a separation of **an** and **ki** [745].

The text, designated as 'The Song of the Hoe', is the only one that indeed mentions Enlil as the god who separated **an** and **ki**. As is amply discussed in the literature, this text is a scholarly word play on the syllable /**al**/ 'hoe' [746]. Despite the uncertainties about the text, it preserves a mythological tradition [747]. In 'The Song of the Hoe', Enlil and his city Nippur

---

742 The text does not mention who produced Enki and Ninki, primaeval gods who are mentioned before the birth of Enlil and of Enki.

743 See the edition of GEN, note at the comment on lines 8-9, in the Appendix; the texts are IAS nos. 113, 136 and 203.

744 See e.g. table 1 in chapter 2.

745 Although the first lines of 'Enki and Ninmaḫ' have been partly damaged, we may safely conclude that this text also starts with the separation of **an** and **ki**, because of the ablative postposition **-ta** after **an-ki** in lines 1-2: '**an-ki-bi-ta**', a postposition that is used in case of a motion away from something.

746 For literature references: see in the Appendix Text editions, no. 9, the note at the comment on line 8 of this text.

747 Michalowski 2010, 199 warned: '(...) the philological virtuosity displayed in HH (Hymn of the Hoe; *JL*) is the primary motivation for the narrative and, therefore, one must be careful not to over-interpret the text and to ascribe to it profound mythological messages, as has often been done in past scholarship.'

play a central role [748]. Undoubtedly this is the (main) cause of the mention that Enlil was the god who separated **an** and **ki**, like an echo of the UGN texts of several hundred years ago. But the mythologem of Enlil as the separator of **an** and **ki** has not been generally accepted in the Sumerian mythology: at least this cannot be proven from the texts.

The introduction of 'Gilgameš, Enkidu and the netherworld' is silent about who really separated **an** and **ki**. It is just stated, in a passive way, that **an** and **ki** were separated. After that, the text makes clear that An has taken the heaven, and it underlines the chthonic character of Enlil: Enlil has taken the earth.

As far as now known, there is no OB narrative/mythological text that reports about the origin of **an-ki**. All texts start with the mere presence of this unit. On the other hand, in the Old Babylonian god list TCL XV 10, line *i*: 37, there is for the first time mention of a goddess called **ᵈama-tu-an-ki** '(divine) mother who has borne **an-ki**'.

The god list TCL XV 10 is an Old Babylonian list, that in certain aspects stands in a long tradition, from Early Dynastic times on. The principles according to which the god lists were composed are not the same for all these lists; they are arranged in a varying combination of theological, genealogical, geographical and lexical orders [749]. The order in TCL XV 10 is a theological one, and the list shows substantial extension compared with the previous god lists. One remarkable addition is **ᵈama-tu-an-ki**. In no prior text from the third millennium – the genuine Sumerian period – is there mention of a '(divine) mother who has borne **an-ki**'. My tentative conclusion is that this idea, this concept of a mother for **an-ki**, originates from the speculative, theological thoughts of the post-Sumerian clergy. This part of the Mesopotamian cosmogony, the origin of **an-ki**, does not belong to the genuine Sumerian cosmogonic ideas. We have to conclude that it belongs to the mental legacy of old of the Old Babylonian Semitic clergy. The later list **an** = *anum*, I: 28, identifies Namma, the primaeval waters [750], with this **ᵈama-ù-tu-an-ki**. In the list TCL XV 10 Namma is mentioned in line *i*: 36, just before the entry **ᵈama-tu-an-ki**.

How important had Namma been up till then? As far as is now known, there is only one report of a temple dedicated to Namma, viz. the temple built by Lugalkisalsi of Uruk in the ED III period [751]. Namma is present in several god lists and in the ED **zà-mì** hymns [752],

---

748 The text certainly had importance: 'The Song of the Hoe' belonged to the core of the school curriculum – the so-called Decad – in Nippur in the Old Babylonian period (Tinney 1999, 168-170).

749 See ch. 3.1. Lambert 1957-1971, 473: 'In form these lists (i.e. the god lists, JL) are just strings of names, but not lacking some order. There are two kinds of arrangement, and this applies to all the god lists from ancient Mesopotamia: (i) theological and (ii) lexical. Theological ordering rests on the conception of a pantheon organized like a tribe. The older members normally have precedence, coming first in the lists, while their offspring make up the lesser members and follow. (...) Lexical ordering follows the principles of arrangement (or lack of them) found in lists generally.'

750 Van Dijk (1964, 58) calls Namma 'la Terre-Mère' and 'la Mère du ciel et de la terre'. But Namma is no Mother-Earth, she represents water. Another detail is unclear: how is it possible that Namma is at the same time 'la Terre-Mère' and 'la Mère du ciel et de la terre' ?

751 Neumann 1981; George 1993, 167, no. 1360.

752 Biggs 1974, 50, lines 140-141: **šu:EŠ-gi še gu ti / ᵈnamma zà-mì**: '**šu:EŠ** reed, barley and flax to live, Namma be praised' (a tentative translation of the present author; the uncertainty of the translation is due to the unknown order of the cuneiform signs). The **šu:EŠ** reed may be some special kind of reed. As Biggs already observed (1974, 55 ad lines 140-141): 'The occurrence of **šu EŠ gi** in [IAS] no. 21 [*iv*: 1] and its duplicate SF 23 iii 14 suggests that these signs belong

but not in the Temple Hymns [753]. Perhaps the Ur III king Ur-Namma, by choosing Namma as the theophoric part of his name [754], drew attention to (or caused a revival of interest in) Namma. In the Old Babylonian period Namma was still venerated in Ur [755]. In some Sumerian god lists Namma was mentioned as '(divine) mother who has borne **an-ki**'. In the Semitic/Akkadian literature the goddess Namma was no longer important [756]. The possible Semitic origin of the idea: 'sea' as primaeval mother, will be discussed in ch. 4.7.1.

About the netherworld Lambert writes [757]: 'Obviously this realm had been set up at some point during the creation and organization of the universe, but the ancient texts so far known seem not to deal with it.' As we have seen, there may be hints in some texts to the **kur**, though indeed not to its creation. Initially there was also no creation of **an-ki**; only from the Old Babylonian god list TCL XV 10 do we know of Namma as the mother of **an-ki**.

4.2 *Theogony*

4.2.1 *The Early Dynastic period*

4.2.1.1 *The Enki-Ninki gods*

One of the oldest texts that is discussed in this study – IAS 114 – is an informative one with respect to the Sumerian theogony. When the text starts, after a break of an unknown number of lines, **an** and **ki** are still together, at least that is my interpretation of **[an⌉-ki niĝin$_2$** (*i*:1') '**an**/heaven and **ki**/earth together'. After **an** and **ki** have placed a sign – presumably a hint at the extispicy in the next columns –, Enki and Ninki appear, seemingly out of nothing, but we do not know what has been related in the broken part of the text. Their names, Lord and Lady Earth, strongly suggest that they developed from **ki** alone [758], but that is not necessarily so, because all the gods, except An, came into being by sexual reproduction. There is no text available at this moment to clarify this dilemma [759].

Shortly later than the IAS 114 text appeared the text Ukg 15. In this text there is no mention of a real creation or the origin of gods. While **an** and **ki** in their union are shouting,

---

together here as well.' Lines 140-141 may express the praise for Namma, because she facilitates the growth of reed, grain and flax, which are necessary for the sustenance of man. There is no indication of Namma as primaeval origin.

753 See also this study, ch. 3.3.2.

754 It is not known why Namma was chosen as the theophoric part of the name Ur-Namma. In the third-millennium texts there are only two examples of a name composed with Namma: **lú-^{d}namma** (CDLI P355945, BM 98148, *rev*. l. 3 under seal 1, an administrative text from Ĝirsu, Ur III period), and another (or the same?) **lú-^{d}namma** (CDLI P104534, AUAM 73.2212, l. 4) (origin unknown; Ur III period).

755 Renger 1967b, 157: a field belonging to Namma was mentioned in UET 5, 107:18, and a temple for Namma in UET 5, 274:4.

756 Stolz 1999, 738.

757 Lambert 1980, 58-59.

758 There are several ED texts in which the chthonic character of Enki and Ninki is alluded to, but these texts do not explain the origin of these gods; for a survey of these texts, see Cavigneaux and Krebernik 1998-2001, 446.

759 If Enlil was introduced into the pantheon of Nippur at the end of the 4th millennium BCE – perhaps as a kind of counterpart of Uruk's An [see ch. 3, Excursus 1: Enlil and Ninlil] – it seems very unlikely that the Nippurian clergy would have made An the father of Enlil.

the text relates that Enki and Ninki do not yet exist; thereafter it is likewise stated that Enlil and Ninlil do not exist. These lines suggest that after the cosmic 'marriage' of **an** and **ki** first Enki and Ninki would appear, and after them Enlil and Ninlil, the same order as was described in IAS 114.

The other ED texts are too damaged to be able to say anything about the primaeval pair Enki and Ninki, if they are mentioned at all.

According to IAS 114, this primaeval pair Enki and Ninki successively brought forth 'the seven' and Enlil.

4.2.1.2 *The 'seven' of Enki and Ninki*

In the god list **an** = *anum* [760] there is a group of gods who have been named 'the ancestors of Enlil'. In earlier god lists, members of this group – always varying in number – are mentioned together, and in relation to Enki-Ninki or Ninki alone. The Fara list SF 1 starts with An and Enlil; there is no Enki-Ninki group, but only three **nin**-gods are grouped around Ninki, and one of them is Ninlil [761]. The presence of Ninlil – written **^{d}nin-KID** – in this small 'ancestor' group may be confusing, because she is not known as an 'ancestor' of Enlil; the other two gods indeed belong to the 'ancestor' group. In the Abu Ṣalābīḫ list the restored first three lines report: An, Enlil and Ninlil [762], while Enlil and Ninlil are also listed in the Enki-Ninki group[763]. In the ED god lists SF 23-24 and the AṢ list however, the 'ancestor' group starts with Enki and Ninki, immediately followed by Enlil and Ninlil. After these two pairs another five (Fara) or six (AṢ) pairs belonging to this 'Enki-Ninki' group are listed. The presence of Ninlil in these 'ancestor' groups may be caused by systematism: all the 'ancestors' are mentioned as **en-nin** pairs.

In IAS 114 *i*: 3'-4' we read that 'Enki and Ninki have brought forth the seven'. These seven are not mentioned by name in this text. According to this narrative text, Enlil – and Enlil alone, not Ninlil – has been brought forth *after* the creation of the seven. At least Enlil has been created by Enki and Ninki, and thus belongs to their offspring, but obviously not to the 'group of seven'. Who are 'the seven', produced by Enki and Ninki in IAS 114? In a first reaction one could think of the seven demons, but these creatures appear to be produced by **an** and **ki** [764]. In the discussion of this text in chapter 2, it was argued that at

---

760 The god lists are included in an Addendum at this study.

761 Krebernik 1986, 174: in column *obv. vi*: 25-28. Different line numberings at Deimel 1923, 2 (*obv. vi*: 24-27) and at Mander 1986, 81 (*obv. vi*: 27'-30').

762 The appearance of Ninlil immediately after An and Enlil, while no other god is mentioned together with his/her respective partner, may be caused by the high position of Enlil in the Sumerian pantheon.

763 Mander 1986, 29: nos. 273-288 in the composite list.

764 Black and Green 1992, 162a. According to Lambert (1975b, 198): 'Demons were gods in Babylonian terminology.' Wiggermann (2000, 1866) wrote about these 'seven': 'Most misfortune was caused by evil. Evil belonged to the different and threatening world outside (...). Some, perhaps all, evil belonged to the earlier unorganized cosmos (before the appearance of Enlil) and was an untamed by-product of creation. Its demons had no cult, received no offerings, did not profit from man's pious activities, and preyed upon him without restraint. (...). The demons that attacked the people were summarized as the "Seven Evil Spirits"; (...).' Bottéro (2001, 63) mentioned that, although the divine determinative was affixed to their names, they were never inserted into god lists. The oldest attestation of an **udug-ḫul** in a Sumerian text was found in an incantation from Ebla (Krebernik 1984, 122-124; texts TM.75.G.1722 and TM.75.G.2459). This text gives no information about the origin of **udug-ḫul**. It is only in Old Babylonian Sumerian

least part of the text IAS 114 is dealing with extispicy. In my opinion it is therefore less probable that these seven are 'the seven **udug-ḫul**, evil demons', who can be found especially in exorcistic texts. Because of the ancestry of these seven from the pair Enki-Ninki, it is also very unlikely that they are the Anunna [765].

The hypothesis of the present author is that the 'seven' belong to the gods who later were called 'ancestors of Enlil'. As already stated, the ED god lists name maximally six **en**/**nin**-pairs, these prefixes being combined with: **giriš**$_x$ 'butterfly', **bulug**$_3$ 'growth', **utua** 'stud', **gukkal** 'fat-tailed sheep', **á** 'power', **an** 'heaven'. Thus some names concern animals; two are abstract concepts, that could be connected with the vital force and the development of the young earth; this development may also be symbolized by the butterfly. The pair that conclude the range of Enki-Ninki gods, ^{d}**en-an** and ^{d}**nin-an**, only appear in the AṢ god list. They are the counterpart of ^{d}**en-ki** and ^{d}**nin-ki**. It seems as if all these gods symbolically represent the development of the 'spheres of concern' of the chthonic god Enlil within the close unit of **an-ki**. They only seem to have the function of preparing the birth of Enlil. This preparation may be primarily conceived of as the development of **ki**/earth, starting with the primaeval chthonic pair Enki and Ninki, a development from a unit consisting of only inanimate material, like stones, to an inhabitable and agreeable place with living animals, as may be derived from some of their names. A striking characteristic in the names of these 'ancestors' is the absence of floral motifs. At the climax of this development of **ki** Enlil was born. That may also have been the moment when Enlil separated **an** and **ki**, as some AṢ texts mention; only then is the primaeval unit broken [766].

These 'seven ancestor' gods were not venerated in any way, e.g. there are no temples or offerings known for them. In the text IAS 114 these gods are not named; in my

---

literature that we find more details about this kind of demon. In 'A *šir-gida* to Ninisina' [ETCSL 4.22.1], l. 46, there is only a mention of **udug-ḫul**. Especially from the OB Sumerian exorcistic incantations UḪF (Geller 1985) the following can be inferred (in the cited lines the transcription is Geller's, the translation is by the present author).

a. *About the origin of the evil demons*

l. 247: **a an-né ri-a-meš dumu ki-in-du tu-da-meš**	they have been begotten by An, they are children borne by earth.
l. 359: ⌜**a du$_{10}$-ga an-na**⌝ ^{d}⌜**udug-ḫul a-ru**⌝**-a**	An's good seed that has produced the evil demon.

l. 367: this line is only a slight variation of line 247, with an identical translation.

l. 378: **ki-a** ⌜**dím**⌝**-bi a-an-na-ke$_4$** *	created on earth by the seed of An

(* Ms M has: <u>**ú**</u>**-an-na-ke$_4$**)

b. *Other information about the evil demons*

l. 390: [^{d}**en-ki-ke$_4$ še]š-gal-**⌜**a**⌝**-[meš]**	they are Enki's elder brothers.

c. *The relation between 'the seven' and demons*

There is – to the best of my knowledge – no third-millennium Sumerian text in which 'the seven' are associated with demons. The first mention of this is in Geller's UḪF ll. 401-403; 415; 421-424; 469 and 489. The conclusion may be, that these demons were created by **an** and **ki**, when both still formed a primaeval unit, and before the birth of Enki. That the creation of these demons took place *before* the appearance of Enlil – as Wiggermann (2000, 1866) says –, cannot be proven, at least the lines discussed in this study do not show this. For some general information about demons and ghosts, see Farber 2000, 1896-1898. It is not unlikely that the relation between the Mesopotamian gods and these demons is comparable with that between the Vedic *devas* and *asuras* (see e.g. Eliade 1958, 417).

765 See ch. 4.2.1.3.

766 See ch. 2.1.1b.

opinion this underlines the relative unimportance of these **en**/**nin**-gods. On the other hand, according to the god lists they do belong to the actual Sumerian pantheon.

If my explanation of the function of the 'seven' of Enki and Ninki is correct, there remains a question that has not yet been answered: why was this Enki-Ninki group necessary in preparation for Enlil's birth? One explanation may be that it had to do with the origin, the 'creation', of Enlil. Then the 'seven' may be considered as a kind of introduction of a new god into the Sumerian pantheon, to give an indication of his fields of concern. From the names of the 'seven', it may tentatively be concluded that Enlil originally was the supreme god of cattle breeders; at least in my opinion, it is conspicuous that no **en**/**nin**-pair was formed with e.g. grain or barley [767].

4.2.1.3 *The 'seven' of Ningal and Ninḫursaĝa*

While in the short text Ukg 15 'the seven' of Enki and Ninki are not mentioned, the Barton cylinder and IAS 174 (a possible parallel to the Barton cylinder text) report that someone – whose name presumably was written in the now broken parts – 'has poured the seed of a septuplet into her womb'. The expectant mothers are named as Ninḫursaĝa (Barton text) and Ningal (IAS 174), respectively. It is not likely that this Ningal is the spouse of Nanna, but rather one of the names for the mother goddess [768]. An explanation for this episode may be the following. Ninḫursaĝa and Ningal [769], in their function as the mother goddess, represent **ki**. The text relates about primaeval times; **an** and **ki** are present. Therefore it is nearly inevitable that in this case **an**, in the capacity as the god of heaven, is the partner of **ki**, who is named by a name of the mother goddess, in a cosmic 'marriage'. The outcome of this intercourse are 'the seven', a septuplet. In both texts the seven are not mentioned by name. Since the parents are **an** (presumably) and **ki** – presented as Ningal or Ninḫursaĝa, who evolved from **ki** analogously to An's evolution from **an** –, these seven must be different from the seven of Enki and Ninki. This last group of seven, I have argued, may be the 'ancestors' of Enlil. For a possible identification of the seven of **an-ki**, the following text citations may be helpful [770].

1. The AṢ **zà-mì** hymn ll. 39-40 read:

39	**ki-sar nun tu-tu**	Garden that brings forth princes,
40	**ama ^{d}nin-gal zà-mì**	mother Ningal be praised.

---

767 See the Appendix God Lists.

768 Krebernik 1993-1997 (Krebernik does not mention the name Ningal in § 3. Namen); Zgoll 1998-2001, 353, § 3.

769 Krebernik 1993-1997, § 6.2, 512: 'Auch als Gattin des Himmelgottes An dürfte die Muttergöttin die Erde repräsentieren.' The present author does not agree with Krebernik that 'Im Nippur-Zylinder (*i.e. the Barton cylinder; JL*) ist vielleicht eine kosmische Hochzeit zwischen Enlil und Ninḫursaĝa als Erde geschildert.' (Lit.: idem). As is substantiated in our text, it is assumed that An is the partner of **ki** = Ninḫursaĝa/Ningal.

770 Due to their lineage, these seven might be the seven **udug-ḫul** 'evil demons', but the context of the text of the Barton cylinder does not fit in with an exorcistic incantation. Thus the identification of 'the seven' in the Barton cylinder with evil demons is very unlikely (see also § 4.2.1.2).

The important elements are: 1. the begetting of **nun** 'princes' [771], and 2. the supposed equation of **ki-sar** and mother Ningal. A special form of **ki**, viz. **ki-sar**, which is the mother goddess Ningal, brought forth several **nun**, 'princes'.

2. In the text NBC 11108 we read:

12	**[diĝir-ga]l a-nu[n]-[ra a]d$^?$ [nu]-um-di-di**	For the great gods, the Anunna, one was not singing$^?$:
13	**diĝir-an-⌈na⌉ ⌈an⌉-[ki]-a / nu-ù-⌈ma⌉-su$_8$-⌈su$_8$⌉-ge-éš**	the gods of heaven still did not stand in heaven and on earth.

These lines – if correctly completed – mention the Anunna, the great gods, the gods of heaven who have a place in heaven and on earth.

3. The first lines of 'The Debate between Grain and Sheep' are:

1	**ḫur-saĝ-an-ki-bi-da-ke$_4$**	When Heaven $^{(l.\ 2)}$ had caused the mountain range of **an-ki**
2	**u$_4$ an-né diĝir-^{d}a-nun-na im-tu-dè-eš-a-ba**	to bring forth the Anunna gods

These lines say that the intercourse between **an** and a part of **ki**, viz. the mountain range of the primaeval unit **an-ki**, results in the production of the Anunna. These lines may also be interpreted as: the god of heaven An and Ninḫursaĝa – 'Lady of the mountain range' –, representing the mother goddess, bring forth the Anunna.

4. Line 82 of 'Lugalbanda and the Anzud bird' reads [772]:

^{d}a-nun-na diĝir-ḫur-saĝ-ĝá: the Anunna, gods of the mountain range.

5. 'A hymn to Nibru and Išme-Dagan' reads [773]:

diĝir ^{d}a-nun-na an-ki-ke$_4$-ne: the Anunna gods of heaven and earth.

6. In 'A *šir-namšub* to Nanna for Ur-Namma' we read [774]:

[^{d}a]-/nun\-na diĝir-an-ki-ka ad mi-ni-in-gi-gi-ne	the Anunna, the gods of heaven and earth, take counsel.

These six citations may lead to the following conclusion. There is a group of gods, the Anunna, whose father is An. The name of their mother varies: **ki**, **ḫursaĝ**, Ninḫursaĝa, Ningal, but all these names represent the mother goddess. This conclusion may be used to explain the character of the septuplet in IAS 174 and in the Barton cylinder. This septuplet is the result of the intercourse between **an** and Ningal/Ninḫursaĝa. With the aid of the above citations, ranging from the ED time up to and including the OB period, I tentatively

---

771 **nun** 'princes': I interpreted the reduplicated stem **tu-tu** as the indication for the plural form of the object **nun**. According to Thomsen is the verbal class of **tu.d** not known (Thomsen 1984, 320).

772 The line has been copied from the text ETCSL 1.8.2.2.

773 Lines 107-108 of the text ETCSL 2.5.4.23.

774 Line 17 of the text ETCSL 2.4.1.6.

conclude that these seven may be the Anunna. The name Anunna has been used since the second half of the third millennium BCE [775].

Who are the Anunna? The Anunna are the major gods of the Sumerian pantheon, but their number is not a fixed one. Some Sumerian sources mention An as their father. As a *group*, the Anunna are not mentioned in offering lists; they do not form the theophoric part of personal names; they have no temple [776].

4.2.1.4 *Enlil and Ninlil*

After the birth of 'the seven of Enki and Ninki' the universe was now prepared to receive other gods. First of all, Enki and Ninki created Enlil [777]. As I have tried to make plausible in the section 4.2.1.1, Enki and Ninki form a primaeval pair that emanated from **ki**, or at least the pair has only aspects of **ki**; in other words: it is a chthonic pair. This is in contrast to **an** with only heavenly qualities. Enlil is not the son of An. Enlil's ancestry is told in such a way as to underline his chthonic character, as opposed to An. The ED god lists are not uniform in the way in which they give Enlil's position in these lists. SF 1 lists Enlil immediately after An, and shows only a small group of only **nin**-gods around Ninki, among them Ninlil. SF 23-24 include Enlil and Ninlil in the primaeval group, immediately after

---

775 Edzard 1976-1980, 38; Falkenstein 1965a, 127-128.

776 Edzard 1976-1980; Katz 2003, 403. If my supposition that **imin** 'the seven' in the Barton cylinder represent the Anunna gods, then it may be that there is an Ur III-text in which these seven receive offerings; in Yildiz-Gomi PDT 2, 767 – a tablet with offerings to several gods – we read *obv. ii*: 4: **1 udu-niga ki-diĝir-imin** '1 fattened sheep for the place of the seven gods'.
In several administrative ED texts, the name **é-a-nun** – as a personal name – is present (e.g. Jestin TSS: 89 *rev. iii*:4; Deimel WF: 6 *rev. iii*:10), but it is uncertain whether **a-nun** in these names represents the Anunna gods. Theoretically **é-a-nun** as a personal name may mean 'Ea is a/the prince; Ea is the best', but this is not very likely, because the names of the individual gods (mostly) are provided with the **diĝir**-determinative in this period, while the Anunna are not.
In the AṢ god list there is one god associated with **imin**: **[^{d}luga]l-IMIN.GI$_4$** (Mander 1986, 30, no. 321). In the list **an** = *anum* this god appears as **^{d}lugal-imin-gi** (Litke 1998, 207: Tabl. VI: 95), as one of the names for Nergal.

777 On several occasions Wiggermann wrote that the **du$_6$-kù**, 'holy mound', created Enlil and the other gods according to third-millennium mythology (Wiggermann 1992, 286; Wiggermann 1996, 214). To the best of my knowledge there are no texts which describe the birth of gods by the **du$_6$-kù**. In the text IAS 114, discussed in ch. 2.1.1a, Enlil was born from the primaeval pair Enki-Ninki (ll. *i*: 9'-11'). The 'ancestor' group of Enlil expanded in due time to 21 pairs in **an** = *anum*. Indeed, in **an** = *anum* Lugaldukuga is mentioned as **a-a-^{d}en-líl-lá-ke$_4$** (Litke 1998, 37, I: 147), but the discussion about 'ancestors' and 'developmental stages' with respect to An and Enlil has made clear, that the god lists are not consistent in this respect. Moreover, the ED god lists from Fara and Abū Ṣalābīḫ, and the OB god list TCL XV 10 do not seem to be familiar with Lugaldukuga, so most likely he is a later invention. The **du$_6$-kù**, as the abode or residence of the Anunna (l. 27 of 'The Debate between Grain and Sheep'), has to be distinguished from the **ḫur-saĝ-an-ki-bi-da** (l. 1 of the same debate) which brought forth the Anunna (ch. 2.1.6 of this study; Black and Green 1998, 72). Wiggermann (1992, 286) wrote that 'Another cosmic entity that is brought forth when Heaven and Earth are separated is **ḫur-sag**, the foot hills.' To the best of my knowledge no text exists in which this mytheme is present. With regards to the line 1 of the text 'The Debate between Grain and Sheep' I argued that **ḫur-saĝ = ki**. The same equation can be deduced from the lines 12-14 of 'The Debate between Winter and Summer' (see: Translations of 'The Debate between Grain and Sheep', comments at the lines 1-2).

Enki and Ninki. The AṢ god list mentions Enlil and Ninlil twice: first immediately after An – at least according to the restored text –, and secondly in the Enki-Ninki group, after Enki and Ninki. But as we know, the order of the gods in these lists is ruled by theological and/or lexical arguments, and an order determined by theological considerations does not necessarily need to coincide with family birth order. The position of Enlil after Enki and Ninki only means that within this group of offspring Enlil is the most important god.

One salient point of difference between the god lists and the narrative texts is the fact that the lists, by mentioning Ninlil together with Enlil, at least suggest that Ninlil has also been created by Enki and Ninki [778], while the narrative texts preserve silence in this respect. In the Excursus 'Enlil-Ninlil' I have set up a hypothesis about the provenance of Enlil and Ninlil. With respect to Enlil, the tentative conclusion was that he was introduced into the northern Sumerian pantheon as a supreme god at the end of the 4th millennium BCE. But in fact, the Sumerian pantheon already had a supreme god at its disposal, viz. An. The Sumerian theologians/priests found an – in my opinion – elegant solution for this dilemma. The primaeval cosmological unit **an-ki** consisted of two parts, **an** and **ki**, joined together. The parts of this unit developed separately into a divine-heavenly section and a divine-chthonic one, respectively. From **an** originated, besides heaven, the supreme god of heaven, An. The section **ki** gave rise to the physical earth and the pair Enki and Ninki; this pair ultimately created the 'newcomer' Enlil who became the supreme chthonic god [779].

There is one other text from the ED period that mentions an ancestry of Enlil, viz. the text 'Lugalzagesi 1' [780]; in these lines An is mentioned as 'beloved father of Enlil'. But this attestation of An's paternity of Enlil must be treated carefully: it is the only such attestation, and moreover paternities of gods as found in texts may vary. E.g. in general it is Enki who is named as the father of Nanše, but according to a passage in 'Enki and the world order' Enlil is the father of Nanše [781]. Sometimes the designation 'father' or 'mother' does not indicate real parenthood, but is just a honorary title. Only when the verb (**ù**-)**tud** 'to give birth, to bring forth' has been used, can one be sure of the ancestry mentioned, as in IAS 114 where 'Enki and Ninki have brought forth Enlil'.

As for Ninlil, my hypothesis is that this goddess was syncretized from the original goddess Ninkid of Tummal [782] and from the city goddess Sud of Šuruppak [783]. At some time – probably before the ED period – she became the spouse of Enlil. As such she may be mentioned at the beginning of the Abu Ṣalābīḫ god list after An and Enlil (after restoration of the damaged text). Ninlil's ancestry has not been described in any known ED text [784]. The mention of the name Ninlil at different places in the ED god lists is rather confusing. In the Fara and Abu Ṣalābīḫ god lists, Ninlil is *inter alia* present in the group of Enlil's ancestor gods. In the Fara list SF 1, Ninlil is included in a small group of only **nin**-gods belonging to Enlil's ancestors. The presence of Ninlil together with Enlil in the primaeval

---

778 The tendency of lists to regularize may be the cause of it.

779 That earth is Enlil's domain may be demonstrated e.g. by the lines 11-12 of the AṢ **zà-mì** hymns (Biggs 1974, 46): **^{d}en-líl a-nun ki mu-ĝar-ĝar** 'Enlil has allocated pieces of land [= cult places; *JL*] to the Anunna'.

780 Steible & Behrens 1982, 319: Lugalzagesi 1, *iii*: 13-18.

781 Benito 1969, 110, ll. 417-420 (ETCSL 1.1.3, ll. 418-421).

782 During the third millennium the names of Enlil and Ninlil were written as **^{d}en-E$_2$** and **^{d}nin-KID**, respectively.

783 See ch. 3, Excursus 1 'Enlil and Ninlil'.

784 In later texts, Ninlil is the daughter of the grain goddess Nunbaršegunu/Nisaba/Nanibgal/Ašnan. Ninlil's father may be Ḫaja, husband of Nisaba (Krebernik 1998-2001b, 456, § 3.3).

Enki-Ninki group of the god list SF 23-24 and the AṢ god list may be caused by a certain systematism: the group of ancestors was composed of **en/nin**-pairs, and that may have been decisive for the incorporation of Ninlil, being the spouse of Enlil, together with Enlil in the Enki-Ninki group of gods.

4.2.1.5 *Enki*

Presumably Enki of Eridu is a genuine Sumerian god, perhaps even in oldest times the principal god [785]. At a certain point Enki lost his position in favour of the newcomer Enlil, who then became *de facto* the supreme god of the Sumerian pantheon. This is probably the situation at the moment that the text of IAS 114 was written. In my opinion, the lines *i*: 9'-13' reveal more than the simple mention of the fact that Enlil and Enki had been born:

1. The birth of Enlil was mentioned first, immediately followed by that of Enki. This order may be interpreted as an indication, that Enlil was more important than Enki in the divine hierarchy of that moment.
2. Eridu's Enki descended from the primaeval pair **an-ki** [786]. This implies that his father was An. To be born of **an** and **ki**, two different primaeval entities – one of which developed into the god of heaven, *the* supreme god of the Sumerian pantheon, though a *deus otiosus* – may give a god a higher status than a lineage from the pair Enki and Ninki. That pair may be considered as an emanation from **ki** alone. Their relatively little importance may be reflected by the absence of any known cult for these gods. Enki's ancestry may be a reflection of his high position in the oldest and genuine Sumerian pantheon.

4.2.1.6 *An*

The ED texts tell about **an-ki** and the preparation of the first and cosmic intercourse or 'marriage' between **an** and **ki**. As will be demonstrated in the Excursus 'The animate *vs* inanimate class of **an** and **ki**' [787], the class of **an** varied between inanimate and animate. The tentative conclusion was that, when **an** is preparing his 'marriage' with **ki**, but surely from the moment onwards that **an** has intercourse with **ki**, this **an** became animate. In other words: it seems that the god of heaven An was present from that moment onwards. Nothing has been said about the 'how' of this transition.

4.2.2 *The Ur III period*

4.2.2.1 *Enki and Ninki and 'the seven'; Enlil; Enki (of Eridu)*

The texts that have been dated in this study as 'Ur III' texts – NBC 11108 and two debate texts –, and also the Weidner god list, keep silence about the pair Enki-Ninki; there is no report either about the birth of 'the seven of Enki and Ninki', nor about Enlil's birth or about that of Eridu's Enki. Instead with respect to theogony a whole category of gods is mentioned: the Anunna gods.

---

785 See ch. 3, Excursus 2 'Enki'.

786 The epithet of Enki, **^{d}am-an-ki**, used in later literature, may refer to this ancestry of Enki; see also ch. 2.2.3, 3.2.4 and 3.2.5 with respect to this epithet.

787 See Appendix Text editions.

4.2.2.2 *The Anunna*

The possible identification of this group of gods with 'the seven of Ningal' or 'the seven of Ninhursaĝa' has been discussed in ch. 4.2.1.3.

In the text of NBC 11108 (ll. 12-13) we read: 'For the great gods, the Anunna, one was not singing$^{?}$; the gods of heaven still did not stand in heaven and on earth.' In other words: the Anunna gods do not stay or live anywhere, because they have not been born. Before these lines, the absence of any cult or rite for An, Enlil and Inanna was described. The cosmic 'marriage' between **an** and **ki** has not yet been celebrated. The suggestion from the text is, that the appearance of the Anunna is one of the main happenings after this 'marriage'.

The first two lines of 'The Debate between Grain and Sheep' I have translated as: 'When Heaven had caused the mountain range of **an-ki** to bring forth the Anunna gods, (...)'. In the rest of this debate text it does not become clear who these Anunna gods are, as they are not mentioned by name. In view of the expression 'the mountain range of **an-ki**', we may tentatively conclude that the action took place in primaeval times, before the unit **an-ki** was separated; **an** was considered inanimate [788]. In the present edition of this text [789], I have argued that **ḫur-saĝ** in these lines is the prefiguration of the mother goddess Ninḫursaĝa.

4.2.3 *The Old Babylonian period, and later*

Although the god list **an** = *anum* is of a later period than the Old Babylonian god list, this text has been included here because of the additional information it contains, so that the picture of Sumerian Beginnings can be described more clearly and more completely.

From the Ur III period onwards the information about the Beginnings is greatly reduced. Instead of a more extensive story about the Beginnings with respect to cosmogony and theogony, the introductions to the texts concentrate on other aspects, and focus on the main theme of the text [790].

All narrative/literary texts from this period pass over in silence the birth of the first gods, with one exception. In 'Enki and Ninmaḫ' we read the following: 'after the gods of heaven have been brought forth' (l. 4). Presumably 'the gods of heaven' refer to the Anunna. In l. 17 of this text we read: 'Namma, the primaeval mother, procreator of a multitude of gods'.

4.2.3.1 *Namma*

In the Old Babylonian Sumerian literature the goddess Namma appears in two prominent roles:

1. As the mother of **an-ki**; no male partner of Namma has been mentioned: Namma is now the primaeval beginning and origin of all (see § 4.1.3).
2. As the mother of some unnamed gods and of Enki:
   a. In the text 'Enki and Ninmaḫ' (l. 17) Namma is named the 'primaeval mother, procreator of a multitude of gods'. This multitude of gods, mentioned without any name, may be identical with 'the seven' – also unnamed – in the third-millennium texts of the

788 See the Excursus 'The animate *vs* inanimate class of **an** and **ki**' in the Appendix Text editions.

789 See Appendix Text editions, no. 6.

790 In this respect it has to be noted, that the main theme of the ED texts in this study could not always be determined.

Barton cylinder and IAS 174, where presumably An and **ki** (= Ninḫursaĝa/Ningal) created these seven. My tentative conclusion was that these gods may have been the Anunna, and the same conclusion may apply to 'the gods of heaven' and to this 'multitude of gods' [791]. Then it is also likely, that An was involved in the procreation of this multitude.
b. Namma as the mother of Enki appears *passim* in 'Enki and Ninmaḫ'. In this period the father of Enki is still An [792]. There is no text available in which An and Namma together are presented as 'the parents who brought forth Enki'. Namma as the mother of Enki is explicitly mentioned in the god list **an** = *anum*, before her equation with **^{d}ama-ù-tu-an-ki** 'the mother who bore **an-ki**'.

That Namma bore Enki may be explained as follows. Namma is the personification of the primaeval waters, and as such she fits the position of the mother of Enki – the god of the **abzu**, the subterranean waters – even more than **ki**. In the ED text IAS 114 *i*: 12'-13' was stated that Enki originated from the primaeval pair **an-ki**, the oldest couple. Possibly this is a recognition of Enki's position as former supreme Sumerian god (after An) before the 'arrival' of Enlil in the Sumerian pantheon. In the OB time, Namma is the primaeval mother, producing **an-ki**. In order to bestow Enki the oldest possible parents: his father is still An, but his mother has become Namma.

In summary: in the OB texts Namma has become the primaeval origin of the universe **an-ki**, and she has taken the position of **ki** in the creation of Enki and, presumably, of the Anunna [793]. The origin of Namma herself was not described in any extant text.

4.2.3.2 *The 'ancestor' groups*

With respect to the ancestor groups several remarks can be made. The first observation is that the Enki-Ninki group as ancestors of Enlil no longer occurs in cosmogonic introductions. However they do appear as **ama-a-a^{d}en-líl-lá** 'ancestors of Enlil' in 'The Death of Gilgameš' [794]. The size of this group is relatively small and it is incomplete in comparison with those of the god lists TCL XV 10 (16 pairs **en-** and **nin**-gods) and **an** = *anum* (21 pairs). In 'Šulgi and Ninlil's Boat' the 'ancestors of Enlil' are only mentioned as a group [795].

The second observation is, that there is an increase in the number of Enlil's ancestors in the OB god lists in comparison with the 'seven' of the ED period. And finally: there appears, for the first time, a rather modest number – only two – of ancestors of the god An. These phenomena are – in my opinion – not separate developments, but they are mutually related.

In ch. 4.2.1.2 I have tried to find an explanation for the appearance of the 'seven' of the Enki-Ninki group. Presumably the 'created' Nippurian god Enlil had become a supreme god in the Sumerian pantheon in the late Uruk period and so he may have needed an introduction; this resulted in the 'seven'. That kind of introduction was no longer necessary in the OB period: Enlil's position in the Sumerian pantheon was unchallenged. To find an

---

791 See ch. 4.2.1.3 and 4.2.2.2.

792 See e.g. 'Enki and the world order' l. 68 (Benito 1969, 89; ETCSL 1.1.3).

793 There exists to my knowledge no text in which An and Namma are a couple. For an hypothesis about how Namma could have been introduced in the position of primaeval origin: see ch. 4.7.1.

794 Kramer 1944, 10: version B, ll. 14-18; Cavigneaux and Al-Rawi 2000, 23, ll. 14-18; 'The Death of Gilgameš, Another version from Nibru', ll. 14-18, ETCSL 1.8.1.3.

795 Klein 1990, 106, l. 65; 'Šulgi and Ninlil's barge: a *tigi* (?) to Ninlil (Šulgi R)', l. 65, ETCSL 2.4.2.18.

explanation for the ancestor lists of Enlil and An in the OB period, the practice of making genealogical lists – like the Sumerian King List – may be helpful.

The Sumerian King List, a genealogy of Sumerian kings, might trace back to Sargonic times [796]. The intention of the post-Ur III list may have been to give the Sumerian kings firm grounds for their rule and power, despite the fact that the periods of office listed for some kings are unreal [797]. Another example to demonstrate the importance of lineage in the first half of the second millennium is the genealogy of Ḫammurapi with his long list of ancestors [798]. What is important is that there *existed* a tradition of making genealogical lists. As we have seen in the god lists, Enlil was the only god with a list of ancestors. God lists were very traditional, and although hardly any examples from the Ur III time remain, they appeared to have been continued in the OB period. In the time between the ED and the OB period, the idea of the 'seven' of Enki and Ninki may have developed into real ancestors by analogy with the development of the genealogical lists for worldly rulers. Once this idea had become fixed, the theologians/clergy must have concluded that the original ED group of 'seven' was too limited. Nearly all the names of the **en**/**nin**-gods of this group were made up of names of animals [799], which was not compatible with the supreme position of Enlil. Therefore the ancestor group as it appears in the OB god list TCL XV 10 has been extended from about 7 to 16 pairs. Although they are called ancestors, the names of these ancestors seem to indicate rather special fields of concern from which Enlil's power had evolved.

At the same time these theologians may have thought that it was less befitting to leave An without any ancestors. A modest start was made on the list TCL XV 10. The two names in this list that appear after An, viz. **an-šár-gal** and **^{d}en-uru-ul-la**, simply might be explained as epithets of An. But the list **an** = *anum* clarified these names: the two 'epithets' are now included in a more extensive group called **ama-a-a-an-na-ke$_4$-ne** – 'ancestors of An' – in this list.

Lambert pointed out a problem with respect to these ancestor lists [800]. All pairs, except the first one, seem to originate from a marriage between a brother and sister. This kind of marriage was a taboo in Mesopotamia. Lambert suggests a solution for this problem: 'die Namen in den Listen bezeichneten nicht verschiedene Wesen – entstanden aus einem früheren Paar und das nächste schaffend –, sondern galten für Entwicklungs-phasen einer Gottheit, (...).' The present author has the same opinion: the 'ancestor' groups of An and of Enlil are 'developmental stages'. In the case of An, these stages represent the development of the inanimate **an** in the primaeval unit **an-ki** to become the animate god of heaven An.

4.3 *Anthropogeny*

4.3.1 *The Early Dynastic period*

As far as I was able to study texts from the Early Dynastic period – some of them are written in the UD-GAL-NUN orthography and not yet translated –, I could not find any

796 Steinkeller 2003.
797 Michalowski 1983; Wilcke 1989; Haul 2004, 260-261.
798 Finkelstein 1966.
799 See the Appendix 'God lists' for the translation of their names: most of the names refer to animals.
800 Lambert 1974, 14-16 (quotation: p.15).

trace of human beginnings, not even a hint in the direction of the creation of any human being or the necessity for this.

4.3.2 *The Ur III period*

In the text of tablet NBC 11108, with its central theme: the preparation of An for his forthcoming marriage with **ki**, the absence of a special category of people has been mentioned, viz. the supreme purification priest of Enlil and the priestess of An were absent. Still nothing has been said about the creation of man.

In 'Ninurta's exploits' [801] we are told about a plan by the gods – who are not named or defined in more detail – (l. 339): 'there was talk of a community of men for the ploughing of the Land.' The reason for this plan may have been indicated in the preceding lines: the gods were assigned to a task in Sumer and they took charge of the hoe and basket; this indeed was their orderly duty (ll. 336-338). In other words: the gods had to work. This Ur III text expresses the suggestion for the creation of man to take over the work of the gods. Only the idea has been described, not the realization of it.

In 'The Debate between Grain and Sheep' the protagonists are, of course, Grain and Sheep. However, in the introduction we meet also a very primitive mankind (ll. 20-25): 'Mankind of those faraway days had not got to know eating bread, had not got to know a garment or something to dress in! The people moved on all fours; like sheep they ate grass with their mouths [802]; the water from the gardens they were drinking.' In this text there is nothing left to guesswork with respect to the developmental stage of humankind: they are behaving like animals, there are no references to a human nature [803]. The self-interest of the Anunna gods – 'they could not be satisfied in their eating and drinking' – lead or even forced them 'to inspirit humankind' (line 36). This means that people became civilized in such a way that the cults of the gods could be founded. This story cannot be considered as a genuine creation of mankind; therefore we have to consult some Old Babylonian texts.

An analogous idea seems to arise from the extent of the first column of 'The Sumerian Flood Story'. The text begins by telling that Enki would like to stop forsaking his creatures; the origins and the reason for this creation have been lost – if they were ever told. From the

---

801 Appendix Text editions, no 8.1.a.

802 The first line of the text 'How grain came to Sumer' is nearly identical to the line 24 of this debate (Bruschweiler 1987, 54-55; ETCSL 1.7.6): **uĝ$_3$-e udu-gin$_7$ ú ka-ba mu-ni-ib-«ni»-[gu$_7$]** 'men used to eat grass with their mouths like sheep'.

803 Another indication for the absence of civilization may be found in line 17 of this debate: 'Uttu had not been born. A cap was not worn.' Uttu represents here the goddess associated with weaving. Because of the absence of sheep and thus of wool, it was not possible to make woollen products. The present author agrees with Bauer (1982a, 378a) that this line does not only mean: because of the absence of Uttu and of wool, it was not possible to manufacture a (woollen) cap, the headgear of a king. The second part of this line alludes to the absence of kingship itself, and thus it points to an era that did not have a king and ultimately had no experience of civilization. Pettinato (1971, 35, note 118) has shown how the relation between kingship and civilization was expressed in Mesopotamian texts. From the viewpoint of settled, civilized people, living in towns – and scribes belong to them – nomadic groups, travelling with and living together with their herds, were at least less civilized. See also Michalowski (1983, 243-244) for the various constituting elements of the Mesopotamian society, among others the Amorite tribes.

lines *i*: 4'-10' one may conclude that the people roamed around (l. 4') [804], there are no cities and no temples (ll. 5'-6'); in short, people lived a nomadic life, and their religious life seems not to have satisfied the gods (ll. 7'-9'). The lines 11'-12' – 'After An, Enlil, Enki (and) Ninḫursaĝa had formed the black-headed people' – do *not* relate to the *creation* of the black-headed people. By analogy with line 36 of 'The Debate between Grain and Sheep' – 'they [the Anunna] have inspirited mankind' –, Enki's wish has been fulfilled: people have become (more) civilized and they now may be called 'black-headed people' [805].

4.3.3 *The Old Babylonian period, and later*

The first part of the cosmogonic introduction of 'Gilgameš, Enkidu and the netherworld' (ll. 1-10) mentions in a few words the arrival of humankind on earth (l. 10): 'the name of mankind has been established on it'. It is hardly possible to tell the appearance of human beings on earth in less words. Moreover, nothing is said about how or why, or by whom this came about; everything is veiled because the announcement is made in a passive way.

One of the most famous Sumerian stories about the creation of man is to be found in 'Enki and Ninmaḫ'. The gods complain of their life because of the work they have to do for their daily ration. On request of his mother Namma, Enki thinks of a solution. First he creates birth goddesses (**se$_{12}$-en-se$_{12}$-šár**). Next the clay on top of the **abzu** has to be thoroughly mixed, and eight birth goddesses – all identified by name – take some of this mixed clay and give it to Namma, who will 'mould it into form'. This piece of clay gets a human form, at least that is what may be understood from these lines, because the next event will be the birth of this creature, thought of by Enki, with Namma as its mother. That the clay foetus became a full-term baby in the womb of Namma can be inferred from l. 36, where we read that the birth goddesses assist Namma while she is giving birth [806].

Two main points arise from this text: 1. the first man was made from clay, and from nothing other than clay, originating from the top of the **abzu**, and 2. Namma is the mother of this first-born man [807].

---

804 Line 5': **ĝissu-bi ní ga-ba-ab-dúb-bu** 'I will calm down their shadow'. I have tentatively interpreted this sentence as: I – Enki – would like that these people get a place where they can settle permanently, so that their roaming, expressed as the unrest of their shadow, will stop.

805 Civilization, in terms of the Sumerian society, has been expressed as: people have become settled, have built cities and temples; they have irrigated the land and they have installed the cult for the gods (ll. *i*: 5'-10').

806 Line 36: **tu-tu-a-zu ḫa-ra-gub-bu-ne** 'they assist you while you are giving birth'. The verb '**tu(d)**' means: 'to give birth, to bear a child'. It is assumed that the verb here has its literal meaning. Only in Lagaš has the verb **tu(d)** been used also for the creation of statues (PSD A III *sub* **alam**, 160-170 *passim*), whereas the ususal verb for that creation is **dím**.

807 Frymer-Kensky (1987, 129-131) draws a parallel between the creation of man from clay on the one hand, and potting and sculpting on the other hand: the 'nipping off', moistening, mixing and casting of the clay. The creation of man in 'The Song of the Hoe', she assigns to a second tradition, viz. the sprouting of man from the earth like plants. However, the line concerned (l. 19) says, that Enlil 'placed the first of mankind in a brick-mould'. This can be interpreted as another image of processing clay, viz. making bricks. In Sumerian and Akkadian texts there is no tradition about the creation of man described as the emerging of plants (Frymer-Kensky 1987, 130); that this tradition did not exist in my opinion, is one of the topics of this study, see e.g. my translation and explanation of the lines 1-4 of 'Enki's Journey to Nibru' (*vide infra*). There is a difference between the use of a metaphor as a figurative expression, and the description of a process analogous to another one. The descriptions of the creation of man from clay,

The creation of man with the aid of clay and the subsequent birth of this first man by Namma was called by van Dijk *formatio* [808]. According to van Dijk, this creation story differs profoundly from that in 'Enki's Journey to Nibru'. His interpretation of the first lines of this text was, that man was created by An, and that man emerged from the earth after the earth had been fertilized by the heaven [809]. Van Dijk designated this kind of creation as *emersio*. Up to the present day the current view among Sumerologists has been that the Sumerian expression **ki dar**, used in 'Enki's Journey to Nibru' line 3, describes the creation of man by means of '*emersio*' [810]. However, a careful analysis of the relevant passage cast doubt on the existence of such a different creation tradition. To see whether this doubt is justified, the first lines of this text will be analysed and commented on [811]:

1	**u_4-ri-a nam ba-tar-ra-ba**	On that remote day, when the fate was determined,
2	**mu ḫ̮é-ĝál an-ù-tud-da**	the year that he (= *Enki; l.4*) brought about abundance,
3	**ùĝ-e ú-šim-gin_7 ki in-dar-ra-ba**	(and) that he (= *Enki*) has made it (= *abundance*) break through the earth for the people like green plants,
4	**en-abzu lugal ^{d}en-ki-ke_4** *(etc.)*	(then) the lord of the **abzu**, king Enki, *(etc.)*

that were used in Sumerian and Akkadian mythological texts, are comparisons with other similar processes in which things are made from clay. Just as bricks, bowls, figurines or images of gods are made of clay, so in the same way the prototype of mankind was formed. Frymer-Kensky gives examples of the use of metaphor when she says: 'The second Sumerian tradition about man's creation also draws on a fundamental metaphor of human existence, the parallel between man and plants.' (p. 130).

808 In chapter 1.2.1.5 van Dijk's theories about anthropogeny in Sumerian texts were discussed.

809 Van Dijk 1964, 23-24. The lines 1-3 of 'Enki's Journey to Nibru' are the most important ones on which he based the *emersio* theory. Two other texts are: 'The Song of the Hoe' (the lines 18-20) and 'Gilgameš, Enkidu and the netherworld' (line 10). His explanation of the lines 1-3 of 'Enki's Journey to Nibru' is based, *inter alia*, on a different reading in the first line: **a-ri-a** instead of **u_4-ri-a**. This **a-ri-a** should be a derivation from the verb **a-ri** 'fertilize', with An as subject, as is proved by the second line **an ù-tu-da**. Based on these suppositions van Dijk says: 'L'homme est né de la terre fécondée par le ciel comme l'étaient les plantes dans cette «année d'abondance»'. The translation of van Dijk was: 1. Lorsque le destin fut fixé de tout ce qui était engendré (par An), 2. qu'An eut engendré l'année d'abondance, 3. que les hommes brisèrent la surface de la terre comme les herbes.

810 Van Dijk (1964, 23) was the first scholar to use the expression '*emersio*', and after him several scholars (see ch. 1.2).

811 The transcription and the translation of the lines 1-4 in ETCSL 1.1.4 are as follows:

1. **ud re-a nam ba-tar-ra-ba**	In those remote days, when the fates were determined;
2. **mu ḫ̮é-ĝál an ù-tud-da**	in a year when An brought about abundance,
3. **ùĝ-e ú-šim-gin_7 ki in-dar-ra-ba**	and people broke through the earth like green plants –
4. **en abzu lugal ^{d}en-ki-ke_4**	then the lord of the *abzu*, King Enki, (*etc.*)

Al-Fouadi's translation of these lines is (1969, 77): 1. In those remote days, when destiny was determined, 2. In a year (full of) abundance, which An had created, 3. When people sprang up from the earth like herbs (and) plants, 4. (Then) the lord Enki, lord of the Abzu....

Comments

Line 2 Al-Fouadi considered **an** as the subject (*idem* in ETCSL), and accordingly **an** has been translated as An, the god of Heaven. **an** in this line as subject is very unlikely, because 1) **an** is missing the ergative ending ***e** which, if **an** is the subject, would be necessary; and 2) the position of **an** – if the subject – in this sentence is very unusual; as subject it should come after **mu** and before **ḫé-ĝál**.
I propose therefore another transcription of this line, viz. **mu ḫé-ĝál an-ù-tud-da**, in other words: **an** belongs to the prefix chain of the verbal part instead of being the subject of this line as is generally believed [812]. Echos of the fact that Enki is responsible for the abundance can be found in 'Enki's Journey to Nibru':
1) l. 49: **eridug^{ki} ^{d}en-ki-ke$_4$ ki-áĝ-ĝá-ni** l. 50: **é-engur-ra šà-bi ḫé-ĝál sù-ga**
49. 'Enki's beloved Eridu, 50. E-engura whose inside is full of abundance'; and 2) in the line 88: **^{d}en-ki u$_5$-a-ni mu ḫé-ĝál sù-ga** 'When Enki embarks, the year will be full of abundance.' [813]

Line 3 The ***e**-ending in **ùĝ-e** in line 3 does not indicate the ergative, as it has been interpreted in the literature, but is here a representation of the directive [814].
A nice parallel for the expression that 'abundance breaks through the earth like green plants' is found in 'A praise poem of Šulgi' [815]: **ḫé-ĝál ú-šim-gin$_7$ ki dar-ra** 'abundance, sprouted from the earth like green plants'.

Line 4 The subject of the verbs in ll. 2-3 is rendered in l. 4: **en-abzu lugal ^{d}en-ki-ke$_4$**.

***

812 Although **an** as a prefix to the verbal chain does not occur frequently, it does so in all epochs, except for the texts of Gudea. The combination of this prefix with the verb **ù-tud** seems to be a hapax (Heimpel 1974). For the function and meaning of the prefix /**a**/ (e.g. as an expression of a general truth), see Attinger 1993, 268, ch. 3.2.5.3 *sub* b) Fonctions. Recently Jagersma discussed the use of the ***a**-prefix in texts from northern and southern Sumer, from the Old Sumerian period up to and including the Ur III-period (2010, §§ 24.4 - 24.6, 535-549). In the examples given by Jagersma also no verbal forms with an ***a**-prefix from Lagaš are present. The copies of 'Enki's Journey to Nibru' originate from the Old Babylonian period and from various places, including Nippur and Ur. The original text may date from the Ur III-period. The use of the ***a**-prefix in **an-ù-tud-da** might be an example of a fossilized expression, originating from northern Sumer. Jagersma wrote (2010, 544) that the ***a**-prefix in northern Sumerian had an additional usage, which is completely absent from texts from southern Sumer: this prefix was also used as a passive marker. A passive marker of the ***a**-prefix in the present text seems unlikely; an analysis of the prefix **an-** as **a-n(i)-** and a translation such as 'in the year that abundance was brought about' is less probable because of the absence of a locative marker after **mu**.

813 An attestation of 'An who brings abundance' could not be found in the literature. More examples in this respect are available for Enki: 1. 'Enki and the world order' [ETCSL 1.1.3]: l. 41: **[^{d}en-ki] en-ḫé-ĝál-la** (...) 'Enki, Lord of abundance, (...)'; ll. 329-330: **^{d}en-ki-ke$_4$ (...) ùĝ-e ḫé-ĝál-la šu mu-un-di-ni-ib-peš-e** 'Enki (...) enhances the abundance for the people'. 2. 'Enmerkar and the lord of Aratta' [ETCSL 1.8.2.3]: l. 150: **^{d}en-ki en-ḫé-ĝál-la** (...) 'Enki, Lord of abundance, (...)'.

814 In the prefix-chain * **in** is no reference to this directive.

815 'A praise poem of Šulgi' (Šulgi P), ETCSL 2.4.2.16, segment C, l. 10. Another example of the figurative usage of this expression: A *balbale* to Nanna [ETCSL 4.13.04], ll. 1-2: **lu-lu-a-bi ú-šim ki dar-re-dam** 'they (*the cows with their calves; JL*) are as numerous as the grasses which break through the earth'.

The conclusion may be justified by saying that it was not 'people' but 'abundance' that broke through the earth like plants, and that this text of 'Enki's Journey to Nibru' does not represent an account of the way man was created.

Another creation story of humankind can be found in 'The Song of the Hoe'. Initially here no reason was given for the action of Enlil, who 'placed the first of mankind in a brick mould' (l. 19). The similarity with the story in 'Enki and Ninmaḫ' is that only clay was used to form a prototype of mankind. But instead of growing in a divine womb, the first man has to take his 'human form' from a brick mould. The metaphoric description of the 'gestation', i.e. the stay in the brick mould, where the brick mould is a metaphor for the womb, keeps going on: when the time for the birth of the first man came, he 'split open the earth towards Enlil' (l. 20). The first-born freed himself from this mould and from his clay envelope, being figurative language for a baby's birth from the womb [816]. The Anunna, apparently very content with this first-born 'black-headed' man, requested of Enlil a whole people of black-headed ones. To fulfil the Anunna's wish, 'Ninmena now establishes procreation'. From that moment onwards the normal human procreation took place. Finally, as if in passing, the reason for the creation of man is told: they have to 'provide for the daily rations for the gods' (l. 31).

Except for line 20 in 'The Song of the Hoe', the Sumerian literature shows no examples in which a human being 'split open the earth'. That unique example has been explained in the above as 'leaving the brick mould, breaking the clay envelope'. Therefore, in my opinion, the origin of man, anthropogeny, in the Sumerian texts has never been described as '*emersio*' as modern scholars have called it.

As a conclusion to the texts about anthropogeny, the text known as KAR 4 may be discussed. After a very short cosmogonic and theogonic introduction, the text describes how 'the designs in heaven and on earth have been fixed'. The gods are asking themselves: 'And now? What else can be done?' Then two of the Anunna gods propose a plan to create mankind: 'We shall slaughter the gods Alla and Illa, so that their blood makes mankind grow.' Here we observe a clean break with the Sumerian tradition as has been discussed up

---

816 The name Uzumua/Uzuea in 'The Song of the Hoe' (ll. 6 and 18) and the use of the expression **ki .. dar** (l. 20) may seem to give evidence for the creation of man via *emersio* 'sprouting from the earth (like plants)'. A literature search showed that e.g. the verb **mú** 'to grow' was not exclusively used with respect to plants growing up from the earth. Combinations of **mú** with fire (**izi**), combat (**du**$_4$), horn (**si**), beard (**sun**$_4$), house (**é**) were found. The expression **ki .. dar** literally means 'to split the earth'; mostly, but not exclusively, it is used in combination with plants. In one example we find a river seeking for its course ('Lugalbanda in the mountain cave' [ETCSL 1.8.2.1], l. 8: [^{íd}]⌜**buranuna**⌝**íd-**⌜**ḫé**⌝**-ĝál-la unug**^{ki}**-ga-**⌜**ke**$_4$⌝ **[ki]** ⌜**in**⌝**-dar-ra-a-ba** 'when the Euphrates, the river full of abundance of Uruk, had split the earth'. In 'The Death of Gilgameš' [ETCSL 1.8.1.3] segment H, l. 8 we read: **u**$_4$**-bi-a murub**$_4$**-**^{íd}⌜**buranun-na**⌝**-ka ki bí-in-dar** 'Then, in the middle of the Euphrates, the earth was broken (burst open)'. 'The Song of the Hoe' is generally considered as a playful text (references: Kinnier Wilson 1965, 273; Wilcke 1972-1975 [Hacke]; Farber 1997; Farber 1999a; Vanstiphout 2004, 82-85; Michalowski 2010). The basic mythemes in this text, like the creation of man with the aid of clay, are serious in my opinion. The name Uzumua (or Uzuea) and the expression **ki .. dar** are most likely used metaphorically (see also ch. 2.1.9); in my interpretation they have not the intention to indicate *emersio* – another way of man's creation than *formatio* – as some scholars believe (ch. 1.2).

to now. As could be shown in the above texts, in the Sumerian stories about the creation of man always clay and only clay was used, nothing else. As described in detail in ch. 2.1.10, 'the slaughter of a god is (...) a motif that is foreign to Sumerian mythology' [817]. The same motif is used in the Akkadian poems *atra-ḫasīs* [818] ['They slaughtered Wê-ila, who had intelligence, in their assembly. Together with his flesh and blood Nintu mixed clay'] and *enūma eliš* [819] ['From his (= *Qingu's*) blood he (*Ea*) created mankind on whom he imposed the service of the gods, and set the gods free'].

4.4 *Sumerian Beginnings: Conclusion*

Now that the various beginnings – cosmogony, theogony and anthropogeny – have been described and analysed for three separate epochs – ED, Ur III and the OB period –, I will formulate a tentative description of the diachronic development of the Sumerian thoughts on these beginnings, as expressed by the various texts, from the Early Dynastic period onwards, up to the end of the Old Babylonian period.

4.4.1 *Cosmogony*

From the Early Dynastic period onwards the cuneiform texts provided evidence of the Sumerian ideas concerning cosmogony. The presence of the unit **an-ki** 'heaven-earth' was taken as the starting-point: there were no questions about any possible beginning state *before* the appearance of **an-ki**, at least not noticeably in the texts. This changed only in the Old Babylonian period in a god list, but it was never expressed in Sumerian literary narrative texts. In that list Namma, a goddess already known earlier, was introduced as the primaeval waters, as the mother who has borne **an-ki**. This means that from now on water was conceived as the ultimate source. As will be discussed elsewhere in more detail [820], the function of Namma as the primaeval mother of **an-ki** may have been borrowed from the West-Semitic goddess Ti'amat, the sea [821].

In the third millennium much attention was paid to the unit **an-ki** and especially to the preparation of **an** and/or **ki** for their intercourse or cosmic 'marriage'. This aspect has been lost in the second millennium texts, which only mention the splitting of **an-ki**. The concept of this separation was mentioned as early as the ED period, but to my knowledge only in some UD-GAL-NUN texts. In these texts Enlil was the active god who separated heaven and earth [822]. This role of Enlil was repeated in one Old Babylonian text only, in 'The Song of the Hoe'; the other OB texts relate in a passive way that **an** and **ki** were

---

817 Steinkeller 1992, 247.

818 Lambert and Millard 1999, 58, tablet I: 223-226.

819 Lambert 2008, 52, tablet *vi*: 33-34.

820 See chapter ch. 4.7.1, and also ch. 3.5 *ad* 1.

821 Namma is present in an ED **zà-mì** hymn (Biggs 1974, 50, ll. 140-141); for the translation of these lines and comments, see ch. 4.1.3, footnote 753. This hymn describes her favourable influence on several plants. The goddess Namma there represents the underground waters, not the sea.

822 Therefore I disagree with the conclusion of Komoróczy (1973, 40) that '(...) according to the earliest Sumerian concept sky and earth divide into two parts automatically, without any external power, e.g. divine interference'. Wang (2011, 228) mentions the Abū Ṣalābīḫ **zà-mì** hymns as a source attesting the separation of heaven and earth by Enlil, but I could not find convincing evidence for that.

separated, and 'Gilgameš, Enkidu and the netherworld' adds to this that 'An took heaven and Enlil took earth'.

The cosmogonic introductions of ED texts mention also the presence of water – in holes or in wells, later also in irrigation channels – as a blessing for the earth. As these texts also deal with the intercourse of **an** and **ki**, it may be concluded that this 'water for the earth' represents the seed of An. Water was present from the very beginning, as part of the earth. The mention of it shows how important water was for the Mesopotamian people. In the cosmogonic introductions of second millennium texts the benevolence of water is not mentioned anymore [823].

The **kur**/netherworld seems also to be present from the very beginning. Sometimes, as in NBC 11108 l. 1 – 'An is looking to the **kur**' –, there is only an allusion to this realm, but a creation story for the **kur**/netherworld is not (yet) known [824].

In summary there are some main points of difference between third-millennium introductions and those of the second millennium with respect to the unit **an-ki**: 1. third-millennium texts amply describe the preparation of the two protagonists **an** and **ki** for their cosmic 'marriage', but not what happened with **an-ki** afterwards, except in some UGN-texts; and 2. second-millennium texts omit the intercourse or marriage of **an** and **ki** and instead they begin with their separation. What may be the reason for the omission of the cosmic 'marriage' between **an** and **ki** in the second-millennium texts? One tentative hypothesis is that the older literary texts did not exist anymore. Michalowski suggested that, at the founding of the **eduba** in Ur, Ur-Namma and his son Šulgi 'threw out most of the older literary texts, which were almost exclusively mythological (...)' [825]. Other possibilities are: the long introductions of the third-millennium texts and the content of the new texts did not match, or the cosmic marriage of **an** and **ki** was a story too well-known to be told again.

The final result of the cosmogony is that the most important realms of the universe are divided: An resides in heaven, Enlil is master of the earth, Ereškigal is queen of the netherworld.

4.4.2 *Theogony*

Without a primaeval **an-ki** there is no theogony in Sumerian religious thought, at least until the Old Babylonian period. The theogony has never been described as comprehensively as in the Early Dynastic text IAS 114. In the third millennium, when the texts only mention an inanimate unit **an-ki** in the beginning, the first god who evolved within this unit was the god of heaven, An. He was present in all his magnificence and glory at the moment of his 'marriage' with **ki**/earth. Within the undivided unit **an-ki** there developed additionally the primaeval divine pair Enki-Ninki 'Lord Earth and Lady Earth', presumably from **ki** alone, since they have only chthonic characteristics [826]. This pair Enki-Ninki brought forth seven gods. Finally Enlil was born of his parents Enki and Ninki.

As the god lists from that time have shown, the number of seven descendants, created by Enki and Ninki as mentioned in the IAS 114 text, was not fixed. These 'seven'

---

823 One reason therefore may be that in due course the irrigation systems for the necessary water supply had much improved, so that better water management was possible.

824 For the possibility of Ninurta creating the netherworld: see ch. 2.3.1.1, footnote 261.

825 Michalowski 2007, 176.

826 See ch. 4.2.1.1.

appeared to represent several aspects of earth and earthly life, sometimes also abstract ones, as can be derived from their names [827]. During the third millennium the identity of 'the seven' of Enki and Ninki seems to have changed. Initially they could be interpreted as brothers and sisters of Enlil, because they all had the same parents. In literary texts 'the seven' are seldom present any longer, but what is more important is that in the god lists of the OB period and later they are listed among the ancestors of Enlil, as they were named in these lists. These 'ancestors' I have called 'developmental stages'. From now on, parallel to the ancestors of Enlil, there also appeared 'ancestors' of An, who had never been mentioned in the narrative texts before. This aspect cannot be considered as a genuine motif of the Sumerian theogony.

The oldest text also tells us that **an** and **ki** have brought forth Enki, the Enki of Eridu. As is discussed elsewhere in this study [828], the mother goddess Ninḫursaĝa developed from **ki**, analogously to An's evolution from **an**. An alternative way to describe Enki's lineage – using the divine names – may be that his parents are An and Ninḫursaĝa. At that moment in the theogony the highest gods of the Sumerian pantheon were present: An, Enlil, Enki and Ninḫursaĝa.

That the unit **an-ki** was necessary for the creation of the first gods appears to be the case, not only from the ED texts (Barton cylinder, IAS 174), but also from the Ur III/OB text 'The Debate between Grain and Sheep', where **an**/An and **ḫursaĝ**/Ninḫursaĝa create the Anunna, the highest gods in heaven. Perhaps more gods were created in this primaeval phase, but the texts do not tell these stories. Once the most important gods are present, viz. An, Enlil, Enki, Ninḫursaĝa, and Namma (from OB times on), the other gods can originate from these first ones, and now it is time for the separation of **an** and **ki**. For this act some texts held Enlil responsible, as we saw in the preceding section 4.4.1.

If there are statements in OB texts about the creation of the first gods, these are very condensed. In principle there are no changes with respect to the story as told above, with one exception. Now in the OB period Namma had become the primaeval element who bore **an-ki**, there also appeared texts where Namma was the mother of Enki (of Eridu), instead of **ki** as stated in the ED text IAS 114. Why Namma? Namma is written with the sign **ENGUR**, as logogram for **engur** which has the meaning 'underground/subterranean waters'. Since olden times Enki had been the god of the **abzu**, another word for these subterranean waters. Perhaps this was the reason for giving him Namma as mother, once she had become a primaeval mother. Thus since the OB period Enki's father was An and his mother Namma. These parents – his father the supreme god of heaven, his mother the primaeval mother – may also have been a recognition of Enki's position as the formerly supreme Sumerian god (after An) before the 'arrival' of the newcomer Enlil [829].

4.4.3 *Anthropogeny*

In search of the Sumerian ideas concerning anthropogeny, it has become very clear from the texts that the beginning of humankind and how man came into existence on earth were not the questions that were dealt with in third-millennium texts. Only at the end of this millennium, in the Ur III period, did the texts contain some hesitant and cautious remarks in which the gods expressed the 'necessity for men' to take over the heavy tasks of the gods.

---

827 See Appendix 'God lists'.

828 E.g. in chapter 4.2.1.3.

829 See ch. 3, Excursus 1 'Enlil and Ninlil' and Excursus 2 'Enki'.

This has always been the reason for the origin of humankind: the gods complained about their life and finally refused to work for their daily rations; other creatures should do that.

'The Debate between Grain and Sheep' [830] (ll. 20-25) contains no creation story of man; mankind appeared to have been present as beings whose behaviour was just like that of animals. The gods somehow inspirited these primitive beings to make them civilized people who were able to perform the necessary tasks for the gods.

Only in the OB period did a real story about the creation of man appear: a creation starting from the real beginning. Man – or rather: his prototype – was formed using clay. The choice of clay as the raw material for man may have been influenced and stimulated by the other end of man's lifespan: his death and especially what happened after his death. Buried bodies disappeared after a while and were 'dissolved' into the earth; literally, they turned into clay. On the other hand, one had experienced the property of clay, that it could be moulded into nearly any desired model: a clay tablet, the image of a god or of a suppliant (for the temples), the image of an animal. It may have been the combination of both observations and experiences, i.e. the versatility of clay and the apparently universal basic substance of living beings, that gave rise to the idea that humans originated from clay.

Next this 'moulded' human prototype was placed in the womb of a mother goddess; in the story of 'Enki and Ninmaḫ' this mother goddess may be Namma [831]. In fact, this action is not expressed in these actual words, but it may be deduced from the continuation of the story after the prototype had been designed: the birth goddesses should assist Namma while she is giving birth. This could be interpreted as follows: after the moulding of the clay, the model was placed in the womb of the mother goddess who was to bear the first human. But there is the somewhat enigmatic sentence: 'let Ninmaḫ act as your companion' (l. 33). Ninmaḫ was one of the names for the mother goddess [832]; other names are e.g. Ninḫursaĝa and Ningal. This goddess may be seen as derived ultimately from **ki** 'earth'. Why should Ninmaḫ now act as a companion? The text does not make clear what this companionship exactly included. It is most likely that this companionship is a function of a midwife [833]. The presence of a mother goddess on the occasion of the birth of the first human being emphasizes the importance of this event. That importance is also accentuated by the presence of seven additional goddesses who 'assist you while you are giving birth' (ll. 34-36) [834].

The observation that, at the first stage of the actual birth process, there is a loss of water (*amniotic fluid*) may have been the direct cause for the supposition that Namma – 'water' – has been involved in the birth process of the first man. As we have seen in the god list TCL XV 10, at that time – the Old Babylonian period – Namma was considered to be the primaeval mother of **an-ki** [835]. This function of Namma may have been associated and

---

830 Van Dijk (1953, 36) has argued that the 'debate' texts may date from the Ur III period.

831 In 'Enki and Ninmaḫ' l. 17, Namma was mentioned as 'the primaeval mother, procreator of a multitude of gods'.

832 Black and Green 1998, 132-133; 141. The supposition of Bottéro and Kramer (1993, 195) that Ninmaḫ might be the spouse of Enki seems unlikely.

833 Black and Green 1998, 141.

834 For a discussion of these additional goddesses, see the Appendix Text editions, no. 8, comments on line 34 of 'Enki and Ninmaḫ'.

835 The idea of 'sea' as the primaeval mother of **an-ki** may have a West-Semitic origin. See also ch. 4.7.1. In that section, especially in the Excursus 'Namma vs Ti'amat' the birth incantation YBC 4603 will be discussed which links 'sea' with amniotic fluid and with the birth of a child. In this way Namma may be linked with the creation of the first man.

also linked with the birth of the first man. The following is a tentative hypothesis about what the Sumerians may have thought about how a foetus is provided with blood [836]. During the pregnancy the menstruation stops, and the blood that is normally lost during those periods may now be used for the growing foetus.

Once the first human being had been born, most likely some birth rituals and/or rituals with respect to the newborn baby are described, at least that is what we may conclude from the broken text at the end of the first episode of the story 'Enki and Ninmaḫ' (the lines 38-43) [837].

In search of an explanation for why the mythological ideas about the origin of man were only described from OB times onwards, one has to consider the increasing attention to human concerns at that time. For the religious aspects of Mesopotamian life, Jacobsen has formulated a development [838]:

> 'In our sources for religion in the fourth and third millennia B.C. just considered the individual seems almost totally immersed in his community as part of which he suffers and rejoices. But with the beginning second millennium the personal fortunes of the individual worshiper, his fears of personal misfortune, anxieties in illness and suffering, begin to be voiced; and the fear that may torture individual existence takes its place with the earlier central, conditioning fears, adding a personal dimension to the relation with the divine. The new development has its beginnings in the concept of the 'personal god', (...).'

In addition to the increasing evidence for a personal god, there also developed the feature of a 'private letter to a god' [839].

Outside the sphere of religion the increased attention to the personal life of individuals, had become noticeable for the first time already in the so-called 'Reform of Uru'inimgina' [840],

---

836 In the OB Akkadian poem *atra-ḫasīs* divine blood was necessary to create the first man.

837 Some fragments of the lines 38-43 in 'Enki and Ninmaḫ' recur in the lines 395-402 of 'Enki and the World Order' (ETCSL 1.1.3) which describe the usual birth rituals.

838 Jacobsen 1970, 44-45.

839 Edzard 2004a, 591-2; 594-596.

840 Perhaps the first steps were made even earlier, by Enmetena (Edzard 1976, 145-147).
With respect to the transcription of the name **URU-KA-gi-na** and its meaning, Bauer (1998, 475-477) and Selz (1998) have reviewed the literature and also given their suggestions. Until now, **uru** has been interpreted as the logical subject of the name phrase. In my opinion there is another solution possible for the interpretation of the name: it cannot be excluded that **uru** represents the directive ***uru-e**. The expression **inim gi.n** is known, meaning 'to establish the word, to confirm'. The transcription of the name **URU-KA-gi-na** then may be **URU-inim-gi-na,** and it may be translated as: 'he who has kept the word towards the city'. Perhaps the word **lú** has been omitted in the name (just as the *regens* of a genitive construction may be missing [Thomsen 1984, 92, § 167]). This interpretation of the name may fit very well with the efforts, e.g. the reforms, of **URU**'inimgina for the city (and thus indirectly for the city god). The text Ukg 4, especially the lines *xii*: 23-28, seems to support my suggested interpretation for the name (play on words 27-28): 23. **nu-siki nu-ma-kuš** 24. **lú-á-tuku** 25. **nu-na-ĝá-ĝá-a** 26. **^{d}nin-ĝír-su-da** 27. **uru-inim-gi-na-ke$_4$** 28. **inim-bé ka e-da-kéš** 23. That the orphan and widow 25. will not be subjugated 24. to a powerful man, 27. has **URU**'inimgina 28. established in a treaty 26. with Ninĝirsu. Krispijn's view (2005, 154-155) is that the vernacular Sumerian (the 'dialect'

and later in more detailed 'laws': those known as the laws of Ur-Namma [841], of Lipit-Ištar, and the code of Ḫammurapi. Also in economic life there was, from the Ur III period onwards, an increase in personal interests [842]. In general: the individual became more and more important. This new interest is also reflected in the stories about the creation of man, an interest observable only since Ur III times.

4.5 *Akkadian mythological texts about Beginnings*

As it is the first intention of this study I have focussed the attention until now on Sumerian texts, with sometimes a small digression to relevant bilingual or Akkadian texts. From the Old Babylonian time onwards no new genuinely Sumerian texts were composed, i.e. by people whose mother tongue was Sumerian. But as we have seen with respect to the god lists, the theologians further developed the Sumerian mythological-religious concepts. We were then able to find the god list **an** = *anum* at the end of a long development, the origin of which we could trace back to the Early Dynastic period and which was concluded in the Middle Babylonian period. Old concepts were continued, but there could also be observed some renewal or extension of ideas. Examples of this extension are: the 'ancestor' group of Enlil, started already in ED times but with only a few 'ancestors', growing to the 42 'ancestors' in **an** = *anum*; the 'ancestors' of An increased from two in TCL XV 10 to 21 in **an** = *anum*; Namma as the mother of **an-ki** was only introduced in OB times, in the list TCL XV 10.

In this chapter I will examine if the Sumerian ideas about the Beginnings were continued after the Sumerian culture had been preserved predominantly in the scribal schools. Therefore I have chosen two main Akkadian texts – *atra-ḫasīs* and *enūma eliš* – and one bilingual text, KAR 4 [843]. The presence of the three aspects – cosmogony, theogony and anthropogeny – will be studied, and compared with the previous results for the Sumerian texts.

4.5.1 *Cosmogony – Theogony*

4.5.1.1 *enūma eliš* [844]

Unlike *atra-ḫasīs* [845], *enūma eliš* does have a cosmogonic-theogonic introduction; the most relevant lines are [846]:

---

form), attested in some personal and geographical names, was even more prominent in the Lagaš texts than Bauer suggested. He also proposes that the 'Lagaš' pronunciation for **URU** may have been **uru**, and the 'official' *eme-gir* pronunciation **iri**. The tentative conclusion may be: the 'vernacular' pronunciation of **URU-inim-gi-na** is Uru'inimgina, and the 'official' pronunciation is Iri'inim-gina.

841 Wilcke 2002.

842 Edzard 2004b; Renger 2004.

843 Edzard and Wilcke (1977, 86) claim to have a monolingual, Sumerian, Old Babylonian version from Isin (tablet IB 591) of the KAR 4 text. Until now (2013) the text of this tablet has not yet been published.

844 I wish to thank Prof. S. Maul (Heidelberg) for his courtesy in having discussed with me the present parts of *enūma eliš*.

845 The story of *atra-ḫasīs* starts with gods who are 'men' and who have to work: cosmogony and theogony are not found in this text (see the Appendix Text editions, no. 8.1.c).

1	*e-nu-ma e-liš la na-bu-ú šá-ma-mu*	When upwards the heaven had not been named,
2	*šap-liš am-ma-tum šu-ma la zak-rat*	downwards the earth had not been called by name,
3	**ZU.AB**-*ma reš-tu-ú za-ru-šu-un*	there were the primaeval Apsû, their begetter,
4	*mu-um-mu ti-amat mu-al-li-da-at gim-ri-šú-un*	(and) *mummu* [847] Ti'amat, who gave birth to their totality;
5	**A**^{meš}-*šú-nu iš-te-niš i-ḫi-iq-qu-ma*	they mingled their waters into one,
6	*gi-pa-ra la ki-iṣ-ṣu-ru ṣu-ṣa-a la še-'e-ú*	but they neither joined pasture nor layered reed marshes.
7	*e-nu-ma* **DIĜIR.DIĜIR** *la šu-pu-u ma-na-ma*	When the gods had not (yet) appeared, not one,
8	*šu-ma la zuk-ku-ru ši-ma-tú la ši-i-mu*	neither they had been given a name nor the destinies had been ordained,

---

846 The Akkadian text – tablet I: 1-10 – has been quoted from Talon 2005; in line 6 Talon's transcription *še-'u-ú* (derived from the verb *še'û* 'to look for' *etc.* [CAD Š II, 355]) has been changed in *še-'e-ú*, derived from *šê'u* 'to pad, layer, to upholster' [CAD Š II, 363]. Recently a survey of the editions and the score of the extant texts of *enūma eliš* were published by Kämmerer and Metzler (2012). To make things easier for the reader I will give here Lambert's translation of these lines: 1. When the heavens above did not exist, 2. And earth below had not come into being - 3. There was Apsû, the first in order, their begetter, 4. And demiurge Tiāmat, who gave birth to them all; 5. They had mingled their waters together 6. Before meadow-land had coalesced and reed-bed was to be found - 7. When not one of the gods had been formed 8. Or had come into being, when no destinies had been decreed, 9. The gods were created within them: 10. Laḫmu and Laḫamu were formed and came into being. 11. While they grew and increased in stature 12. Anšar and Kišar, who excelled them, were created.

847 There are different opinions in the literature about the meaning of *mummu* (Michalowski 1990a: no translation; Lambert 2008, 37: demiurge; Foster 2005, 439: matrix). In my translation *mummu* has been conceived as an epithet of Ti'amat [CAD M II, 197 *mummu* A 1. 'craftsman, creator', *ad* 1b) 'as epithet of Tiāmat'; AHw 672 *mummu* I: 'etwa "lebenbewirkende Kraft"?'].
Van Dijk (1971, 443) has developed some philosophical notes with respect to *mummu*, which will be cited here in order to understand why he included Mummu in the 'Eridu' theology (van Dijk 1964, 10; see also ch. 1.2.1.2 in this study): 'In diesen Kreisen [i.e. the Babylon clergy influenced by the 'Eridu' theology; JL] wurde (...) ein Grundbegriff der sumerischen Vorstellungswelt entwickelt: sumerisch **mum** akkadisch *mummu*, "die *forma intelligibilis* der Materie, die in aller Ewigkeit *in ratione seminali* bestanden hat". Da die Materie nur zum Gegenstand sinnlicher Wahrnehmung gemacht werden kann, wenn sie geformt ist, und da diese Materie außerdem nicht durch einen Schöpfungsakt entstand, sondern monistisch mit den Göttern identisch war, war auch diese *forma intelligibilis* nicht geschaffen: sie war in aller Ewigkeit das allem Bestehenden innewohnende Ordnungsprinzip, (...); dieser Begriff wurde die am besten geeignete Bezeichnung für die "objektive Wissenschaft", das, was man wissen kann und was zu wissen ist. Zugleich war er im monistischen Denken das notwendige Prinzip in der Entwicklung aller Phänomene. Die Sumerer kannten keine Schöpfung aus dem Nichts: das "göttliche Universum" entwickelte sich selbst und aus sich selbst. Zusammen mit dem Wasser und der Erde war dieses *mummu* ein *tertium necessarium*, "ein drittes notwendiges Prinzip", das formende Prinzip.'

9	*ib-ba-nu-ú-ma* **DIĜIR.DIĜIR** *qé-reb-šú-un*	(then) gods were created in their inner part:
10	^{d}*làḫ-mu u* ^{d}*la-ḫa-mu uš-ta-pu-ú šu-mi iz-zak-ru*	Laḫmu and Laḫamu came into existence, were called by name.
11	*a-di ir-bu-ú i-ši-ḫu*	By the time they had grown in age and stature,
12	*an-šár u* ^{d}*ki-šár ib-ba-nu-u e-li-šu-nu at-ru*	Anšar and Kišar were created; they excelled them.

Comments

Line 2 *am-ma-tíš* // **GIM** *er-ṣe-tú* (Lambert 1996, 74, commentary at line 58; CT 41, pl. 44, *obv.* 7). CAD A II, 75 reads: *ammatu* B 'earth'. In the lexical part of this lemma *ammatu* is equated with *dannatu* (CAD D, 87, *ad* 3) 'ground, terrain', and (in AHw 160, *ad* 2) 'Baugrund'. This means that *ammatu* is not the same as **ki**, but is like **ki**.

The verbs used in these sentences have to be considered in more detail:

Line 4 *(w)ulludu* 'to give birth' (CAD A I, 287 ff); idem in line 16 [848];
Line 7 *šūpû* 'to bring forth; to make manifest, make appear (CAD A II, 20 *ad* 5);
Line 9 *banû* N-stem 'to be built, created' (CAD B, 84, *ad* 7);
Line 10 *šutāpû* (among others) 'to come into existence' (CAD A II, 201, *ad* 7);
Line 12 *banû* N-stem 'to be built, created' (CAD B, 84, *ad* 7).

Only the verbal forms of the lines 4 and 16 indicate a real parenthood.

***

The current explanation – of nearly all scholars – of these lines is as follows. Cosmogony begins with water. The mingling of the waters of Apsû and Ti'amat represents a sexual intercourse. The first result of this intercourse is the divine pair Laḫmu and Laḫamu [849], who were created 'within them' (l. 9). This interpretation could be observed e.g. in the publications of van Dijk, Jacobsen and Lambert [850]. This 'within them', in the words of Lambert, 'means of course within these mingled waters' [851]. There is uncertainty about the parents of the next pair, Anšar and Kišar: are they Apsû-Ti'amat or Laḫmu-Laḫamu [852] ?

In the following discussion I would like to suggest a new interpretation 1) of the lines I:1-12 and 2) of the lines IV:137-138 and V:62.

848 Line 16: *ù* ^{d}*a-num tam-ši-la-šú ú-lid* ^{d}*nu-dím-mud* 'Anu, too, begot his image: Nudimmud.'
849 The character of Laḫmu-Laḫamu was the subject of a polemic debate between Wiggermann (1981-1982) and Lambert (1985b).
850 Van Dijk 1964, 10; Jacobsen 1976, 168; Lambert 2008, 18-19.
851 Lambert 2008, 18.
852 Lambert 2008, 26; Seri 2012, 9-10.

1. *The lines I:1-12*

The division of the lines 1-12, in my opinion, is as follows [853]:

i) First episode: the lines 1-6 form a unit.

ii) Second episode: the lines 7-10 *idem*. In line 7 a *new* episode starts (*enūma*...), analogous to the lines 1-4.

iii) A third episode starts from line 11.

Ad i) The first critical point in this discussion is: to what or to whom does *-šu-un* refer in line 3 (*za-ru-šu-un*) and *-šú-un* in line 4 (*gim-ri-šú-un*) ? Until now in the opinion of all scholars both these *-/šun/* refer to the gods. But the gods are only mentioned in the second episode (line 7). In my opinion it is much more likely that *-/šun/* in lines 3 and 4 refers to *šamāmū* and *ammatum* – heaven and earth –, which are mentioned in lines 1 and 2. If this supposition is true, this means that in lines 3-4 it is stated that Apsû and Ti'amat brought forth – expressed by the verb *(w)ulludu* – heaven and earth. The expression *gimrišun* 'their totality' most likely indicates that *šamāmū* ('heaven') and *ammatum* ('earth') are still forming a unit [854]. Line 6 of *enūma eliš* adds that nothing else happened with respect to the earth: Apsû and Ti'amat neither joined any pasture nor layered reed marshes. What else can this addition mean than that the earth remained 'empty' ? Line 6 has only significance if the earth really existed. But the word *ammatu* modifies the concept of 'earth'. The word *ammatu* expresses that at that moment only the hard ground is present, empty, without e.g. rivers, greenery. The earth has not yet its present appearance, which is described by the word *erṣetu*. In my opinion the lines 1-4 indicate that heaven and earth – with the restriction made above – were born as a unit.

The usual translation of *gimrišun* 'them all' and the concomitant current interpretation that 'all' refers to the gods seem very unlikely. Recently Lambert and Seri were of the opinion that only one pair of gods, viz. Laḫmu and Laḫamu, were brought forth with certainty by Apsû and Ti'amat (l. 10) [855]. If *gimrišun* refers to more than *šamāmū-ammatum* alone, then this would be because Apsû and Ti'amat might have been thought of as the *origin* of the primaeval universe and of all the gods, but not as the *real parents* of all those gods.

The unit heaven-earth does not play any part in the continuation of the composition, until Marduk will use this unit for the creation of the present universe (the lines IV:138 and V:62).

Ad ii) In line 7 a new episode starts with the word *enūma*, parallel to the lines 1-4. This episode deals with the gods, who initially were also absent. This absence is expressed in lines 7-8, partly with an analogous formulation as in lines 1-2 to indicate the initial absence of heaven and earth. But another formulation was also used: the gods had not (yet) appeared, for which the verb *šūpû* was used. This verb does not express the appearance by birth. Immediately after this, in line 9, we read that gods were formed, rendered by the verb *banû*. This verb has also not the connotation of production by sexual intercourse. The first

---

853 For the most recent analysis of the poetic structure of *enūma eliš*, see Kämmerer and Metzler (2012, ch. 1.7, pp. 55-72), who concluded that the 'formal principle of quatrains or at least four-line stanzas characteristic for our poem' (Vanstiphout 1987b, 53) does not exist.

854 The examples given in CAD G, 76-77 *ad gimru* 1. show that *gimru* can also refer to one object in its totality.

855 Lambert 2008, 26; Seri 2012, 9-10.

gods who appear are the pair Laḫmu-Laḫamu; the verb *šutāpû* 'to come into existence' also does not imply: via sexual intercourse.

These gods were formed *qerebšun* 'in their inner part' (l. 9). As already mentioned: this *qerebšun* 'within them' was explained by Lambert as '(...) of course within these mingled waters' [856].

The next development is the creation (a form of *banû*) of Anšar and Kišar. Anšar and Kišar are the parents of the god Anu (ll. 14-15). Although the enumeration of An's 'ancestors' is relatively short, in general it corresponds to what the god list **an** = *anum* tells us.

2. *The lines IV:137-138 and V:62*

The next discussion point is the translation and interpretation of the lines IV:137-138 and V:62 [857]. These lines read:

IV:137	*iḫ-pi-ši-ma ki-ma nu-un maš-ṭe-e a-na ši-ni-šu*
IV:138	*mi-iš-lu-uš-ša iš-ku-nam-ma šá-ma-mi uṣ-ṣal-lil*
V:62	*[meš-la-šá] uṣ-ṣal-li-la er-ṣe-ti uk-tin-na*

In lines IV:138 and V:62 the verb *ṣullulu* has been used. The first meaning of this verb is 'to roof', or in the words of AHw 'überdachen, zum Dach machen' [858]. Therefore my translation of the respective lines is [859]:

IV:137	he split her like a fish for the drying place into two parts
IV:138	he placed one half of her and roofed the heaven (with it);
V:62	[one half] he used as a roof and fastened it to the earth.

In both sentences the verb *ṣullulu* expresses that an existing entity is covered with a roof: these existing things are heaven and earth, respectively. The unit *šamāmū-ammatum* was born of Apsû and Ti'amat. The two lines IV:138 and V:62 indicate that Marduk forms the present world – heaven/sky and earth as we can observe it, the earth now written as *erṣetu* –

---

856 Lambert 2008, 18.

857 In the translation of Lambert (2008, 48-49):
IV:137-138: He split her into two like a dried fish: One half of her he set up and stretched out as the heavens.
V:62: [(Thus) the half of her] he stretched out and made it firm as the earth.

858 AHw 1110, *ṣullulu* 'überdachen, überdecken (ad 3. zum Dach machen)'.
CAD Ṣ, 239, *s.v. ṣullulu* A. The first quotation *ad* 1.b) concerns line IV:138, with translation: 'he set up half of her and roofed the sky (with it).'
CAD R 299, *s.v. retû* 1e) line V:61 is cited, with the translation of V:61-62. V:61 'he placed half of her (so that) she was wedged in the sky', V:62 'with the other half he roofed the earth'.

859 IV:138 *mi-iš-lu-uš-ša*. Strictly speaking is *mišlušša* the locativus-adverbialis (**mišlum-ša*). AHw mentions (661a, *mišlu*) at this example: 'für Akk[usativ]', also von Soden 1995, 109, § 66f; and I interpreted it likewise; literally: 'he placed her for the half and made it to roof the heaven'.
V:62 *[meš-la-šá] uṣ-ṣal-li-la er-ṣe-ti uk-tin-na*. The preceding lines describe how Marduk used several parts of Ti'amat's body, e.g. from her eyes he let flow the Euphrates and the Tigris (55), on her breasts he heaped up mountains (57). With one half of Ti'amat's body prepared in this way, he roofed the earth and gave earth its appearance.

with the aid of Ti'amat's body [860]. The apparently amorphous mass of the unit heaven-earth of the beginning was now tranformed by Marduk to give both heaven and earth their actual appearance. Especially the lines V: 47(?)-62 show how Marduk is using various parts of Ti'amat's body to cover the earth, which had been without any distinguishing features until now (see also line I:6). This may be called: the creation of the world by Marduk [861].

The whole action, viz. the splitting of Ti'amat's body and the subsequent covering of heaven and earth, each with one half of her body, may also include the well-known mytheme of earlier Sumerian mythologies: the separation of the primaeval unit **an-ki**. But now it is Marduk who takes over the role of Enlil in this act and who executes the separation of heaven and earth [862].

Conclusion

As a result of these considerations I tentatively conclude that *enūma eliš* starts with the birth of the unit *šamāmū-ammatum* or heaven-earth from the primaeval waters Apsû and Ti'amat. It appears that the cosmogony as described in **an** = *anum* is different and more 'conservative' than the cosmogony in *enūma eliš*. In **an** = *anum* it is still Namma – without any partner, as in the OB god list TCL XV 10 – who gave birth to the unit **an-ki** 'heaven-earth'. This **an-ki** unit was active in the creation of gods, as was discussed in the previous sections with respect to the Sumerian Beginnings. In *enūma eliš* the primaeval water consists of a pair – Apsû and Ti'amat – which brought forth *šamāmū-ammatum* 'heaven-earth'. But the Akkadian unit of heaven-earth differs from the **an-ki** unit in the Sumerian cosmogonic mythology. The unit *šamāmū-ammatum* is not active in the theogonic process: it is the pair Apsû and Ti'amat that created the first gods Laḫmu and Laḫamu.

In *enūma eliš* some 'ancestral' pairs of An are mentioned, viz. Laḫmu-Laḫamu and Anšar-Kišar, who are also present in the god list **an** = *anum*. The order of the 'ancestors' of

860 The formulation of van Dijk (1983, 10) with respect to this episode also suggested that in his opinion the heaven and the earth already existed: '(...) Ti'āmat, coupée en deux, devient le firmament céleste et la surface de la terre; (...).' Jacobsen (1984, 16) obviously has the same opinion, in view of his comment: 'After her (= *Ti'amat; JL*) death at Marduk's hand she continues to exist. Her watery body is up there behind the vault of heaven, held back by it and guarded.'

861 Compare this creation with the text of CT 13 36: 17-18 (Neo-Assyrian period), where Marduk wove a raft over the primaeval waters and poured soil on this raft, thus creating dry land.

862 The episode in which is described how Marduk moulded heaven and earth, was concluded as follows (text: STT 1,12, line *obv.* 15'-16'; CDLI P338328):

V:65	*ip-⌜te!⌝-eq-ma* AN-*e ù* KI-*tim* ⌜KU⌝ *x* [*xx*]*x*	And thus he fashioned the heaven and the earth [...],
V:66	[...] *ri-kis-su-nu-ma* ⌜*ma-a-da*⌝ *kun-nu-ni*	[he? ...] that their bond was very lasting.

Comments:
1. V:66 Other transcriptions: a. [...] *ri-kis-su-nu-ma e-pi-iš kunnuni* (coll. W.G. Lambert in CAD R, *sub riksu* 6a, 352b); b. [...] *ri-kis-su-nu ma-*⌜*x*⌝*-iš kun-nu-ni* (Talon 2005, 58); c. [...] *ri-kis-su-nu ma-*⌜*di!-iš*⌝ *kun-nu-ni* (Kämmerer and Metzler 2012, 236).
2. *kun-nu-ni*: the ending *-ni* points to the subjunctive mood of the verbal form.

An in *enūma eliš* is different from that in **an** = *anum*, but as we have seen (ch. 3), the order of the 'ancestors' of both An and Enlil is not fixed, but rather variable [863].

4.5.2 *Anthropogeny*

In studying Akkadian accounts of the creation of man, I will discuss both *atra-ḫasīs* and *enūma eliš*, but also the bilingual text KAR 4.

In *atra-ḫasīs* or 'The Babylonian Story of the Flood' the creation of man is extensively described. After the rebellion of a number of the gods against Enlil in protest against the heavy work they had to do, Ea proposed creating man to carry out the toils of these gods. One of the rebellious gods had to be sacrificed [864], and from his flesh and blood, mixed with clay, man should be created. The mother goddess nipped off pieces of this mixture and with the aid of birth goddesses these pieces were moulded into male and female human beings. The formation of the womb which had to be opened by Nintu after the gestation period is lost in a text gap. The plan for man's creation is first described, followed by a detailed account of the realization of it [865]; as the plan contains the most important principles for the creation of man, they may suffice to describe the whole process. The announcement – by the god Enki – of the intention to create man and the plan is phrased in the following lines of tablet I:

208	*ilam iš-te-en li-iṭ-bu-ḫu-ma* [866]	Let them slaughter one god, so that
209	*li-te-el-li-lu ilū*^{meš} *i-na qé-⌜er⌝-bi* [867]	the gods become freed (from their duties) [868] thereby.
210	*i-na ši-ri-šu ù da-mi-šu*	Let Nintu (l. 211) with his flesh and blood
211	^{d}*nin-tu li-ba-al-li-il ṭi-iṭ-ṭa*	mix clay,
212	*i-lu-um-ma ù a-wi-lum li-ib-ta-al-li-lu*	so that god and man are mixed

---

863 For a recent discussion of the mythological motifs used in *enūma eliš*, see Katz 2011. I do not agree with her interpretations – which are in accordance with the usual ones until now – 1. that the introduction of *enūma eliš* goes back to a time 'before heaven and earth came into being' (p. 127), and 2. that 'Marduk used the corpse of Tiāmat to create the whole universe' (p. 131). On the contrary, this introduction starts with the birth of heaven and earth, although earth has not yet its final appearance. Marduk used Ti'amat's body to *finish* the appearance of heaven and earth by covering both with a half of Ti'amat's body, as may be clear from the present translation and interpretation.

864 A late 3rd millennium Akkadian seal may illustrate this episode: to the right some gods are filling and carrying baskets, some are building a house/temple; in the middle a god is raising his hands (Enlil?), and to the left a god is killing another god (by courtesy of Wiggermann; Wiggermann 2011, 301, fig. 2).

865 For the Akkadian text I refer to Lambert and Millard 1999, 58 (ll. 208-217; the planning) and 58-64 (the ll. 233-295 give a detailed account). The translation is by the present author.

866 The use of the verb *ṭabāḫu* 'to slaughter, to slit the throat' [CAD Ṭ, 1] instead of e.g. *dâku* or *nêru* indicates that the way of killing the god was such that his blood could be caught.

867 For the reading *i-na qé-⌜er⌝-bi* instead of *i-na ṭi-⌜i⌝-bi*: see George and Al-Rawi 1996, 187 *ad* 92.

868 Oshima (2012, 415, note 56) suggested the meaning: 'to be cleared (from their duties)' [CAD E, *elēlu*, Dt-stem)].

213	*pu-ḫu-ur i-na ṭi-iṭ-ṭi*	together in the clay.
214	*aḫ-ri-a-ti-iš* u_4*-mi up-pa i ni-iš-me*	Let us hear in future days a drum;
215	*i-na ši-i-ir i-li e-ṭe-em-mu li-ib-ši*	let there be the spirit from the god's flesh:
216	*ba-al-ṭa it-ta-šu li-še-di-⸢šu⸣-ma*	let it (*the spirit*) make its characteristic recognizable to the living, and
217	*aš-šu la mu-uš-ši-i e-ṭe-em-mu li-ib-ši*	let the spirit be there in order not to forget this.

Comments

Line 214 *uppu* 'drum': Lambert has given a possible explanation for the use of a drum as a cultic instrument [869]. Abusch's 'drum [= heart (beat)]' [870] is less likely because it does not fit in the context, which is all about the spirit of the god. I agree with Lambert that the drum had a cultic use. The drum may accompany the daily meals of the gods for which man has to care; that was the reason why man was created.

There is also talk of the spirit of a god – *eṭemmu* –, which spirit transformed this new creation into a human being. At the death of a person this spirit goes to the netherworld. I think that this is an opportunity *par excellence* to remember the presence of the god's spirit. Rituals, perhaps accompanied by the sound of a drum, are necessary and are performed to keep the memory of the deceased person – in fact the memory of his soul or spirit, his *eṭemmu* – alive. Then the drum in l. 214 may also be an allusion to and may belong to funerary rituals.

Line 216 As Lambert has already mentioned, *ba-al-ṭa* is an adjective, not a noun [871]. Therefore a translation as 'life' seems not to be justified [872]. I agree with Lambert, that *eṭemmu* (l. 215) is the subject in lines 216-217.

***

Several discussions have dealt with lines I: 214-217 [873]. My interpretation of the preceding lines is as follows. For the creation of man, clay was necessary for the human physical body, the part that returns to dust or clay after the man's death [874]. It was and will be dead material. Then the flesh and blood of a slaughtered god were necessary. This god provided the clay body with his *eṭemmu*: 'spirit' or 'ghost'. This spirit had a two-fold function, explained in lines 216-217. It is this *eṭemmu* that is the subject of these lines. The first

---

869 Lambert and Millard 1999, 152 *ad* 214: 'The drum called *uppu* had a cultic use, (...). Perhaps at the time of the composition of this epic the daily meals of the gods were introduced in the sanctum to a beating of the drum.'

870 Abusch 1998, 365.

871 Lambert and Millard 1999, 152 *ad* 216, 229.

872 Katz 2005, 58.

873 In this respect I only mention Lambert and Millard 1999, *esp.* p. 22 with the explanation of these lines; Abusch 1998, 365; Katz 2005, 57-59; Katz 2007b, 173 (Abusch and Katz have a different interpretation).

874 For references to Akkadian texts with the expression 'return to dust/clay': CAD Ṭ *ṭīdu*, *ad* c, 108-109.

function is, that the *eṭemmu* of the slaughtered god makes the clay body alive; that is the proclamation of the *ittu* 'characteristic' or the 'sign' of the spirit [875]. The *eṭemmu* has a second function after man's death. The ghost of the deceased person goes to the netherworld; the memory of this ghost has to be kept alive by the surviving relatives. Due to the (recurring) funeral rituals man is obliged to remember how he has come into being and that the part that remains after his death has a divine origin; this part remains in the netherworld and has to be honoured by the living relatives of the deceased person [876].

In conclusion: the creation of man in *atra-ḫasīs* exists of a combination of lifeless clay and the *eṭemmu* of the slaughtered god: his spirit makes human life possible. The spirit of the god was not created, but is present forever.

As we have seen [877], the creation of man in KAR 4 *obv.* 19-20, has been mentioned only briefly: 'we shall slaughter the gods Alla and Illa, so that their blood makes mankind grow.' Only the blood of two gods – for man and woman? – seems to be necessary; clay is absent. Also in *enūma eliš* it is just a casual remark that pays attention to the creation of man. In tablet VI: 5-6 we read: 'I (= *Marduk*) will compose blood vessels and form the body's frame, I will bring into being Lullû, whose name shall be 'man' ', and in VI: 33: 'From his blood he (*Ea*) created mankind' [878], while the blood was originating from the sentenced god Qingu. The conclusion is that here only divine blood was used.

4.6 *Akkadian Beginnings: Conclusion*

The Akkadian texts discussed above show, that the Sumerian ideas about the Beginnings were gradually developed further during the second millennium BCE, with differences in some main points [879].

Namma, introduced about or shortly after 2000 BCE as the primaeval water who bore **an-ki**, was finally replaced by the pair Apsû and Ti'amat, the primaeval waters who brought forth *šamāmū* and *ammatum*, heaven and earth. This initial unit *šamāmū-ammatum*

---

875 In the opinion of Katz it is Nintu who 'announces life' as man's sign (Katz 2005, 58), but Nintu has only brought the necessary elements – clay and god's flesh and blood – together (ll. 225-226).

876 Katz (2007b, 173) wrote that the *eṭemmu* was 'an everlasting reminder to the gods of the circumstances that led to the creation of man.' But in my opinion, in the netherworld the *eṭemmu* functions as a reminder *not* to the *gods* but to the surviving relatives of the deceased person: *they* are to remember forever how man was created, by honouring his *eṭemmu*.

877 Chapter 2.1.10 and the Appendix Text editions, no. 10.

878 Lambert 2008, 52 (with a slight variation in the translation). The relevant lines of *enūma eliš* read (Akkadian text: Talon 2005):

VI:5	*da-mi lu-uk-ṣur-ma eṣ-mé-ta lu-šab-ši-ma*	'I (= *Marduk*) will compose blood vessels * and form the body's frame'.
VI:33	*ina da-me-šú ib-na-a a-me-lu-tú*	From his blood he (= *Ea*) created mankind.

* *Comment*: For the translation *damu* 'blood *vessels*' in this line: see Stol 2000, 11; Ziegler 2005. *kaṣāru(m)*: AHw 456-457, 'knoten, fügen, sammeln'; CAD K 257 ff., a.o. 'to tie, bind together, to join; to make compact'.

879 There is a totally different concept of the beginning, expressed in the Akkadian OB text YOS 11, 5:1, and which has been repeated in later times: *^{d}a-nu ir-ḫi-a-am ša-me-e ša-mu-ú er-ṣe-tam ul-d[u-n]im* 'Anu inseminated the heaven, the heaven gave birth to the earth'. For references: Westenholz 2010, 303, note 42.

was waiting for the action of Marduk who created the *actual* heaven and earth, and who – instead of Enlil – separated both.

In Sumerian texts An's ancestors and some other gods developed from the primaeval unit **an-ki**. In *enūma eliš* Apsû and Ti'amat – and not the unit *šamāmū-ammatum* – produced some of the ancestors of An.

Clay is the common principle of the Sumerian creation story of man and of the OB Akkadian story *atra-ḫasīs*. In this Akkadian myth, in addition to clay the blood and flesh of a slaughtered god were also necessary; thus it implied a violent action [880]. On the other hand: in this story it is not stated that a mother goddess became pregnant. The text is very concise at the point after the clay has been mixed. Nintu summoned the great gods, who spat upon the clay. Nintu then addressed these gods with the words: 'You command me a task, I have completed it.' [881]. Thereafter normal human reproduction was established.

Other Akkadian stories about the creation of man only mention the necessity of using blood (KAR 4; *enūma eliš*); however clay was not mentioned any more.

One element is common to both the Sumerian and the Akkadian stories about the creation of man, viz. the purpose of this creation. Man had to free the gods from their burden and to take over their toil; man had to provide them with food and drink, he had to serve the gods.

4.7 *The Beginnings: Sumerian vs Semitic origins*

As has become clear from the above overview of Mesopotamian history (ch. 1.1): there were several periods in which there were occasional intensive contacts between the Sumerians and people of another ethnic identity. During the Uruk expansion this contact reached as far as the present Syria and Anatolia in the north-west. The contact with the Semitic Kiš civilization was perhaps initially confined to the Nippur-Kiš region, but could be observed in later times also in Mari and Ebla. In the Sargonic period, Sumer was ruled by the Akkadians. Keeping these considerations in mind, let us now return to the main subjects of this study: the Sumerian Beginnings with respect to cosmogony, theogony and anthropogeny. Is it possible to discern whether these ideas are genuinely Sumerian?

4.7.1 *Cosmogony – Theogony*

We have no information about the ideas with respect to these topics in the 4th millennium, simply because there are no texts. The first texts written in the Sumerian language with an introduction about beginnings date from the Early Dynastic time; they tell only about **an-ki** (heaven - earth) as the primaeval origin of cosmos and gods. This idea was maintained until the OB period. I suppose that the concept of heaven impregnating the earth with water may have had its origin in Neolithic times, when agriculture had begun and irrigation channels were not yet known, and the water necessary for the farming lands was provided by rain (in the lower hills) or by the flooding of Euphrates and Tigris. It is also possible that nomadic

---

880 Michalowski (1991, 136) calls this creation of man 'a form of ritual reversal', since: ' (...) the sacrificial cult of animals, intended to feed the gods, is grotesquely reversed so that the divine ones offer up one of their own.' And further: 'Then follows another ritual reversal, the mixing of blood with clay to produce primeval man. Humanity is here definitely subjugated to the gods, although the ritual bond established at its creation provides the symbol of the dialectical relationship between humans and their masters.'

881 Lambert and Millard 1999, 59, lines 231-237.

people, who were dependent on rainwater for the pasture of their herds, may have had the same idea. In the third-millennium texts discussed in this study [882], we have seen that during the phase of courtship and the preparation for the 'marriage', **an** is inanimate, representing heaven. During the intercourse **an** represents the god of heaven, An [883]. In the OB text 'Enki and the World Order' it is Enki who fills the Euphrates and/or Tigris with his semen [884]. This act may be seen as a repetition of the primaeval insemination of the earth by An. In other words: the picture of a god who fecundates or irrigates the land may be an old and long-lived theme [885].

A second point concerns Enlil and the separation of **an-ki**. From the oldest god lists it may be concluded that Enlil already in the ED period had been accepted as a supreme god in the Sumerian pantheon. So far we have found only a few third-millennium UD-GAL-NUN texts from ED Abū Ṣalābīḫ, in which it is expressly stated that Enlil has separated heaven and earth [886]. This was mentioned later, in only one OB text: 'The Song of the Hoe'. In 'Gilgameš, Enkidu and the netherworld' the separation of **an** and **ki** has been nuanced: in the first episode (ll. 1-10) it is been mentioned in an neutral, passive way [887], in the second episode An and Enlil both are active [888]. Therefore we may conclude that Enlil's separation of **an** and **ki** was not a generally accepted idea. These data may lead to the following hypothesis.

The idea of **an-ki** as the primaeval origin goes back to before the third millennium. My supposition is that this idea was already in existence before Enlil had been accepted as a member of the Sumerian pantheon, which may have taken place at the end of the Uruk period when the Semitic influence from the Kiš region started to increase. This primaeval origin appeared to be deeply rooted in Sumerian thought, as can be deduced from the fact that it was present in the Sumerian texts during the third millennium. Obviously, the introduction of Enlil into the Sumerian pantheon did not *de facto* imply certain cosmological consequences, viz. Enlil's separation of **an** and **ki**, for the Sumerian mythology, since this action was only mentioned in some UD-GAL-NUN texts and not in third-millennium texts written in the normal orthography. Therefore these considerations may lead to the tentative conclusion that **an-ki** as the primaeval origin was a genuine Sumerian idea, which the Sumerians probably shared with other peoples, and that Enlil's action was not a genuinely Sumerian one which earlier – before Enlil's entry into the Sumerian pantheon – has been accomplished by another Sumerian god.

---

882 For a survey: see Table 1 in ch. 2.

883 See the Excursus 'The animate *vs* inanimate class of **an** and **ki**' in the Appendix Text editions.

884 Cooper (1989, 87 note 2) discusses the uncertainties about the Euphrates and Tigris in the lines 251-257 of this text [ETCSL 1.1.3].

885 Dupré 1975, 253: 'The idea of divine world parents, of the sacred wedding, or of conflict between heaven and earth, are further ways in which basic experiences of the divine being are conceptualized chiefly among the planters.' And further (p. 255): 'The paideumatic experience (...) of the Sky and All-father among the pastoral nomads, of sexuality and fertility among most of the planters affect and shape the whole of the religious system.'

886 Perhaps the fact that the texts concerned were written in UD-GAL-NUN-orthography and that they originated from Abū Ṣalābīḫ was the cause of the relative unfamiliarity with Enlil's act in southern Mesopotamia, i.e. Sumer; at least: this is not visible in other third-millennium Sumerian texts.

887 GEN, lines 8-9: 'After heaven had been separated from earth, after earth had been delimited from heaven,'.

888 GEN, lines 11-12: 'At the time that An had taken heaven, Enlil had taken earth,'.

After the Ur III period the cosmogonic primaeval principle changed, from **an-ki** into Namma, the primaeval ocean as the one who gave birth to **an-ki**. A scanty indication of this new mythology could be observed in the OB Sumerian god list TCL XV 10, which was fully expressed in the list **an** = *anum*. In the Akkadian *enūma eliš* Namma as the one and only primaeval origin had been taken over by a pair, the male Apsû and the female Ti'amat. In the next Excursus first the relation between Namma and Ti'amat will be examined.

Excursus: Namma vs Ti'amat

A survey of the available information from different kinds of texts about Namma on the one hand, and Ti'amat on the other hand shows the following. Namma is known during the third millennium, but not yet in her quality as mother goddess [889]. At that time she did not seem to be an important goddess [890]. In the Ur III period no cultic feasts for Namma are known [891], while the first ruler of the Ur III dynasty – Ur-Namma – was bearing her name. Richter only mentions a sanctuary for Namma at Ur in the Old Babylonian period [892]. In the oldest Semitic pantheon – as described by Roberts [893] – Namma or an equivalent, like Ti'amat, is absent. In Eblaitic texts [894], with ᵈAma-ušum as protagonist, Namma is present in an as yet unclear context.

The name of Namma's Semitic counterpart is Ti'amat – 'sea, ocean'. As Jacobsen has shown: Ti'amat's ultimate identity was a personification of the Mediterranean sea and its powers [895]. Ti'amat could have entered the Mesopotamian mythology since the Akkad period (via the campaigns of Sargon to the West?). Support therefore may be found in an Old Akkadian tablet from Ešnunna, on which the name Ti'amat appeared in *a-ba-ra-ak ti-àm-tim* 'steward of Ti'amat' [896]. Further evidence for an early familiarity with Ti'amat in Mesopotamia comes from the name of one of the spouses of the Ur III-king Šu-Suen, viz. the **lukur** Ti'amat-bāštī. This woman originates obviously from Niniveh, and is possibly a sister of Tiš-atal [897]. It is at that time – at the end of the third millennium – that there is an increasing influence of (West-) Semitic immigrants in Mesopotamia. Subsequently Ti'amat was syncretized with the Sumerian goddess Namma, who in turn received more attention than in the preceding times: her name was also used in personal names from the Ur III period onwards (e.g. Ur-Namma).

***

---

889 See also ch. 4.1.3.

890 See ch. 3.3.2.

891 At least: Sallaberger (1993) does not mention such a feast.

892 Richter 2004, 506.

893 Roberts 1972. Myers (2002) also has no indications for Ti'amat or Namma in the Sippar pantheon.

894 The Eblaitic texts are ARET 5: 20-21, *xii*: 3 / *xiii*: 2. These texts have a parallel in IAS 278 (Edzard 1984, 39).

895 Jacobsen 1968, 105. CAD T 150, *s.v. tâmtu, o.a.* '1. see, ocean, from OAkk. on; and 4. (a deity) from OAkk. on' (with ref. on pp. 156-157).

896 Westenholz 1974-1977, 102. The alternative transliteration and translation of Durand (1993, 43) are: '*a̲(b) pá-a̲-ra-ak ti-àm-tim (underlinings JL)* = *a(b) parrâk ti'âmtim* = O Père! toi dont la tâche est d'être la barrière contre les flots de la Mer.' The suggestion that *a̲* might be used for *ab* is not persuasive. Moreover, the *a̲* is not present in the OA text.

897 Wilcke 1990; Sallaberger 1999, 161. Ti'amat has two forms: sea and monster. When Ti'amat is part of a personal name, it is hardly conceivable that the monstruous aspect of Ti'amat is meant.

How has Namma, or perhaps better: primaeval water, become *the* primaeval origin? A tentative hypothesis is the following.

The idea of 'sea' or 'water' as the primaeval origin of the universe is a relatively late concept – from the Old Babylonian period – in the Mesopotamian mythology. This may have originated from the increasing interest – from Ur III times – in human existence (ch. 4.4.3), in personal affairs, as evidenced e.g. in the appearance of a personal god and of laws. It is not surprising that in such a context questions about the origin of man and why he is on earth have arisen. In this respect, there exists an interesting and unique Old Babylonian tablet containing an Akkadian incantation which accompanies the birth of a human [898]. The lines which are relevant for our discussion are:

YBC 4603
obv.

1	*i-na me-e na-a-ki-im*	In the fluid of the intercourse
2	*ib-ba-ni e-ṣé-em-tum*	bone is made;
3	*i-na ši-i-ir [še]-er-ḫa-nim*	with the flesh of the muscles
4	*ib-ba-ni [l]i-il-li-du-um*	the offspring is made.
5	*i-na me-e* **a-ab-ba** *ša-am-ru-tim*	In the water of the violent and awe-inspiring (l. 6) sea,
6	*pa-al-ḫu-ú-tim*	
7	*i-na me-e ti-a-am-tim ru-qú-ú-tim*	in the distant water of the ocean,
8	*a-š[a]r ṣe-eḫ-ru-um ku-us-sà-a / i-da-a-šu*	where both arms of the little one are bound:
9	*qé-er-bi-is-sú la-a uš-na-wa-ru*	its interior does not illuminate
10	*i-in ša-am-ši-im*	the eye of the sun,
11	*i-mu-ur-šu-ú-ma* **ᵈasar-lú-ḫi** / *ma-ri* **ᵈen-ki**	but Asarluḫi, the son of Enki, looked at him.
rev 4	**ka-inim-ma-munus-ù-tu-da-[kam]**	Incantation formula for a woman who is giving birth.

Comments

Line 3 *šerḫānu* is a variation of *šer'ānu* 'sinew, tendon, vein, muscle' [899]. The copy of the cuneiform text [900] renders as the last syllable: *nim*, but van Dijk transcribes *ni-im* [901].

***

[898] YBC 4603: van Dijk 1973; van Dijk *et al.* 1985, plate LXXVII, no. 86. The lines 12-15 describe the intervention of Asalluḫi (= the physician), and lines 16-27 the assistance of the mother goddess (= the midwife) at birth. Veldhuis (1999, 39-41) discussed the poetic language of this incantation. With reference to this childbirth incantation, Woods (2009, 221-222) describes the relation between cosmography and the birth metaphor, where amniotic fluid is compared with a cosmic ocean.

[899] CAD Š II, 308-309, especially the lexical part.

[900] Van Dijk *et al.* 1985, plate LXXVII, no. 86.

[901] Van Dijk 1973, 503.

This incantation describes in a poetic way the development of the foetus after the conception up to and including its birth. In particular lines 5-11 picture the situation of the foetus in the womb, viz. in the amniotic fluid which is represented by the metaphor of a violent and awe-inspiring ocean in which no light can enter. At the same time there is an allusion to the Engur (ll. 9-11), the abode of Enki, viz. to the line 13 of 'Enki and Ninmaḫ': '(Enki, in) the deep Engur, the subterranean water – a place the inside of which not a single god can observe –'. In a subtle way we find in these few lines of this incantation a relation between amniotic fluid - sea - Engur. From here to Namma, who combines the aspects of ocean and mother, seems only a small step. My suggestion is, that the idea of a foetus in the amniotic fluid – after all an ancient experience – may be the source of inspiration for Namma as the primaeval mother of man and subsequently also of **an-ki** [902].

Primaeval water as a cosmogonic beginning, personified by Namma – not Ti'amat – was present in the OB god list TCL XV 10. In the OB Sumerian literary text 'Enki and Ninmaḫ', Namma is also the mother of Enki and of 'a multitude of gods', presumably the Anunna. Moreover, Namma is the 'mother' of the first human individual, at least she bore the clay model of the first human until his birth [903]. These are all OB attestations, from a time when the people were supposed to speak another language than the Sumerian one. But the use of the Sumerian language and of the Sumerian names of gods may indicate that the scribal schools were still Sumerian-oriented and preserved the Sumerian culture. However, it is the above-mentioned Akkadian text that in a clear way relates amniotic water and sea, thus the idea of Namma-Ti'amat as mother. The god list **an** = *anum* – in its composition not even Akkadian but rather Sumerian [904] – still mentions Namma as the mother of **an-ki**. In the nearly contemporaneously composed *enūma eliš* the primaeval sea also appears as the origin, but now split up into a male part Apsû and a female part with the name Ti'amat [905].

In the second half of the second millennium BCE the influence of West-Semitic mythology became prominent and influential [906]. My tentative conclusion is that the idea of

902 This suggestion I have made already in my undergraduate thesis (Lisman 2005, 88). Recently Katz (2011, 128-129) expressed the same idea. It is not known whether or not the Mesopotamian idea of water as the primaeval origin was influenced by the Egyptian mythology. In the Egyptian beliefs, there is a world of pre-existence as origin of the creation. This world may be described as a watery chaos and is called Nun (see Epilogue, § 1.1). This water contains a pre-existent form of Atum, the creator god, who comes into full existence by autogenesis. In my opinion there is only a difference in nuance between Namma and Nun as primaeval origin: the principle is that water is the primaeval beginning in both mythologies.

903 'Enki and Ninmaḫ', ll. 31-36 (ch. 2.1.8).

904 Litke 1998, 6.

905 The idea of primaeval water as the first beginning is also present in the bilingual text known as 'The creation of the world by Marduk', a Neo-Assyrian text from the first half of the first millennium (CT 13, 35-38; Foster 2005, 487-489; Horowitz 1998, 129-132). After summing up what was absent, the text reads (l. 10): **[niĝin]-kur-kur-ra-ke$_4$ a-ab-ba «a-ba»** / *nap-ḫar ma-ta-a-tú tam-tum-ma* 'All the world was sea'. The lines 17-18 of this text tell that Marduk creates dry land by pouring soil on a raft; this text is several hundreds of year younger than the text of *enūma eliš*.

906 For the influence of the West-Semitic mythology on the Mesopotamian one: see the chapter Epilogue. Lambert (1991; reprint of a paper from 1965) mentioned the impact of the (West)-Semitic culture on the Babylonian civilization; in the attendant postscript (p. 111) he wrote: 'The archive from mid-third-millennium Ebla (...) shows that much of the Ugaritic pantheon was firmly established in Syria long before the Amorites appeared, and they had comparatively little impact on it. The third-millennium texts from Mari favour the same conclusions.'

Ti'amat and Apsû as the primaeval origin – as it is known from *enūma eliš* – has a (West)-Semitic origin [907]. After the OB period the cosmogonic origin developed from a monistic principle (Namma) into a dualistic principle: Ti'amat and Apsû.

4.7.2 *Anthropogeny*

In this section the texts about the creation of man in Sumerian and Akkadian language will be analysed with respect to the origin of the ideas about that creation. Is it possible to distinguish between Sumerian and Akkadian/Semitic ideas in this respect? Several aspects will be discussed.

1. It is the general opinion that the OB corpus of Sumerian texts – at least partly – consists of text traditions, which originally were written in the Ur III period (or even earlier). 'Enki and Ninmaḫ' presumably belongs to that OB corpus of Sumerian texts. 'The Song of the Hoe' is probably of OB provenance.
2. Both Sumerian texts mention only the use of clay for the creation of man [908].
3. In the OB Akkadian text *atra-ḫasīs*, the creation of man is performed with the aid of clay and the flesh and blood of a slaughtered god.
4. In the bilingual MB text KAR 4, only the blood of two slaughtered gods was used for the creation of man.
5. In the MB text *enūma eliš* only divine blood was used.

---

907 Komoróczy (1973, 31-32) mentioned that the theomachy of *enūma eliš* – the fight of Marduk against the creatures of the primordial chaos – originates from Ugaritic mythology (with references to Lambert and Jacobsen). Already in a Mari-letter (A.1968; Durand 1993) was written: 2' ${}^{\text{ĝiš}}$**tukul-[meš]** 3' *ša it-ti te-em-tim am-ta-aḫ-ṣú* 4' *ad-di-na-ak-kum* 'the weapons with which I defeated *tâmtu*/the sea I gave to you'. In other words: the mytheme of the fight against *tâmtu*-the sea-Ti'amat was already known during the 18th century BCE. The word *tâmtu*/the sea in this letter may have been used metonymically, e.g. for the West-Semitic enemies.
Several authors (Yeivin 1971; Yadin 1971; Grafman 1972; Kaplan 1976) have tried to identify some of the scenes pictured 1. on an early-Akkadian seal [PSBA 14, 210, no. 6]; 2. on the ʿAin-Samiya goblet from the Middle Bronze Age I period (ca. 2100-2000 BCE); and 3. on a clay plaque from Khafajeh dating from the Isin-Larsa period, with the battle between Marduk and Ti'amat in *enūma eliš*. Ad 1. Amiet (1953, 150) described the early-Akkadian seal as an example which '(...) annonce déjà celui du dieu vainqueur des scènes mythologiques akkadiennes.' The figure supposed to represent Marduk is just a naked 'hero'. Ad 2. The scene on the ʿAin-Samiya goblet, although seemingly fitting in with the Marduk-Ti'amat combat, more likely represents an unknown mythological scene, but possibly a predecessor or forerunner (or West-Semitic parallel) of that combat described in *enūma eliš*. Ad 3. The Khafajeh clay plaque shows 'a god killing a fiery cyclops' (Frankfort 1970, 113, fig. 122). In all likelihood this scene represents the fight of Ninurta against Asag (see e.g. van Dijk 1983, Tome I, frontispiece). The present author doubts if these pictures of about 700-1000 years before *enūma eliš* could have described a scene from this text. In this respect I would like to refer to Wiggermann (1993-1997, 229) who wrote: 'Before Ee (= *enūma eliš; JL*) a connection of monsters with the early cosmos cannot be proved, (...).' In conclusion: it is very unlikely that the above-mentioned scenes 1-3 represent the combat of Marduk - Ti'amat.

908 As has been proposed before (chapter 4.4.3), it may be that in 'Enki and Ninmaḫ' the use of divine blood has been implicitly assumed. For Namma is bearing the first human foetus, and at least according to human principles, the menstruation ceases and that blood may now be used for the growth of the foetus.

The motifs in the Sumerian texts are:
1. The gods, worn-out by working hard for their daily rations, decide – without rebellion – that a new being, man, should be created;
2. The creation of man is effected with the aid of clay alone.

The motifs in the Akkadan/Semitic texts are:
1. The gods who have to toil hard rise in revolt against the higher gods;
2. Violence occurred before the creation of man: the slaughter of one or two gods was necessary in order to be able to use the divine blood for the creation of man. In the OB text *atra-ḫasīs* clay and the flesh of the slaughtered god was also necessary. In the later texts (KAR 4; *enūma eliš*) the divine blood was sufficient.

Several scholars have discussed the Akkadian art in Sargonic times [909]. These authors were united about one aspect of Akkadian art, viz. the fighting between gods was typically Akkadian. Westenholz concluded [910]: 'Most of the distinctive features make their appearance with abrupt suddenness under Sargon, and many disappeared just as suddenly after Akkade's fall.' In the words of Steinkeller [911]: 'We know, for example, that, while not entirely free of conflict, the relations within the Sumerian pantheon were generally peaceful, with co-operation rather than rivalry being the norm of behaviour. And, if a conflict did occur, it was rarely if ever solved through violence, mediation being the usual way out. Thus, the motif of theomachy, which is so conspicuous and prevalent on Sargonic seals, is completely unknown in Sumerian mythology.'

Komoróczy wrote [912]: 'Old Mesopotamian mythology (...) did not know the theomachy: in the Sumerian texts the well-ordered world developed without its gods being compelled to fight with each other. Although we find the fights of different deities also there, but these fights do not constitute a part of the cosmogonic process (...).'

Of course it is clear that in OB times we cannot speak of 'Sargonic art'. But the feature of violence, characteristic of that Sargonic art, is reflected again in the second-millennium Akkadian texts about anthropogeny.

Based on these considerations, the tentative conclusion may be that the story of man's creation as told in 'Enki and Ninmaḫ' – although the oldest copy stems from the OB period – represents Sumerian ideas about this subject: the Sumerian goddess Namma is one of the protagonists, there is no question of violence, and only clay is used to create man. Sumerian ideas were still living on and written down in the scribal schools in the OB times.

The Akkadian texts concerning anthropogeny are suffused with the spirit of violence; this violence, viz. the slaughter of a god, and the explicit use of blood have been shown to be typical Akkadian/Semitic features [913].

---

909 Nissen 1993; Steinkeller 1992; Westenholz 1999, ch. 6, 85-89.

910 Westenholz 1999, 85.

911 Steinkeller 1992, 246-247. Nissen (1993, 104) was more careful: 'In spite of what I said before about differences between the ethnic groups of the Sumerians and the Akkadians, I see no reason to reintroduce the concept of ethnically bound art.'

912 Komoróczy 1973, 32.

913 Hruška (1974, 274) has a different opinion: 'Obwohl die einzelnen keilschriftlichen Quellen zur Menschenschöpfung in zwei Sprachen geschrieben sind und ihr Inhalt mit der Zeit ab und zu Änderungen und Neuerungen aufweist, *kann man doch nicht von zwei grundverschiedenen Überlieferungen von der Erschaffung des Menschen sprechen.*' (*ital. JL*).

4.8 *Some final remarks with respect to former theories*

In ch. 1.2 the theories of several authors with respect to Mesopotamian Beginnings have been described. In this section some general comments on the main points of these theories will be made.

4.8.1 *Absence of a diachronic study*

One disadvantage of the former studies is the lack of a diachronic study [914]. One refers to Sumerian and Akkadian texts, thereby mixing up texts of both languages and of all periods. When presented in this way, it may seem as if all the mentioned prime elements can be found in texts of every period. My study has tried to show that a diachronic study is possible.

4.8.2 *Van Dijk* [915]

4.8.2.1 *The cosmic and the chthonic motif*

Van Dijk contrasts the two 'beginning' motifs, the *cosmic* motif and the *chthonic* motif, which should indicate a pluralism of the Sumerian ideas about 'the Beginning'.

In the cosmic motif, heaven and earth 'fertilize each other' [916], or in other words by van Dijk, too: **an-ki** should be androgynous. This system corresponds with the *emersio* of gods and man, and may belong to the milieu of desert-dwellers.

In the chthonic motif, the **abzu** and mother earth form the principle of life; the *formatio* of man corresponds with this system, which may be attributed to the sedentary and agricultural population.

As I have shown in this study, these two systems are not opposed to each other. There was a gradual development and extension of the – originally Sumerian – ideas from ED times onwards, starting with the primaeval unit **an-ki**, consisting of the male principle **an** and the female principle **ki**. The principle of the primaeval waters, viz. Namma, who brings forth the unit **an-ki** – and thus the origin of the universe – was introduced during or shortly after the Ur III period. After the Old Babylonian period, the Akkadian/Semitic mytheme of the watery beginning from one single entity – Namma –, was developed into the beginning in which a sexually active couple was present, viz. the male Apsû and the female Ti'amat.

4.8.2.2 *Emersio vs formatio*

The *emersio*-type of man's creation [917] – 'man is sprouting from the earth' – could not be demonstrated with the aid of texts. On the basis of the first lines of 'Enki's Journey to Nibru', van Dijk concluded that An had been involved in the process of man's creation [918], which could not be proved in my studies.

---

914 Van Dijk (1971, 432) agrees with Oppenheim that such a diachronic study is hardly possible, probably the reason why he never tried to do it.

915 See ch. 1.2.1.6.

916 Van Dijk 1964, 14, 58 *ad* 2: 'Ciel et Terre se fertilisent mutuellement'. Enki and Ninki are also called 'êtres androgynes' by van Dijk (1964, 12). He derived this appellation from Eliade1954, 78-80. The indication **en-/nin-** is sufficient to prove that they are not androgynous beings.

917 This theory was first formulated by van Dijk 1964, 23 (see ch. 1.2.1.5) and it is still held today.

918 Van Dijk 1964, 23-24.

4.8.2.3 *Nomadic vs sedentary ideas*

Van Dijk's attribution of his cosmic and chthonic system to two different population groups – cosmic ~ nomads; chthonic ~ sedentary and agricultural population – cannot be proven by my study [919]. The first observed mytheme, **an-ki** as the primaeval unit, may have been developed by nomads, but we do not know how long this idea had already existed before it was written down. In the earliest sedentary communities – without irrigation canals – people were also dependent on rainfall, just like the pastoral people. The first attestations originate from the ED period, in which many of the Sumerian population were sedentary. Water, viz. Namma, as the primaeval origin may have originated from Ur III times onwards in a Mesopotamian community, in which human affairs were becoming more and more important, and whose population was predominantly sedentary. The idea may have come into being by analogy with the process of pregnancy and the birth of a human baby. The origin of the universe, realized by the mingling waters of Apsû and Ti'amat, presumably has a (West)-Semitic origin; but the present author would not like to give the label 'nomadic' or 'sedentary' to this idea [920].

4.8.3 *The concept of 'Time' in Mesopotamian Beginnings* [921]

4.8.3.1 *Lambert's ideas*

On several occasions, Lambert discussed 'Time' as a cosmological principle in Sumerian and Babylonian thoughts [922]. In the summary of his last contribution we read: 'Sumerian and Babylonian thinkers (...) looked for one element from which everything else in the universe somehow evolved. Earth and water were the two elements most commonly assumed to have had this function. (...) A third, and much less common principle was Time: eternal time, which is a remarkable sophistication, seeing that time is immaterial.' From Lambert's earlier publications it is known that he based the idea of Time as a primaeval principle on the presence of the pair *^{d}du-rí - ^{d}da-rí* 'ever and ever' in the so-called 'ancestor' list of An [923]. As has been discussed before [924], the 'ancestor' list of An is a theo-

---

919 According to Staal (1986), fertility myths seem more prominent in agrarian, sedentary communities, where the change of the seasons is the all-controlling factor, rather than in an nomadic community, whose life has a discontinuous character in several respects. This may be based on the dependence of those agrarian communities on rainfall, which can be seen as 'heaven impregnating earth'. On the other hand, in my opinion nomadic people are dependent on rainfall as well, for their transhumance – the seasonal move with small live stock between summer and winter pastureland – is the proof of this. Therefore it seems to me very difficult to connect fertility myths exclusively to either agrarian/sedentary or to nomadic people.

920 See also Epilogue § 2.2, Mesopotamian cosmogonies *ad* World Parents creation.

921 The theme of the 56th RAI 2010 at Barcelona was 'Time and History in the Ancient Near East'. At the time of writing of this section, only the abstracts of the contributions are available to me.

922 Lambert: 1980-1983a, 220; 1975a, 54; 2008, 31-32; 2010 56th RAI.

923 The god list **an** = *anum*, I: 12-13. Lambert (2008, 32) also referred to 'a second-millennium example of Dūri-Dāri' [a short incantation in Elamite language] with 'two lines in dialectal Sumerian': *obv.* 7 **tu-ri ta-ri en mu-ul-li** 8. **ta-ri ta-ri en mu-ul-li**; his translation and comments: 'Dūri-Dāri, lord Enlil, Dāri-Dāri, lord Enlil. Most probably "lord Enlil" is abbreviated for "lords of mother-father of Enlil", and here the motif of eternal time is made the ancestor of Enlil (...)'. Although I am not able to place these lines in context or to translate this incantation, I still would like to make some comments. Remarkable is the absence of the **diĝir**-determinative at **mu-ul-li**, while in *rev.* 11 of this text one of the names for Enki/Ea (? [see **an** =

logical construct. A few names on this list – Anšar-Kišar; Laḫmu-Laḫamu – are also present in *enūma eliš*, albeit in the reverse order. Both in *enūma eliš* and in the list **an** = *anum* the primaeval origin is water, not time. Time has never been, in any Sumerian or Akkadian text, the primordial beginning of the universe. The appellations *dūri* - *dāri* each may be translated as 'the eternal one' [925].

4.8.3.2 *Time expression in Sumerian texts about 'Beginning'*

As far as the first lines of the texts concerning the 'Beginning' – the texts discussed in this study – have been preserved, all the time indications begin with **ud/u$_4$** [926]. This sign originally represented a picture of the rising sun. This indicates that the Sumerian conception of 'Time' has had its origin in the perpetual 'movement' of the sun. My translation of the word **ud** is 'sun, day'. Sometimes the expression **u$_4$-ri-a** is translated as 'at that time' [927]. It must be clear that the translation 'time' at that instance is our modern interpretation; the literal meaning is: 'on that remote day', and it does not reflect any philosophical idea. In fact there is a contradiction between the Sumerian phrase indicating the beginning – **u$_4$-ri-a** 'on that remote day' – and the absence of **ud/u$_4$** 'sun', with the aid of which a day and thus time could be measured, at that moment. In Sumerian texts there is no mention of 'Time' before the Beginning, or of the impossibility of measuring 'time' until the appearance of the sun god Utu. Therefore it may be concluded that Sumerians implicitly accepted that things happened 'on that remote day' – like the cosmic 'marriage' and/or the preparation for it – before time was measurable, viz. before the birth of Utu.

4.9 *Summary – Conclusions*

The results of my research as presented in this study may suggest that the Sumerian ideas about the Beginnings (gradually) developed along one line of tradition. However, there may have existed several Sumerian mythological traditions because of the following reasons:

1. The time covered by the Sumerian texts in this study was about one millennium.
2. The various Sumerian cities – each with its own city god – may have had their own specific tradition.
3. Several foreign civilizations influenced the Sumerian culture in varying degrees in the various parts of Sumer.

---

anum II:159 ^{d}**nun-ur$_4$-ra**]) is written with the determinative as ^{d}**nun-ùr-ri**. The question is: does **mu-ul-li** really represent Enlil, for the usual dialectal spelling is **mu-ul-lil$_2$**, or could it be a participle of *ullû* 'to elevate, raise extol' ? And why, if it is 'dialectal Sumerian', is **en** written instead of /**umun**/ for 'lord'? Lambert (1980-1983a, 220b) also suggested that (Ḫ)alma and (Ḫ)al(l)ama, in **an** = *anum* I:20-21 written as ^{d}"**ALAM**, may express eternal Time. According to Lambert's explanation, 'The reference to time is based on the equation *almû* = *ullû* ("eternal") in Malku = šarru VIII 110 and on the Hebrew ʿ*ôlām* "eternity" '. Note, however, that neither in CAD A I 364 *s.v. almû*, nor in PSD A III 160-170 *s.v.* **alam**, the connection between **ALAM** and eternity could be found.

924 Chapter 4.2.3.2 of this study.

925 For a possible explanation with respect to the presence of the pair *dūri* - *dāri* in the list **an** = *anum*, see ch. 3.2.8.1 note 425.

926 Ukg 15: **u$_4$-da**; Barton cylinder: **u$_4$-rí-a**; GEN: **u$_4$-ri-a**; Enki & Ninmaḫ: **u$_4$-ri-a-ta**; KAR4: **u$_4$**.

927 E.g. Rubio, 56th RAI: column *iii* of AO 4153 (= Ukg 15). By courtesy of Rubio, who has sent me copies of the sheets he showed on the occasion of this Rencontre.

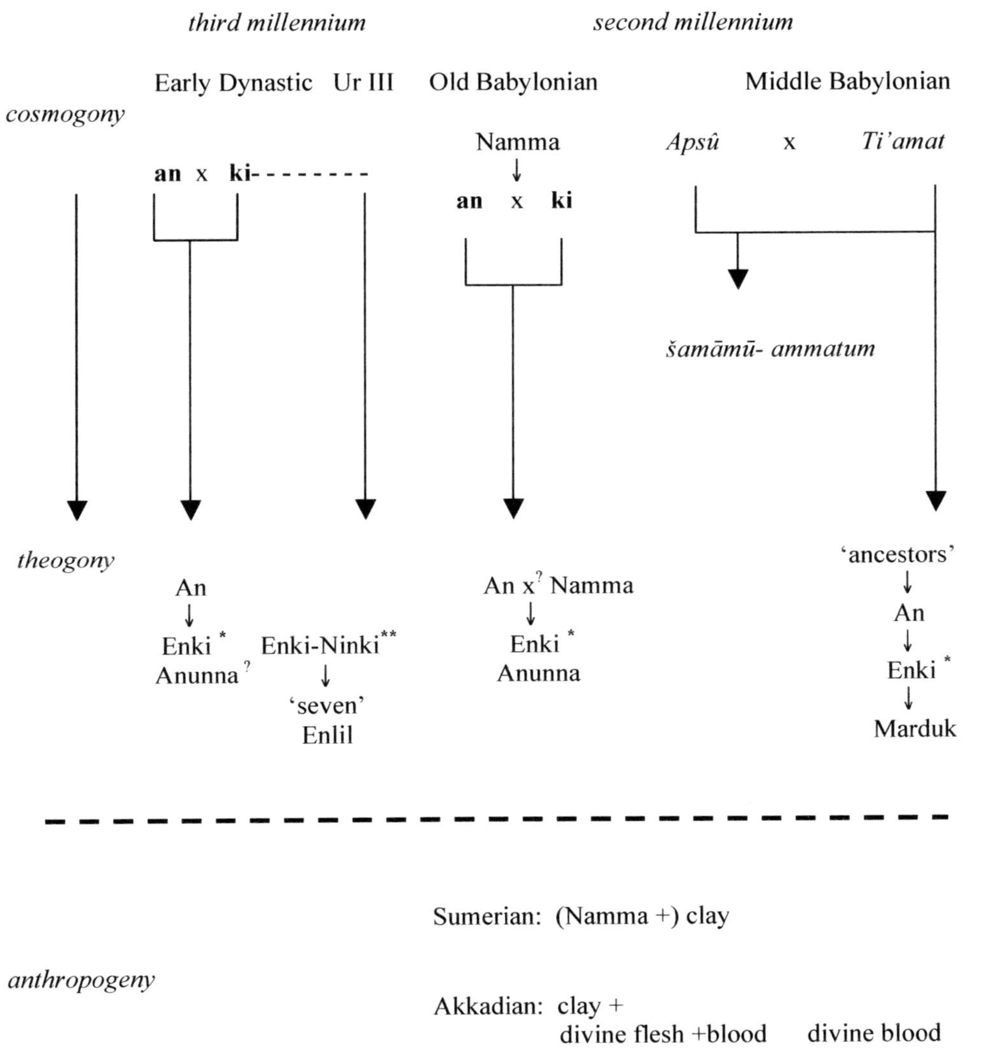

* Enki of Eridu

** Enki and Ninki: ancestors

x? An x? Namma: the question mark indicates that I do not know of any text which records that An and Namma *together* produced Enki and the Anunna.

Table 2:
Schematic summary of Mesopotamian Beginnings in the 3rd and 2nd millennium BCE

However, for the study of a possible differentiation between various mythological traditions in Sumer we have not enough texts at our disposal. The texts we have are for the greater part from Nippur. Moreover, the extant texts are sometimes damaged at crucial passages, or they are partly incomprehensible (e.g. the UD-GAL-NUN texts). In my opinion it is therefore not justified to do a study into the existence of a possible diversity of the Sumerian mythological tradition with the aid of these few texts.

The main conclusions of this study, schematically summarized in table 2, are:

Third millennium BCE

- The Sumerian idea about the Beginning was, that there was a primaeval universe, viz. the unit **an-ki** 'heaven-earth'.
- **an-ki** – as An and Ninḫursaĝa – created Enki (of Eridu) and probably the Anunna.
- Presumably from **ki** alone, the pair Enki and Ninki developed, who produced 'the seven', 'developmental stages' or 'spheres of concern' (later called 'ancestors') of Enlil; finally Enki and Ninki became the parents of Enlil.
- Initially, Enki of Eridu was the most important god, before Enlil was introduced and accepted as the supreme god of the Sumerian pantheon, presumably during the late Uruk period.
- Enlil divided the unit **an-ki**.
- The creation of man was not mentioned.

Beginning of the second millennium BCE

- The primaeval origin was water, viz. Namma, who bore **an-ki**.
- Presumably An and Namma created Enki (of Eridu) and the Anunna.
- Only the god lists mention 'ancestors' of Enlil.
- Heaven and earth were separated.
- The creation of man according to the Sumerian view: Namma produced the first man from a clay model.
- The creation of man according to the Akkadian/Semitic view: man was created from clay mixed with the flesh and blood of a slaughtered god.

Second half of the second millennium BCE

- The Akkadian/Semitic ideas of the primaeval beginning were expressed as a pair, Apsû (male) and Ti'amat (female), both representing primaeval water, who produced the unit *šamāmū-ammatum* 'heaven-earth'.
- Within these primaeval waters several pairs of gods, as ancestors of An, came into existence (Laḫmu-Laḫamu and Anšar-Kišar).
- Marduk created the actual universe – by covering heaven and earth each with the half of Ti'amat's body – and separated the unit *šamāmū-erṣetu*.
- Man was created with the aid of divine blood [928].

---

[928] Van Wolde (2009a,b) discussed the word ברא *bārāʾ* in Genesis 1.1 - 2.4a and she concluded that the verb in this context does not mean 'to create' but 'to separate'. This conclusion was criticised and finally renounced by Baasten (2009) and Becking & Korpel (2010). I fully agree with these authors in this respect. Baasten wrote that *bārāʾ* does not mean 'to create from nothing' (2009, 5). Becking & Korpel (2010, 3) referred to Gesenius, who offered as one of the meanings of *bārāʾ* 'hervorbringen'. To prevent a possible misunderstanding as a result of the translation of *bārāʾ* as 'to create' in Genesis 1.1 - 2.4a, I suggest the following translation for the verb *bārāʾ* in

4.10 *Analysis and interpretation of the texts*

In this section, at the end of the text interpretations and conclusions, I would like to justify briefly the way in which the texts used in this study were analyzed. Formally speaking, 1) not all the texts edited in this study can be attributed with certainty to a specific genre; and 2) the 'stories about the Beginnings' are mostly contained within just a few lines, as an introduction to the main text. In any case, these 'stories' – no matter how short they may be – could be conceived of as myths.

There are several definitions and approaches with respect to mythology and myths. On the one hand there is Bottéro's definition of mythology [929]:

> 'An intellectual procedure which consists of responding to the great questions about the origins and the meaning of the universe and our existence, as well as the role and the activity of the gods, who are considered to have directed everything. This procedure worked not by rational and conceptual analysis, in order to find the truth, but by imaginative activity that gave answers that were no more than probable but were considered to be sufficient.'

Lambert wrote about the Mesopotamian myth [930]:

> '(...) der mesopotamische Mythos (...) umfaßte Religion, Philosophie und Wissenschaft für den frühen Menschen und diente dazu, die Welt und des Menschen Rolle darin zu erklären, das heißt, dem geistigen wie dem gefühlsbestimmten Verhältnis des Menschen zu den großen kosmischen Mächten, die ihn umgaben, Ausdruck und Sinn zu geben.'

Lévi-Strauss devoted a chapter to 'The Structural Study of Myth' [931]. One of the consequences of his working hypothesis is, that 'Myth, like the rest of language, is made up of constituent units.' [932]

When we analyse the Sumerian Beginnings according to this principle, it appears that several main themes are present:

---

this context, viz. 'to bring out'. My translation of Genesis 1.1 - 1.2 is as follows: 'When God began to bring out the heaven and the earth – the earth was without form and void, darkness was upon the face of the deep, and the Spirit of God was hovering over the face of the waters – God said: ...'. In my opinion there is a striking similarity between the lines 3-6 of *enūma eliš* and Genesis 1.1 - 1.2. In both episodes the earth, as it appeared in the beginning, was void and not complete, but it got its actual appearance by the actions of Marduk and God, respectively.

929 Bottéro 1995, 298.

930 Lambert 1974, 3.

931 Lévi-Strauss 1969, ch. XI, 206-231. A critical analysis of the structuralistic ideas of Lévi-Strauss was presented by De Ruijter (1979). One of the shortcomings of Lévi-Strauss' study, as he called it, is that all the examples on the basis of which he makes generalisations have not been derived from Semitic, pre-Hellenic or Indo-European cultures (De Ruijter 1979, 89, with reference to Ricoeur). Doty (2000, ch. 9, 266-302) discusses Lévi-Strauss' structuralism. A summary of criticisms of Lévi-Strauss' approach to mythology is given (pp. 282-283). Dupré (1975, 258) and Ong (2004, 161-162) also mention objections to Lévi-Strauss' theories and structuralistic approach. Larsen (1987) discussed Lévi-Strauss' presentation (in his book 'La pensée sauvage') of the dicho-tomous interpretation of the world of human thoughts.

932 Lévi-Strauss 1969, 210.

- the intercourse and/or the cosmic marriage of **an** and **ki**;
- the division of the unit **an-ki** into **an** and **ki**;
- the creation of the gods;
- the creation of (the first) man.

Within each main theme several sub-themes – mythemes – may be discerned, e.g.:

- the preparation of each of the two partners for their cosmic 'marriage';
- the moulding of the first man with the aid of clay.

As far as I can see, all these mythemes endeavour to find and to explain origins. The basis for this was the visual experience of their environment and/or their experience of life. To give only a few examples:

- Water makes the earth productive. Rain is water from heaven. The experience at human level of fertility is the union of man and wife [933]. Thus the union of heaven and earth must have been the beginning of the universe.
- In the beginning – **an** and **ki** are alone – there was darkness, for the sun was not yet present. During the night there is at least some light, especially due to the moon. Thus, in Sumerian thought, the moon came into being prior to the sun; in mythological terms: Utu, the sun god (male), is the son of Nanna, the (male) god of the moon, son of Enlil and Ninlil [934].
- When a dead body has been buried, it will become dust, clay. So the basic constituent for the formation of man must also have been clay.

In recent years more and more the degree of applicability of the structural analysis in the study of Sumerian and Akkadian texts has been discussed. Michalowski wrote in this respect [935]:

> 'The written culture must be seen (...) as a limited as well as a limiting form of discourse, with its own rhetoric and its own ideology. (...) Thus, one must see this written mythology as quite different from the kinds one encounters in oral societies. As in such cultures, myth never transmits pure information but is a repository of rhetoric and cultural style as much as it is of values and traditions. All these elements are different, however, if the texts are available to a limited group within a culture.'

As we have seen in the Sumerian Beginnings, there was a change in the concept of the origin of the universe: the cosmogonic beginning, initially represented by the cosmic marriage of **an** and **ki**, was later replaced by the primaeval water(s) as the primordial origin. This happened in a period when interest in the human being had developed, and as a consequence the 'birth' of the universe and that of a human being may have been conceived as analogous processes.

---

933 The cuneiform sign **a** has the meaning: 1. water, 2. semen.

934 Michalowski (2002, 414-415) wrote about the peculiarity of this relationship between sun and moon – both male gods – in the Mesopotamian religion, because in other societies 'one finds that most often the two great lights are of different sex, (...).' He also refers in this respect to Lévi-Strauss' chapter 'The Sex of the Sun and the Moon' in his *Structural Anthropology, vol. II.*

935 Michalowski 1996, 191-192.

The inclusion of the (visual) experience of everyday life in mythological stories may lead to different stories by different peoples about the same subject [936]. An example of this is provided by the myths about Beginnings. These are not identical for the various peoples: their environmental experience – in the broadest sense – has a determining influence [937].

In conclusion: attention has been focused on the philological aspects, on the diachronic aspects, the influence of a changing population in Mesopotamia, and the advancing scientific and cultural development of Mesopotamian people. I have tried to make a logical and coherent analysis of the material available, with a restricted use of the various analytical methods as discussed before, taking into consideration the critical reviews, some of which are referred to here.

***

936 Quirke (2008, 62) made the following remarks with respect to the study of creation stories: 'Certain preliminary observations may be made to encourage innovative research in the comparative study of creation stories. In the first place, the objects of this comparative research involve human subjects that are difficult to reduce to the level of similar and different, however heuristically and didactically useful this may be for introducing an unfamiliar society. Differences and similarities may mislead when taken separately. A tendency to focus exclusively on similarities between different societies or periods resulted in postwar structuralist anthropology, and in prewar diffusionist archaeology and history [C. Renfrew, P. Bahn, *Archaeology: Theories, Methods and Practice.* London and New York: Thames and Hudson, 2000, 34]. More recently, the differences between societies have been identified as more appropriate objects of research, in the continuing specialisation within both archaeology and anthropology. Although this is dominant in contemporary research, it is not necessarily much more productive than the opposite extreme. Similarities and differences may be interesting in comparative study, but they should be understood as interpretative and partial aspects of the complexities of human social interaction.'

937 See the Epilogue of this study.

Epilogue

Beginnings and creation myths worldwide compared with the Mesopotamian Beginnings

'Il ne suffit pas d'identifier avec précision chaque animal, chaque plante, pierre, corps céleste ou phénomène naturel évoqués dans les mythes et le rituel (...), il faut aussi savoir quel rôle chaque culture leur attribue au sein d'un système de significations.'
[C. Lévi-Strauss, *La Pensée Sauvage*; 1962, 73]

In this chapter the Sumerian/Akkadian Beginnings, as outlined before, will be compared with creation myths of the rest of the world. This will be done in two parts. First of all, the creation stories of the Sumerians/Akkadians will be compared with those of other people in the Near East. In the second part I will investigate which of the Mesopotamian ideas about the Beginnings can be found in other parts of the world, outside the Near East.

1 *The Ancient Near East*

1.1 *Egypt* [938]

The only people with a literary legacy nearly contemporary with that of Sumer are the Egyptians. The extant texts may be divided as: texts dealing with (rites for) the afterlife and funerary rites, and those dealing with festival rites. These texts, and the periods from which they originate, are:
1. Pyramid texts; Old Kingdom (from *ca.* 2400 - 2150 BCE);
2. Coffin texts; Middle Kingdom (from *ca.* 2050 - 1850 BCE);
3. Book of the Dead; New Kingdom (from *ca.* 1600 - 100 BCE).

My main interest is in the Pyramid and Coffin texts, because the Sumerian texts studied also date from that period. The sources from the Old and Middle Kingdom show a homogeneous picture with respect to the origin of the world; this applies to sources within the entire country of Egypt. The texts show a particular unity with respect to the creator; within copies of the same spell there may be modifications, different wordings, but the cosmogonic principle is constant [939]. In fact, these ancient Egyptian texts are not narratives about genesis; at least: such texts – if they ever have existed – do not survive. Quirke prefers to speak of 'creation motifs' instead of 'creation stories'. In the Pyramid and Coffin texts there is not one fixed order of cosmogonic mythemes. This may be, at least partly,

938 The following text is based on Bickel 1994 and Quirke 2008, with notes to other references if necessary. The Egyptian 'Beginnings' are summarized in a very concise way; only the main points are described. For the role of birds in Egyptian creation myths: see Bailleul-LeSuer 2012, 131-134.

939 This is especially remarkable for the Coffin Texts, which were also found in South Egypt. The Pyramid Texts all come from Memphis and surroundings.

inherent in the non-narrative character of these texts. The Egyptian 'Beginnings' may be summarized as follows.

i. Nun

In the Egyptian beliefs, there is a world of pre-existence as origin of the creation. This pre-existing world, called Nun, may be described as a watery and dark chaos, both an element and a place. Nun is not a creator god, but it is the energy source that is the determining factor with respect to the beginning of creation. Nun is the milieu in which the creator god Atum is '*in inertia*'; at a certain point Nun stimulates Atum into consciousness. Nun continues to exist in the inner part of the created world and around it.

ii. Atum

Atum's 'awakening' – the transition from his pre-existent form into full existence – is described as a spontaneous impulse from himself, by his will: autogenesis. The Egyptian texts tell of the solitude of Atum in this primaeval period. Atum is the origin of the visible world, and as such he is the most important god. He is sometimes associated with the sun god Ra (or Re). Atum and Ra are two complementary aspects of the creator god, active at different points in the creation. Atum is the god who created everything 'in the beginning', but he does not intervene in the created world. The sun god Ra is the visible manifestation of Atum, who appears every day.

Some Pyramid texts mention clearly the concept of the *Primaeval Hill*, rising up from Nun and with which Atum is identified. This picture is considered one of the oldest cosmogonic notions.

After his autogenesis, Atum creates the constituent parts of the universe, with the aid of substances that he takes from himself. Creation is possible with the aid of his semen, saliva, breath, tears, word, will, or intellectual power. The substance sent out by the creator is immediately the object that has to be created.

iii. Ennead [940]

The Pyramid and Coffin texts mention how other gods were created. Atum brought forth *Shu* (male; god of air) and *Tefnut* (female; goddess of moisture[?] [941]). From this pair originate *Geb* (male; god of earth) and *Nut* (female; god of heaven), and from them: *Osiris*, *Isis*, *Seth* and *Nephthys*. These gods together – from Atum to Nephthys – are called: the Heliopolitan Ennead [942].

---

940 The concept of the Ogdoad – four pairs of gods at the beginning – seemed also to be known already in the Old Kingdom: a Pyramid text (PT § 2270) mentioned Hermopolis as 'the town of the Eight' (Khemenu).

941 This interpretation, just like the interpretation of Assmann (2004, 158) as the goddess of fire, is very uncertain (by courtesy of prof. O.E. Kaper).

942 Bickel (1994, 128) remarks with respect to 'creation': 'Bien que les principaux constituants du monde soient des émanations du dieu unique (= Atum; *JL*), nous avons constaté que l'apparition de tous les éléments (à l'exception peut-être de l'humanité) est accompagnée d'un acte volitif. La cosmogonie égyptienne n'est pas une processus «automatique», mais une série de décisions prises par l'autogène et suivies d'actes créateurs conscients. C'est pour cette raison que nous utilisons dans cette étude les termes créateur et création le plus souvent à propos d'Atoum et de son œuvre cosmogonique.'

iv. Heaven and Earth

The creation of heaven and earth does not have the same interest as it has in other civilizations. It has been described in only a few Coffin texts, and it does not seem to be a fundamental cosmogonic idea. It is mentioned as it were 'casually', without saying how Atum has done it. Other texts relating about the appearance of heaven and earth do not mention their origin nor their creator, as if heaven and earth had originated from an inherent force.

Bickel summarizes the separation of heaven and earth as follows [943]:

> '1. La séparation comme phénomène lié à la constitution de l'univers se situe dans la phase de transition entre la préexistence et l'existence. La conception d'une éclosion commune d'Atoum, du ciel et de la terre suppose que ces trois éléments étaient liés dans la substance de la préexistence et qu'ils se sont séparés les uns des autres et par rapport au Noun au moment de l'autogenèse et de la prise de conscience du créateur. La créateur peut aussi être considéré comme celui qui a provoqué la séparation des deux éléments. (...)
> 2. La séparation du ciel et de la terre par Chou s'inscrit dans la phase de structuration et de maintien de l'univers créé. Elle est décrite de façon très imagée et les textes préfigurent clairement un tableau iconographique [944] répandu bien plus tard. Le but principal de cette image, à laquelle s'attachent de nombreuses conceptions et espérances funéraires, est d'expliquer le monde tel qu'il est. Cette image constitue l'étiologie de l'œuvre du créateur. Elle fournit en même temps la garantie de la stabilité du cosmos.'

v. Creation of man

The Pyramid texts contain hardly any information about the origin of man [945]. In one Coffin text it is said that 'people are the tears of my eye' [946]. There is no mention of 'the first man'; Atum created 'the species' man.

The notion of the god Khnum modelling a human being, like a potter with the aid of a potter's wheel, has been attested already in Pyramid texts. One spell in a Coffin text refers to the same activity of Khnum. On a bas relief in the temple of Amun-Ra in Luxor (in the room of theogamy; 14th cent. BCE), Khnum is present, creating a Pharaoh and his *ka* on a potter's wheel [947]. Although the texts never mention the use of clay with respect to the creation of a human being, the use of a potter's wheel may at least suggest it.

vi. Summary

The Pyramid texts already deal with the origin of the universe. Atum, inspired by Nun, creates his own existence, and after that two children, Shu and Tefnut. The period of the Coffin texts only extends the cosmogonic knowledge, but there is no essential difference

---

943 Bickel 1994, 197.
944 See e.g. Baines and Málek 1993, 214, fig. with Geb, Shu and Nut.
945 There is a passage in a Pyramid text (Faulkner 1985, 226: utterance 571, § 1466) that mentions that the Pharaoh 'was fashioned by his father Atum before the sky existed, before earth existed, before men existed, before the gods were born, before death existed.'
946 Bickel 1994, 93: CT 1130 vii 465a.
947 For a reproduction of this relief: see Wasilewska 2000, 139, fig. 8.1.

between the ideas of the Old Kingdom and those of the Middle Kingdom. There are some additions. One is that the origin of all is not a voluntary autogenesis of Atum, but a process of exchange of life energy between Atum, Shu and Tefnut; these three protagonists form a symbiotic unity. Other extensions concern the ways in which Atum is able to create.

Atum was a very theological entity, hardly to be approached by the faithful, in contrast to his children who were responsible for the daily routine. This picture of an inactive and distant creator has led to changes in the theological thoughts from the New Kingdom (*ca.* 1500 BCE.). One of these new ideas was the concept of more creator gods. From now on the creator god was, after his creation act, also responsible for what he had created. Gods were now closely connected with one city or region. As a consequence, every city or its temple could be considered as the place where the creator – who had nearly always an aspect of the sun – had ascended from Nun [948]. The primaeval mound, originally restricted to Heliopolis, became one of the most widespread concepts.

The concept of the Ogdoad – four pairs of gods at the beginning – became also more widespread in the New Kingdom than it had been.

Both Bickel and Quirke dispute the idea that the Egyptian religion was made up of local and regional concepts. This idea was based on sources from the New Kingdom. The Pyramid texts and the Coffin texts – thus texts from the Old and Middle Kingdom – form one homogeneous picture. What can be observed are variations within several editions of one spell, but the unique creator was always Atum, emerging from Nun.

In the Old Kingdom Heliopolis was an important religious centre for the whole country. The religious principles were conceived during the first dynasties at the Pharaonic court, and Heliopolis was in the neighbourhood of it. It is likely that Heliopolis and Memphis continued to exist as a theological centre and as 'editorial places' of the greater part of the Coffin texts during the Middle Kingdom.

The changes that took place from the beginning of the New Kingdom – e.g. other creator gods besides Atum – may have been stimulated by the people who tried to come into closer contact with the gods. Therefore the local gods were the appropriate ones. These gods, originally descendants from Atum, became in due course themselves creator gods [949].

vii. Comparison of the Egyptian and Mesopotamian Beginnings

1. In the oldest known Egyptian cosmogonic system – mentioned in the Pyramid texts, 3rd millennium BCE – water (Nun) was the primaeval origin. This may be explained by the dependence of Egypt on the annual flooding of the Nile. At the same epoch, the Sumerian texts only show **an-ki** – Heaven-Earth – as the primaeval beginning. As a similar dependence on inundation can be observed for Sumer at that time, it may be remarkable that water was not conceived by the Sumerians as the beginning of all. Therefore the tentative conclusion may be, that the idea of **an-ki** as the primaeval beginning did not originate in Sumer itself. It may have been introduced by the Sumerians who brought this idea with them on their arrival in South Mesopotamia from a region where they did not depend on inundation but on rainfall for their sustentation.

---

948 Not every god originated from Nun; e.g. Ptah created with words.

949 For the detailed hypothesis with respect to the evolution of the Egyptian religion: see Bickel 1994, 297-298.

2. When water, being the goddess Namma, became the primaeval origin in Sumerian texts – from the beginning of the second millennium BCE onwards – it is noteworthy that Namma is the mother of **an-ki**. In the Egyptian cosmogonic system, Nun only stimulates Atum, who is already present in an inert phase, to become active.

3. Whereas Atum produces everything only with the aid of his own 'emanations' (of every kind), the Sumerian theogony requires the action of **an** and **ki**, or in terms of gods: of An, the god of Heaven, and of a mother goddess, e.g. Ninḫursaĝa.

4. Heaven-Earth (**an-ki**) are the first generation after Namma in the Sumerian cosmogony. In the Egyptian cosmogony Heaven-Earth (Geb-Nut [950]) are the second generation, after Shu and Tefnut (air-moisture$^{?}$).

5. In Egyptian texts the creation of man is mentioned, but no more than that. The texts do not describe why man was created. Perhaps the scant attention paid to 'normal' human beings in this respect may be explained by the very humble position of the common man in comparison with the Pharaoh. The Sumerian texts explicitly mention that man was created by the gods in order to support the gods.

For both the Egyptian and the Sumerian religion it is interesting, that only in the first half of the second millennium any attention was paid – at least in texts – to the origin of man.

6. The primaeval waters Nun and Namma have a different character. Nun is a watery and dark chaos [951], who is never described as the parent of Atum. Nun only stimulates the appearance of Atum by autogenesis. Namma, originally a minor Sumerian goddess representing the subterranean waters, in all likelihood has obtained her function as the primaeval mother under the influence of the Semitic Ti'amat, who represents the 'sea'. In *enūma eliš*, the primaeval waters are the male Apsû and the female Ti'amat, whose interaction is necessary for the generation of *šamāmū* and *ammatum*, heaven and earth.

Conclusion

The 'Uruk expansion' extended its influence as far as Egypt, as can be inferred from the copying of ceramic shapes, the imitation of iconographic themes and the adaptation in the management of goods [952]. This influence proved to be unidirectional: from Mesopotamia to Egypt. On the other hand, in the religious-theological-philosophical sphere, the differences as outlined above are so fundamental, that any kind of influence – be it unidirectional or mutual – at first glance seems out of the question, at least before the Akkad period.

1.2 *North-West Mesopotamia and Anatolia: Ḫattian-Hurrian-Hittite mythology*

The Hittite mythology has elements adopted from the Ḫattian, Hurrian and Mesopotamian mythology. The texts do not contain stories about cosmogony or how gods came into

---

950 The usual gender for Heaven is male and for Earth female; in the Egyptian mythology the gender of both is reversed.

951 In the Ogdoad, the male Nun is accompanied by a female partner: Naunet.

952 Watrin 2004-2005; Wilkinson 2002.

being[953]. The text CTH 344, known as 'The Kingdom in Heaven' or 'The Song of Kumarbi', includes some theomachy (Alalu was defeated by Anu, who in turn was defeated by Kumarbi), but no real 'Beginnings'. 'The Song of Ullikummi' contains two short references to the separation of heaven and earth, but only as 'informal remarks': 'And when they came and cut heaven and earth apart with a copper cutting tool (...)', and further: 'Let them bring forth the primaeval copper cutting tool with which they cut apart heaven and earth.' [954]

A text, called by Collins 'The creation of Man by the Fate Deities' [955], describes 'a ritual in which a female suppliant stands over a pit and recounts the story of the creation of man as part of a rite to regain or to ensure her own ability to conceive.' The prayer starts as follows:

> 'When they took heaven and earth, the gods divided (it) up for themselves, and the upperworld deities took heaven for themselves, and the underworld deities took the land beneath the earth for themselves. So each took something for himself.'

This refers to the moment that the gods came into being, and that heaven and (the land beneath the) earth became divided among the gods. Unlike the Mesopotamian concept in which Enlil, or An and Enlil together, are responsible for the separation of heaven and earth, in the Hittite mythology it appears to be a joint operation of cosmic and chthonic deities. This is the only reference in this text to primaeval times. The creation of man is mentioned thereafter, in the briefest way it could have been done: '... the mother goddesses (**diĝir-maḫ-meš**), who create human beings.' [956]

1.3 *Ugarit*

Several authors discuss the Ugarit literature in general [957]. Smith has given a comprehensive survey of the various opinions with respect to the cosmogonic interpretation of the Baal cycle [958]. One of the difficulties in these discussions is the definition of 'cosmogony'. His conclusions are: 'The Baal cycle does not describe primordial events such as the

---

953 Güterbock 1980-1983; Hoffner 1990; Lebrun 2000. Kümmel (1973, 76-77): 'Obwohl neuerdings neben Fassungen des Gilgamesh-Epos auch Fragmente des Atrachasis-Epos in Boğazköy nachgewiesen werden konnten, fehlt z. B. das mythische Motiv der Schöpfung und Sintflut in hethitischen Texten völlig, (...)'.

954 'The Song of Ullikummi', Tablet 3, lines 42 and 52-53, respectively. Güterbock 1952, 29; Hoffner 1990, 59, §§ 61, 63.

955 Collins 2002, 231. The text has been edited by Otten and Siegelová 1970.

956 Otten and Siegelová (1970, 38) comment: 'Unklar ist bei der Wendung «welche den Menschen *ša-am-(ma)-ni-eš-kán-zi*» die Zeitbeziehung, d.h. ob damit auf den Schöpfungsakt zu Anfang der Menschheitsgeschichte Bezug genommen wird (...) oder ob nicht vielmehr die Gottheiten bei jeder einzelnen Menschwerdung bis zur Geburt tätig sind.' This depends on the interpretation of the verbal form: as a 'historic present' or as a form with an iterative meaning.

957 The authors who contributed in this respect to Sasson (ed.) *Civilizations of the Ancient Near East* (2000) are: Crenshaw, Parker, Smith, van der Toorn. Further literature: Caquot 1980; Clifford 1994, ch. 5; Margalit 1981. The conclusion of these authors is that no myth concerning 'Beginnings' can be demonstrated in the extant texts.

958 Smith 1994, 75-87. Besides cosmogonic interpretations, Smith reviewed other explanations of this cycle, like ritual and seasonal theories, and historical and political views.

creation of the cosmos, but rather its maintenance through the power of the storm-god. The Baal Cycle concerns a more "recent" series of divine events compared to either the opening tablet of Enuma Elish or the opening chapters of Genesis, and in this sense the Baal Cycle differs sharply from the well-known "cosmogonies".' And further: 'The Baal Cycle therefore appears to be fundamentally different from any of these texts [959] in its expression of "cosmogony".' [960] With respect to Yamm, Smith noted: 'Like Tiamat, Yamm may represent a deity from an older theogonic tradition. Creation stories regularly describe the primeval chaos as watery.' And also: 'Given the absence of an attested Ugaritic tradition regarding creation, Yamm perhaps constituted the primeval waters in older traditional material or was related to these waters in some manner.' [961] Pardee criticized Smith [962], because he did not explain why Yamm, rather than an entity designated by the root *thm*, is Baal's enemy. The question is, according to Pardee, 'why these watery forces were seen as the enemy of the creator deity (probably ᵓIlu at Ugarit, though no cosmological myth is yet attested).'

Another discussion Smith has dealt with is the provenance of the conflict between gods, or between a god and a monster [963]. The mytheme of fighting gods is not uncommon in Mesopotamian texts: e.g. Ninurta's battle against the Anzu bird. According to Lambert, this element of conflict may have been used in the Marduk - Ti'amat fight in *enūma eliš* [964]. Others, including Jacobsen [965], argued that the Baal - Yamm story served as a model for the intrigue of *enūma eliš*. In any case: Sumerian texts do not contain 'cosmic sea' as the enemy. Smith concludes: 'Therefore, it is plausible that the complex development of the rendering of Marduk and Tiamat in Enuma Elish involved primarily East Semitic elements, but possibly West Semitic ones as well.' [966]

Finally, Smith and Pitard wrote [967]: 'While the Baal Cycle presents episodes of cosmic conflict (the battles with Yamm and Mot), one of the most noteworthy aspects of these stories is that they do not culminate in creation. It is thus not too surprising that there is no indication that these stories are set in a distant past. Rather the cycle seems to be set in the vague near present.'

Two Ugaritic gods may have been involved in creative processes in the primaeval time, as may be deduced from their epithets [968]. The god El has several epithets: *bny bnwt* 'builder of beings'; *ab adm* 'father of mankind'; *ab šnm*, most likely 'father of the lofty ones'. The

---

959 These texts are *atra-ḫasīs* and *enūma eliš*.

960 Both citations: Smith 1994, 82.

961 Both citations: Smith 1994, 85.

962 Pardee 1997, 242.

963 Smith 1994, 110-112 (with previous bibliography).

964 Lambert 1986b.

965 Jacobsen 1968, 106-107.

966 Smith 1994, 111-112. In a recent publication, Katz (2011) studied recurring mythemes in Mesopotamian literature. One of such mythemes is the battle, of which the fight between Marduk and Ti'amat is one example that has presumably a West-Semitic origin. Katz also points to the two different characters of Ti'amat: she is both a goddess (and mother) and a monster; both characteristics of Ti'amat are present in *enūma eliš*.

967 Smith and Pitard 2009, 45. Kirk (1970, 221-222), also commenting on the Baal cycle, interpreted these stories as the theme of the disappearing fertility god (= Baal).

968 Margalit 1981, 138.

goddess ʿAṯirat, the wife of El, is known as *qnyt ilm* 'creatrix of the gods'. These epithets are the only available allusions to Beginnings in Ugarit texts.

My summarizing conclusion is that at this moment there exist no Ugaritic texts about 'Beginnings' comparable with the Sumerian or Akkadian ones as discussed in this study.

2 *Creation myths worldwide* [969]

2.1 *Several distinguishable types of myth*

Leeming, following Long in this respect, distinguishes several main types of creation:

i. Creation from nothing, ex nihilo creation

This is the most common type. The central fact of the creation from nothing, or *ex nihilo*, creation myth type is a supreme deity, existing alone in a pre-creation emptiness or void, who consciously creates an organized universe on his own. This type of creation can be found in the Hebrew and Christian religion (Genesis 1), and in the Egyptian, Rig Veda (X:129), Mayan, Tuamotuan, Zuñi, Maori and Polynesian myths. In the Hebrew/Christian texts, and in the Maori and Maya myths the supreme god creates through his word.

ii. Creation from Chaos

Leeming defined this type as: '(...) creation from a pre-existing, undifferentiated or chaotic state represented by primal elements or sometimes by a primal object such as a cosmic egg (...)' (p. 1), and also '(...) the potential for creation already exists in some form of material that is eternal, (...)' (p. 9). The assumption is that at the creation from chaos the material already existed before the creator. As a consequence the creator in this type of creation is less omnipotent than in the ex-nihilo myth. Examples are found in the Chinese mythology with primaeval vapor containing *yin-yang* principles. Myths with a cosmic egg as primaeval principle are well-known in Egyptian mythology, and they have also a broad geographical distribution in the area surrounding the Black Sea [970]. Leeming classified the creation of humans as told in the Sumerian text 'Enki and Ninmaḫ' [971] as 'creation from chaos', the clay used for man's creation being the chaos.

iii. World Parent creation

'The world parent myth involves the breaking apart of a static primeval state. In one form of the world parent myth the beginning consists of the eternal union of the parents, a union that has to be broken in order for creation to take place. Another sort of world parent myth, sometimes a second part of the first, involves a stage of creation in which it is the body of a world parent that is itself separated, usually by an act of dismemberment. The body parts of this sacrificed deity parent actually become the world.' (p. 16). Heaven and earth are often identified as world parents. This kind of myth is widespread.

---

969 This survey of the various types of creation myths depends on the studies of Long (1963) and Leeming (2010, part I, 1-29). Only in particular cases will the exact references with page numbers be given, to avoid an overload of footnotes.

970 Haarmann and Marler 2008, 45.

971 Translation and edition of this text in ch. 2.1.8 and in the Appendix 'Text editions' no. 8 of this study.

iv. Emergence creation

This type of creation has been defined as 'creation by way of a hole in the earth' (p. 1). The creation of men is emphasized, '(...) the focus is on a process by which humans emerge in stages into this world from under the earth.' (p. 21). Emergence creation is characteristic of the culture of Native North Americans, particularly in the south-west region.

v. Earth-Diver creation

Leeming's definition (p. 1): '(...) creation by means of diving into the depths of the primordial waters.' And further: 'It is a myth type that stresses the creation of Earth as opposed to the larger cosmos. Animals often play an important role in the creation, as do the primeval waters and often an evil force that balances the good in a dualistic tension.' This type of myth can be found in many parts of the world, but is particularly important in Central Asia, India and native North America (p. 24).

2.2 *Mesopotamian cosmogonies*

The Mesopotamian cosmogonies may be best characterized by the following creation types: 'creation from chaos' and 'the world parent creation'.

Creation from Chaos

In this type of myth '(...) the potential for creation already exists in some form of material that is eternal, (...)' [972]. This material may be water. Another favoured image of chaos is the cosmic egg, that appears in all parts of the world (e.g. in Egypt) [973]. In the section 'Primordial waters in creation', Leeming explains the reasons for the ubiquity of water as the primaeval motif: 'All cultures naturally recognize water as a necessary source of life and survival, making it a useful symbol of creative fertility. Large masses of water are uncontrollable and, therefore, aptly representative of chaos. In tandem these two symbolic functions lead us to the idea of potential, as yet unformed creation. Perhaps most importantly, the waters speak to the larger metaphor of creation as birth. We are born of the maternal waters and so, in these myths, is creation itself.' [974]

It is obvious that the mythemes in which primaeval water(s), like Namma alone or Apsû and Ti'amat, produce the unit heaven-earth belong to the type 'Creation from Chaos'. Although in many myths of the world the primordial waters are the place where creation begins, there are no mythical notions that show any resemblance to the Mesopotamian myths in which cosmic waters – the Sumerian Namma or the Akkadian pair Ti'amat-Apsû – give birth to heaven-earth.

---

972 Leeming 2010, 9

973 According to Haarmann and Marler (2008, 45), the origin myth of the world egg belongs to the mythic repertoire of Uralic peoples in eastern Europe and of various peoples of the Ancient Orient. The world egg myth has a broad geographical distribution in the areas surrounding the Black Sea.

974 Leeming 2010, 341. Eliade (1958, § 60, 188) also discusses water symbolism: 'In cosmogony, in myth, ritual and iconography, water fills the same function in whatever type of cultural pattern we find it; it *precedes* all forms and *upholds* all creation.'

World Parent creation

The older Sumerian idea – **an-ki** (Heaven-Earth) is the primaeval origin – may be assigned to the class of 'World Parents creation'. Eliade gives a survey of the civilizations in which the creation motif of a primaeval pair Heaven-Earth as 'Beginning' occurs, e.g. in all civilizations of Oceania, and in several tribes in Africa and America [975]. '(...) this formula covers a large part of the beliefs concerning agriculture.' In the words of Long [976]: '(...) the symbol of completeness in high agricultural communities is often expressed by the union of earth mother and sky father.'

For the continuation of the creation it is necessary that after the coital connection between Heaven and Earth both parents are separated [977]. This can be effectuated in several ways, mostly by the offspring of the primaeval couple. This separation is sometimes violent. There is no myth in which the separation was performed just as the Sumerian texts tell us. In the oldest texts, from the Early Dynastic period [978], it is Enlil who separated **an-ki**; in an Old Babylonian text [979] An has taken heaven and Enlil has taken earth. In the Egyptian mythology, the situation is different: the father (Shu; Air) of Geb (Earth; male) and Nut (Heaven; female) carries out their separation [980].

2.3 *Mesopotamian anthropogeny*

For a survey of the world-wide aspect of the creation of man it may suffice to refer to the first chapter of Frazer's book about creation and evolution in primitive cosmogonies [981]. He established *inter alia*: 'The simple notion that the first man and woman were modelled out of clay by a god or other superhuman being is found in the tradition of many peoples.' To this extent the Sumerian anthropogeny is not a unique story. On the other hand, the use of divine blood and/or flesh of a slaughtered god to create man – as we have seen in the Akkadian myths of man's creation – is hardly to be found in the world mythology. Schipper describes one example in which a Being sprinkles two lifeless figures of men, scratched in the ground, with his own blood; then they came to life (Vanuatu, Melanesia).

---

975 Eliade 1958, § 84, 240-242. Citation: p. 241. Komoróczy (1973, 38) said that the idea of **an-ki** in the Sumerian cosmogony was clearly an analogical explanation. 'It was brought about by a visual experience, *viz.* the sight of the sky merging in the horizon with the level land.' In note 74 he mentioned: 'Although it is seldom a fortunate thing to base too much on the ethnographic parallels at the analysis of ancient mythology (...), now I still have to refer to such a «parallel», the value of which lies just in its negative character. It is generally known that among the peoples of Oceania there existed such cosmogonic ideas, according to which in olden times sky and sea formed an inseparable whole. There, where the horizon is on the plane of the sea, *this* idea is the natural one.'

976 Long 1963, 21.

977 Leeming 2010, 346-347.

978 See ch. 2.1.1b.

979 'Gilgameš, Enkidu and the Netherworld'; ch. 2.1.7.

980 The usual gender for Heaven is male and for Earth female. An is the god of Heaven, the personification of **an**. Enlil belongs to a younger generation than **an-ki**.

981 Frazer 1968. Citation: pp. 5-6. Frazer described one example (Maori, New Zealand; p. 7) in which Tiki, like the Babylonian Bēl, kneaded red clay with his own blood to create the first man. In recent literature this example could not not be found. Leeming (2010) describes the creation of man, e.g. in the sections 'Creation from Chaos' (9-16) and 'Emergence creation' (21-24).

In one other example (Maya, Guatamala) a god used parts of his own flesh to complete the first men [982].

3. *General conclusion*

In my study I have given a diachronic survey of the Mesopotamian ideas about 'Beginnings'. One may conclude that these ideas changed in due course, from the Early Dynastic period (early third millennium BCE) up to and including the second half of the second millennium BCE. Heaven-Earth as the primaeval origin were replaced by primordial water(s) who brought forth Heaven-Earth and the first gods. Creation of man was not a mythological subject until the Old Babylonian period.

The studies of Eliade, Frazer, Long and Leeming, and Schipper gave us the opportunity to compare the Mesopotamian mythologies about 'Beginnings' with world-wide ones. The Mesopotamian stories appeared not to be unique in the basic ideas about 'Beginnings'; most mythemes have a universal value, with one exception: the slaughter of a god to create the first man in the Akkadian version of man's origin.

***

982 Schipper 2010, 74 (Vanuatu) and 88-89 (Maya).

‘Mythical thinking is not concerned primarily with logic.
On the other hand, it is not illogical or prelogical.’
[C. Long, *ALPHA. The Myths of Creation*; 1963, 12]

Appendix

Text editions

1a. IAS 114 [983]

column i

	UGN-orthography	'Normal' orthography	
	... Unknown number of lines broken		
1'	[ ] [UD⌝ **ki LAGAB**	[ ] [an⌝ **ki niĝin$_2$**	[ ] **an**/heaven and **ki**/earth together
2'	**ĝiškim NAM$_2$-⌜LU$_3$⌝**	**ĝiškim nam-⌜ĝar⌝**	have placed a sign.
3'	**^{d}GAL-ki ^{d}nin-ki**	**^{d}en-ki ^{d}nin-ki**	Enki and Ninki
4'	**7 àm-ta-tud**	**7 àm-ta-tud**	have brought forth the seven.
5'	**a UNUG šim-LAGAB**	**a ki šim-gin$_7$**	Water for the earth, like something fragrant,
6'	**rig$_7$-dug$_3$ ŠA-NAM$_2$**	**rig$_7$-dug$_3$ na$_5$-nam**	is really a gift.
7'	**a pú-šè**	**a pú-šè**	Water for a well
8'	**rig$_7$-dug$_3$ ŠA-NAM$_2$**	**rig$_7$-dug$_3$ na$_5$-nam**	is really a gift.
9'	**^{UD}GAL-NUN**	**^{d}en-líl**	Enlil
10'	**AMA-a**	**tud-a**	has been brought forth
11'	**^{d}GAL-UNUG ^{UD}nin-ki**	**^{d}en-ki ^{d}nin-ki**	by Enki and Ninki.
12'	**^{d}GAL-UNUG AMA-a**	**^{d}en-ki tud-a**	Enki has been brought forth
13'	**UD-GIŠGAL *ù* ki-EREN-TUKU**	**an-uru$_{16}$ *ù* ki-dilmun**	by the mighty An and the luxuriant earth.
14'	**^{UD}ŠEŠ-KI AMA-a**	**^{d}nanna tud-a**	Nanna has been brought forth
15'	**^{UD}KIŠ-NUN**	**^{d}en-líl**	by Enlil
16'	**^{UD}nin-KID**	**^{d}nin-líl**	and Ninlil.

[983] The cuneiform text of this tablet was published by Biggs (1974). It was classified as a literary text written in the so-called UD-GAL-NUN (UGN) orthography. Text IAS 114 has some duplicates: from Abū Ṣalābīḫ IAS 248, and from Fāra SF 37 and SF 38 (Krebernik 1998, 340, *ad* SF 37 [with corrections: Biggs 1971; Krebernik 1984]; Lambert 1981, 82, note 3). For the decoding of the UGN-signs, the lists published by Krebernik (1984, 267-286; 1998, 298-302) have been used.

Editions / Translations [984]

1. W.G. Lambert, Studies in UD.GAL.NUN. *OrAnt* 20 (1981) 83-85 (lines 3'-4'; 9'-12').
2. M. Krebernik, Die Texte aus Fāra und Tell Abū Ṣalābīḫ. In: J. Bauer, R.K. Englund, M. Krebernik, *Mesopotamien. Späturuk-Zeit und Frühdynastische Zeit.* OBO 160/1. Freiburg: Universitätsverlag; Göttingen: Vandenhoeck & Ruprecht, 1998; 322, note 806 (lines 9'-11'; 14'-16').

***

Comments

1' The remaining parts of the first two visible signs strongly suggest the reading **UD ki**, in normal orthography **an ki**. As appears from the following lines, the text is dealing with primaeval times, when gods were born. If the reading of **UD/an ki** is correct, this reminds us of other texts where **an** and **ki** are interacting in the beginning (Ukg 15, Barton cylinder, NBC 11108, 'Debate between Tree and Reed' [985]). In this context a plausible reading for the UGN-sign **LAGAB** is **niĝin$_2$**, that may express the situation in which **an** and **ki** have not yet been separated.

2' It seems unlikely that a sign is missing in this line. In this period (Fara; ED III), the sign **LU$_3$** is normally twice as large as the remains of it in this line. As a consequence, a possible reading as: **IGI-DUB-[x] nam-LU$_3$** (for **x** e.g. **šè**) is very unlikely.

Next we have to look at the sign combination **IGI-DUB**. With respect to **IGI-DUB** two transcriptions are possible: 1. **agrig**, and 2. **ĝiškim**. The presence of an **agrig** 'steward, housekeeper' in primaeval times is not very likely. Instead **IGI-DUB** has been transcribed as **ĝiškim**: 'sign'. The possibility that **IGI-DUB** is written in UGN-orthography cannot be excluded, but until now such notation is not known.

Lambert [986] has proposed the following equivalences for the UGN-sign **lù**: **ĝar** and **ĝál**. One of these must be applied in this line, because the normal meaning of **lù** (to disturb, to cover completely, to mix [987]) does not fit. Choosing for **ĝar** a striking parallel to the resulting Sumerian line **[an-ki niĝin$_2$ ĝiškim nam-ĝar]** is found in an Neo-Assyrian Akkadian text [988]: *šamê u erṣetim ištēniš gis-kim-ma ub-ba-lu-ni* 'heaven and earth all together bring forth omens' [989]. Krebernik has already observed that at a particular place the tablet IAS 114 is dealing with omens, viz. 'Enki nahm die Opferschau vor' [990].

---

984 In these text editions a few times is referred to 'Krispijn 2004', especially in connection with the analysis of verbal forms; it concerns a syllabus of Krispijn's Sumerian grammar used during his university classes. This grammar is present in the NINO library (Leiden), but it was not formally published.

985 This study.

986 Lambert 1981, 86, *ad* 6).

987 ePSD.

988 Virolleaud 1911, 111, l. 39.

989 Instead of the translation *ištēniš* 'alike' (in CAD G, 98 *ad* '*giskimmu*') the present author prefers 'all together', 'jointly', meanings given in CAD I-J 279. A short comment is added there: 'SB instructions to the interpreter of omens'.

990 Krebernik 1984, 278, *ad* *ME = šu: IAS 114 *iii*: 14' ff. However, this column *iii* shows more examples of extispicy (e.g. ll. *iii*: 2', 5').

The translation '... have placed a sign' has to be interpreted as: '... have brought forth an omen.'

3'-4' The 'seven' who have been brought forth by Enki and Ninki are probably the ancestors of Enlil (see the chapter 3 'God lists' in this study).

For a beginning of the study of the technical terminology for creation in the Ancient Near East: see Lambert 1998.

5'-8' The idea that water is a gift for **ki**/earth has also been expressed, though in other words, in Ukg 15 column *i*: there water makes earth appear luxuriant, it makes earth materialize in all its fertility. The remark of Krecher [991] – '*IAS* 114 I 5'-8': compare 'Enki and Ninḫursanga' [45]. 49 // 56. 60' – is rather cryptic; there is no literal correspondence between these lines of IAS 114 and those of 'Enki and Ninḫursaĝa', except in a general sense that water can provide abundance, and in that sense can be considered as a gift.

9'-11' These lines explicitly mention Enki and Ninki as the parents of Enlil. According to Lambert [992], Enlil's ancestry is only described in a list: there is no narrative known about his origin. The ancestors of Enlil were discussed in more detail in the chapter 'God lists'.

Lambert [993] combines the lines 11' and 12', and thus he considers Enki and Ninki (line 11') as the parents of Enki(g): 'Enki and Ninki bore Enki(g)'. After having said that he did not understand the lines 5'-8', he writes in the next paragraph:

> 'If our understanding of the phrases cited is correct, this passage [= *the lines 3'-14'; JL*] first describes how Enki and Ninki were responsible for bearing 'seven', then that Enlil was born (presumably of the same parents), then that Enki(g) was born to the same pair, and after the obscure 114 *i* 13 (...) Sîn is born (114 *i* 14).'

12'-13' After the remarkable and unique mention of the parents of Enlil in the preceding lines, this text has another 'surprise': it tells us that **an** and **ki** are the parents of Enki. And although there is no difference in spelling between the Enki of line 11' and the one of line 12', it may be assumed that in line 12' the Enki of Eridu is meant, whose birth is mentioned between those of Enlil and Nanna. Like Enlil's ancestry, Enki's ancestry has also not been recorded in any other Sumerian narrative. Occasionally one of his parents is mentioned, e.g. An as his father, Namma as his mother [994].

With regard to **ĜIŠGAL**, the ePSD gives as translations 'station, attendant'; both are not suitable in this context. **ĜIŠGAL** has not been included in the UGN-lists of Krebernik. But for **ĜIŠGAL** also a sound value /**uru**/ has been proposed [995]. In this respect

---

991 Krecher 1992, 303. The respective lines are [ETCSL 1.1.1]: 45. **iri-zu a ḫé-ĝál-la ḫu-mu-ra-na$_8$-na$_8$** 49. **dilmun**[ki] **é-gún- kar-ra-kalam-ma-ka ḫé-a** 56. **iri-ni a ḫé-ĝál-la im-ta-na$_8$-na$_8$** 60. **iri-ni é-gún-kar-ra-kalam-ma-ka na-nam** 45. May your city drink water in abundance from them. 49. May Dilmun become a storehouse on the quay of the Land. 56. Her city drank water in abundance from them. 60. Her city was indeed a storehouse on the quay of the Land.

992 Lambert 2000, 1830-1831.

993 Lambert 1981, 84 *sub* 5, 85.

994 An as his father: 'Enki and the world order', ETCSL 1.1.3, ll. 61-80; Namma as his mother: 'Enki and Ninmaḫ', ETCSL 1.1.2, e.g. ll. 17-18.

995 Borger 2010, 61, no. 80, with reference to Sjöberg and Bergmann, 1969, 140 *ad* l. 488; van Dijk: A *balbale* to Ninazu [= ETCSL 4.17.1], 1960, 57 and 66 *ad* 8.

ĜIŠGAL is homophone with **EN** = /**uru**/ = mighty. The epithet **EN** is a common one, not only for An but in general for gods. I propose that in this case **ĜIŠGAL** has been used as an UD-GAL-NUN-sign for the orthographic sign **EN** with meaning /**uru**/.

As far as I know the expression **EREN-TUKU** has not been used elsewhere in the Sumerian literature. If orthographically used, **EREN-TUKU** could be translated as 'having cedars'. A search in the Sumerian literary text corpus for **erin** shows that in more than 30% of the results it concerns **šim-^{ĝiš}erin-na** 'cedar essence'. If this might be applied to the present text, I have to suppose that the expression **EREN-TUKU** has been used rather elliptically. A tentative translation of **ki-EREN-TUKU** then might be: 'the fragrant earth'. But if we have a closer look at several cosmogonic introductions edited in this study – Ukg 15, NBC 11108 (?), 'Debate Tree-Reed' –, it will be seen that there is reported on the embellishment of **ki**/earth, which then appears luxuriant. Therefore the combination **EREN-TUKU**, at least with respect to **EREN**, might be an UGN-writing for the orthographic combination **NI-TUKU** = **dilmun** 'luxuriant' [996]. If this explanation is true, the relation between the UGN-sign **EREN** and the orthographic **NI** is unclear. Of both mentioned possibilities for a translation of **EREN-TUKU** – fragrant and luxuriant, respectively – I prefer the last one as being more in line with the other descriptions of **ki**/earth in the cosmogonic introductions treated in this study [997].

***

[996] See also the comments concerning **dilmun** at Ukg 15 *i*:3. In this line a variant spelling for **NI-TUKU** – viz. **MUNUS-ḪUB$_2$** – can be observed.

[997] The relation between **eren** 'cedar' and divination is demonstrated in 'an Old Babylonian prayer of the divination priest' (Goetze 1968). As magical texts belong to the oldest literary tradition of Mesopotamia (van Dijk et al. 1985, 1), the literal meaning of **eren** (cedar) cannot be excluded in the text IAS 114, because of the relation of this text with *omina*, as mentioned in the commentary at line 2'.

1b. IAS 136, 113, 203 [998]

IAS 136

iii	UGN-orthography	'Normal' orthography	
1'	**^{UD}GAL-NUN**	**^{d}en-líl**	Enlil,
2'	**an UNUG-ta bad**	**an ki-ta bad**	who has separated heaven from earth,
3'	**ki an-ta bad**	**ki an-ta bad**	who has separated earth from heaven.

IAS 113

ii	UGN-orthography	'Normal' orthography	
5	**^{UD}KIŠ-[NU]N**	**^{d}en-líl**	Enlil,
6	**GAL nu-nám-NAGAR**	**en nu-nam-nir**	Lord Nunamnir,
7	**GAL du$_{11}$-TUKU DU$_{6}$-GAG-GAG**	**en du$_{11}$-ga nu-gi$_{4}$-gi$_{4}$**	the lord who does not revert to an order,
8	**UD UNUG-ta LAGAB**	**an ki-ta bad**	who has separated heaven from earth,
9	**ki UD-ta LAGAB**	**ki an-ta bad**	who has separated earth from heaven,
10	**UD UNUG-ta LAGAB**	**an ki-ta bad**	who has separated heaven from earth.

IAS 203

ii	UGN-orthography	'Normal' orthography	
3'	**[^{UD}GAL/KIŠ]-NUN**	**^{d}en-líl**	Enlil,
4'	**[U]D [k]i-ta [LAGAB]**	**an ki-ta bad**	who has separated heaven from earth,
5'	**ki UD-ta LAGAB**	**ki an-ta bad**	who has separated earth from heaven.

---

998 The cuneiform text of these tablets was published by Biggs (1974). They were classified as a literary text written in the so-called UD-GAL-NUN (UGN) orthography. For the decoding of the UGN-signs, the lists published by Krebernik (1984, 267-286; 1998, 298-302) have been used.

2. Ukg 15 (AO 4153) [999]

Obverse

column iii

1	**u$_4$-⸢da⸣ im-ma**	One day, in a bygone year,
2	**ul-[lí-a] im-m[a]**	in the long-ago, in a bygone year,
3	**u$_4$ nu-zal**	the sun did not get up early,
4	**ì-ti nu-è-è**	the moon did not appear all along.

column ii

1	**an en-nam šul-le-éš al-gub**	An, as **en**, was standing there as a youthful man.
2	**an ki téš-ba šeg$_{12}$ am$_6$-gi$_4$-gi$_4$**	An/heaven and Ki/earth, in their unity, were shouting.
3	**u$_4$-ba en-ki nun-ki nu-se$_{12}$**	Then Enki and Ninki did not exist;
4	**^{d}en-líl nu-ti**	Enlil did not exist,
5	**^{d}nin-líl nu-ti**	Ninlil did not exist.

column i

1	[...]	[...]
2	**[k]a-<a>-muš ḫa-mu-ni-se$_{11}$-se$_{11}$**	He has lowered the inlets of the irrigation channels in it,
3	**ki-e dilmun-na dalla ḫa-mu-ak-e**	in order to make earth appear in luxuriance:
4	**kiri$_6$ duru$_5$-am$_6$ te-me-nam**	a garden, moist and cool;
5	**ki-buru$_3$ a šè-ma-si**	water has filled the holes in the earth.

The remaining part of the obverse, and the entire reverse are blank.

---

999 This text is also known as AO 4153; a copy of it has been published by Cros 1910, 180. Afterwards Sollberger published this text again as Ukg 15 (in: *Corpus des Inscriptions 'Royales' Présargoniques de Lagaš*. Genève: Droz; 1956, 57).

Editions / Translations

Editions

1. J. van Dijk, Le Motif Cosmique dans la Pensée Sumérienne. *AcOr* 28 (1964) 39-44.
2. W.H.Ph. Römer, Die Urzeit: Vor der Schöpfung (II). In: W.H.Ph. Römer, D.O. Edzard (eds), *Weisheitstexte, Mythen und Epen. Mythen und Epen.* TUAT III. Gütersloh: Mohn, 1993; 355-356.
3. Å.W. Sjöberg, In the Beginning. In: T. Abusch, *Riches Hidden in Secret Places. Ancient Near Eastern Studies in Memory of Thorkild Jacobsen.* Winona Lake, Indiana: Eisenbrauns, 2002; 229-239.
4. W. Horowitz, *Mesopotamian Cosmic Geography.* Winona Lake, Indiana: Eisenbrauns, 1998; 140-141. Edition of columns *ii* and *iii*.

Translations

1. Th. Jacobsen has translated this text into Danish; see Sjöberg 2002, 229, note 1. Sjöberg only refers to this translation in his comments.
2. W.W. Hallo translated the French text of van Dijk into English, and commented only on the contents: see Hallo 1970a, 65-66.

***

Comments

Following the order in which the columns are presented in the copy there seems to be a chronological inconsequence in the story. In column *i* someone takes care for the earth, while column *iii* describes prehistoric times, when nothing is present. In other words, the logical order of presentation of occurrences seems to be reversed: not going from column *i* to column *iii*, but from column *iii* to column *i*. This is supported by the formulation in column *iii*: 1. **u$_4$-⌜da⌝**... 2. **ul [la / lí-a]**..., constructions related to those which are usual at introductions of mythologies that refer to primaeval times. The question arises of whether the text of AO 4153 has been miscopied by the scribe; for instance: he started at the reverse of the original tablet, so that the sequence of the columns on the copy has a reverse order. Arriving at the prehistory the copyist became aware of his mistake, and perhaps he abandoned his task...? As a consequence this copy – tablet AO 4153 – has been written on one side only, and moreover on that side only partially.

In this study column *iii* is considered as the first column, followed by column *ii* and column *i*, respectively. However, the already existing numeration of the columns has been maintained.

Meyer-Laurin doubts – on orthographic grounds – whether the text of this tablet may be considered as a pre-Sargonic one [1000].

***

1000 Meyer-Laurin 2012, 229-230. See also my comments at the lines *i*: 3 and *i*: 5.

column iii

1-2 The opening phrase **u$_4$-da im-ma** seems to be unique in the Sumerian literature; at least, no duplicate could be found in the Sumerian literature.

The reference by van Dijk [1001] to line 5 of 'Gilgameš, Enkidu and the Netherworld' in view of **ul [?] im-ma** does not seem applicable, because in that line is written: **ul (..) mí-zi (dug$_4$ ..)** instead of: **ul (..) im-ma**.

On the translation of **im-ma** as 'last year' has Sjöberg commented in detail [1002]. However, such a translation as '*last* year' seems to be most unlikely at the beginning of a myth that evokes a remote past of which no one can have any memory, and the time of which was definitely not 'last year'.

With respect to the completion of **ul** to **ul-la** or **ul-lí-a**, some references to the literature may suffice [1003]. Both expressions refer to a far future, but **ul-lí-a** also may denote a long-distant past. This last expression seems the most likely one at the beginning of a myth.

Restoration of the text as **ul-[ḫé$^?$]** (Horowitz 1998, 140: 'In the firm[ament$^?$]') is not very likely, in view of the aspect of time which is operative in these lines.

3. For the translation of **u$_4$ nu-zal** as 'the sun did not get up early' has been chosen because of the 'antithetic parallelism' between the lines 3 and 4. There are many examples in the literature in which **ud/u$_4$ zal** has been translated with 'the day dawned' or 'the day passed' (depending on the context), but the present text strongly suggests the parallelism between sun and moon [1004].

Michalowski and Katz discussed the use of negations in Sumerian literature, which are prevalent in cosmological introductions of stories [1005].

4. **è-è** may be interpreted as the iterative form of the *ḫamṭu* stem: 'the moon did not appear *all along*'. The reduplication of the *ḫamṭu* stem could also be caused by the plurality of the absolutive: 'the moons did not appear'. In that case the different phases of the moon may be meant.

1001 Van Dijk 1964, 43, *ad* 11-12.

1002 Sjöberg 2002, 237, and note 15 and 16.

1003 On **ul-la**: see e.g. van Dijk 1976, 126, 130; Behrens 1978, 23, line 37; 106-108; 115; Civil 1994, glossar; AHw, 1408, *s.v. ulla* B. On **ul-lí-a**: see CAD Ṣ, 116-117, *s.v. ṣâtu*, lexical part. References to literature in ETCSL: via ePSD **ul** 'distant'.

1004 Some results with translation 'the day dawned/broke' for **ud/u$_4$ zal**: 'Inanna and Šukaletuda', ll. 110, 126, 239, 280 (ETCSL 1.3.3); 'Gilgameš, Enkidu and the Netherworld', ll. 47, 48, 91, 92 (ETCSL 1.8.1.4); 'Lugalbanda and the Anzu bird', l. 44 (ETCSL 1.8.2.2); 'Enmerkar and the Lord of Aratta', ll. 308, 391 (ETCSL 1.8.2.3); 'Enmerkar and En-suḫgir-ana', ll. 42, 48, 87 (ETCSL 1.8.2.4); 'The building of Niĝirsu's temple' (Gudea, cylinders A and B), ll. 485, 577, 928 (ETCSL 2.1.7); 'A praise poem of Šulgi' (Šulgi C), l. 44 (ETCSL 2.4.2.03); 'A hymn to Ḫendursaĝa' (Ḫendursaĝa A), l. 31 (ETCSL 4.06.1); 'A hymn to Inana as Ninegala', l. 3 (ETCSL 4.07.4); 'The Debate between Copper and Silver', l. 103 (ETCSL 5.3.6).

1005 Michalowski 1991; Katz 2007a, 578-579.

column ii

1. This text may refer to the primal '*hierós gámos*', viz. that of **an** and **ki**. This justifies the transcription of **EN** as **en**, in the meaning of 'high priest; lord'. The order: name of a god followed by a plain substantive as epithet like **en** or **lugal**, is amply attested [1006]. On the other hand: the adjective **EN** = **uru**$_{16}$ 'mighty, exalted' is not unusual for An [1007].

2. **šeg**$_{12}$ **.. gi**$_{4}$: 'to make noise, shout, roar' [1008]; perhaps used here with a sexual connotation: 'to court, make love' [1009]. In the expression /**šeg**/.. **gi**$_{4}$ the noun part /**šeg**/ may be written as **šeg**$_{10}$, **šeg**$_{11}$, en **šeg**$_{x}$, where /**šeg**/ is a compound sign: **KA** x another sign [1010]. In the older literature (ED III, Gudea cylinders, Ur III [1011]) /**šeg**/ is also written by the sign **sig**$_{4}$ = **šeg**$_{12}$, undoubtedly because of the similar sound.

Horowitz places an **x** instead of **šeg**$_{12}$, and does not translate the verbal part at all. In a footnote he states: 'In this context, a meaning of the verb **gi**$_{4}$**-gi**$_{4}$ 'to be joined together' would be appropriate.' [1012]

The prefix **AN** in **AN-gi**$_{4}$**-gi**$_{4}$ has been transcribed as **am**$_{6}$, whereas van Dijk, Horowith and Sjöberg write **an**. Since this tablet is supposed to be written in presargonic times [1013], the sign **AN** may be used for **am**$_{6}$ [1014]. The subjects are **an**/heaven and **ki**/earth. This plural subject may be represented by the collective pronomen ***b** before the *ḫamṭu* stem. The consequent analysis of **am**$_{6}$**-gi**$_{4}$**-gi**$_{4}$ is: **a-m-b-BḪ**_reduplicated_-∅ [1015]; the reduplication of the *ḫamṭu* stem has been interpreted as the iterative aspect of the action.

3. For the writing of **nun-ki**: see Sjöberg 2002, 237 note 14, but especially Alster 1970. The transcription of van Dijk (1964, 40) – **eridu**^{ki} – seems not meaningful in this line; later he corrected this in **nun-ki** [1016].

4-5. The name of Enlil, which is now written **^{d}en-líl** in transcription, should possibly be transliterated as **^{d}en-E**$_{2}$, as was usual throughout the third millennium [1017]. To judge from

---

1006 Both possibilities – **en** before and after the name of a god – occur.

1007 More examples and discussion: Falkenstein 1959, 33; Falkenstein 1962, 69-72; van Dijk 1960, 66, 92; PSD B 96^{a} *ad* 2.1.2; 'An **adab** to An for Lipit-Eštar', l. 31 (ETCSL 2.5.5.3); 'An **adab** to An for Ur-Ninurta', l. 1 (ETCSL 2.5.6.5); NBC 11108, l. 1: ch. 2.1.4 and Appendix Text editions no. 4 in this study.

1008 Black 2003, 40.

1009 Michalowski (1998, 240) supposes likewise.

1010 Borger 2010, 56 no. 42, and 57 no. 47; ePSD *via* **šeg** and **šeg gi**.

1011 To be found via ePSD at the 'card' **šeg gi** at the bottom: 'see ETCSL: **šeg** $_{11}$ = loud noise'.

1012 Horowitz 1998, 140, note 40.

1013 Van Dijk 1964, 39, note 110.

1014 Sollberger 1961, 7, *sub* 12 **AN**; Krecher 1967b, 19. For the writing **am**$_{6}$ in the prefix-chain, see also Edzard 2003b, 92 and note 12; Jagersma 2010, 21-22.

1015 If the god An had been meant here, with **ki** 'earth' as inanimate, and the plural of the personal pronominal element had been used, the transcription may be: **<u>an</u>-gi**$_{4}$**-gi**$_{4}$, whereby the /**eš**/ as pronominal suffix of the plural subject has been omitted; final consonants are not always written, and that may explain the absence of an ending **-(e)š**. The animate or inanimate class of **an** (An, heaven) will be dicussed in the Excursus at the end of the Text editions. To the best of my knowledge there is no example in this kind of 'myth of origin' in which **ki** is considered as animate.

1016 Van Dijk 1976, 128, note 22.

the enlargement of the photograph of the tablet AO 4153 [1018], there seems to be a (slight!) difference between the sign after ^{d}**en** in *ii*: 4 and that after ^{d}**nin** in *ii*: 5; this last one is **KID** = **líl** [1019].

column i

1. This line was completed by Sjöberg as [**an-e**]. Because of the text in column *ii*: 1 and also in NBC 11108 line 1, this line might be read as: **an-uru$_{16}$-né**. As appears from the text of Ukg 15, there are no other protagonists present than **an** and **ki**, so An as subject in this line seems obvious. However in the Barton cylinder, column *ii*: 13-15, in all likelihood ^{d}**íd-maḫ** and **diĝir-íd** are responsible for, or concerned with, the water management.

2. Van Dijk assumes that the text begins with a conversation in which instructions are given: 'Que ... dans ... les [rep]tiles descendent;'. In an explanation van Dijk justifies his translation of **ka-muš** as 'reptile' [1020]. Sjöberg did not translate this line at all.

Because of the context a translation of **ka-muš** as a reptile or the like seems less meaningful. The first sign of this line could be **saĝ** or **KA**. The expression **saĝ .. SIG** 'to lower the head, to humble oneself' does not fit here. With respect to the contents of especially the lines 4 and 5 and the verb used in line 2 – **SIG** = **se$_{11}$** 'to be / make low' – it is more likely that the first sign is **KA**. With an emendation to **ka-<a>-muš**, the line tells about the efforts of – most probably – An to lower the inlets ('the mouth') of irrigation canals [1021]. This means: the dams between a big water reservoir and the canals connected with it are lowered, so that water can flow out from that reservoir into irrigation canals. In this way the earth can be made or make herself resplendent.

The verbal form **ḫa-mu-ni-se$_{11}$-se$_{11}$** in this context is affirmative with a plural *ḫamṭu* basis (plural, because of more than one inlet through which the water can flow out), rather than a precative form with a *marû* basis [1022], or with the words of Civil: the modal prefix /**ḫe**/ has an epistemic function rather than a deontic one [1023].

3. The suffix ***e** in **ki-e** was interpreted as a directive [1024].

For a discussion about **SAL-ḪUB$_2$** (or **MUNUS-ḪUB$_2$**), Sjöberg refers to 'Civil *apud* Wiggermann' [1025] and he concludes with a translation 'lavishness(?)'. But in his

---

1017 Steinkeller 1999, 114, note 36.

1018 CDLI no. P315470.

1019 For a discussion of the reading of the signs involved: Biggs 1966, 84, n. 85; Englund 1988, 131-132, n. 9; Jacobsen 1989, 267-269; Steinkeller 1995c, 700, nos. 129 and 142.

1020 Van Dijk 1964, 41.

1021 **a-muš**: (a type of irrigation canal), UrIII (Sjöberg 1984+: PSD A-I, 116). **SIG** (also **se$_{11}$** [Borger 2010, 223 and 449: no. 881]): to be/make weak/low/thin/narrow (ePSD).

1022 According to Thomsen 1984, 314, the verbal class of **SIG** is not known. However, in 'Ninurta's exploits: a *šir-sud* (?) to Ninurta' [ETCSL 1.6.2], l. 549 there is the form **ḫé-ta-sig-ge**, and in 'A hymn to Nungal' [ETCSL 4.28.1], l. 80 a verbal form **im-sig-ge-en**. Therefore I tentatively conclude that the *marû* form of **sig** is regular, and that in this line **se$_{11}$-se$_{11}$** represents a plural *ḫamṭu*-stem. The analysis of the verbal form may be: **ḫe-(i)-m-b-ni-en-/se$_{11}$-se$_{11}$/-ø**; here ***b** + **ni** refers to the earth: 'in it' and ***en** refers to An (animate).

1023 Civil 2000b, § 3, 31-35.

1024 Meyer-Laurin (2012, 229) considered the ***e** as an ergative ending.

treatise of **SAL-ḪUB₂**, Wiggermann also says that 'the sign DILMUN (NI.TUK, SAL.TUK) occasionally takes the shapes NI.ḪUB₂ or SAL.ḪUB₂, (...)' [1026]. Borger states that **SAL-ḪUB₂** may also be an older spelling for **NI-TUK** = **dilmun** [1027], with its Akkadian equivalent *šūpû* [1028]. Because of the context a meaning like 'luxuriance' seems most appropriate. Now the syllable ***na** after **SAL-ḪUB₂** can easily be explained as being composed of a linking ***n** plus a locative [1029].

dalla .. ak seems to be a hapax, and a synonym of **dalla .. è** [1030]. An emendation to **dalla-è ... ak**, analogous to **pa-è .. ak**, is also possible.

4. The transcriptions of this line by both van Dijk and Sjöberg are the same: **sar-àm te-me-nam**. According to van Dijk: '***am** dans **sar-àm** et ***temen-am** remplace ici un locatif qui a été incorporé dans la forme verbale **a šè-ma-si** (...)' [1031]; consequently his translation of lines 5 and 4 is: 'un trou dans la terre remplit d'eau les rigoles du jardin (et) l'enclos'. Sjöberg says (line 4): 'she was green (like) a garden, it was cool'.

The interpretation given here is different. **SAR** = **kiri₆**: garden. **A-AN** can be transcribed as follows: 1. = **šeĝ₃**: rain; 2. = **àm**; as a copula it is in use from the Old Akkadian period, whereas the present text is older; 3. = **duru₅-am₆**: he/she/it is moist [1032]. This last meaning, 'to be moist', refers to the earth, just like the next expression **te-me-nam**. With respect to **te-me-(e)n** as a long form for **te-en** 'cool', the discussion of Sjöberg may suffice [1033].

In line *i*: 4 we read the final result of what has been described in the lines *i*: 2-3, viz. humidity and coolness for the earth. The state of being moist and cool forms part of the luxuriance of the earth. Most likely the circumstances 'being moist and cool' were experienced as positive and agreeable in Mesopotamia with its very dry and warm periods. An alternative order for the signs **te-me-nam** is **me-te-nam**: 1. it is fitting; 2. it is from him/her/itself. Of the two given possibilities, **te-me-nam**, 'being cool', seems the more appropriate one in this context.

***

1025 Sjöberg 2002, 232 *ad* 3; Wiggermann 1988. For a discussion of Wiggermann's study, see Michalowski 1990b.

1026 Wiggermann 1988, 231.

1027 Borger 2010, 116, no. 380.

1028 CAD Š III, 328, lexical part; *šūpû* (among others) 'brilliant, shining, splendid'.

1029 Another possibility might be to analyse ***na** as **ani** + **a**: animate possessive suffix followed by a locative. But in that case one is halting between two opinions: on the one hand **ki** is considered as animate because of ***ani**, the possessive pronoun for animates; on the other hand as inanimate, for **ki** has been supplied with the suffix ***e**, the directive or 'dative for inanimates'. The suffix ***e** in **ki-e** as ergative suffix is very unlikely in this sentence. See e.g. the Barton cylinder *i*: 14, where **ki** is subject, but written without an ergative suffix.

1030 See also Sjöberg 2002, 232. The analysis of the verbal form **ḫa-mu-ak-e** may be: **ḫe-(i)-m-b-ni-/*marû-stem*/-ē**. Here ***b + ni** refers to the indirect object with the locative: **dilmun-na̲**.

1031 Van Dijk 1964, 41-42.

1032 The enclitic copula /**am**/ was written as ***AN** = ***am₆** in the Old Sumerian period, but thereafter as ***A-AN** = ***àm** (Sollberger 1961, 7 *sub* 12 **AN**; Krecher 1967b, 19).
The present writing **kiri₆ duru₅-am₆** instead of 'SAR (kiri₆?)-**àm** (Meyer-Laurin 2012, 229) agrees with a pre-Sargonic dating of this text.

1033 Sjöberg 2002, 232-233.

3a. The Barton cylinder

column i

u$_4$-rí-a u$_4$-rí-šè	On that remote day, until that remote day,
na-nam	it was indeed;
ĝi$_6$-rí-a ĝi$_6$-rí-šè	in that remote night, until that remote night,
na-nam	it was indeed;
mu-rí-a mu-rí-šè	in that remote year, until that remote year,
na-nam	it was indeed.
u$_4$ na-du$_7$-du$_7$	Then a gale was really blowing unceasingly,
nin na-ĝír-ĝír	there were really flashes of lightning continuously.
èš-nibru^{ki}	Near the sanctuary of Nippur
u$_4$ na-du$_7$-du$_7$	a gale was then really blowing unceasingly,
nin na-ĝír-ĝír	there were really flashes of lightning continuously.
an-né ki-da	An/heaven is shouting $^{(l. 13)}$ together with Ki/earth;
gù am$_6$-dab$_6$-e	
ki an-da ⌞gù⌟ am$_6$-dab$_6$-e	Ki/earth is shouting together with An/heaven.
[...]	[...]

[*about 7 lines broken*]

column ii

igi-⌜zi⌝-gal-an-n[a]	With the true, great Queen of heaven,
nin-gal-^{d}en-líl	the older sister of Enlil,
^{d}nin-ḫur-saĝ	Ninḫursaĝ,
igi-zi-gal-an-na	with the true, great Queen of heaven,
nin-gal-^{d}en-líl	the older sister of Enlil,
^{d}nin-ḫur-saĝ-ra	Ninḫursaĝ,
ĝìš mu-ni-du$_{11}$	he has had intercourse;
ne mu-ni-sub$_5$	he has kissed her;
a-maš-imin	the seed for a set of septuplets
š[à] mu-ni-ru	he has poured into her womb.
ki muš-ĝír-da	Earth chatted cheerily $^{(l. 12)}$ with the muš-ĝír-snake:
BAL-bal am$_6$-da-za	

13	**^{d}íd-maḫ**	'Exalted Divine River,
14	**níĝ-tur-zu a mu-da-de$_6$**	your small things have brought along water;
15	**[p]a$_5$-ra diĝir-íd-da-ke$_4$**	in the canals, the god of the river
16	**[...] mu [...]**	[...] has $^{?}$ [...]'
17	[...]	[...]

[*about 6 lines broken*]

Editions

1. G.A. Barton, No. 1. The Oldest Religious Text from Babylonia. In: G.A. Barton, *Miscellaneous Babylonian Inscriptions*. New Haven: Yale University Press; London, Humphrey, Milford: Oxford University Press, 1918; 1-20 [1034].
2. J. van Dijk, Le Motif Cosmique dans la Pensée Sumérienne. *AcOr* 28 (1964) 35-39 (columns *i* en *ii*).
3. B. Alster, A. Westenholz, The Barton Cylinder. *ASJ* 16 (1994) 15-46.

***

Comments

Column i

1-6. **na-nam**: for the meaning of the **na**-prefix at the beginning of Sumerian literary texts, see Jagersma and other scholars [1035].

7-11. Both van Dijk and Alster & Westenholz omit the translation of the **na**-prefixes. The reduplication of the *ḫamṭu*-stems has been interpreted as the iterative aspect of the actions.

9. About the reality of the sanctuary of Nippur, van Dijk writes: 'Cela ne veut pas dire que le sanctuaire de Nippour existait déjà. C'est une licence poétique. L'auteur veut dire: l'endroit où le sanctuaire sera construit un jour' [1036].

12-14. In the expression **KA dug$_4$** there are two possible transcriptions for **KA**: 1. **inim dug$_4$**, 'to talk, to speak about a matter/question'; 2. **gù dug$_4$**, 'to shout' [1037]. The second expression seems to fit better into the picture as outlined in the lines 7-11.

1034 Barton's translation and interpretation of this text are not discussed in the present edition, since the understanding of the Sumerian language and myths have changed a good deal since 1918.

1035 The modal prefix ***na** can have a prohibitive and an affirmative meaning: Jagersma 2010, ch. 25.5 and 26.3; Thomsen 1984, 194-199, § 371- 383; Attinger 1993, 289-291, ch. 3.2.6.4; Streck 2002, 256, § 5.1; Edzard 2003a, 118-120, ch. 12.11.9 and 12.11.10.

1036 Van Dijk 1964, 37, note 103. According to Alster (1976b, 19 and note 28) are 'temple hymns' more concerned with the celestial archetypes of these temples than with the concrete temples on earth, which are only secondary representations of the celestial 'houses'.

1037 Attinger 1993, 526-536, ch. 5.3.63.

In these lines both **an** (Heaven) and **ki** (Earth) are supposed to be of the inanimate class [1038]; therefore the present translation for **an** is Heaven instead of the god An. Consequently the transcription of the prefix **AN** is **am**$_6$ (ll. 13 and 14) [1039].

Column ii

1. The epithet **égi-zi-an-na**, in Ur III and older times written as **igi-zi-an-na**, seems to be almost exclusive to the goddess Ninḫursaĝa [1040].

7. With respect to the 'he', Alster and Westenholz suppose: ' "He" is hardly "Earth", since it is most likely that Heaven and Earth represent the original male and female couple, and that the male counterpart of Ninḫursag (...) was the son of Heaven and Earth, presumably Enlil himself' [1041]. Enlil the son of **an** and **ki**? That is not supported by any mythological text. But Enlil is indeed mentioned as the sexual partner of Ninḫursaĝa, viz. in 'The Debate between Winter and Summer' [1042]. Another god, Enki, is the sexual partner of Ninḫursaĝa in the story entitled 'Enki and Ninḫursaĝa' [1043]. Both gods, Enlil and Enki, are not under consideration for a possible partner of Ninḫursaĝa in the 'Barton' story, which tells about the primaeval cosmic marriage between **an** and **ki**. On the other hand, it cannot be excluded that An, the sky god, had intercourse with Ninḫursaĝa. This supposition is also based on the first lines of 'The Debate between Grain and Sheep' [1044].

9. **maš-imin** is 'a set of septuplets', not 'seven twins' (translation of Alster and Westenholz) [1045].

10. A verbal form **mu-na(!)-ni(!)-ru**, as written by Alster and Westenholz, seems less probable, compared with the other short-cut prefix-chains in this text. Presumably attempts have been made to erase **na** [1046].

1038 See also the comments at Ukg 15, *ii*-2. Van Dijk (1964, 37), Alster & Westenholz (1994, 18) and Attinger [1993, 500, *sub* e) Le comitatif, 1^0] all transcribe **AN** as **an**. Attinger adds: 'an et ki sont personnifiés'. A more detailed study about the inanimate or animate class of **an** and **ki** is given in the Excursus 'The animate *vs* inanimate class of **an** and **ki**' at the end of this Appendix.

1039 For the transcription **AN** = **am**$_6$, see Sollberger 1961, 7, *sub* 12 **AN**, and also Edzard 2003b, 92 and note 12. The analysis of the verbal form **am**$_6$**-dab**$_6$**-e** is: **a-m-b-da-b-e** (= *marû*-stem) **-ē**.

1040 Steinkeller 2005, 303.

1041 Alster and Westenholz 1994, 33, *ad* ii 7.

1042 'The Debate between Winter and Summer' [ETCSL 5.3.3], line 12: **ḫur-saĝ-gal-gal-la ĝiš bí-in-dug**$_4$ **kur-re ḫa-la ba-an-šúm** 'He (= *Enlil*) copulated with the great Mountain Ranges (= *Ninḫursaĝa*), the Mountain (= *Enlil*) gave her its share '. See also Steinkeller 2007, 229-230 for the translation of **ḫur-saĝ** and **kur**.

1043 See 'Enki and Ninḫursaĝa' (ETCSL 1.1.1), ll. 73-74. For a detailed analysis of 'Enki and Ninḫursaĝa': see Katz 2007a and 2008.

1044 For a discussion with regard to this, and the possible prefiguration of **ḫur-saĝ** for the goddess Ninḫursaĝa: see the edition of 'The Debate between Grain and Sheep' in this Appendix, and a more detailed analysis of the 'Barton' text in ch. 2.1.3.

1045 For a discussion of the possible identity of this set of septuplets, see ch. 4.2.1.3. For **maš-imin** as 'a set of septuplets': see Gudea Cyl. B, *xi*: 11: **dumu-maš-imin-**^{d}**ba-ú-me** 'they are the septuplets of Bau', preceded by the names of seven children of Bau.

1046 Alster and Westenholz 1994, 18; comments on this line: p. 33.

11. The interpretation of the **muš-ĝír** – a fanged snake [1047] – is difficult. This snake is also spelled as **muš-mir** [1048]. The change **ĝ** - **m** in words, between the Sumerian main dialect and the Emesal, is known [1049]. According to Heimpel the **muš-mir** concerns a 'Gürtelschlange, (...) wahrscheinlich eine Riesenschlange'; Klein refers to MSL VIII/2 for the equation **muš-mir** = *šibbu* 'girdle-snake' [1050].

Alster and Westenholz comment on **muš-ĝír**: '**muš-ĝír**, 'scorpion' is here apparently an epithet either of Ninḫursaĝ's consort, or of Ninḫursaĝ herself.' In their comment on line 13, both scholars suppose that '**ᵈíd-maḫ** is here perhaps an epithet of Ninḫursaĝ' [1051]. In my opinion neither **muš-ĝír** nor **ᵈíd-maḫ** is an epithet of Ninḫursaĝa [1052].

12. **BAL-bal .. za**: Black has reviewed onomatopoeic expressions or ideophones [1053]. The present expression **BAL-bal .. za** seems to be a hapax [1054], with concomitant uncertainty about the reading of the first **BAL**: **bal** or **bùl**, and also about the precise meaning. As far as the text has been understood, there begins a new episode from line 11, in which Earth is talking with a snake about the distribution of water via rivers ('small things'; l. 14) and canals (l. 15). Since water for the earth in Mesopotamia means fertility, I suppose that Earth is rejoicing and speaks gratefully and happily; therefore **BAL-bal .. za** is translated as 'to chat cheerily'.

The transcription of **AN** in **AN-da-za** depends on the character of **muš-ĝír**; if this snake belongs to the animate class, **AN** is **an**; if the snake is of the inanimate class, which supposition seems to be the most probable one since animals in general belong to this class, **AN** is **am**$_6$.

13-15. In ch. 2.1.3 an interpretation of these lines is given, starting from the hypothesis that **muš-ĝír** is the addressee, who is called **ᵈíd-maḫ** in line 13. However, if the **muš-ĝír** and **ᵈíd-maḫ** are considered as distinguished entities and **ᵈíd-maḫ** is the subject of the lines 13-14, then as a consequence the translation should be: 'The Exalted Divine River, being aware of small things, has brought along water.' The absence of a case ending at **ᵈíd-maḫ** makes both interpretations possible. If **ᵈíd-maḫ** is the subject of the verb **de**$_6$ (l. 14), the

---

1047 According to ePSD. Sjöberg and Bergmann (1969, 118, comments at l. 336) discuss the word **muš-ĝír**, but give only a partial translation: '...-snake'; their review provides no clue for the supposed animal in the present text. In the Ebla list with animals (MEE 4 no. 116) several snakes are mentioned, but no **muš-ĝír** (Sjöberg 1996b). Cooper (1983, 256-257, comments at l. 267) posits that the reading **ĝír** in **muš-ĜIR**$_2$ is uncertain. A search in the Sumerian literature only results in a few lines – the same as those mentioned by Sjöberg and Bergmann – with **muš-ĝír**: 'The Debate between Grain and Sheep' [ETCSL 5.3.2], line 128; 'The Victory of Utuḫeĝal' [ETCSL 2.1.6], lines 1 and 59; 'A hymn to Nungal' [Nungal A; ETCSL 4.28.1], line 102; 'A praise poem of Šulgi' [Šulgi E; ETCSL 2.4.2.05], lines 208 and 230.

1048 In 'A praise poem of Šulgi' [Šulgi E; ETCSL 2.4.2.05], line 212 has **muš-mir** instead of **muš-ĝír**. The spelling **muš-mir** has also been found in another 'praise poem of Šulgi' (ETCSL 2.4.2.04, line 173).

1049 Schretter 1990, 43-49; see also p. 213 *sub* nr 249 /**mer**/.

1050 Heimpel 1968, 508, nr. 98. Klein 1981, 76, line 173; comments p. 100. See also CAD Š-II, 375, *šibbu A*. In ePSD **mir** = 'type of snake'. In an EDIIIa lexical list from Fara (TSŠ 46, *viii*: 3), **muš:ĝír** is present (see also Sjöberg 2000, 412).

1051 Alster and Westenholz 1994, 33 (*ad* ii:11 and ii:13, respectively).

1052 In ch. 2.1.3 the possible meaning of the **muš-ĝír** was discussed in more detail.

1053 Black 2003.

1054 Black 2003, 48, nr 13.

ergative case ending is obscured by the final **zu** of the complex **^{d}íd-mah̯-níĝ-tur-zu**. If **^{d}íd-mah̯** is the addressee, there is no need for a case ending. In both cases the 'small things' are most likely the rivers. The water of the rivers in turn is spread among the canals [1055] by the god(s) who rule(s) the rivers (= **diĝir-íd** in l. 15).

For the transcription of **DU** = **de$_6$** (and not **túm**): see Sallaberger (2005b).

With respect to the comitative infix ***da** in **mu-da-de$_6$**, Balke observes the following: 'Daneben erscheint ein Komitativpräfix -da- oftmals in Kontexten, die weder eine personale noch impersonale Referenz des Dimensionalpräfixes aufweisen, d.h. kein explizites Komitativkomplement im Satz enthalten'. Examples are e.g.: **de$_6$** 'to bring'; **de$_6$** + comitative 'to bring along' (...) [1056].

***

[1055] For the reading **[p]a$_5$** in line 15, I am indebted to Th. Krispijn (personal communication).

[1056] Balke 2006, 112-113; see also p. 31 with examples 46 and 47 for the verb **de$_6$**, and note 150.

3b. Addendum to the text of the Barton cylinder

IAS 174

column i

	UGN-orthography	‘Normal’ orthography	
	... Unknown number of lines broken		
1’	[ ] **NUN-ni-⌜sub$_5$**	**[ne $^?$] mu-ni-⌜sub$_5$**	He kissed her.
2’	**ki-ná**	**ki-ná**	Together with her (l. 3’)
3’	**NUN-GAL-ak**	**mu-da-ak**	he made up the bed (l. 2’),
4’	**ĝiš$_3$ Ḫ̮I-NUN-ŠID**	**ĝiš$_3$ ḫ̮e-mu-du$_{11}$**	to have intercourse with her.
5’	**a MAR**	**a šà**	He has poured (l. 6’) the seed
6’	**NUN-ŠA-ru**	**mu-na$_5$-ru**	into her womb (l. 5’).
7’	**[a⌟-maš-imin**	**[a⌟-maš-imin**	The seed for a set of septuplets
8’	[**x]-[ru⌟**	[**x]-[ru⌟**	[he has] poured [into her womb].
	... Unknown number of lines broken		

column ii

	... Unknown number of lines broken		
1’	[ ]	[ ]	....
2’	**BUR$_2$-BUR$_2$**	**BUR$_2$-BUR$_2$**	She murmured (ll. 2’-3’).
3’	**mu-za**	**mu-za**	
4’	**MAR-^{d}nin-gal**	**šà-^{d}nin-gal**	Inside Ningal
5’	**ušum mu-lu**	**ušum mu-lu**	stirred the snake.
6’	**DAL-DAL**	**DAL-DAL**	She babbled (ll. 6’-7’).
7’	**NUN-za**	**mu-za**	
8’	**⌞KU$^?$⌟ a nin-[ ]**	**⌞KU$^?$⌟ a nin-[ ]**	....
	... Unknown number of lines broken		

Comments

A possible parallel to the text of the Barton cylinder column *ii* has been found in the fragment IAS 174, written in the UGN-orthography [1057]. Preliminary transcriptions, translations and comments have been given by Krebernik [1058].

Column i

4' Krebernik considered **ḪI** = $\mathbf{du_{10}}$ as a gloss for the sign **ŠID**, the UGN-spelling for $\mathbf{du_{11}}$. A more likely transcription for **ḪI** may be **ḫe**; the modal proclitic is usually written as **ḫé** instead of **ḫe** [1059].

Column ii

2'-3' and 6'-7' For the compound verbs $\mathbf{BUR_2}$**-**$\mathbf{BUR_2}$ **za** and **DAL-DAL za**: see Black [1060]. The precise translation for 'making noise' could not be determined due to the deficient context.

4' The goddess Ningal is particularly known as the goddess of Ur, as the spouse of Nanna and as the mother of Utu [1061]. In the Abu Ṣalābīḫ **zà-mì** hymns, Ningal is mentioned after Nanna and Utu, thus most likely in her well-known functions [1062]. In this text – with its striking similarity with the text of the Barton cylinder – there seems to be no specific role for Ningal as the goddess of Ur. As 'Great Divine Lady' she represents in this text the mother goddess, just like Ninḫursaĝa in the Barton cylinder.

***

1057 The cuneiform text of this tablet is published by Biggs (1974). It has been classified as a literary text written in the so-called UD-GAL-NUN (UGN) orthography. For the decoding of the UGN-signs, the lists published by Krebernik (1984, 267-286; 1998, 298-302) were used.

1058 Krebernik 1984, 277-278, *sub* *MAR; Krebernik 1993-1997, § 4.3, 508.

1059 In a few texts from Ebla **ḫe-mu-DU** is present (ARET 13, 5 [CDLI P241971]; ARET 13, 9 [P241801]; ARET 13, 13 [CDLI P241853]; ARET 13, 15 [CDLI P241831].

1060 Black 2003. In his overview of the attested compound verbs (p. 37), **bur-bur za** was registered as 'other ideophones (meanings uncertain)'. **dal-dal za** is missing in this summary.

1061 Zgoll 1998-2001.

1062 Biggs 1974, 47, ll. 39-40.

4. NBC 11108

Obverse

1	**a[n]-uru$_{16}$-né an mu-zalag$_{2}$ / ki mu-kikki kur-šè igi m[u]-[íl$^{?}$]**	The mighty An lighted heaven, earth he darkened, he looked at the netherworld.
2	**buru$_{3}$ a nu-bal ninda nu-ĝar ki-daĝal / ⌜uru$_{4}$⌝ nu-ak**	From the depths no water was drawn; bread was not put down. Cultivation of the wide earth did not happen.
3	**išib-mah̬-^{d}en-⌜líl⌝-lá nu-ù-ĝál / [š]u-luh̬-kù-ge šu nu-ù-⌜ma-du$_{7}$⌝**	The supreme purification priest of Enlil was not there, a holy purification rite was not perfected.
4	**[igi-z]i-an-na-ke$_{4}$ šu nu-ù-tag / [zà$^{?}$]-mí nu-di**	The priestess of An did not play an instrument, did not sing a song.
5	**[an k]i téš-bi-a mu-lug**	An/heaven lived together with Ki/earth,
6	**[nam-dam-š]è ⌜nu-ù⌝-tuku**	(but) as wife he had not taken her.
7	**⌜u$_{4}$⌝ nu-⌜zalag⌝ ĝi$_{6}$-àm mu-lá**	The day did not dawn, the night had spread over her (= **ki** earth).
8	**an-né da-ga-an-na / kiri$_{3}$-zal mu-ni-íb-guru$_{17}$**	An was wearing luxuriance in the residence,

Reverse

9	**ki-gub-ú-šim-ma / ní nu-mu-[gíd]-gíd-e**	(but) on a verdant place he still did not lie down.
10	**me-^{d}en-⌜líl-lá-ke$_{4}$⌝ kur-kur-ra / ⌜šu nu-ù-du$_{7}$⌝**	The ritual of Enlil had not been completed in the lands.
11	**[k]ù-⌜in-nin⌝ [nin]-⌜an⌝-na-ke$_{4}$ / ni[dba] ⌜šu nu-mu⌝-ši-te$^{?}$**	The holy lady, the lady of heaven, did not receive$^{?}$ the food offerings.
12	**[diĝir-ga]l a-nu[n]-[ra a]d$^{?}$ [nu]-um-di-di**	For the great gods, the Anunna, one was not singing$^{?}$:
13	**diĝir-an-⌜na⌝ ⌜an⌝-[ki]-a / nu-ù-⌜ma⌝-su$_{8}$-⌜su$_{8}$⌝-ge-éš**	the gods of heaven still did not stand in heaven and on earth.

Editions / Translations

Editions

1. J. van Dijk, Existe-t-il un 'Poème de la Création' Sumérien? In: B.L. Eichler (ed.), *Cuneiform Studies in Honor of Samuel Noah Kramer*. AOAT 25. Neukirchen-Vluyn: Neukirchener Verlag, 1976; 125-133.
2. W.H.Ph. Römer, Die Urzeit: Vor der Schöpfung (I). In: W.H.Ph. Römer, D.O. Edzard (eds), *Weisheitstexte, Mythen und Epen. Mythen und Epen.* TUAT III. Gütersloh: Mohn, 1993; 353-354.
3. W. Horowitz, *Mesopotamian Cosmic Geography.* Winona Lake: Eisenbrauns, 1998; 138-139.
4. Å.W. Sjöberg, In the Beginning. In: T. Abusch (ed.), *Riches Hidden in Secret Places. Ancient Near Eastern Studies in Memory of Thorkild Jacobsen.* Winona Lake: Eisenbrauns, 2002; 239-244.

Translation:

1. M. Dietrich, Die Kosmogonie in Nippur und Eridu. *JARG* 5 (1984) 159-160. [1063]

***

Comments

1. A possible transcription for **EN,** besides **en** ('lord'), is **uru**$_{16}$ or **urun** [1064]: (inter alia) 'mighty, exalted'. The present translation – the mighty An – is different from those in the previous editions [1065]. It is not a Sumerian habit to write a name, followed by a simple noun as the epithet; in that case the order is (usually) reversed. Rather a name is followed by an adjective, in this case **uru**$_{16}$ [1066].

The last legible sign, as copied by van Dijk, in the upper part of line 1 is **zalag**$_2$ [1067]. Horowitz reads **ká[r$^?$]** instead of /**zalag**/, but the characteristic '*gunû*'-character is not visible. The opposition mostly used in this respect is /**zalag**/ *vs* /**kikki**/ or /**kukku**/ [1068].

[1063] The text of Dietrich is a literal translation into German of van Dijk's edition; therefore Dietrich's translation will not be discussed.

[1064] Civil 1989, 55; Borger 2010, 79, no. 164.

[1065] Van Dijk: 'An, (étant) Bēl'; Römer: 'An, der Herr'; Sjöberg: 'An, the En'; Horowitz: 'Anu$^?$, the Lord'. Michalowski (1993a, 159 and note 46) suggests a reading **⌜en⌝-né an mu-za[lag]**, with a 'remote possibility' that **en-né** 'is an abbreviation of a syllabic spelling of **én-é-nu-ru**, "incantation formula", and that the beginning has to be rendered as: "The earth was lit up, *etc.*" '. But where is 'earth' in the first part of this line?

[1066] The adjective /**uru**/ is not unusual for An. Examples are: 'An **adab** to An for Lipit-Eštar' [ETCSL 2.5.5.3], l. 31; 'An **adab** to An for Ur-Ninurta' [ETCSL 2.5.6.5], l. 1; Ukg 15 *ii*:1 (included in this study). Other examples and discussions with respect to /**uru**/: Falkenstein 1959, 33, *ad* 15; Falkenstein 1962, 69-72; van Dijk 1960, 66, 92; PSD B, 96^a *ad* 2.1.2 (Hymn to Nippur; UET 6/1, 118 *ii*: 11-13).

[1067] Both van Dijk and Sjöberg read **zalag**, while the sign as copied by van Dijk is **zalag**$_2$. Both signs sometimes alternate in different editions of the same text, *e.g.* Cooper 1978, 82, l. 145.

[1068] Krispijn: personal communication. For the transcription of **MI,** there are several possibilities of which Borger has given a survey of the readings, varying from /**mi**/, /**me**/, /**gig**/, /**gigi**/, to /**ku**/ (Borger 2010, 179, no. 681; see also Krecher 1967a, 98, note 14). Borger explicitly mentions, that **MI** = /**kukku**/ has not been found in lexical lists; for **MI-MI** there is more unity; it is

According to Sjöberg, the reading **m[u]-zala[g]-⌜ge⌝** is almost certain; but the photo and the copy of the text have no indication for it. Moreover, a *marû* stem before two immediately following *ḫamṭu* stems seems less probable: the order is rather reversed [1069].

An is supposed to be the subject, not only of **zalag$_2$**, but also of the verbs **kikki** and **igi .. íl**. On the other hand, van Dijk and Römer consider the second part of line 1 as being intransitive or passive, but the verbal forms do not fit with this view.

Of all possible combinations **igi** + verb, it seems that **íl** is the best choice, because of the use in Sumerian literature [1070].

2. Because of the opposition **a bal** and **NIĜ$_2$ ĝar**, **NIĜ$_2$** may be read as **ninda**; **ĝar** 'to put down', is to be understood as 'make available'.

The traces in the copy of van Dijk do not contradict the proposed reading **APIN** [1071]. For the transcription of **APIN** there are two possibilities in this case. The first one is: **absin$_3$** = furrow; the expression '**absin$_3$ ... ak**' might mean: to make furrows, to plough. The complete expression for 'to plough', in combination with **AK**, however reads: **APIN-DU$_3$ AK** [1072]. The direct object of **AK** in idiomatic phrases is often a verb alone [1073]. A second possibility is to transcribe **APIN** = **uru$_4$** 'to cultivate'; then we may read: **ki-daĝal uru$_4$ nu-ak** 'cultivation of the wide earth did not happen' [1074].

---

equated with /**kukku**/. According to ePSD, **MI** = '1.**kukku$_5$** (to be) dark; 2. **giggi** = (to be) black; **MI-MI** = **ku$_{10}$-ku$_{10}$**, (to be) dark)'. As a result I have chosen **MI = kikki**, as Sjöberg did already (see also discussion at Sjöberg 2002, 241, and note 21).

1069 Streck 1998, 184-186 (ex. 7-8), 191-192: the conclusion § 2.5.

1070 I have checked the presence of the various combinations **igi + verb** with **kur** in the Sumerian literature. For **igi + ĝar**, **bar** or **lá** there are no examples with **kur**, for **igi .. tab** and **kur** there is only one example. The meaning of **igi íl** 'to lift the eye; to look at', might fit in the context, but there seems to be not enough space for the sign **íl**. There are several examples for the combination of **igi ĝál** and **kur**. But **kur-šè mu-ĝál** seems grammatically impossible, because the verbal prefix-chain must have a prefix ***ši** (one should expect the form **im-ši-ĝál** in that case; Jagersma, personal communication). The sign following **igi** is **mu** (Sjöberg 2002, 241). The combination **igi** + **íl** is chosen here, because a literature search showed that in this case a noun + terminative ***šè** is not always accompanied by a ***ši** in the prefix-chain; e.g.: 'The cursing of Agade' [ETCSL 2.1.5], l. 152; 'A praise poem of Šulgi (Šulgi A)' [ETCSL 2.4.2.01], l. 45; 'The temple hymns' [ETCSL 4.80.1], l. 241.
Referring to this line, Katz (2003, 13, note 36) writes: 'This line is interesting because it describes the situation with *an* and *ki* in positive terms of reality, but with regard to the *kur* it comments that it is invisible (or non-existent).' This remark has been based only on the edition of van Dijk, who in the second part reads: **kur-šè igi nu$^?$-ĝ[ál]**. Since Sjöberg (2002, 241) has written that the sign following **igi** is certainly **mu**, and not **nu**, this comment of Katz on the invisibility or non-existence of the **kur** cannot be based on this passage.

1071 A reading **á** instead of **APIN** (Horowitz 1998, 138) seems less probable; moreover, the meaning of the expression **á AK** is obscure. Horowitz translates: '... service was not done'. This meaning for **á AK** was not found in PSD , at ePSD or at Attinger 2005.

1072 PSD A-III, 78 *ad* 8.20. For the expression **APIN ... AK**: see also 'The farmer's instructions' [ETCSL 5.6.3], l. 30. Other expressions for the making of furrows are: **ab-sin$_2$ / absin$_3$ .. ur$_{11}$ / ra / gub** (PSD A-II, 148-149 *ad* 2.1).

1073 PSD A-III, 76 ff.

1074 The expression **uru$_4$ .. AK** was not found in PSD, at ePSD or at Attinger 2005.

3. **išib** (*išippu*; CAD I, 242^{b}): purification priest. According to Renger [1075], the **išib** belongs to the high-rank 'Beschwörungspriester', incantation priests or exorcists. One of his tasks is to perform purification rites. A notable detail: besides the **išib** there also exists an **išib-maḫ**, about whom Renger says: '**išib-maḫ** bezieht sich in den erwähnten Fällen auf Eridu' [1076].

Sjöberg reads in the second part of this line: ... **šu nu-ù-ma-<u>ni</u>-du$_7$**, but there seems to be not enough place for the sign **ni**. Moreover he did not translate this part of the line, without any comment.

4. For the first signs of this line, van Dijk proposed the restoration **[nu-gi]g**$^{??}$ [1077]. But in view of the style of writing applied at the tablet NBC 11108, it seems unlikely that there is enough space for the signs **nu-gig** [1078]. Another possible candidate for restoration may be **égi-zi** (the traces do not contradict this supposition). Besides the use as epithet, almost exclusively used for Ninḫursaĝa [1079], there exists also a general function **égi-zi (-an-na)**, the priestess in question being a very important cultic functionary [1080]:

> 'In spite of the seeming rarity of **égi-zi**, it now becomes clear that this office was in reality quite common. Such at least was the situation in Ur III times, from which numerous attestations of it survive. However, in Ur III sources the title in question is spelled differently: **igi-zi** or **egi (ŠE$_3$)-zi**.'

At the beginning of line 4 there is enough space for the writing **igi-zi**, which seems the most probable way of writing here, because the tablet NBC 11108 dates from the Ur III time, as suggested by van Dijk [1081].

If, in view of the clear traces of the sign **mí**, the restoration of the second part of this line with **zà** is correct, then this line refers to 'making music'. According to Karahashi, **šu**

1075 Renger 1969, 122-126.

1076 Renger 1969, 125, and note 637. The position of the **išib-maḫ** in relation to the **išib** was not discussed by Renger.

1077 Van Dijk 1976, 129 (l. 4); 131 comments.

1078 Renger (1967a, 179-184) and Henshaw (1994, 206-213) have described several meanings and functions of **nu-gig**. The conclusion of Henshaw with respect to the function of a **nu-gig** (*qadištu*) is: 'The *qadištu*'s roles seem to have been: to exalt the god, to take part in the procession, to sing the *inhu*-song. There is nothing of a sexual role.' (Henshaw 1994, 209 *sub* 4.11.9). Henshaw also noticed, that **nu-gig** is one of the titles of Inanna, a subject that most recently was dealt with by Zgoll (Zgoll 1997a; Zgoll 1997b, 304-305, *ad* Zeile 3). Her conclusion (1997a, 194): '**[nu-gig]** ist ein Hoheitstitel, mit welchem ein unüberhörbarer Anspruch auf Land und Herrschaft verbunden ist, besonders auch die Vorherrschaft über den Himmel (**nu-gig-an-na**); so ließ sich an einigen Stellen ein Bezug auf An ("Hierodule Ans") ausschließen und die Bedeutung "Herrscherin über den Himmel" begründen.' Besides, Zgoll mentions that **nu-gig** was also used as an indication or title for certain priestesses (Zgoll 1997a, 182, 184; Zgoll 1997b, 304, *ad* Zeile 3 *sub* 3). Glassner (1992, 76) noticed: 'Le titre **nu.(u$_8$.)gig.an.na**, habituellement traduit par "hiérodule céleste", ne peut faire référence à la déesse comme modèle céleste des courtisanes d'ici bas. On l'a vu, en effet, le sumérien **nu.gig** comme son équivalent akkadien *qadištum* ne font aucunement allusion, dans les sources du troisième millénaire ou du début du second, à une manière de prostitution ou d'hiérodulie.'

1079 Steinkeller 2005, 303. See also the Barton cylinder, col. *ii*: 1-6 (this study).

1080 Steinkeller 2005, 301, 303 (quotation).

1081 Van Dijk 1976, 128.

tag is 'to touch', and when used with musical instruments it means 'to play' [1082]. Although no musical instrument is mentioned, it may be concluded from the context that 'making music' was meant here. With respect to **zà-mí dug$_4$**, see Attinger [1083]: 'adresser un/des éloge(s); chanter des louanges; faire l'éloge', which has been translated here as 'to sing a song' [1084]. In the expression written here – **zà-mí nu-di** – the non-finite *marû*-form of the verb **dug$_4$** has been used.

5. The chosen phrasing **an ki** is different from those of van Dijk and of Horowitz (**an-ki**; Sjöberg places x x at the beginning). Instead of the unit **an-ki**, as plural absolutive, it is more plausible, in view of the singular verbal form, that there is also a singular absolutive, in this case **an** [1085]. This **an** is at the same time the subject in the next line.

The last sign of this line, as copied by van Dijk, is **lug** ('to live in its natural habitat, said of an animal' [1086]), not **dib** (van Dijk) or **dab$_5$** (Horowitz) [1087].

6. **nam-dam-[a-ni]-šè tuku** *aḫāzu*: 'to take a wife, to marry' [1088]. This expression was only used from the point of view of a man, who takes a woman as spouse. Therefore the translations of van Dijk, Römer and Sjöberg, saying that both partners marry each other, are less probable [1089]. Sjöberg distinguishes two times **tuku** on the cast, as van Dijk already had supposed, but the copy of van Dijk shows just one **tuku** [1090].

7. Van Dijk has interpreted the first (broken) sign as /**itud**/ 'moon'. Due to the broken text this reading cannot be excluded. In my edition I have followed Sjöberg who writes [1091]: 'The reading ⌜**u$_4$**⌝ ⌜**nu**⌝-⌜**zalag**⌝ is almost certain.'

The phrasing **ĝi$_6$-àm mu-lá**, which has been followed in the present edition, is that of Sjöberg. It accentuates the contrast day *vs* night. Van Dijk's spelling of the verbal form **àm-mu-lá** is gramatically not possible [1092].

1082 Karahashi 2000, 166-167.

1083 Attinger 1993, 755-761, ch. 5.3.206.

1084 The suggestion of Horowitz (1998, 138) for this line: [x] x **an.na.⌜ke$_4$ šu⌝ nu.ù.⌜du$_7$⌝ [di.k]u$_5$**$^?$ ⌜**nu**⌝.**di**: '[the .] .. of Anu was not performed, [judgi]ng$^?$ [was not] judged', is very unlikely. The readings **du$_7$** instead of **tag**, and **ku$_5$** instead of **mí**, are nearly to be excluded. Sjöberg (2002, 239) has no doubts about the reading **tag**.

1085 Plural pronominal suffixes are written in this text, see e.g. line 13.

1086 Civil 1972, 386^b *sub* P. 98:67. For **lu.g** = *rabāṣu* 'to lie down', see: CAD R, 10, and Steinkeller 1984.

1087 Mittermayer (2006) clearly shows the difference between these signs: **LU** 164-165, no. 416; **DIB** 166, no. 419.

1088 CAD A I, 173-177.

1089 The example, quoted by Sjöberg (2002, 242, note 25), from 'Enki and Ninmaḫ' ll. 6-7, in justification of a possible completion of the beginning of line 6, is rather unfortunate, because in the lines in question **nam-dam-šè** was not written, but **nam-NIR-PA** (see also the edition of 'Enki & Ninmaḫ' in this study).

1090 The possibility of a reading **kin** instead of **tuku** (Horowitz) seems less likely because of the positive reading **tuku** by Sjöberg on the cast.

1091 Sjöberg 2002, 242, *ad* 7. My analysis of the verbal form **nu-ù-tuku** is: **nu-i-<u>en</u>-/*ḫamṭu-stem*/-Ø** or **nu-i-<u>b</u>-/*ḫamṭu-stem*/-Ø**. A conclusion with respect to the pronomen-prefix – the animate class (***en**) or the inanimate class (***b**) – is not possible. An analysis of ***nu-ù-** as **nu-ù-** is not possible, because the prefix ***u** cannot be used together with the negative proclitic ***nu** (Jagersma 2010, 517, ch. 24.2.1).

8. **da-ga-an**, a syllabic spelling of /**daggan**/, seems to be a hapax in its present presentation [1093]. In the opinion of van Dijk [1094], the meaning 'chamber' does not fit here: 'Il s'agit plutôt d'un topos cosmique: **dag-an** signifierait "demeure céleste".' But finally he did not translate this word at all: 'le Ciel montra sa face resplendissante dans le *Dagan*' [1095]. Römer and Sjöberg followed the interpretation of van Dijk. According to Civil, **dakan** is an early Semitic loanword in Sumerian [1096]. Due to a different phrasing and transcription – **da-ga-an-na-ka-ni** – Sjöberg is not able to give a complete translation of this line.

Most probably the verb **guru**$_{17}$ is a *ḫamṭu*-stem, followed in the next line by a *marû*-stem, expressing simultaneity [1097]. The ***b** before the *ḫamṭu*-stem [1098] represents the ergative, so indicating the inanimate class of **an**. This line relates that **an** has adorned himself for a festal occasion, viz. the approaching marriage between **an** and **ki**.

9. Line 9 expresses that the expectation, raised in line 8, has not yet been fulfilled. **an** has prepared himself for a marriage, but nevertheless: he does not stretch out on the appropriate place. Therefore the translation of this line might start with: 'but....'.

The first part of this line – in the present transcription **ki-gub-ú-šim-ma** – was interpreted in different ways [1099]. In my opinion: **ú-šim** 'greenery' is an apposition to **ki-gub**, e.g. abode, resting place [1100], and the final ***a** is a locative ending. This 'verdant place' describes the bed, consisting of grass and probably fragrant herbs, that has been prepared (or: has yet to be prepared) for the celebration of the marriage between **an** and **ki** [1101].

1092 Jagersma, personal communication. Horowitz has even a third phrasing: **ĝi**$_6$**.a an mu.lá** 'in night, heaven stretched forth'. This seems very unlikely, because heaven also stretches forth during the day. The evident 'subjects' are day and night.

1093 The examples found via ePSD show spellings like **da-ga-n(a)**, **da-gán**, **dag-ga-n(a)**. The Akkadian word is *dakkanu* (AHw, 151-152) or *takkanu* A (CAD T, 74). Behrens (1978, 189) gives references to the literature, where this word has been discussed. Krecher (1987, 88, note 39) discusses two words, written **da**, one of them ending on /**g**/, which is mostly used as reference or allusion to sexual overtures; **da-ga-na** then means 'in his/her nearness'. This is not applicable here, because **da-ga-an-na** is written in this line, unless we emend **da-ga-«an»-na**, with the consequent translation 'An wears luxuriance in her nearness for her'.

1094 Van Dijk 1976, 131.

1095 While van Dijk wrote 'le *Dagan*', Dietrich interpreted it as the Syrian god Dagan (Dietrich 1984, 160).

1096 Civil 2007, 18, *ad* CaCaC 006.

1097 Streck 1998, 184-186, examples 7-10.

1098 My analysis of the prefix-chain **mu-ni-íb-** is: **(i)-m-b-ni-b-**, ***b-ni** referring to **da-ga-an-na**.

1099 For **ki-gub** van Dijk writes **ki du** ('là où il allait'), Sjöberg **ki-du** ('Ki-Earth'), and Horowitz **ki DU** ('earth'). Sjöberg (2002, 243, and also note 27): '(...) I have preferred to understand **ki-du** as "(the) earth". **ki-gub**, Akk. *manzāzu*, can hardly be considered.'

1100 For **ki-gub** *manzāzu*: see AHw, 638, *s.v. mazzāzu(m)*; CAD M I, 234-239.

1101 [*1*] A nice parallel for the description of a nuptial bed as a bed strewn with flowers, spread with herbs, is given in the Old Babylonian text 'A *kunĝar* to Inana', AO 6967, ll. 40-41 (Sefati 1998, ch. 15, 247-256, especially pp. 248 and 250; comments pp. 255-256). In the transcription and translation of this text in ETCSL 4.08.20:
40. **ĝiš-nú gi**$_4$**-rin-na-ĝu**$_{10}$**-a dè-ma-ab-gub-bu-ne** (source: **un-na-ab-gub-bu-ne**)
41. **ú za-gin**$_3$ **dur**$_5$**-ru-ĝu**$_{10}$ **dè-ma-ab-barag**$_2$**-ge-ne**
40. Let them erect for me my flowered bed.
41. Let them spread it for me with herbs like translucent lapis lazuli.
[*2*] In 'Enlil and Sud' we find (Civil 1983; ETCSL 1.2.2):
148. **e**$_2$ **ki-nú-a /ĝiš\-nu**$_2$ **gi-rin-/na** ^{ĝiš}**tir šim** ^{ĝiš}**erin-gin**$_7$ **[...]**

On the authority of Sjöberg I write **nu-mu-gíd-gíd-e**, instead of one verbal stem **nu-mu-gíd-e** as van Dijk did [1102]. According to Thomsen [1103], **gíd** is a regular verb; this implies that the verbal form in this line represents a reduplication of the *marû* stem: 'he *still* does not...'. The *marû* stem expresses the simultaneity with what happens in line 8, i.e. simultaneity in the past [1104].

10. For a recent summary and discussion of the conception of **me**, see Berlejung [1105].

11. In spite of the severe damage of this line, I have tried to restore it, starting from the suggestions of van Dijk [1106]. Several alternatives are proposed here.

Instead of **é-an-na-ke$_4$**, I suggest **nin-an-na-ke$_4$**; the expression **in-nin-é-an-na-(k)** has not been found in Sumerian literature [1107]; on the basis of this literature research I suggest the restoration **kù-in-nin nin-an-na-ke$_4$**.

For the last sign of the second part of this line (van Dijk: **te**$^?$; Horowitz: **túm**$^?$; Sjöberg: ?) I would like to accept van Dijk's suggestion: **te**. Karahashi wrote about the compound verb **šu–ti/te** 'to receive' [1108]: 'The received is the semantic object of the composite predicator **šu–ti** and receives an oblique case marking', e.g. the locative **-a**. Therefore a locative marker may be necessary after **nidba**.

12. The first identifiable signs are **a** and **nu[n]**. The sign in front of the **a** might be **[ga]l**. The usual epithet for the Anunna-gods – assumed that they are mentioned indeed in this line – is **diĝir-gal-gal**, but more than one **gal**-sign is not very likely (not enough space).

I have tried to complete this line in relation to the general tenor of the lines 3, 4, 10 and 11: rituals. The traces of the first sign after **nu[n]** might be those of **ad** (since in general

---

149. **/^{d}en-lil$_2$\ nita$_3$-dam-a-ni ĝiš$_3$ /bi$_2$$^?$\-[...] /mu\-ni-in-hi-li-[...]**
148. In the sleeping quarters, in the flowered bed …… like a fragrant cedar forest,
149. Enlil made (?) love to his wife and took great pleasure in it.
[*3*] A description of a resting-place (probably a bedroom) as a verdant place is found in an Ur III text, from the year that Amar-Suen has become king (CDLI P102380, obv. *i*: 23): **eša šà é-ki-tuš é-ú-šim**.

1102 Horowitz 1998, 138-139, and notes 36 and 37: **ki DU ú.šim.ma ní nu.mu.[x].šér.e**, '(but) earth, bringing forth plant life did not glow on its own'. For the expression **ní šér** no examples could be found in the literature. Moreover, in view of his commentary, Horowitz himself seems not satisfied if **ki DU ú.šim.ma** is to be construed as 'participle plus noun' which is 'awkward in Sumerian' (note 36).

1103 Thomsen 1984, 304.

1104 Streck 1998, 184-185, exs. 7-8.

1105 Berlejung 1998, 20-25.
ePSD: **me**, (inter alia) '(cultic) ordinance'.

1106 Van Dijk 1976, 129: **[k]ù$^?$ i[n-ni]n$^{??}$-é-an-na-ke$_4$ n[idba$^{??}$ š]u n[u-mu-š]i-te$^?$**.

1107 In the literature I have found the following: A. with **é-an-na**: 'Enmerkar and the lord of Aratta' [ETCSL 1.8.2.3], l. 233: **diĝir nin-é-an-na-ka**; ll.484 and 624: **^{d}inana nin-é-an-na-ra**; 'The Death of Ur-Namma' [ETCSL 2.4.1.1], l. 201: **nin-gal-é-an-na**. B. with **an**: 'Inana and An' [ETCSL 1.3.5], l. B3: **nin$^?$-an-na-ke$_4$**; l. D38: **nin-gal-an-na-ke$_4$**; 'Enmerkar and the lord of Aratta' [ETCSL 1.8.2.3], l. 229: **nin-gal-an-na**; 'A *šir-namursaĝa* to Ninsiana for Iddin-Dagan' [ETCSL 2.5.3.1], passim: **nin-gal-an-na**.

1108 Karahashi 2000, 168.

the signs are written rather large, there seems to be hardly any space for more than the suggested signs) [1109].

Van Dijk suggested the equation **di-di** = *dâlu* 'se mettre en mouvement' [1110], but the meaning of *dâlu* does not seem to fit in this context [1111].

For **DI-DI** = **sá-sá** = *kašādu*: see CAD K, 271vv. Then the consequent translation should be: 'The great gods, the Anunna, had not arrived'. First: this statement duplicates for a great part that of the following line, an argument that is also valid for van Dijk's translation. That kind of duplication is not in line with the rest of the text. Second: the lexical attestations for this equation are availbale only from the Middle Babylonian period.

13. Hitherto the transcriptions and translations of the first part are: **diĝir-an-na diĝir-ki-a ...** 'the gods of heaven (and) the gods of earth ...'. However, it is very unlikely that after **ki** a genitive ***a(k)** should have been written [1112]. On the contrary, a locative postposition would be written [1113]. The present transcription therefore is different from the former ones: **diĝir-an-na an-ki-a** [1114].

The reduplication of the verbal stem **sug$_2$** (or **su$_8$**), which itself is already a plural form, has been interpreted as being the expression of an iterative meaning.

***

1109 For the 'correct' form **^{d}a-nun-na-ke$_4$-ne-ra**, in case my suggestion for a completion is right, is not enough space. But compare also the lines 11-12 of the AṢ **zà-mì** hymns (Excursus 1 Enlil and Ninlil, § 3.2, note 583): **^{d}en-E$_2$ a-nun ki mu-ĝar-ĝar**, where **a-nun** is written even without the dative ***ra**. For **ad dug$_4$** 'to sing', see Krispijn 1990, 14-15. Another expression in relation to making music is **balaĝ di** (Attinger 1993, ch. 5.3.20, pp. 451-453), but for the sign **balaĝ** seems to be not enough space.

1110 Van Dijk 1976, 132, with translation of l. 12 (p. 129): 'Les gr[ands dieux], les Anunna, ne s'étaient pas mis en mouvement.'

1111 *dâlu* CAD D 58: 'to wander around aimlessly, to wander about in despair, to prowl, to disport oneself, to move (unnaturally)'. See also the examples given by Barré 2001, 178-179.

1112 Thomsen 1984, 90, § 161.

1113 Thomsen 1984, 98, § 182; Krecher 1986, 45.

1114 As an example of 'gods who stand in heaven': see 'An *adab* to Enlil for Išme-Dagan' [ETCSL 2.5.4.08], l. 4: **diĝir-an-na an-na bí-su$_8$-ug** 'he made the gods of heaven stand by in heaven'.

5. The Debate between Tree and Reed[1115]

1	**ki-ùr-gal-e ní pa bí-íb-è bar-dul-le-eš nam-sa$_7$**	The large surface of the earth introduced herself; then she has embellished herself as with a *bardul*-garment.
2	**[ki]-daĝal-e kù ^{na4}za-gìn-bi bar-ba àm-mi-íb-si**	The vast earth has filled her exterior with precious metals and lapis lazuli.
3	**$^{[n]a4}$esi ^{na4}nir$_7$ ^{na4}gug sù-du-áĝ-ĝá šu-tag ba-ni-ib-du$_{11}$**	With diorite, nir-stone, cornelian and suduaĝa she has adorned herself.
4	**[ki]-ú-šim-e ḫi-li gú bí-íb-è nam-nun-ba mu-un-gub**	The earth, the fragrant vegetation, covered herself with attractiveness. She stood in her magnificence.
5	**[k]i-kù-ki-sikil-la an-kù-ra ní-bi mu-na-ab-sa$_7$**	The pure earth, the virgin earth, has beautified herself for the holy An.
6	**an-an-maḫ-e ki-daĝal-la dùb im-ma-ni-ib-nir**	An, the exalted heaven, had intercourse with the vast earth.
7	**a-ur-saĝ-ĝiš-gi-bi-da-ke$_4$ šà-ga ba-ni-in-de$_5$**	He poured the seed of the hero's Tree and Reed into her womb.
8	**ki-šár-áb-zi-dè a-du$_{10}$-ga-an-na da bí-íb-ri**	The whole earth, the fecund cow, took the good seed of An under her care.
9	**ki-ú-nam-ti-la-ke$_4$ šà im-ḫúl ù-tu-ba mu-un-gub**	The earth, life-giving vegetation, innerly happy, devoted herself to the production of it (*i.e. the vegetation)*.
10	**ki-kiri$_3$-zal-e ḫé-ĝál im-gùr kurun$_2$ lal$_3$ ir sù-ud**	The earth, full of joy, bore abundance, while juice and syrup gave out their smell.

1115 The Sumerian text was composed from the tablets AO 6715, Ni 4463, Ni 4598, and from the texts given by van Dijk 1964, 44-45, and Sjöberg 2002, 244. Römer (1993c, 357) and Vanstiphout (1990, 305, note 23) give a survey of the Sumerian tablets with the text of this debate. Because van Dijk had an unpublished manuscript of Civil and the collations of Nougayrol at his disposal, I have not written a score. The present version of the text has been copied from Sjöberg (the most recently published text), with only a few adaptations in the transcriptions. Wilcke (2007, 43-44) has given a score of the lines 1-12. The text of this debate most likely dates from the Ur III-period (van Dijk 1953, 35-36).

Editions / Translations

Editions
1. J. van Dijk, Le Motif Cosmique dans la Pensée Sumérienne. *AcOr* 28 (1964) 44-57.
2. Å.W. Sjöberg, In the Beginning. In: T. Abusch (ed.), *Riches Hidden in Secret Places. Ancient Near Eastern Studies in Memory of Thorkild Jacobsen.* Winona Lake: Eisenbrauns, 2002; 244-247 [1116].

Translations
1. S.N. Kramer, *History Begins at Sumer: Thirty-Nine Firsts in Recorded History.* Philadelphia: University of Pennsylvania Press, 1981; 303-304.
2. W.H.Ph. Römer, Der Prolog des Streitgespräches zwischen Holz und Rohr (Z. 1-29). In: W.H.Ph. Römer, D.O. Edzard (eds), *Weisheitstexte, Mythen und Epen. Mythen und Epen I, TUAT III.* Gütersloh: Mohn, 1993; 357-360.
3. J. Bottéro, S.N. Kramer, Le Prologue du Tournoi 'Arbre contre Roseau'. In: J. Bottéro, S.N. Kramer, *Lorsque les Dieux Faisaient l'Homme. Mythologie Mésopotamienne.* [Paris]: Gallimard, 1993; 479-481 (ll. 1-8).
4. M. Jaques, *Le Vocabulaire des Sentiments dans les Textes Sumériens. Recherche sur le Lexique Sumérien et Akkadien.* AOAT 332. Münster: Ugarit-Verlag, 2006; 262, S403: lines 7-11.
5. C. Wilcke, Vom altorientalischen Blick zurück auf die Anfänge. In: E. Angehrn (ed.), *Anfang und Ursprung. Die Frage nach dem Ersten in Philosophie und Kulturwissenschaft.* Berlin, New York: Walter de Gruyter, 2007; 17 (translation), 43-44 (score).

***

Comments

1. **ki-ùr** has been dealt with by van Dijk. Eventually, he did not translate it, perhaps because the multitude of possibilities he had listed obstructed a definitive choice. From the context it is clear that the surface of the earth was meant [1117]: **ki** - earth adorns herself for her approaching intercourse with An.

Sjöberg discusses whether it should be written **bar dul-le-eš** or **bar-dul-le-eš** [1118]. He did not take up a clear position. There may be some arguments against the second reading. The determinative **túg**, although not obligatory, is absent; moreover, the most current spelling for this kind of garment is **bar-dul$_5$** [1119]. The second reading was followed in this edition, because the first reading resulted in a translation like: 'She embellished the exterior by way of covering', which in my opinion is less expressive.

1116 Sjöberg gives the Sumerian text and comments on earlier translations, but he himself does not present a translation.

1117 **ki-ùr** = *duruššu*: AHw, 178: '(planiertes) Fundament'; CAD D, 198: 'basis, base, foundation, habitation'. Behrens 1978, 145, note 281 gives a short survey of the meanings of **ki-ùr.**

1118 Sjöberg 2002, 246. Attinger discussed words with an ***eš**-ending (L'adverbiatif; Attinger1993, 168, § 105 a)), with the meaning 'de manière', 'à la manière de', 'comme'.

1119 **bar-dul$_5$** A: PSD B, 119-121.

Places in the lexical lists with **SIG**$_7$, provided with glosses, are gathered by Pettinato, and one of them is **SIG**$_7$ = /**sa**/ = *banû* [1120].

2. The copy of AO 6715 has **za-gìn-bi bar-ba;** in Ni 4463 we read: **za-gìn-ba bar-bi** and also **àm-mi-íb-si** (Sjöberg: **àm-mi-ib-si**). Van Dijk supposes **gá** at the end of this line; the copy of Ni 4463 has **si**, which is not incompatible with the rest of the concerning sign in AO 6715.

The passive translation of this line in PSD [1121] ('precious metals and gems were inlaid in the outer façade of the vast place') neither agrees with the Sumerian text, nor with the idea behind this text: it is earth who is actively adorning herself (see also ll. 1, 3-5).

3. The exact nature of the precious stone **nir**$_7$ cannot be given. ^{na4}**ZA-MIR** = **nir**$_7$ is the old spelling for ^{na4}**ZA-TUN**$_3$ = **nír** = *ḫulālu* [1122]. According to Borger it might be agate or chalcedony (?) [1123]. With respect to **sù-du-áĝ-ĝá**, also written as **sud-áĝ**, and mostly without the **na**$_4$-determinative, – the Akkadian *elmēšu* – it is not sure to which kind of stone it refers. Borger supposes that it might be amber$^?$ (Bernstein$^?$), Attinger thinks of electrum (?) [1124]. Antimony, as Bottéro and Kramer translate it, is very improbable, because its usual spelling has been derived from **ŠIM** (see ePSD).

Attinger discusses the expression **šu-tag du**$_{11}$, and he mentions also this line in the examples. It is remarkable that in this sentence, with **ki** as subject, at least in one text the subject pronomen is of the animate class [1125].

A parallel of the embellishment of **ki** can be found in an Old Babylonian text [1126], where an elaborate description of Inanna's preparation for a celebration of marriage with Dumuzi is given, e.g. with the precious stone lapis lazuli.

4-10. The lines 4-6 and 8-10 show similar constructions: **ki** and **an** are attended with appositions; e.g. [**ki**]**-ú-šim-e** (l. 4) and **an-an-maḫ-e** (l. 6).

4. Van Dijk was pretty certain about his supplement of the beginning of this line with [**an-e**] [1127]: 'Le sujet ne peut être qu'An'. His opinion was based on the text AO 4153 *ii*: 1:

1120 Pettinato 1971, 52. *banû* B (CAD B, 90): 'to grow, to be pleasant'; *D-stem*: 'to beautify, adorn, to improve, to decorate, to prepare carefully'. For the transcription of **SIG**$_7$ as **sa**$_7$: see Borger 2010, 152, no. 564. At ePSD the spelling **sag**$_{12}$ (**sa**$_7$) is given, but a spelling **sag**$_{12}$ is not found at Borger. See also Lambert (1998, 193) for the meanings of *banû* and *bunnû*.

1121 PSD B, 98 *sub* 3.2.5 of **bar** A.

1122 Sjöberg 1974, 173 *ad* 15. Wilcke (2007, 43) reads **nír** instead of **nir**$_7$.

1123 Borger 2010, 440, no. 851.

1124 Borger 2010, 376, no. 584. Attinger 1993, 725, § 857 d 1^0. *elmēšu* in AHw, 205: 'wohl ein kostbarer, gelbscheinender Stein'. *elmēšu* in CAD E, 107-108: '(a precious stone)'. The fact that *elmēšu* in the lexical list **ḪAR / ur**$_5$**-ra** = *ḫubullu* is mentioned under the pigments of mineral origin and not under the stones, indicates that *elmēšu* has a characteristic colour.

1125 Text AO 6715: **... ba-ni-in-du**$_{11}$; text Ni 4463: **... ba-ni-ib-du**$_{11}$. Attinger 1993, 720-725, ch. 5.3.181; this sentence is mentioned in § 857 d 1^0, p. 725. See also the Excursus 'The animate *vs* inanimate class of **an** and **ki**' at the end of this Appendix.

1126 'A *kunĝar* to Inana': Sefati 1998, ch. 15, 247-256, especially ll. 11-24 [Text: AO 6967; ETCSL 4.08.20].

1127 Van Dijk 1964, 45, and comment *ad* 4 p. 49.

an en-nam šul-le-éš al-gub, in his translation: 'An, l'En, se dressa en jeune héros' [1128]. He translated this line of 'Tree and Reed' as follows: '[Le Ciel] revêtit d'attraits les herbages, se dressa dans sa majesté'. Kramer and Römer shared the opinion of van Dijk that An is the subject of this sentence [1129]. But in the translation of Bottéro and Kramer, the earth is the subject [1130]. In my opinion **ki**/earth is the subject of the lines 1-5, which show a bride preparing her wedding dress. In addition to a literal meaning – **ki** is adorning herself –, this sentence has also a metaphorical meaning: the embellished earth represents at the same time the marriage bed. This becomes the more clear, when we consider NBC 11108, and especially line 9: '(but) on a verdant place he [= An] still did not lie down' [1131].

The /**n**/ before the verbal stem in **mu-un-gub** can be explained as a mark for the locative case, as an alternate form of the locative infix /**ni**/ [1132], and referring to **nam-nun-ba̱**.

5. Van Dijk translates: 'La Terre pure se montra enjolivée au pur An dans une place immaculée'. He comments: 'nous traduisons «dans une place immaculée». On pourrait traduire aussi: «la terre, la pure terre». Grammaticalement cette traduction l'emporte, parce que après **ki-kù** il manque le **-e** de l'agentif.' [1133] This seems to be a strange and unrealistic situation: earth who presents herself *in a virgin place*.

The suffix /**a**/ in **sikil-la** might have a determining character [1134]; one should expect the ergative ending /**e**/.

6. Van Dijk, who discusses two possible transcriptions – **an an-maḫ-e** and **an diĝir-maḫ-e** –, has given one argument in favour of the first reading [1135]: that reading outlines the characteristics of 'the exalted heaven' on the one side, and 'the vast earth' on the other side. I may add a second and grammatical argument for the first option, *viz.* the /**b**/ before the verbal stem. In the transcription **an diĝir-maḫ-e** 'An, the exalted god', the god An (animate) should have been the subject; then we should expect /(**e**)**n**/, the pronominal prefix for the animate class, before the stem. Therefore the preferred transcription is **an-an-maḫ-e** with the concomitant translation: 'An, the exalted Heaven...'. The /**b**/ refers to the inanimate 'Heaven', not to An.

Cooper [1136] and ePSD have for **dùb nir** (Akkadian *rakābu*) 'to ejaculate', but a translation more close to the Sumerian expression – a rather plastic description – is 'to have intercourse'. The consequence of this action is mentioned in the next line.

The same idea – Heaven copulates with Earth – is expressed in 'Ninurta's exploits' (Lugale), line 26: **an-e ki sig$_7$-ga ĝiš im-ma-dug$_4$** 'Heaven copulated with the verdant Earth' [1137].

1128 The translation of AO 4153 *ii*:1 (Ukg 15) in this study reads: 'An, as **en**, was standing there as a youthful man'.

1129 Kramer 1981, 303; Römer 1993c, 358.

1130 Bottéro and Kramer 1993, 480.

1131 See the edition of NBC 11108 in this study.

1132 Delnero 2007, *esp.* 122-123.

1133 Van Dijk 1964, 46, and 49-50. Römer 1993, 358, follows van Dijk in his translation.

1134 Thomsen 1984, 64, § 80.

1135 Van Dijk 1964, 50 *ad* 6.

1136 Cooper 1989, 87. See also Karahashi 2000, 81 *s.v.* **dùg nir**.

1137 Van Dijk 1983, l. 26; ETCSL 1.6.2, l. 26.

7. The *e at the end of the complex **a-ur-saĝ-ĝiš-gi-bi-da-ke$_4$** represents the deictic /**e**/; or -**ke$_4$** is the Old Babylonian form for the earlier -**kam** [1138]. In the verbal chain the pronominal prefix for the animate, /(**e**)**n**/, was used by the scribe; that means that in this case he was thinking of the god An, not of Heaven, who was ejaculating.

8. Instead of **ki-šár** it may be read **ki-du$_{10}$** [1139], 'the good earth'. Because the result of this insemination can be seen everywhere on earth, **ki-šár** has been chosen in the present edition. For the expression **da ri**: see Selz 1995b.

9. Van Dijk discusses **gub** + locative as 'to serve', and translates it as 'se tenir prête, se dévouer' [1140]. In my opinion **šà im-ḫúl** points to the inner make-up of the earth, and not to the process of bringing forth [1141].

10. **DIN** = **kurun$_2$** most likely represents here (sweet) juice of fruit, which is a natural product, and not some kind of beer, which is the result of human activity. In the introduction to this debate humanity does not yet exist.

Jaques has interpreted this line differently [1142]: 'Ki: dans la magnificence, l'opulence s'est levée, fruits, miel, parfum répandus.' In this text, however, both **an** and **ki** are provided with appositions [1143]. The *-**e** at the end of the complex **ki-kiri$_3$-zal-e** is not a directive ending, but the ergative ending.

***

1138 Black & Zólyomi 2007, 22, § 4.2.1.C.

1139 Except Bottéro & Kramer and Jaques, all scholars read **ki-du$_{10}$**. Jaques (2006, 262, *ad* S403) does not translate **ki-šár**: **ki-šár áb zi-dè a du$_{10}$-ga an-na-da bí-ib-ri** 'en Kišar, la vache légitime, la bonne semence y fut éjaculée/versée par An'.

1140 Van Dijk 1964, 52. According to Thomsen 1984, 304: **gub** + **da** 'to stand by, to serve'.

1141 Van Dijk 1964, 46: 'La Terre se dévoua toute à donner une heureuse naissance aux herbes de vie (= à la végétation)'. Probably this translation was the consequence of his reading **šà-ḫúl** (van Dijk) instead of **šà im-ḫúl** (Sjöberg).

1142 Jaques 2006, 262, S403.

1143 In the present transcription these appositions are indicated by the use of hyphens after **an** and **ki**, respectively: e.g. l. 6 **an-an-maḫ-e**; l. 10; **ki-kiri$_3$-zal-e**.

6. The Debate between Grain and Sheep

Published sources

Siglum	*Signature*	*Publication*	*Provenience*
A	CBS14005	MBI 8	Nippur
B	Ni 4462	ISET 2, 71	Nippur
C	Ni 9885	ISET 1, 193	Nippur
D	Ni 4215	ISET 1, 138	Nippur
E	CBS 6983	SEM 54	Nippur
F	Ni 2505 +	SRT 25	Nippur
	Ni 4094	SLTNi 20	Nippur
G	Bodleian CI	BE 31, 15	Nippur
H	CBS 2228	HAV 6	Nippur
U		UET 6, 33	Ur

1		**ḫur-saĝ-an-ki-bi-da-ke$_4$**	When Heaven [l. 2] had caused the mountain range of **an-ki**
	A *obv*	ḫur-saĝ-an-ki-bi-da-ke$_4$	
	B	[ ]-ke$_4$	
	U *obv*	ḫur-saĝ-an-ki-bi-da-ke$_4$	
2		**u$_4$ an-né diĝir-^{d}a-nun-na im-tu-dè-eš-a-ba**	to bring forth the Anunna gods
	A	u$_4$ an-né diĝir-^{d}a-nun-na im-tu-dè-eš-a-ba	
	B	[ ]-a-ba	
	C	[ ] ^{d}a-nun-na [x]	
	U	u$_4$ an-né diĝir-^{d}a-nun-na im-tu-dè-eš-a-ba	
3		**mu ^{d}ezina$_2$ nu-ub-da-tu-da-aš nu-ub-da-an-sig$_7$-ga**	– because Ezina had not been engendered together with them, had not grown up with them;
	A	mu ^{d}ezina$_2$ nu-ub-da-tu-da nu-ub$^{!}$-da-an-sig$_7$-ga	
	B	[ ]x-da-aš nu-ub-da-an-sig$_7$-g[a]	
	C	mu ^{d}ezina$_2$ nu-ub-da-[x] / nu-ub-da-an-[ ]	
	U	mu ^{d}ezina$_2$ nu-ub-ta-tu-da-aš nu-ub-da-sig$_7$-g[a]	
4		**kalam-ma gu-^{d}uttu nu-ub-da-an-dím-ma-aš**	because in the Land the thread of Uttu had not been fashioned together with them,

	A	kalam-e$^{?}$ gu-^{d}uttu nu-ub-da-an-dím-ma-aš	
	B	[ ]-[m]a gu-^{d}utt[u] nu-ub-[d]a-an-dím-m[a-]	
	C	kalam-ma gu-^{d}utt[u] [ ]	
	U	kalam-ma gu-^{d}uttu nu-ub-da-dím-ma-à[m]	
5		**^{d}uttu-ra temen nu-mu-un-na-si-ga-aš**	(and) for Uttu no poles (*for the loom*) had been driven into the ground –,
	A	^{d}uttu-ra temen nu-mu-na-si-ga-aš	
	B	[x]-ra temen nu-mu-un-⌜na-si⌝-[ ]	
	C	^{d}uttu-ra temen nu-[x]	
	U	^{d}uttu-ra temen nu-mu-un-na-si-ga-à[m]	
6		**u$_{8}$ nu-è-a sila$_{4}$ nu-šár-ra**	when the ewe had not appeared, so lambs were not present in great numbers,
	A	u$_{8}$ nu-è-a sila$_{4}$ nu-šár-ra	
	B	u$_{8}$ nu-[ ] sila$_{4}$ nu-šár-ra	
	C	u$_{8}$ nu-[x] [ ]	
	U	u$_{8}$ nu-è-a sila$_{4}$ nu-šár-ra-àm	
7		**ùz nu-è-a máš nu-šár-ra**	when the goat had not appeared, so kids were not present in great numbers,
	A	ùz nu-me-a máš nu-šár-ra	
	B	ùz nu-[x] máš nu-šár-ra	
	C	[x]	
	U	ùz nu-è-a máš nu-šár-ra-àm	
8		**u$_{8}$-e sila$_{4}$-min-bi nu-ub-tu-ud**	the ewe did not give birth to her two lambs,
	A	u$_{8}$-e sila$_{4}$-min$^{!}$-bi nu-ub-tu-ud	
	B	u$_{8}$-e [x]-bi nu-ub-tu-ud	
	U	u$_{8}$-e sila$_{4}$-min-bi nu-ub-tu-ud	
9		**ùz-e máš-eš$_{5}$-bi nu-ub-tu-ud**	the goat did not give birth to her three kids;
	A	ùz-e máš-eš$_{5}$-bi nu-ub-tu-ud	
	B	ùz-e [x]-eš$_{5}$-bi nu-ub-tu-ud	
	U	ùz-e máš-eš$_{5}$-bi nu-ub-tu-ud	
10		**mu-^{d}ezina$_{2}$-^{d}kù-sù-u$_{8}$-bi-da-ke$_{4}$**	the names of Grain, the holy blade, and of Ewe,
	A	mu-^{d}ezina$_{2}$-kù-su$_{13}$-u$_{8}$-bi-da-ke$_{4}$	
	B	mu-^{d}[ezi]na$_{2}$-^{d}kù-sù-u$_{8}$-bi-da-ke$_{4}$	
	D	[ ] [x]	
	U	mu-^{d}ezina$_{2}$-^{d}kù-sù-^{d}u$_{8}$-bi-da-ke$_{4}$	

11		**^{d}a-nun-na diĝir-gal-gal-e-ne nu-mu-un-zu-uš-àm**	did even the Anunna, the great gods, not know!
	A	^{d}a-nun-na diĝir-gal-gal-e-ne nu-mu-un-zu-uš-àm	
	B	^{d}a-nun-[x] diĝir-gal-gal-e-ne nu-mu-un-zu-u[š]-[ ]	
	D	[ ] diĝir-ga[l]-⌜gal]-⌜e]-[ ]	
	U	^{d}a-nun-na diĝir-gal-gal-e-ne nu-mu-un-zu-uš-àm	
12		**še-muš$_5$ u$_4$-30-àm nu-ĝál-la-àm**	The 30-days *šeĝuš*-barley was not there!
	A	še-muš$_5$ u$_4$-30-àm nu-ĝál-la-àm	
	B	[x] nu-ĝál-la-[ ]	
	D	[ ] ⌜u$_4$-30-àm nu-[ ]	
	U	še-muš$_5$ u$_4$-30-kam nu-ĝál-la-àm	
13		**še-muš$_5$ u$_4$-40-àm nu-ĝál-la-àm**	The 40-days *šeĝuš*-barley was not there!
	A	*missing*	
	D	[ ] ⌜u$_4$-40-àm [ ]	
	U	še-muš$_5$ u$_4$-40-kam nu-ĝál-la-àm	
14		**še-muš$_5$ u$_4$-50-àm nu-ĝál-la-àm**	The 50-days *šeĝuš*-barley was not there!
	A 13	še-muš$_5$ u$_4$-50-àm nu-ĝál-la-àm	
	D	*missing*	
	U	še-muš$_5$ u$_4$-50-kam nu-ĝál-la-àm	
15		**še-di$_4$-di$_4$ še-kur-ra še-á-dam-kù-ga nu-ĝál-la-àm**	The small barley, the foreign barley, (and) the barley from the holy fields around the dwelling-places were not there!
	A 14	še-di$_4$-di$_4$ še-kur-ra še-á-dam-kù-ga nu-ĝál-la-àm	
	D	[ ] še-kur-ra še-[x]	
	U	še-di$_4$-di$_4$ še-kur-ra še-á-dam-kù-ga nu-ĝál-la-àm	
16		**túg níĝ mu$_4$-mu$_4$-bi nu-ĝál-la-àm**	A garment (or) something to dress was not there!
	A 15	túg níĝ mu$_4$-mu$_4$-bi nu-ĝál-la-àm	
	D	[x]-bi nu-[ ]	
	E	[-l]a-à[m]	
	U	túg níĝ mu$_4$-mu$_4$-bi nu-ĝál-la-àm	
17		**^{d}uttu nu-ub-tu-ud men nu-íl**	Uttu had not been born. A cap was not worn.

	A 16	^{d}uttu nu-ub-tu-ud men nu-íl	
	D	[-u]b-tu-ud men [x]	
	E	[-u]b-tu-ud men nu-íl	
	F	[x]	
	U	^{d}uttu nu-tu-ud men nu-íl	
18		**en-^{d}niĝir-si en-kal-kal nu-ub-tu-ud**	Enniĝirsi, the very mighty lord, had not been born;
	A 17	en-^{d}niĝir-si en-kal-kal nu-ub-tu-ud	
	D	[ ] ⌜en-kal-kal [ ]	
	E	[ ] niĝir en-kal-kal nu-tu-ud	
	F	diĝir en niĝir [ ]	
	U	en-^{d}niĝir-si en ^{d}KAL-KAL nu-tu-ud	
19		**^{d}šakan$_2$ bar-rim$_4$-ma la-ba-ra-è**	Šakan had not appeared in the dry area.
	A 18	^{d}šakan$_2$ bar-rim$_4$-ma la-ba-ra-è	
	D	[x] la⌝-[ ]	
	E	^{d}šakan$_2$ bar-rim$_4$-ma la-ba-ra-è	
	F	^{d}šakan$_2$ bar-rim$_4$-ma [x]	
	U	^{d}šakan$_2$ bar-rim$_4$-ma la-ba-ra-è-a	
20		**nam-lú-ulu$_3$-u$_4$-ri-a-ke$_4$-ne**	Mankind of those faraway days
	A 19	nam-lú-u$_{12}$-lu-u$_4$-ri-a-ke$_4$-e-ne	
	D	[x]	
	E	nam-lú-ulu$_3$-u$_4$-ri-a-ke$_4$-ne	
	F	nam-lú-ulu$_3$-u$_4$-ri-a-[ ]	
	U	nam-lú-ulu$_3$-u$_4$-ri-a-ke$_4$-ne	
21		**ninda gu$_7$-ù-bi nu-mu-un-zu-uš-àm**	had not got to know eating bread,
	A 20	ninda gu$_7$-ù-bi nu-mu-un-zu-uš-àm	
	E	ninda gu$_7$-ù-bi nu-mu-un-zu-uš-àm	
	F	ninda «a» gu$_7$-ù-bi nu-mu-un-[ ]	
	U	ninda gu$_7$-ù-dè nu-mu-un-zu-uš-àm	
22		**túg níĝ mu$_4$-mu$_4$-bi nu-mu-un-zu-uš-àm**	had not got to know a garment (or) something to dress in!
	A 21	túg-ga mu$_4$-mu$_4$-bi nu-mu-un-zu-uš-àm	
	B	túg-ga mu$_4$-m[u$_4$] [ ]	
	E	túg-ga mu$_4$-mu$_4$-bi nu-mu-un-zu-uš-àm	
	F	túg-ga mu$_4$-mu$_4$-bi nu-mu-un-[ ]	
	G	túg níĝ mu$_4$-mu$_4$-bi nu-mu-un-zu-uš-àm	
	U	túg níĝ mu$_4$-mu$_4$-bi nu-mu-un-zu-uš-àm	

23		**ùĝ ĝeš-gen$_6$-na-a šu-bi mu-un-ĝen**	The people moved on all fours;
	A 22	ùĝ ĝeš-gen$_6$-a-na šu-bi mu-un-ĝen	
	B	ùĝ ĝeš-g[e]-[ ]	
	E	[ù]ĝ ĝeš-ge-na-a šu-ba mu-un-ĝen	
	F	ùĝ ĝeš-ge-na-a šu-bi mu-[ ]	
	G	ùĝ su ge-en-na-àm / sù-bi mu-un-ĝen	
	U	kalam-ma ĝeš-ge-na su-bi mu-un-ĝen	
24		**udu-gin$_7$ ka-ba ú mu-ni-ib-gu$_7$**	like sheep they ate grass with their mouths,
	A 23	udu-gin$_7$ ka-ba [ ]-in-íb-gu$_7^{!}$	
	B	udu-gin$_7$ k[a]-[ ]	
	E	[ ]-gin$_7$ ka-ba ú mu-ni-ib-gu$_7$	
	F	udu-gin$_7$ ka-ba ú mu-ni-ib-[ ]	
	G	udu-gin$_7$ ka-ba ú mu-ni-ib-gu$_7$	
	U	udu-gin$_7$ ka-bi ú mu-ni-ib-gu$_7$	
25		**a-mú-sar-ra-kam i-im-na$_8$-na$_8$-ne**	the water from the gardens they were drinking.
	A 24	a-mú-sar-ra-[ ] im-na$_8$-na$_8$-ne	
	B	[x] [ ]	
	E	[.. m]ú-sar-ra-kam i-im-na$_8$-na$_8$-ne	
	F	a-mú-sar-ra-kam i-im-na$_8$-na$_8$-n[e]	
	G	a-mú-sar-ra-ka i-im-na$_8^{!}$-na$_8^{!}$	
	U	a-ḪAR-ḪAR-ra-ke$_4$ i-im-na$_8$-na$_8$-ne	
26		**u$_4$-ba ki-ulutim$_2$-diĝir-re-e-ne-kam**	At that time, near the place of residence of the gods,
	A 25	u$_4$-ba ki SIG$_7$-[-r]e-e-n[e]	
	E	[ ] ki-ulutim$_2$-diĝir-re-e-ne-kam	
	F	u$_4$-ba ki-ulutim$_2$-diĝir-re-e-[ ]	
	G	u$_4$-ba ki-ulutim$_2$-diĝir-re-e-ne-[ ]	
	U	u$_4$-ba ki-SIG$_7$-ALAM-AL[AM] [ ]-re-e-ne-ke$_4$	
27		**é-bi-du$_6$-kù-ga u$_8$ ^{d}ezina$_2$-bi mu-un-se$_{12}$-eš-àm**	in their abode, the holy hill, Ewe and Grain were living.
	A 26	é-bi-du$_6$-[ ]-bi kur$^{?}$ [x]	
	E	[b]i-⌜du$_6$-kù⌝-ga u$_8$ ^{d}ezina$_2$-bi / mu-un-se$_{12}$-eš-àm	
	F	é-bi-du$_6$-kù-bi u$_8$ ^{d}ezina$_2$-bi mu-[ ]	
	G	é-bi-du$_6$-kù-ga u$_8$ ^{d}ezina$_2$-[ ] / mu-un-se$_{12}$-eš-⌜àm]	
	U	é-bi-du$_6$-kù-ga ^{d}[ ] [^{d}e]zina$_2$ [ ] ⌞un⌟-SIG$_7$-e	

28		**èš-ninda-gu$_7$-diĝir-re-e-ne-ka mi-ni-ib-de$_5$-de$_5$-ge-eš-a**	After they [*the Anunna*] had gathered them in the dining hall of the gods,
	A 27	èš [ ] g[u$_7$] [ ] bi	
	E	[ ] gu$_7$-diĝir-re-e-ne-⌞ka⌟ [ ] / [ ]	
	F	èš-ninda-gu$_7$-diĝir-re-e-ne-kam mi-ni-ib-de$_5$-de$_5$-[ ]	
	G	èš-ninda-gu$_7$-diĝir-re-ne-[ ] / mi-ni-ib-de$_5$-de$_5$-[ ]	
	H	[ne-ke$_4$$^?$ / [ ]-ni-ib-de$_5$-de$_5$-ge-eš-a	
	U	èš-ninda-gu$_7$-diĝir-re-[ ]	
29		**ḫé-ĝál-u$_8$-^{d}ezina$_2$-bi-da-ka**	of the abundance of Ewe and Grain
	A 28	[x]	
	E	[ ] ⌜u$_8$⌝ ^{d}e[zina$_2$]-[ ]	
	F	ḫé-ĝál-u$_8$-^{d}ezina$_2$-bi-da-k[a]	
	G	ḫé-[x..x]	
	H	ḫé-ĝál-u$_8$-^{d}ezina$_2$-bi-da-ka	
	U	ḫé-ĝál [ ]	
30		**^{d}a-nun-na-du$_6$-kù-ga-ke$_4$-ne**	the Anunna of the holy hill
	F	^{d}a-nun-na-du$_6$-kù-ga-ke$_4$-[ ]	
	G	^{d}[ ]	
	H	^{d}a-nun-na-du$_6$-kù-ga-ke$_4$-ne	
	U	^{d}a-nun-na-du$_6$-k[ù]-[ ]	
31		**i-im-gu$_7$-gu$_7$-ne nu-mu-un-dè-si-si-eš**	were eating, but all along they could not satisfy themselves.
	F	i-im-gu$_7$-ù-ne nu-mu-un-dè-[ ]	
	H	i-im-gu$_7$-gu$_7$-ne nu-mu-un-dè-si-si-eš	
	U	i-im-gu$_7$-gu$_7$-ne [ ]	
32		**ubur-kù-ga-ne-ne ga-bi níĝ-dùg-ga**	The milk, that sweet substance of their pure udder,
	F	ubur$_x$-kù-ga-ne-ne ga-bi níĝ-[ ]	
	H	ubur-kù-ga-ne-ne ga-bi níĝ-dùg-ga	
	U *rev*	ubur-kù-ga-ne-ne a [ ]	
33		**^{d}a-nun-na-du$_6$-kù-ga-ke$_4$-ne**	the Anunna of the holy hill
	F	^{d}a-nun-na-du$_6$-kù-ga-ke$_4$-[ ]	
	H	^{d}a-nun-na-du$_6$-kù-ga-ke$_4$-ne	
	U	*missing*	

34		**i-im-na$_8$-na$_8$-ne nu-mu-un-dè-si-si-eš**	were drinking, but all along they could not satisfy themselves.
	A *rev* 2	[ ] x [ ]	
	F	i-im-na$_8$-na$_8$-ne nu-mu-un-dè-[ ]	
	H	i-im-na$_8$-na$_8$-ne nu-mu-un-dè-si-si-eš	
	U	*missing*	
35		**ubur-kù-ga níĝ-dùg-ga-ne-ne-šè**	For the sake of the sweet substance of their pure udder
	A *rev* 3	[ub]ur$^?$ ŠE A[ ]	
	F	ubur$_x$-kù-ga níĝ-dùg-ga-ne-[ ]	
	H	ubur-kù-ga níĝ-dùg-ga-ne-ne-šè	
	U	*missing*	
36		**nam-lú-ulu$_3$ zi-šà im-ši-íb-ĝál**	they have inspirited mankind.
	A *rev* 4	[ ]-lú-ulu$_3$ [ ]	
	F	nam-lú-ulu$_3$ zi-šà [ ]	
	G	⌜nam⌝-lú-[ ]	
	H	nam-lú-ulu$_3$ zi-šà im-ši-íb-ĝál	
	U 33	nam.lú.ulu$_3$ KAxX-bi [x]	
37		**u$_4$-ba ^{d}en-ki-ke$_4$ ^{d}en-líl-ra gù mu-un-na-dé-e**	At that time Enki spoke to Enlil:
	A *rev* 5	[ ]-ba ^{d}en-ki-ke$_4$ x-[ ]	
	F	u$_4$-ba [ ]-ki-ke$_4$ ^{d}en-líl-ra gù mu-na-d[é]-[ ]	
	G	u$_4$-ba ^{d}en-ki-ke$_4$ [ ]	
	H	u$_4$-ba ^{d}en-ki-ke$_4$ ^{d}en-líl-ra gù mu-un-na-dé-e	
	U 34	u$_4$-ba ^{d}en-ki-ke$_4$ ^{d}en-líl-ra gù mu-u[n]-[ ]	
38		**a-a-^{d}en-líl u$_8$-^{d}ezina$_2$-bi-da-ke$_4$**	'Father Enlil, now that Ewe and Grain
	A *rev* 6	a-a-^{d}en-líl x-[ ]	
	F	a-a-^{d}en-líl u$_8$ ^{d}ezina$_2$ [x-]	
	G	a-a-^{d}en-líl [x...x...x]	
	H	⌞a-a-^{d}en-líl u$_8$-^{d}ezina$_2$-bi⌟	
	U 35	a-a-^{d}en-líl ^{d}u$_8$-^{d}ezina$_2$-bi-da-⌜ke$_4$⌝	

39		**du$_6$-kù-ga um-ma-da-an-se$_{12}$-eš-a**	have resided together at the holy hill,
	A *rev* 7	du$_6$-kù-ga um-ma-da-an-[ ]	
	F	du$_6$-kù-ga um-ma-da-an-se$_{12}$-eš-a [?]	
	G	du$_6$-kù-ga um-ma-da-an-sig$_7$-[x]	
	U 36	du$_6$-kù-ga um-ma-da-an-sig$_7$-ga	
40		**du$_6$-kù-ta ga-àm-ma-da-ra-ab-e$_{11}$-dè-en-dè-en**	let us send them down together from the holy hill'.
	A *rev* 8	du$_6$-kù-ta$^{!}$ ga-àm-ma-da-ra-ab-[x]	
	F	du$_6$-kù-ta ga-àm-ma-da-ra-ab-e$_{11}$-d[è-]	
	G	du$_6$-kù-ta ga-àm-ma-da-ra-ab-e$_{11}$-[x]	
	U 37	du$_6$-kù-ga-ta àm-ma-da-ra-ab-e$_{11}$-dè-en-dè-en	
41		**^{d}en-ki ^{d}en-líl-bi inim-kù-ga-ne-ne àm-dug$_4$-ga**	At the holy command of Enki and Enlil
	A *rev* 9	^{d}en-ki$^{!}$ ^{d}en-líl-bi inim-kù-ga-ne$^{!}$-n[e]-[ ]	
	F	^{d}en-ki ^{d}en-líl-bi inim-kù-ga-ne-ne a [ ]	
	G	^{d}en-ki-ke$_4$ ^{d}en-líl-ra inim-e$^{?}$-ga-ne-ne [ ]	
	U 38	^{d}en-ki ^{d}en-líl-bi inim-kù-ga-ne-ne «a » àm-dug$_4$-ga	
42		**u$_8$ ^{d}ezina-bi du$_6$-kù-ta im-ma-da-ra-an-e$_{11}$-dè**	Ewe and Grain set off down together from the holy hill.
	A *rev* 10	u$_8$ ^{d}ezina$_2$-bi du$_6$-kù-ta im-ma-da-r[a]-[ ]	
	F	u$_8$ ^{d}ezina$_2$-bi du$_6$-kù-ta im-ma-da-ra-[ ]-dè	
	G	u$_8$ ^{d}ezina$_2$-bi du$_6$-kù-ta ma-da-ra-[ ] / -dè-en-dè-en	
	U 39	^{d}u$_8$ ^{d}ezina$_2$-bi du$_6$-kù-ga im-ma-da-ra-an-e$_{11}$-dè	

Editions / Translations

Editions

1. G. Pettinato, *Das altorientalische Menschenbild und die sumerischen und akkadischen Schöpfungsmythen.* Heidelberg: Carl Winter Universitätsverlag, 1971; 86-90 (ll. 1-42).
2. B. Alster, H. Vanstiphout, Lahar and Ashnan. Presentation and Analysis of a Sumerian Disputation. *ASJ* 9 (1987) 1-43.
3. J. Bottéro, S.N. Kramer, Prologue du Tournoi 'Céréale contre Menu-Bétail'. In: J. Bottéro, S.N. Kramer, *Lorsque les Dieux Faisaient l'Homme. Mythologie Mésopotamienne.* Paris: Gallimard, 1993; 511-514 (ll. 1-42).

Translations

1. S.N. Kramer, *Sumerian Mythology. A Study of Spiritual and Literary Achievement in the Third Millennium B.C.* Philadelphia, University of Pennsylvania Press, 1972; 53-54, 72-73.
2. S.N. Kramer, *History Begins at Sumer: Thirty-Nine Firsts in Recorded History.* Philadelphia: University of Pennsylvania Press, 1981; 108-109.
3. H.L.J. Vanstiphout, The Disputation between Ewe and Wheat. In: W.W. Hallo (ed.), *The Context of Scripture. Vol. I. Canonical Compositions from the Biblical World.* Leiden: Brill, 1997; 575-578.
4. M.P. Streck, Die Prologe der sumerischen Epen. *OrNS* 71 (2002) 247-248, 250.
5. C. Wilcke, Vom altorientalischen Blick zurück auf die Anfänge. In: E. Angehrn (ed.), *Anfang und Ursprung. Die Frage nach dem Ersten in Philosophie und Kulturwissenschaft.* Berlin, New York: Walter de Gruyter, 2007; 19.
6. ETCSL 5.3.2: The Debate between Grain and Sheep.

***

Comments

Transcription and translation of (the names of) **u$_8$** *and* **ezina**

The sign **u$_8$** is written without the diĝir-determinative throughout the lines 1-42 of all texts, with the exception of the lines 10, 38 and 42 of text U. According to Edzard the sign combination ^{d}**u$_8$**, including the diĝir-determinative, indicates the goddess Laḫar [1144]. In this respect Lambert speaks of a god, and he is not sure about the Sumerian reading, not necessarily Laḫar, which is indeed the name of the Akkadian god(dess) [1145]. In the present transcriptions always **u$_8$** is written, which has been translated as Ewe.

The Sumerian reading of the complex **ŠE-TIR** is /**ezina**/; the Akkadian equivalent is *ašnan.* Krecher justifies the reading /**ezina**/, and speaks of a 'Getreidegöttin' [1146]. According to Edzard *ašnan* is a goddess, but in the Old Babylonian epoch there is also

[1144] Edzard 1965, 94.

[1145] Lambert 1980-1983b, 431.

[1146] Krecher 1966, 132-133. Although he is convinced of the reading /**ezina**/, he maintains the reading ^{d}**ezinu** 'der Bequemlichkeit halber'. CAD A II, 450-452: *ašnan* and ePSD give additional references to lexical lists for this reading; see also: Diri Sippar [MSL XV] ii: 20'.

attestation of a god, written as **^{d}ŠE-TIR**, as counterpart of the female *ašnan* [1147]. In spite of the fact that in the present texts the complex always has been written with the diĝir-determinative, it is still difficult to discern whether the god or the grain itself is meant. About this question Bottéro and Kramer have written: 'Dans le présent tournoi littéraire, c'est bien entre les archétypes de ces deux réalités d'ici-bas, et non entre les divinités qui les patronnaient, que se déroulent discussion et comparaison (...)'[1148]. The proof is, they say, the lack of the diĝir-determinative at **u_8**, whereas this was written at **ezina**. I subscribe to this view. The present translation of **ezina** is Grain, except in line 3 where I write 'Ezina': in lines 2-5 there is mention of the presence or absence of gods.

***

Structure of part I: lines 1-42

In this part three main sections are to be discerned. The first section (ll. 1-25) describes what is happening at the very beginning of time. The lines 1+2 [subordinate clause] and 10+11 constitute the first main sentence: the Anunna, engendered by An, do not even know the names of Ewe and Grain; the lines in between explain why: Ewe and Grain have not yet been created. The following lines (12-25) delineate in more detail what is missing at the beginning. Although mankind is mentioned (line 20) – we are not informed about its creation –, human civilization is lacking and people are behaving like animals.

The second section (ll. 26-36) starts by mentioning that Ewe and Grain are living at the holy hill, without any story about their creation included. Then the Anunna gods enjoy the abundance produced by Ewe and Grain, but they are not satisfied at all. The conclusion therefore is that mankind has to be brought into such a condition ('inspirited') that they could fulfil the needs of the gods. People have to become civilized, which means in this case that they must learn to breed cattle and to cultivate cereals.

The third section tells us that, on the proposal of Enki, Ewe and Grain went from the holy hill down to Sumer. This last aspect however, the goal being Sumer, is not mentioned explicitly in the text here, but it can be concluded from the text in line 55: **kalam-ma zi-šà-ĝál mu-da-an-ĝál-le-eš** 'In the Land they have brought vitality'; i.e. vitality in the form of food, sustenance.

***

Comments on the present translation

1-2. The complex **ḫur-saĝ-an-ki-bi-da-ke_4** [1149] has a prominent position in this composition. Unlike the more usual start of a myth with a temporal clause, beginning with

---

1147 Edzard 1965, 68 *sub* Getreidegottheiten.

1148 Bottéro and Kramer 1993, 511.

1149 The comitative ***da** in **ḫur-saĝ-an-ki-bi-da-ke_4** seems to strengthen the idea that **an** and **ki** are still together, constitute one complex and are not yet separated. Recently Balke (2006, 116-118, ch. 4.4) wrote about the 'koordinierende Konjunktion' **bi-da**: 'Dadurch werden in der Regel zwei Nominalphrasen paarweise (...) derart miteinander verknüpft, daß sie ein geschlossenes Syntagma bilden'. In the literature no other examples of the expression **ḫur-saĝ-an-ki-bi-da** have been found. Sjöberg and Bergmann (1969, 51, note 2) summarize briefly the discussion in

ud/u$_4$[1150], this debate opens with the complex **ḫur-saĝ-an-ki-bi-da-ke$_4$**, placed outside the construction of the temporal clause, as it were to emphasize it. In this respect it is comparable to the anticipatory genitive construction which has a stylistic role, viz. that of topicalization, recently reviewed by Haber [1151]. In all the texts of this composition the complex in line 1 ends in **-ke$_4$**. The translations published so far show, that this ***e**-ending has been interpreted as a locative(-terminative) postposition: '*Upon* the mountain range..'. About this ***e**-postposition the following can be said:

i. It might concern a weak reproduction of a locative postposition ***a**. Falkenstein wrote in this respect: 'Die in nachsumerischen Texten häufige Veränderung der Lokativ-postposition -a > -e, die zu einer Verwischung des Unterschieds zwischen Lokativ und Lokativ-Terminativ der unmittelbaren Nähe geführt hat (...).' [1152]

Balke has another opinion: 'Die Lokativpostposition besitzt, wie aus dem vorhandenen Textbestand eindeutig hervorgeht, ausschließlich die morphologische Form -/a/, (...). Zu dieser Form existieren (...) keine phonologisch oder phonotaktisch bedingten Allomorphe, etwa der Form *-/e/, *-/i/ oder *-/u/.' [1153]

The preference is given to the proposition of Balke, also in view of the clear writing of a locative postposition ***a** in the texts of 'Grain and Sheep'.

ii. It could be a directive postposition; its meaning is approximately the direction 'near to' [1154]. The use of ***e** in this sense seems to be limited to a rather small number of mostly compound verbs [1155]. Not surprisingly, the majority of these verbs express a movement [1156]. In any case the verb /**utud**/ has not been incorporated into the range of examples given by Thomsen. Therefore a directive postposition seems to be excluded.

iii. ***ke$_4$** could be a weak form for ***kam** [1157]. The copula **-am** emphasizes **ḫur-saĝ-an-ki-bi-da** [1158]. Any possible postposition after the genitive ***ak** – either a locative or a

---

the literature about **ḫur-saĝ-an-ki-bi-da**. For the translation **ḫur-saĝ** 'mountain range', see Steinkeller 2007.

1150 Thomsen 1984, 246-248, §§ 489, 490.

1151 Haber 2009.

1152 Falkenstein 1978a, 109, § 30.a.3. He gives as an example: Gudea Cyl. A xiii: 29: **gi$_6$ sub-be$_7$ mu-na-zal-e**, an example that is not convincing, because the verb expresses a movement and consequently the *e represents a directive ending.

1153 Balke 2006, 33, ch. 3.1. In nearly all manuscripts of 'Ewe and Grain' a locative ending is written as ***a**; for instance l. 4: **kalam-ma̱**; l. 19: **bar-rim$_4$-ma̱**.

1154 Falkenstein 1978a, 110-112, § 31; 1978b, 118-129, § 106. Thomsen 1984, 95-96, § 174. Krispijn (2004, 23) en Balke (2006, 171-190, ch. 7) call this postposition 'directive'.

1155 Thomsen 1984, 95-96, § 174.

1156 The example of Falkenstein (Gudea Cyl. A xiii: 29) with the verb **zal** may be considered as such.

1157 Black and Zólyomi 2007, 18, 22. Attinger (1993, 312, § 206 a 1^0) claims: 'Le remplacement de /(C)am/ par /(C)e/ (…) attesté sporadiquement déjà à l'ép. pB (…).' Falkenstein (1978a, 147, note 3) says: 'Nachsumerisch ist die Entwicklung von **-bi-im** > **-bé-e** bezeugt.' Maybe the transition of the writing ***kam** > ***ke$_4$** runs parallel with this development. Wilcke (1998, 464) wrote: 'Der promiskue Gebrauch von auslautendem [-a] und [-e] könnte darauf zurückzuführen sein, daß der Unterschied zwischen Lokativ und Lokativ-Terminativ keine genaue Entsprechung bei den Präpositionen des Akkadischen findet.' In 'Ewe and Grain', l. 25, in two texts (E, F) **-kam**, in one text (U) **-ke$_4$** has been written; in l. 26: E shows **-kam** and U again **-ke$_4$**. Poebel

directive – will be obscured by the enclitic copula ***am**. Such a postposition is very likely, because otherwise line 1, merely as information 'it is the mountain range of **an-ki**', is less meaningful. This leads to the following translation of ll. 1-2: 'It was near the mountain range of **an-ki**. When **an**/heaven had engendered the gods, the Anunna, ...'.

But at least in one text, the text from Ur, the spelling ***ke₄** exists besides ***kam**. In the lines 102 and 104, and probably also in line 106: **ĝá-a-kam** 'it/these is/are mine'.

iv. Another possibility that does justice to the ***e**-ending of the complex is the following. It represents a causative construction, in which the second agent is provided with the dative or directive ending; for inanimate beings this ending is ***e** [1159]. When this possibility will be applied to the lines 1-2, **ḫur-saĝ-an-ki-bi-da-*ak** may be considered as second agent, because of the ***e**-ending of this complex [1160]. **an**, Heaven, as transitive subject supplied with an ergative ending, is the one who let the mountain range bring forth the Anunna gods [1161]. The translation of the lines 1-2 then is: 'When Heaven had caused the mountain range of **an-ki** to bring forth the Anunna gods,'. The prominent position of the '**ḫur-saĝ**'-complex in line 1, even before the temporal clause beginning with **u₄** in the next line, seems to justify this translation; for just *the place where* the Anunna had been born does not seem worth mentioning with such emphasis, at least in my opinion.

The question that has to be asked is, whether the verbal form **im-tu-dè-eš**, and especially the prefix chain, is compatible with this translation. The prefix chain could be analysed as /**i-m-b**/. The pronominal element ***b** refers to an inanimate ergative [1162], which in this case must be **an**; inanimate, because the complex **an-ki** has not been separated in **an** and **ki**? Consequently, the translation of **an** should be 'Heaven', not the god An [1163]. However, in the prefix chain there is no space for reference to the complex **ḫur-saĝ-an-ki-bi-da** [1164]. Perhaps the position of this complex outside the temporal clause, beginning with

---

(1923, 137, § 373) wrote: 'Für das im Auslaut stehende Genitivelement -a(k), bez. -(k), tritt jeztz (= in nachsumerischer Zeit; JL) häufig als einfache Genetivendung -ake, bez. -ke ein, also das Genetivelement vermehrt um das in diesem Falle natürlich bedeutungslose Subjektselement -e.'

1158 The enclitic copula ***am** may be used to emphasize an element to which it has been coupled, e.g. a nominal part of a sentence (Falkenstein 1978b, 75, § 100).

1159 Thomsen 1984, 145, § 284; Attinger 1993, 197-199, ch. 3.2.3.12, and 233-234, § 148; Balke 2006, 185-186, ch.7.3.4.

1160 Also in case ***ke₄** is a weak spelling for ***kam**, the directive postposition could be obscured by the 'imaginary' enclitic copula.

1161 Theoretically, also the reverse is possible: **an** is the second agent and **ḫur-saĝ-an-ki-bi-da** is the transitive subject. But since the sky-god An originates from **an**, this possibility is not very likely.

1162 Neither in the (unpublished) manuscript of Heimpel 1974, nor in the texts of ETCSL, are there other examples of the verbal form **<u>im</u>-tu-xx**. In the corpus of literary Sumerian texts there has been found only one example, where **an** is represented by the inanimate pronominal element **b** in the prefix chain of the verbal form: 'The Debate between Tree and Reed', l. 6: **an-an-maḫ-e ki-daĝal-la dùb im-ma-ni-ib-nir** 'An, the exalted heaven, had intercourse with the vast earth'. In this case too, the 'narrative' enacts at the beginning of time.

1163 See also: 'The Debate between Tree and Reed', ll. 6-7 and comments in the Appendix Text editions no. 5.

1164 Another analysis of the prefix chain is possible: /**i-m-b-i**/, with **b+i** incorporating reference to the locative-complex **ḫur-saĝ-an-ki-bi-da-ke₄** (Krispijn 2004, 80). In that case the subject **an-né** (ergative) has been kept outside the prefix chain, which is highly improbable. The 'complete'

u$_4$, is the cause of the 'incomplete' prefix chain ***im**. But in the case of the incorporation of that complex into the prefix chain, this would not have allowed one to distinguish between a causative or a locative/directive construction; in this respect both translations are possible.

The mytheme that a mountain range (**ḫur-saĝ**) brings forth something through a god is known from the literature, e.g. 'The Debate between Winter and Summer'. After a short introduction with An as protagonist (ll. 1-2) the intentions of Enlil are described (ll. 3-11). Then the next lines follow [1165]:

12	**ḫur-saĝ-gal-gal-la ĝìš bí-in-dug**$_4$ **kur-re ḫa-la ba-an-šúm**	He (= *Enlil*) copulated with the great Mountain Ranges (= *Ninḫursaĝa*), the Mountain (= *Enlil*) gave her its share.
13	**é-me-eš en-te-en ḫé-ĝál-zi-kalam-ma šà-ga mu-ni-in-ri**	Summer and Winter, prosperity and life of the Land, he poured in her womb.
14	^{d}**en-líl-le ki ĝìš dug**$_4$**-ga-ni am-gin**$_7$ **mur im-ša**$_4$	While Enlil copulated with earth, there was a roar like a bull's.
15	**ḫur-saĝ-e u**$_4$ **ki-bi-šè nam-zal ĝi**$_6$ **ḫaš**$_4$ **nam-mi-ib-dug**$_4$	The mountain range passed the day calmly, at night it made love.
16	**é-me-eš en-te-en ì-ḫe-nun-na-gin**$_7$ **šu nam-ta-ab-ús**	Summer and Winter have squeezed their way outside like precious oil.

prefix chain should have been: **im-mi-ib** (**an-né** as inanimate ergative), or **im-mi-in** (**an-né** as animate ergative).

1165 Translations: Kramer 1981, 304; Bottéro/Kramer 1993, 481-483; Vanstiphout 1997b, 584-588; Vanstiphout 2004, 167-179; Steinkeller 2007, 229-230; ETCSL 5.3.3. My translation of line 12 is conform the one of Steinkeller, and different from all other ones with respect to **kur-re**. **kur** in this line refers to *the* Mountain (= Enlil, with ergative ***e**), and not to '*a* mountain' with directive ***e**. The translations of the lines 15-16 in the literature are: 1. Kramer: 15. There Mountain spent the day, rested happily at night, 16. Delivered herself of Summer and Winter like rich cream. 2. Bottéro/Kramer: 15. Sur la Montagne, immobile, il passa tout un jour et la chevaucha (encore) toute une nuit: 16. Ainsi, comme on extrait de l'huile fine, en tira-t-il Été et Hiver. 3. Vanstiphout 1997b: 15. Hursag spent the day at that place, and at night she opened her loins. 16. Emesh and Enten she bore as (smoothly as with) princely oil. (His Dutch translation (2004) is of the same tenor). 4. ETCSL 5.3.3: 15. The hill spent the day at that place and at night she opened her loins. 16. She bore Summer and Winter as smoothly as fine oil.
My comments on the lines 15-16 of 'Winter and Summer': *1. Line 15.* **ki-bi-šè**: Some examples from the Sumerian literary corpus ('A praise poem of Šulgi' [ETCSL 2.4.2.05, Šulgi E], l.19; 'A praise poem of Šulgi' [ETCSL 2.4.2.24, Šulgi X], ll. 143,144; 'A *šir-namursaĝa* to Ninsiana for Iddin-Dagan' [ETCSL 2.5.3.1, Iddin-Dagan A], l. 32) show that this expression may be translated as 'in due course/order'; this resulted here in 'calmly'. *2. Line15.* **ḫaš**$_4$ **dug**$_4$: ePSD translates 'to have sex'. From the translation of Bottéro & Kramer it can be concluded that they consider Enlil as the subject of both the lines 14 and 15. *3. Line 16.* **šu ús**: see Karahashi 2000, 172-173: 1. to send, to ship out; 2. to push open (the door). The second meaning has been used in the present translation. The ***b** in the prefix chain of **nam-ta-ab-ús** may represent the subjects Summer and Winter, although /**emeš**/ and /**enten**/ have no ergative postposition.
The figurative language at a birth 'like **ì-ḫe-nun-na**' is found in 'Enki & Ninḫursaĝa' [ETCSL 1.1.1] *passim*: … **ì-ḫe-nun-na-gin**$_7$ **[**^{d}**A] [**^{d}**B] in-tu-ud** '... like precious oil goddess A gave birth to god(dess) B'.

It is questionable whether 'mountain range' is literally meant, both in 'Grain and Sheep' as in 'Winter and Summer'. In an essay about Nintur Jacobsen discusses the idea **ḫur-saĝ** [1166]:

> 'The name Ninḫursaĝa is etymologically transparent; it means "The Lady of the Foothills". Ḫursaĝ, the term which we have translated "Foothills", has actually a somewhat wider application. (...) As "the Lady of the Foothills" Ninḫursaĝa may be assumed to be the numen of rocky or hilly land, (...), and this is confirmed by the passage Gudea Cyl. B xxii (Frgm. S ii 3-5) in which the parentage of her son, the power in the yearly floods Ninurta/Ninĝirsu, is given thus:
>
> | | |
> |---|---|
> | **ᵈnin-⌜ĝír⌝-[su] e$_4$-zi-ᵈen-l[íl-lá]** | Ninĝirsu, rightful semen of Enlil, |
> | **ḫur-saĝ-e dú-da** | born by Ḫursaĝ (the foothills) |
> | **maš-lulim-e ga-zi-gu$_7$-a** | suckled right with milk by deer |
>
> Here Ḫursaĝ, the personified "Foothills", serves as name for the numinous power in them, without any anthropomorph honorary epithet **nin** – "Lady"–, just as the early designation of Nin-tur$_5$ was simply Tur$_5$ "birth-hut".'

This passage shows that **ḫur-saĝ** is presented as the one who has given birth to a god. In a footnote Jacobsen further mentions the following [1167]:

> 'Similarly the "Disputation Between Winter and Summer" ll. 11-16 tells how Enlil in the shape of a bull copulated with **ḫur-saĝ** engendering Summer and Winter to whom she subsequently gave birth. Here too there is no **nin** and no divine determinative before her name, she is just **ḫur-saĝ**.'

Recently Steinkeller discussed the Sumerian words **ḫur-saĝ** and **kur** [1168]. With respect to line 12 of 'The Debate between Winter and Summer' he noted: 'But the "mountain" (**kur**) is of course Enlil! Apart from the grammar (the agentive case **-e** marking **kur**), this is confirmed by the fact that Ninḫursag is never referred to as "mountain" (*meant is:* **kur**, *JL*).' His translation of this line is: 'With the great mountain ranges (= Ninḫursaĝ) he copulated, the Mountain allotted (his) share (to her).'

Both these examples support my view, that in 'Grain and Sheep' **ḫur-saĝ** also might bring forth gods – in this case on the initiative of **an** –, and that this complex does not refer to the birth-place of the Anunna gods simply as an indication of place. Steinkeller seems to be not consequent in his interpretation of **ḫur-saĝ** [1169]: when Enlil copulates with **ḫur-saĝ**, then the latter represents Ninḫursaĝa; but when An caused **ḫur-saĝ** to bring forth the Anunna, then – according to Steinkeller – '(...) **ḫur-saĝ** denotes a huge mountainous barrier

---

1166 Jacobsen 1973, 281-286; quotation: 281-282.

1167 Jacobsen 1973, 282, note 28.

1168 Steinkeller 2007; for line 12 of 'The Debate between Winter and Summer': see 229-230; quotation: p. 230.

1169 Steinkeller 2007, 230. In the lines 2 and 5 in 'A *tigi* to Nintur' [ETCSL 4.26.1] we read: **ḫur-saĝ-ki-sikil-la ù-tud-da**; it is said that Aruru (l. 1) 'has been borne by the mountain range of the pure earth' (the translation in ETCSL reads: '(Aruru) born in the mountains'). My supposition is that this whole expression represents the primaeval **ki**; mother goddesses seem to have no genealogy, and they all may be conceived as being derived from primordial **ki**.

separating the sky from the earth. This, in fact, is the proper (and expected) context of the creation of the gods, (...)'.

One more important conclusion may be drawn from the lines cited above of 'The Debate between Summer and Winter'. Line 12 mentions that Enlil copulates with **ḫur-saĝ**, while in line 14 Enlil copulates with **ki**. If **ḫur-saĝ** represents the goddess Ninḫursaĝa, then these lines imply that **ki** is identical with the mother goddess Ninḫursaĝa.

In 'The Song of the Hoe' line 46 we read: **ama-diĝir-re-e-ne ᵈnin-ḫur-saĝ-ĝá-ke$_4$** 'The mother of the gods, Ninḫursaĝa', another support for the present explanation of the lines 1-2 of 'The Debate between Grain and Sheep'.

2. For the spelling **diĝir-ᵈa-nun-na**: see Attinger 2001, 139 comment on line 236.

3. In my interpretation ***da** in ***b-da** of the prefix-chain represents the comitative. Jaques proposes a different interpretation of this ***b-da** [1170]: 'Dans Laḫar-Ašnan 3, **-da-** est probablement l'allomorphe de {**ta**} ablatif ('enfanter' = 'faire sortir de la matrice/vulve') plutôt que le comitatif.'

The verb **SIG$_7$** has several meanings: 1. **SIG$_7$** = **sag$_{12}$ (sa$_7$)** = *banû,* 'to be good, beautiful'; 2. **SIG$_7$** = *arāqu*, 'to be green-yellow, pale'; 3. **SIG$_7$** = **se$_{12}$** is the plural form (BḪ and BM) of **ti.l** (*ašābu*) 'to sit, reside, live' [1171]. The form **SIG$_7$-ga** points to the fact that in this line **SIG$_7$** does not represent **se$_{12}$**, the plural form of **ti.l**, a fact that is also apparent from the lack of a plural absolutive which should belong to this verb [1172]. 'To be green-yellow' fits very well in this context: 'becoming green-yellow' as said of the grain goddess Ezina, is an allusion to the growing-up and the ripening process of grain. The lines 3-5 describe the absence of the gods Ezina and Uttu, and of the possibility of weaving, which is not so surprising because the ewes are also absent.

The question of why in this dispute Ezina instead of the grain goddess Nisaba was mentioned, cannot be answered. Michalowski, in his communication about Nisaba [1173], notices that sometimes both goddesses are identical, but mostly they are different.

4. The suffix /**aš**/ belongs to the construction '**mu** (l. 3)...verb-**a-eše**' 'because'.

The /**n**/ before the stem represents the locative infix **ni**, referring to the locative suffix ***a** in **kalam-m<u>a</u>**.

5. Dunham has published about the expression **temen si-g** with respect to Uttu [1174]. The literal meaning is: 'to strike a peg into the ground'. Dunham wrote:

1170 Jaques 2006, 407, note 883.

1171 **SIG$_7$** = I). *banû* B, CAD B, 90-94; II). *arāqu*, CAD A II, 231-232; III). *ašābu*, CAD A II, 386 (lexical part). For **SIG$_7$** = **se$_{12}$** as the plural form of **ti.l**: see Steinkeller 1979, 55, with references to lexical lists and literature. Thomsen (1984, 314-315) also mentions the meaning 'to create' for **SIG$_7$**, but this is not supported by any lexical list or otherwise. The translations of **SIG$_7$** in this line are: Pettinato: 'hervorspriešen lassen'; Bottéro/Kramer: 'faire apparaître'; Alster/ Vanstiphout: 'create'; Kramer: 'fashion'; Streck (2002, 248): 'schaffen'; ETCSL: 'create'.

1172 Bauer (1982a, 381, note 10) supposes that this line needs an emendation: 'In Zeile 3 unserer Rangstreitdichtung fehlt die Erwähnung van **ᵈu$_8$**, dem Mutterschaf, vor **ᵈašnan**, sonst wäre das pluralische Verbum **sig$_7$**, 'wohnen (von vielen)', unerklärlich.'

1173 Michalowski 1998-2001, 576, § 3.

1174 Dunham 1986, 55-57; quotation: 56.

'Since in the Ur III field plans the *temen* was a figure made of pegs with cord stretched between them, very possibly the *temen* of the weaver in "Lahar and Asnan", "Enki and the World Order", and the "Hymn to Utu" refers to a horizontal ground loom. More specifically, perhaps the *temen* ... *sìg* (...) refers to striking in the four pegs.'

That **temen sìg** in this connection results in a horizontal ground loom, where **temen** can be considered as *pars pro toto*, is beyond doubt. Because this is evident, but on the other hand **temen** does not have the meaning of loom [1175], the present translation sticks as closely as possible to the Sumerian text.

6 ff. From line 6 onwards, the text relates about the absence of the ewe (and goat) and of grain; therefore **^{d}ezina$_2$** has now been translated as Grain, one of the protagonists of this debate.

8-9. These lines seem to be a recapitulation of what has been said in lines 6-7, albeit that numbers are mentioned now: from a general to a more specific situation. Lines 6-7 tell us that there are not numerous lambs and kids, so there is no sheep- or goat-*farming*. Lines 8 and 9 are saying, that there is not even one ewe or goat to give birth to the usual number of progeny for a sheep or goat (sheep produce on average a smaller number of young than goats).

10. **^{d}kù-sù** is understood as an apposition to **^{d}ezina$_2$** [1176].
The ***ke$_4$**-ending of the complex is possibly owing to a weak form of ***kam** [1177].

12-14. **še-muš$_5$** has been rendered as '*šeĝuš*-barley' [1178]. According to Bauer this kind of barley, which he translates as 'Bittergetreide', grows on soil of lower quality [1179], and

---

1175 Waetzoldt 1972, 130-132, *ad* 2. Der 'Webstuhl'.

1176 Bauer 1982b; Selz 1995a, 157; Michalowski 1993a, 158-159; Michalowski 1998-2001, 576, § 3. Krecher (1966, 133-134) discusses 'Kusu': 'An den meisten, wenn nicht allen Stellen, an denen Ezinu und Kusu zusammenstehen, dürfte eine Gottheit gemeint sein, (...). Wo vor Kusu das Determinativ fehlt, kann **kù-sù** bloßes (stehendes) Epitheton von Ezinu sein'.

1177 Another possibility is that the ***e**-ending reflects a directive. For the directive at the verb **zu**: see Falkenstein, who writes that, besides the accusative, the directive also has been found with this verb (Falkenstein 1978b, 127, note 6). Falkenstein speaks about a 'dimensionale Rection', which means the directive. Also van Dijk gives examples (van Dijk 1960, 43 *ad* 14). Both scholars do not mention explicitly whether **zu** with directive and **zu** with accusative have different meanings. In the texts of ETCSL there are more than 780 lines with the verb **zu**; of these there are only two examples with **zu** + directive ('Inana and Enki' [ETCSL 1.3.1: l. B12] and 'A praise poem of Iddin-Dagan' [ETCSL 2.5.3.2: l. 62]), both times **níĝ-nam-e zu-a**. In four cases the ***e**-ending was written as ***ke$_4$**; therefore it is not clear whether these endings concern a directive, or that ***ke$_4$** is a weak spelling for ***kam** ('Inana and Enki' [ETCSL 1.3.1: l. B10]; 'The Death of Ur-Namma' [ETCSL 2.4.1.1: ll. 84 and C18-19]; 'Letter from Sîn-idinnam to the god Utu' [ETCSL 3.2.05: l. 26]).

1178 **še-muš$_5$**: Akkadian: *šeguššu*; CAD Š II, 261: '(a cereal)'; AHw, 1208: *še/igu(š)šu(m),* 'eine Art Gerste?'

1179 Bauer 1972, 104; 107 [7 (Fö 170) V:4] and comments on p. 113 *ad* V: 4. His conclusion that the soil is of lower quality is based on the lower tax levied on it.

Powell says that this barley has a low frequency of occurrence [1180]. It appears that **še-muš$_5$** mostly was used at rituals and in medical treatments [1181]. Stol points out that **še-muš$_5$** is a late barley, and remarks: 'Die in "Laḫar und Ašnan" (...) genannten kurzen Anbauperioden van 30, 40, 50 Tagen für **še.muš$_5$** läßt an die 45 Tage für "early planted millet" – aber spät im Kalenderjahr! – denken.' [1182]

The significance of these lines in the present text is not completely clear. Seen in combination with the text of line 15, mentioning several other kinds of barley, does it mean that barley was not growing anywhere? Regarding the kinds of barley in question, these all grow on different soils [1183]. On the other hand, if **še-muš$_5$** here refers to rituals, these lines might mean that certain rituals could not be performed, which is not surprising because nothing was yet in existence.

15. The Sumerian term **á-dam** – not translated by Pettinato, and rendered as 'villages' by Alster & Vanstiphout and as 'ville' by Bottéro & Kramer [1184] – is found in PSD: 'habitation, settlement (and its inhabitants)' [1185]. The term has been identified with the Akkadian *namaššû* and *nam/wû* [1186]. The first two meanings of *namû* A, given in CAD, are applicable in the present context: 1. pasture land on the fringes of cultivated areas as the habitat of nomads, and its population and flocks; 2. pasture land, outlying area around a city. This resulted in the translation: 'fields around the dwelling-places'. With **á-dam-kù-g** are meant: the temple domains belonging to these dwelling-places.

16, 22. The phrasing of the first part of these lines may be: 1. **túg-níĝ-mu$_4$-mu$_4$-bi**, or 2. **túg níĝ mu$_4$-mu$_4$-bi** [1187]. Each phrasing has its own interpretation as a consequence.

Ad 1. **túg-níĝ-mu$_4$** = *lubūšu* [1188]. The examples given in CAD show, that this term is connected predominantly with gods and kings. In that case line 16 mentions, that there are

---

1180 Powell 1984, 59-60, § 4.1-4.4. Powell (1984, 60, § 4.4) doubted the interpretation 'bitter' for **muš$_5$**, because 'Akkadian *marru* 'bitter' is not attested with the reading /**muš**/ for the sign **ŠEŠ**.'

1181 With the exception of the present disputation, in literary texts **še-muš$_5$** is only found in some proverb collections. CAD Š II, 261-262, and ePSD *ad* **še.muš$_5$**, refer to texts with ritual and medicinal use of **še-muš$_5$**; these texts date from the 1[e] millennium. See also: Hallo 1992, 398. The Sumerian Proverbs collection 3 *ad* 3.168 reads: '**še-muš$_5$ níĝ-gig lú-gidim-ma-ka** "Bitter" barley is reserved for the "ghost-man" ' (Alster 1997, I, 108; comments: II, 393); in ETCSL 6.1.03, this line [l. 297] is translated: 'The **muš** barley is reserved for the necromancer.' See also Geller 1990, 108; Hallo 1985, 29. Via CDLI (entry **še-muš/muš$_5$/ŠEŠ**) references are obtained to economical texts from the ED IIIb-period up to the Ur III-period inclusive, and to the lexical lists **ur$_5$-ra** = *ḫubullu*; these texts do not give an insight with respect to the exact nature of **še-muš$_5$**, but they show that it was used as food.

1182 Stol 1993-1997, 349[b].

1183 A ritual interpretation is less plausible; the most frequent ritual days are the days 7, 15 and 20 of every month (see e.g. Sallaberger 1993, Glossar *ad* **u$_4$** 7 /15 /20.

1184 Pettinato 1971, 88; Alster and Vanstiphout 1987, 15; Bottéro and Kramer 1993, 512.

1185 PSD A II, 48-50. In section 2 (p. 49[a]) this line of 'Grain and Sheep' is quoted, with a translation of **še-á-dam-kù-ga** as 'grain of the holy habitations'.

1186 *nammaššû:* AHw, 728; CAD N I, 233, with lexical references. *namû* / *nawû:* AHw, 771; *namû* A: CAD N I, 249, with lexical references.

1187 Alster and Vanstiphout (1987, 32) discuss the deviating spelling **túg-<u>ga</u> mu$_4$-mu$_4$-bi** in the texts A, B, E, and F of line 22.

1188 *lubūšu*: AHw, 561 'Gewand'; CAD L, 236 (with lexical references), 'e.g. clothing'.

no garments for the (statues of the) gods [1189], for the possessive suffix ***bi** has to refer to An and the Anunna, so far the only divine authorities mentioned that are present. Line 22 tells us, that the clothing for the worldly rulers was not yet known; the suffix ***bi** refers to mankind (l. 20).

Ad 2. **túg níĝ** $\mathbf{mu_4}$**-**$\mathbf{mu_4}$. This phrasing – chosen in the present edition – represents a notation for clothing in general. This option seems more suitable in both the context of line 16 and of line 22. The contexts are a summing-up of things that are absent or not yet known, due to the fact that neither Ewe nor Grain were present. The suffix ***bi** now refers to **níĝ**.

17. With respect to **men**, a neutral translation has been chosen: cap [1190]. A cap or head-band, made of some kind of textile, was the original symbol for gods and kings [1191]. The mention in this line that Uttu had not been born might be a general explanation for the fact that there was no clothing at all (l. 16). As a consequence there was also no **men**. It is not clear whether **men** in this line is a cap for gods or a cap for kings. A translation as 'so no royal cap was worn' – as proposed by Alster and Vanstiphout – focuses the attention only on the absence of any form of authority on earth (not so surprising, because mankind did not yet exist in a civilized condition), and it may suggest a causal connection between (the absence of) Uttu and (the absence of) worldly power. Extremely remarkable is the article of dress that according to Bottéro and Kramer is meant by **men**, viz. 'pagne', a loincloth.

18. This line involves some problems, which are also discussed by Alster and Vanstiphout and by Bauer [1192]: how to read the name or names, and who is or are meant? Starting from the verbal form, which has no indication for a plural absolutive [1193], it may be concluded that this line mentions only one god (as is also the case in line 17 with the same verbal form). In three texts (A, D, and E), all from Nippur, it is written **en-KAL-KAL**; only in one text (from Ur) can we read **en** $^{\mathbf{d}}$**KAL-KAL**. The god Kalkal is known as the gate-keeper of the Ekur, the temple of Enlil in Nippur [1194]. However, it seems improbable that this god is meant here: not one text from Nippur shows **KAL-KAL** with diĝir-determinative, and moreover, it is hardly understandable what should be the function of this minor god in this context. The solution is, that **en KAL-KAL** is the epithet of the god at the beginning of this line, with a transcription **en-kal-kal** in consequence [1195].

1189 For the clothing of the statues of gods: see Berlejung 1998, 40-41.

1190 CAD A I 153-157: *agû* A (lexical part: *agû* = **men**): 'crown (of gods and kings)'.

1191 Steinmetzer 1922, 150-154, § 37; Waetzoldt 1980-1983. For illustrations: Boehmer 1980-1983.

1192 Alster and Vanstiphout 1987, 41, note 6; Bauer 1982a, 382, notes 19 and 20.

1193 This should have been: **nu-ub-tu-ud-dè-eš**, analogous to the verbal form in line 2.

1194 Lambert 1976-1980; Lambert and Millard 1999, 47; I: 74-75; Richter 2004, 30, 43, 45. Perhaps the scribe of the text from Ur did not understand the original intention of **en KAL-KAL**, and misunderstood the name as the god Kalkal.

1195 The reduplication of the adjective **kal** is considered as an expression of the superlative (Thomsen 1984, 65, §§ 82, 83). See also Bauer 1982a, 382, note 20. Alster and Vanstiphout neglect the reduplication of **kal** (their translation is: 'the precious lord'). Pettinato writes in his translation: 'En-Kalkal'.

To whom does this epithet belong? In all undamaged texts the **diĝir**-sign is present at the first complex, so the conclusion may be that it is a god's name [1196]. In the literature there are references to the equation: Enniĝirsi is the god Dumuzi. Enniĝirsi is also mentioned in lists, among other names, which are synonyms for Dumuzi [1197]. The next question is: in which capacity is Enniĝirsi-Dumuzi mentioned in this line? The most mentioned aspect of Dumuzi is, that he is a shepherd, or the god of the shepherds; one of his epithets is also **sipa** [1198]. Therefore it seems logical to assume that in this line Enniĝirsi-Dumuzi refers to the god of the shepherds [1199], the more so because previous lines refer to flocks of sheep and goats. Thus Enniĝirsi-Dumuzi in this text may be related to the domesticated animals. And the text goes further (l. 19): Šakan, the god of the wild animals in the steppe, has not appeared there. This means in other words: the ultimate absence of any wild animal. But without wild animals it is certainly not possible to begin domestication [1200].

19. The reading of the second sign is, according to Borger, **šakan$_2$** or **šakkan$_2$**, referring here to Lambert [1201]. Alster and Vanstiphout translate: 'Shakkan did not go out *to* the arid lands'. However, there is no terminative but a locative postposition written behind **bar-rim$_4$**: *in* the dry area.

20-21. Kramer thinks that the Anunna-gods are the subject of these lines, as appears from the explanation between brackets in his translation [1202]: 'Like mankind when first created, They (the Anunnaki) knew not the eating of bread, ...'. Bottéro & Kramer comment on these lines as follows [1203]:

1196 Based on the equation **niĝir-si** = *susapinnu* (AHw, 1063; CAD S, 416), Bauer concludes: 'Wir verstehen den Namen [**en-^{d}niĝir-si**, *JL*] (...) als 'Herr, Brautführer' '(Bauer 1982a, 382, note 19).

1197 Van Dijk 1967b, 259, l. 33 (**^{d}niĝir-si** = **^{d}dumu-zi**); Bauer 1982a, 382, note 19, with references; Emesal vocabularium I:77: **^{d}en-niĝir-si** = **^{d}dumu-zi**; Fritz mentions the presence of **^{d}en-niĝir-si** in several godlists, and he calls the god with this name 'Dumuzigestalten bzw. -beinamen' [Fritz 2003, 61 (godlist of Nippur); 62 (Genouillac-list); 66 (emesal godlist)]; Richter 2004, 30, 312.

1198 See e.g. Jacobsen 1985a, 45; Sefati 1998, 78-79. A recent survey of the literature about Dumuzi: Fritz 2003, 17-45.

1199 Krebernik (2003, 158, note 45): 'Falls hier in der Tat Dumuzi gemeint ist, dürfte er in seinem Aspekt als Hirt(engott) angesprochen sein'. According to Edzard (1965, 52), one important aspect of Dumuzi is: he is a god of the vegetation.

1200 In a sense we see here a parallel between the lines 6-7 (referring to sheep- and goat-*farming*; for ewes and goats are not present *in great numbers*) on the one hand, and the lines 8-9 (the natural, wild way of life) on the other hand. But the vegetation aspect of Enniĝirsi-Dumuzi cannot be ruled out, though in that case the mention of Šakan is more difficult to explain.

1201 Borger 2010, 183, no. 701; Lambert 1986a. In the texts of this disputation **šakan$_2$** is written in different ways. Mittermayer 2005 gives a survey of the development of signs with animal heads. The sign in the texts A and U is found on p. 156 *ad* /**šagan**/. The sign in text E is similar to the first sign at /**šagan**/ on p. 148, but with an extra **TAB**-sign [as at the sign **ANŠE** one line above /**šagan**/]. In text F the sign **ANŠE** (p. 148) is written, provided with the gloss **PA** = /**šag**/ instead of the sign **TAB**.

1202 Kramer 1956, 145; Kramer 1972, 72-73; Kramer 1981, 109.

1203 Bottéro and Kramer 1993, 513-514.

'L'expression sumérienne **nam.lú.u$_{18}$.lu**, qui désigne, au pied de la lettre, «l'ensemble des hommes», «l'humanité», est surprenante en *20*, compte tenu que les hommes proprement dits sont censés n'être venus à l'existence que plus tard (*36*). Il s'agit donc vraisemblablement là d'une image pour désigner «la foule des dieux», «la population divine», peut-être en suggérant discrètement, comme fera le premier vers d'*Atraḫasîs*, qu'elle représentait alors une façon d'humanité surnaturelle, et surtout jouait le rôle laborieux ultérieurement assumé par les véritables hommes. Si l'on prend ainsi les choses, on ne suivra point A. Falkenstein*, selon qui le passage *35* s., dans lequel est mentionnée la «création des hommes», était interpolé, et à retirer du texte originel: solution simple et radicale, mais subjective et risquée, dans la mesure où elle n'est pas inévitable.

Ainsi les dieux n'ayant pas de quoi se vêtir, et seulement des herbes sauvages à manger, menaient donc une vie comme animale, primitive et de pure nature (*20-25*).

* Voir les notes de A. Falkenstein, ZA 58 (1967) 10s, 11 n. 3.'

This comment is remarkable for more than one reason. In the first place their translation of **nam-lú-ulu$_3$** reads 'les hommes (?)', so after all 'people', although with a question mark. Moreover, both scholars do not translate **ùĝ** in line 23, whereas this sign clearly indicates that it concerns people. In the preceding lines it has been related, that only the Anunna-gods, together with An, are present. In other words, the pantheon is relatively empty; there is no reason to think of 'la foule des dieux'. A multitude of gods presented as **nam-lú-ulu$_3$** would be unparalleled in the Sumerian literature. A still more serious objection has to be made to the supposition that the gods would have walked around naked and lived like animals. Such a description of the life of the gods 'in the Beginning' is not supported by any Sumerian literature or parallel passages in other texts. Rather we see here, starting from the real meaning of **nam-lú-ulu$_3$** 'mankind', a striking similarity to the description of the originally primitive life of Enkidu, who ate grass and drank water together with the wild animals at the watering-place, and who lacked any kind of civilization, as described in the Gilgameš-epos [1204].

In view of the above comment of Bottéro and Kramer, these scholars understood that line 36 of 'Grain and Sheep' mentions the creation of man. The expression used in this line '**zi-šà ĝál**' then would be an unique description for the creation of man. Pettinato gives a survey of the usual Sumerian words or expressions for the creation of man, with a critical note for the expression '**zi-šà ĝál**' [1205]. Perhaps Bottéro and Kramer were guided by Falkenstein [1206], who translates this phrase as 'Lebensodem ins Leibesinneren sein (lassen)'.

Van Dijk writes about the lines 20-25, that these describe the situation of mankind after the creation: 'Mais il semble se rattacher à la création par "emersio" '[1207]. The several types of creation of mankind, which van Dijk has suggested, are dealt with in chapter 1 of this book.

1204 George 2003, 544-545, ll. 110-112: 110. *it-ti ṣabâtim (*maš.dà^{meš}) *-ma ik-ka-la šam-mi* 111. *it-ti bu-lim maš-qa-a i-tep-pir* 112. *it-ti nam-maš-še-e mê* (a^{meš}) *i-ṭib lìb-ba-šú*: '(he was) ... feeding on grass with the very gazelles. *Jostling at* the water-hole with the herd, he enjoyed the water with the animals.'

1205 Pettinato 1971, 49-57.

1206 Falkenstein 1967, 10-15.

1207 Van Dijk 1964, 30, note 79.

This disputation does not deal with the creation of mankind. The lines 1-25 describe a general prehistory: only An and the Anunna-gods were present, other gods were absent. Any people there might have been were living in a primitive and uncivilized stage of development.

23. The first sign of this line in text U certainly has to be read as **kalam**, as it is followed – but only in this line – by ***ma**. The choice of a transcription of **kalam** (without ***ma**) as a resultant of the total score – as done by Alster and Vanstiphout – and their translation as '*in* the Land' seem not convincing. Their translation implies that both scholars have taken the plural subject of the lines 20-22 – with the corresponding plural verbal forms in the lines 21 and 22 – also as the subject of line 23. The singular verbal form in line 23, as well as the absence of a locative ending after the first sign (except for text U), indicate that the reading of the first sign in this line most likely has to be **ùĝ**.

With respect to the final spelling **ĝeš-gen$_6$-na-a šu-bi**: this part of the line – insofar it has been preserved – has been written differently in all texts. Falkenstein writes: '**giš-gi-na** gehört mit dem vor allem in Beschwörungstexten häufig belegten **giš-ge-en-ge-na** = *bināti* "Gliedmaßen" (...) zusammen' [1208]. He also refers to the line 23 of 'Grain and Sheep', and translates: 'die Menschen gingen auf Händen (und) Füßen'; in a footnote he notes down: 'Wörtlich "auf ihren Gliedmaßen, ihren Händen" ' [1209].

Wilcke proposed a different translation for this expression [1210]; he based his analysis of this complex phrase on just one text variant (in the present edition called G). In this variant **sù** is written, instead of **šu** (A, E, F) or **su** (U). Wilcke based his translation on the equation **sù** 'to be empty, to be naked'. His final score reads: **un giš-ge-(-en)-na sù-bé mu-un-gin**, translated as [1211]: 'das Volk ging unbekleidet (... wörtl. in der Nackheit der Glieder)'. That people was going about naked, is also the interpretation of Alster and Vanstiphout, Bottéro and Kramer, and of ETCSL [1212].

The following considerations have led to the present final text. In three out of five legible texts there is **šu**, and in two /**su**/; therefore **šu** has been chosen [1213]. According to ePSD **ĝeš-gen$_6$-na** stands for 'limbs'. The extra ***a** (text E, F) has been understood as a locative ending [1214]. This, combined with **šu-bi** [1215], results in a translation: 'on the limbs, its hands', in usual English: 'on all fours'. It is striking that in this line the feet are not mentioned, presumably because one is normally moving thereon, and therefore it is not worth mentioning, or because of the comparison of people with animals which do not exhibit a difference between 'feet' and 'hands'.

1208 Falkenstein 1947-1952, 45.

1209 In the same footnote Falkenstein mentions, that the variant in what is called here text G, is incomprehensible to him. Pettinato follows Falkenstein in his translation; he also refers [p. 90 *ad* 23] to Falkenstein 1965a, 132, note 66; in this note Falkenstein remarks that Kramer (1956, 145) connects the lines 19-24 'irrtümlicher Weise' with the Anunna-gods, but Falkenstein did not translate line 23 any more, except for: 'die Menschen...... '.

1210 Wilcke 1969, 168-170.

1211 A correct Sumerian expression for 'in their nakedness' should have been: **sù-ga-bi**.

1212 Kramer (1956; 1972; 1981) did not translate this line at all, without any comment.

1213 Variant G, the only text with **sù**, will not be discussed, because of its incomprehensibility to me.

1214 The reverse of the order ***a-na** in text A also results in **ĝeš-gen$_6$-na-a**.

1215 For **šu-bi** it is not necessary to repeat the locative ending, as has been done in text E.

There may also be an argument from the viewpoint of narrative style against the mentioning in this line of people going about naked. If line 23 had mentioned it, then this line would only have repeated – though in other words – line 22. This kind of repetition does not occur in this disputation. Moreover, *naked* limbs do not seem so peculiar as to merit mention. Line 23 shows how people were moving like animals [1216]; the next lines even show more of such animal behaviour.

24. This line is nearly identical with line 1 of the text 'How grain came to Sumer' [1217]: **ùĝ-e udu-gin$_7$ ú ka-ba mu-ni-ib-«ni»-<gu$_7$>** 'men used to eat grass with their mouths like sheep'. Also in a hymn to Inanna for Ur-Ninurta of Isin there is a reference to such human behaviour in the prehistory: **udu-gin$_7$ ka ú-gu$_7$ ḫa-ba-kíĝ-kíĝ** 'may he let (them) graze like sheep' [1218].

25. Comparison of the different spellings for **a-mú-sar-ra-k***$^{(*)}$ in this line in the various sources could support the thesis, that in the Old Babylonian period ***ke$_4$** has become a weak form for ***kam**: the texts E and F have ***kam**, text U has ***ke$_4$**, whereas text G takes a position in between with ***ka**.

The various observed translations of **mú-sar** (Pettinato: 'Gräben'; Alster/ Vanstiphout: 'ditches'; Bottéro/Kramer: 'fondrières') do not render the meaning of it: 'flower-bed; (plot of a) garden' [1219].

The spelling of the prefix chain with an extra ***i** is demonstrable since the Old Babylonian period [1220]. The use of a *marû* stem, after and in contrast to a *ḫamṭu* stem in one sentence, expresses a simultaneousness with respect to a certain point of reference [1221].

26. The interpretation of **ki-ulutim$_2$** is slightly confusing, when comparing dictionaries, (literary) texts and comments. In ePSD, **ki-ulutim$_2$** is translated as 'place of creation', with reference to Landsberger [1222]. In this article Landsberger writes: 'Die Örtlichkeit **ki-uludin** (…) findet sich häufig in lit. sum. Texten. Die dafür angesetzte Bedeutung "Geburtsstätte" ist unrichtig'. Landsberger supports his thesis with several examples, and proposes as a

---

1216 Compare also e.g.: **máš-anše-lu-a ^{d}šakán-na nì-zi-gala$_7$-eden-na / níg-úr-limmu an-nì-dagal-ba** 'Das zahlreiche Getier Šakans, die Lebewesen der Steppe, / die Vierfüssler unter dem weiten Himmel'; Römer 1965, 131; 139; ll. 93-94: 'Hymne an Inanna-Dilibad/Ninsi'anna' [ETCSL 2.5.3.1: 'A *šir-namursaĝa* to Ninsiana for Iddin-Dagan', ll. 95-96].

1217 'How grain came to Sumer': ETCSL 1.7.6.

1218 'A *šir-namgala* (?) to Inana for Ur-Ninurta' [ETCSL 2.5.6.1] l. 26: 'May he search out food for them to eat as if for sheep, (...).' (Translation in ETCSL).
Falkenstein (1959b, 106, l. 24; comments p. 128 *ad* 24) transcribes **ur$_4$** instead of **kíĝ**: 'wie den Schafen möge er (den Menschen) Nahrung zum Essen sammeln, (...).'

1219 AHw, 681: *mūšarum* 'Beet'; CAD M II, 233-234: *musarû* B 'garden'; ePSD: **mú-sar** 'garden'. In 'A *šir-namursaĝa* to Ninsiana for Iddin-Dagan' (Römer 1965, 131, l. 95; ETCSL 2.5.3.1, l. 97), there is an enumeration: **pú-^{ĝiš}kiri$_6$ mú-sar ĝiš-gi-sig$_7$-ga**, 'orchards, gardens and green reed-beds', which shows that it is very unlikely that **mú-sar** means 'ditch'.

1220 Heimpel 1974, 179-181. In 'The Debate between Grain and Sheep' it is not used consistently.

1221 Streck 1998, 184-186, examples 7-10. Because of the change of the *ḫamṭu* stem into a *marû* stem, the collective subject ***b** in the prefix chain in line 24 has been replaced by the pronominal suffix 3-pluralis /**ēne**/ in line 25, for there is no collective subject ***b** in combination with the *marû* stem.

1222 Landsberger 1964-1966, 77-79.

translation: 'ein hervorstechender Platz' [1223]. In CAD [1224] it is shown, that **SIG$_7$-ALAM** (= **ulutim$_2$**) is equated with *nabnītu*, to which is given as second meaning: 'habitat, place of growth', but not a signification like 'creation'. Therefore it seems very improbable that the phrase **ki-ulutim$_2$** should mean 'place of creation' or the like. In the few places in literary texts with **ki-ulutim$_2$** [1225], the translation of this phrase should be revised. A replacement of the translations used, like 'birth-place, creation place, place of engendering, place of formation', by 'place of residence' or synonyms hereof, seems more in harmony with the original Sumerian texts.

27-28. As appears from the translations, most scholars consider the Anunna-gods as creators of Ewe and Grain [1226], and also as those who gathered both of them in the dining hall of the gods. The supposition, that the Anunna are the subject of both the verbs in the lines 27 and 28, is debatable. If in both lines the Anunna should be the subject, then it is difficult to explain why the pronomina, used for the gods in the respective verbal forms, are different [1227]. The solution may be as follows. First of all, the verb **sig$_7$** does not have the meaning 'to create, fashion' (see discussion above at l. 3). The verb **SIG$_7$** = **se$_{12}$** is used intransitively [1228], with the absolutive /**eš**/ as suffix pronomen for Ewe and Grain. The /**n**/ before the stem represents the locative infix /**ni**/, referring to the locative in **é-bi-du$_6$-kù-g̱a**. We are not told at all who created both Ewe and Grain.

The verbal form in line 28 shows the collective pronomen /**b**/ for the Anunna-gods, mentioned in line 30; the suffix pronomen /**eš**/ represents Ewe and Grain.

28. With respect to the reading **RI(-g)** = **de$_5$(-g)**, see Sallaberger [1229], who, as opposed to Thomsen, says that the verb belongs to the reduplication class. In this case, **de$_5$-de$_5$-g** is the plural form of the *ḫamṭu*-stem [1230].

1223 His translation of ll. 26-27 of 'Grain and Sheep': 'Damals, an dem hervorstechenden Platz der Götter, deren Haus der heilige Urhügel war, saßen Mutterschaf und Ähre.'

1224 CAD N I, 27: *nabnītu* '1. offspring, progeny, product; 2. habitat, place of growth, 3. living creature, 4. appearance (inclusive lexical references)'. See also ePSD *sub* **uludin$_2$** en **ki-uludin$_2$**.

1225 'Inana and Bilulu' [ETCSL 1.4.4], ll. 138 and 168; 'The rulers of Lagaš' [ETCSL 2.1.2], l. 60; 'The lament for Eridug' [ETCSL 2.2.6], segm. C , l. 11; 'A hymn to Nanna', [ETCSL 4.13.15], ll. 8 and 12; 'A *šir-gida* to Ninisina' [ETCSL 4.22.1], ll. 89 and 97; 'The Debate between Grain and Sheep' [ETCSL 5.3.2], l. 26.

1226 Bottéro/Kramer 1993, 512: 'créèrent'; Alster/Vanstiphout 1987, 17 en Vanstiphout 1997, 575: 'fashioned'; Pettinato 1971, 88: 'ließen hervorsprießen'. [This last verb – *hervorsprießen* – refers to one of the two possibilities for creation, which van Dijk 1964 has proposed: the chthonic motif. The other possibility is called: the cosmic motif. This was discussed in chapter 1.2 of this book.] Kramer does not commit himself to the Anunna as creators, and gives a passive translation: 'Laḫar and Ašnan were fashioned' (Kramer 1956, 145; 1972, 73; 1981, 109).

1227 In line 27: prefix /**en**/ + suffix /**eš**/, and in line 28: the collective pronomen /**b**/ as prefix.

1228 **se$_{12}$** is the plural stem of **ti.l** 'to live' and of **lug** 'to dwell' (ePSD). Since **se$_{12}$** as plural stem of **lug** has been used predominantly in the ED IIIb period (ePSD), I suppose that in this text **se$_{12}$** represents the plural stem of **ti.l**.

1229 Sallaberger 2005a, 233: reduplication class. Thomsen 1984, 312: alternating class; as plural stem probably **ri.ri.g**.

1230 For further analysis of the verbal form: see above, comments *ad* 27-28.

31, 34. The comitative /**da**/ in the prefix chain has been interpreted in the present translation as 'to be able to' [1231]. Alster and Vanstiphout translate in a neutral way: '... but were not sated'.

32, 35. The reading of the first sign in these lines needs some comments. All translations are based on the presumed presence of the sign **amaš**. However, there are some arguments for the reading **ubur** of this sign:

1. In all the texts the sign in the lines 32 and 35 clearly differs from the sign **amaš** in the lines 43, 47 and 48 [1232].
2. In the lines 32 and 35 the milk and the sweet substance, respectively, are more compatible with **ubur** 'breast' than with **amaš** 'sheepfold'. For in the story told so far, there is talk of only one Ewe; moreover, this Ewe is in the dining hall of the Anunna (l. 28) and not in a sheepfold.
3. On the other hand, **amaš** fits very well with the circumstances as described in the lines 43, 47 and 48: after the gods have 'inspirited' mankind, Ewe has been sent down from the holy hill with the intention, that Ewe will be increased to the proportion of a flock of sheep, in order to satisfy the needs of the gods.

35. The previous translations that are discussed here show, that /**anene**/ has been understood as referring to the Anunna-gods. But, as can be seen in line 43, – **u$_8$ amaš-<u>a-na</u>** – the pronomen possessivum /**ane**/ also can refer to Ewe, because both the protagonists of this dispute are considered as persons.

36. The translations of Pettinato and Bottéro/Kramer have nearly the same tenor [1233]: the gods give 'life-breath' to the people. Alster and Vanstiphout have a different view: '(l. 35) And so, for their own well-being in the goodly sheepfold / (l. 36) They gave them to mankind as sustenance' [1234]. In view of their comment, Bottéro and Kramer consider the expression **zi-šà ĝál** [1235] as a wording for the creation of mankind. In the literature, how-

---

1231 Thomsen 1984, 226, § 448; Krispijn 2004, 72, ad 6.

1232 In text F, ll. 32 and 35, the first sign seems to be: **DAG-KISIM$_5$ x kur**, here transcribed as **ubur$_x$**. Such a transcription has not been attested in the literature. If it is read correctly, this **kur** perhaps functions as a gloss. Alternatively, it may be a (bad) writing of the sign **ubur**, which is found in Mittermayer 2006, 68, nr 171, line 2, first sign.

1233 Pettinato 1971, 89: '(l. 35) in reinen Schafpferch ließen sie (= die Anunna-Götter) für ihr Wohlergehen / (l. 36) in den Menschen den Lebensodem vorhanden sein'. Pettinato comments in an earlier section (p. 57) the expression **zi-šà-ĝál**; he supposes that with this expression the second phase of the creation, viz. the introduction of the civilization, was meant. Bottéro/ Kramer 1993, 512: '(l. 35) C'est pourquoi, en leur auguste bergerie, et pour leur bénéfice, / (l. 36) ils octroyèrent aux hommes le souffle-de-vie'.

1234 In this same dipute, l. 55, Alster and Vanstiphout translate **zi-šà-ĝál** as 'well-being'. ETCSL 5.3.2 has nearly the same translation as Alster/Vanstiphout. Hallo and van Dijk (1968, 95 *ad* **zi-šà-gala$_7$**): 'they gave sustenance to mankind'.

1235 ePSD translates **zi-šà-ĝál** with 'encouragement'. In AHw, 1533: *zišagallu* 'etwa "göttliche Ermutigung" ' CAD Z, 138: *zišagallu*: '1. divine encouragement, fortitude of heart; 2. (greeting formula used in addressing a king)'. Line 36 is cited (with two times **ĝál** instead of one time), and translated: 'they (the Anunnaki) caused encouragement to be among men'. In an explanation of the word *zišagallu* it has been added: 'In the Sumerian passages, the word denotes the encouragement conveyed by word of mouth, a specific act or by mere presence, from one god to another, or by a god or king to a human being or person of lesser status.'

ever, no examples have been found, in which this expression has ever been used in connection with the creation of mankind, or to bring a lifeless human being to life. Moreover, 'mankind of those faraway days', although in his primitive stage, was mentioned already earlier in this debate (ll. 20-25). Therefore it is not logical to suppose that only now, in line 36, the creation of mankind is described.

Pettinato commented on the expression **zi-šà ĝál** [1236]. He assumes that with this phrase the second phase of the creation is meant, viz. the introduction of civilization, and at the end of his comment he adds: 'Danach ist mit **zi-šà-ĝál** primär die Arbeit bezeichnet, so daß meine Deutung von der Einführung der Arbeit, als Mittel der Zivilisation, bestätigt wird.' In his review of Pettinato's study, Kümmel notes: '**zi-šà-ĝál** als "Arbeit" leuchtet nicht ein, bringen doch die zum Beweis angeführten Ackerbaugeräte physische Lebenserhaltung, eben "sustenance" '[1237].

In my opinion, in the present context it is said, that the gods elevate the people from their primitive stage and animal life style into a civilized state, in order to enable them to carry out agriculture and cattle breeding. The civilization of mankind is a certain guarantee for the satisfaction of the needs of the gods.

38. The ***ke$_4$** at the end of the complex **u$_8$-^{d}ezina$_2$-bi-da-ke$_4$** has presumably been a weak form for ***kam**. Following Thomsen the genitive construction can be explained [1238]: 'The *regens* of the genitive construction may be missing. This construction is especially frequent with the enclitic copula.' A literal translation may be e.g. 'those/the ones of Ewe and Grain'.

40. In text U (l. 37) the signs **ga** and **ta** have exchanged their place, possibly a scribal error as a consequence of the **du$_6$-kù-ga** in U line 36 (= l. 39 in the complete text).

42. The verbal form needs some explanation. Text U is the only undamaged one, so one should expect there a complete verbal form. There are two possibilities to be considered with respect to the subjects of the verb /**e$_{11}$**/: Enki and Enlil are subjects or Ewe and Grain.

If Enki and Enlil were subjects, then the plural ending ***eš** should have been written in the verbal form: **im-ma-da-ra-an-e$_{11}$-dè-eš**. As in this text it is common practice that plural endings are written, the scribes of the texts F and U must then have made a mistake in this line [1239]. On the other hand, if Ewe and Grain are the subject, no plural ending is necessary, unless Ewe and Grain are considered as 'animate' [1240]. The verbal form is the *marû*-stem, 3-singularis (inanimate plural subject), intransitively used. The preverbal element /**n**/ may be explained as having been used to mark the locative case [1241]: **inim-kù-**

1236 Pettinato 1971, 57.

1237 Kümmel 1973-1974, 32.

1238 Thomsen 1984, 92, § 167.

1239 Text G, with ending **-dè-en-dè-en** (1-pluralis), is clearly a mistake (resulting from the [correct] verbal form in l. 40?).

1240 Falkenstein 1978b, 152, § 110, *sub* a. Plural forms of finite verbs are only applied if the subjects are of the animate class. This rule is valid during the Gudea period, but Falkenstein has not indicated that this rule has been abandoned in later times.

1241 Delnero 2007, 122-124. With respect to the prefix /**n**/, Geller (1998, 92-94, § 3 Intransitive verbs with pre-radical /n/) asserts that in Old Babylonian literary Sumerian texts there are many examples of intransitive verbs with /**n**/ before the stem, in spite of the fact that this /**n**/ seems to be formally excluded on grammatical grounds. His explanation is that this /**n**/ represents a

ga-ne-ne àm-dug$_4$-g<u>a</u> in line 41 (the final /**a**/ here represents both a subordination suffix and a locative ending). No emendation is necessary: the verbal form is correct.

The conclusion is that Ewe and Grain are the, now inanimate, subjects of the lines 41-42: inanimate, because from now on ewe and grain represent the future food for the gods [1242].

***

reflexive element. His examples show verbs that express a movement, in which the final destination has not yet been reached. Attinger (1998) comments on this study of Geller. On the basis of further research on two verbs (**mu$_4$** en **a ... tu$_5$/tu$_{17}$/tu$_{22}$**) he concludes: 'En bref: que /n/ préradical soit toujours un élément réflexif est exclu; qu'il le soit sporadiquement est à la limite envisageable, mais devrait être étudié sur une grande échelle. Un candidat plus vraisemblable pour le réflexif est le préfixe [ba].' In the study of Delnero (2007) there remain verbal forms with an additional /n/, in which this added /n/ cannot be assigned to one of the four functions (discussed by Delnero, pp. 121-124) that this element has been shown to possess. Delnero (2007, 128): 'The high percentage of instances in which the additional /n/ does not have an identifiable function may therefore indicate that in most cases, the added /n/ reflects an erroneous or idiosyncratic interpretation of the form'. The possibility that this /n/ represents a reflexive element with verbs of movement has not been discussed by Delnero.

1242 In the translations of Alster & Vanstiphout (1987), Vanstiphout (1997) en ETCSL (5.3.2), Enki and Enlil are subjects. On the other hand, Pettinato (1971), Bottéro & Kramer (1993) and Kramer (1956; 1972) have the opinion that Ewe and Grain are the subjects.

7. Gilgameš, Enkidu and the Netherworld [1243]

or: Gilgameš and the Ḫalub-tree

Published sources

Siglum	*Signature*
A	N 1452
B	CBS 14068 (SEM 21)
C	3N-T381
D	3N-T903, 132
E	Ni 4507
F	Ni 4249, SLTN 5
G	UET VI, 55
H	N 3474 + CBS 19950 + UM 29-13-438 + CBS 15150 + N 3634
I	3N-T557
J	3N-T905, 198

1		**u$_4$-ri-a u$_4$-sù-rá-ri-a**	On that day, on that distant day,
	A	[ ]-a u$_4$-sù-rá-r[i]	
	C	[ ]-sù-rá-r[i]	
	E	u$_4$-ri-a [ ]	
	F	[ ]	
	G	u$_4$-ri-a u$_4$-sù-rá-ri-a	
2		**ĝi$_6$-ri-a ĝi$_6$-bad-rá-ri-a**	in that night, in that night long past,
	A	[r]i-a ĝi$_6$-bad-rá-r[i]	
	B	ĝ[i$_6$]	
	C	[ ]-bad-rá-r[i]	
	D	ĝi$_6$-[ ]	
	E	ĝi$_6$-ri-a [ ]	
	F	ĝi$_6$-ri-a [ ]	
	G	ĝi$_6$-ri-a ĝi$_6$-bad-rá-ri-a	
3		**mu-ri-a mu-sù-rá-ri-a**	in that year, in that distant year,
	A	[m]u-[r]i-a mu-sù-rá-r[i]	
	B	mu-ri-[r]á-ri-[ ]	
	C	[r]i-a mu-sù-rá-ri-a	
	D	mu-r[i]	
	E	mu-ri-a mu-s[ù]	
	F	[r]i-a m[u-s]ù-rá-[ri]	
	G	mu-ri-a mu-sù-rá-ri-a	
4		**u$_4$-ul níĝ-ul-e pa è-a-ba**	in olden times, when eternal things were manifested,
	A	u$_4$-ul níĝ-du$_7$-e pa è-[ ]	
	B	u$_4$-ul níĝ-du$_7$-e [ ]-a-ba	
	C	[ ]-du$_7$-e p[a U]D-DU-a-ba	

[1243] Henceforth: GEN.

	D	u$_4$ níĝ-d[u$_7$]	
	E	u$_4$-ul níĝ-du$_7$-e [p]a è-a-[ ]	
	F	[ ]-e pa è-a-[ ]	
	G	u$_4$ níĝ-du$_7$-e pa è-a-ba	
5		**u$_4$-ul níĝ-ul-e mí-zi dug$_4$-ga-a-ba**	in olden times, when eternal things were taken care of,
	A	u$_4$-ul níĝ-du$_7$-e mí-zi dug$_4$-ga-a[ ]	
	B	[u]l níĝ-du$_7$-e [ ] dug$_4$-ga-a-ba	
	C	[ ]-ul níĝ-du$_7$-e [ ]-zi dug$_4$-ga-a-ba	
	D	u$_4$-ul níĝ-d[u$_7$]	
	E	u$_4$-ul níĝ-du$_7$-e [ ]-zi dug$_4$-g[a]	
	F	u$_4$-ul níĝ-du$_7$-e [ ]-zi dug$_4$-g[a]	
	G	u$_4$-ul [n]íĝ-du$_7$-e mí-zi dug$_4$-ga-a-ba	
6		**èš-kalam-ma-ka ninda šú-a-ba**	when in the shrines of the Land bread was tasted,
	A	èš-kalam-ma-ka ninda šú-a-[ ]	
	B	[k]alam-m[a]-k[a nin]da šú-a-ba	
	C	èš-kalam-ma-ka [ ]	
	D	èš-kalam-ma-[ ]	
	E	èš-kalam-ma-ka [nin]da šú-[ ]	
	F	èš-ka[lam k]a nin[da]-a-ba	
	G	èš-kalam-ma-ke$_4$ ninda šú-a-ba	
7		**^{im}šu-rin-na-kalam-ma-ka níĝ-tab ak-a-ba**	when in the ovens of the Land fire was blown,
	A	^{im}šu-rin-na-kalam-ma-ka níĝ-tab ak-a-b[a]	
	B	$^{[im]}$šu-rin-[n]a-kalam-ma-[k]a (?) níĝ-tab ak-a-ba	
	C	^{im}šu-rin-na-ka[lam]	
	D	^{im}šu-rin-na-ka[lam]	
	E	^{im}šu-rin-na-kalam-ma-ka níĝ-tab ak-[ ]	
	F	^{im}šu-rin-na-kalam-ma-ka [a]k-a-ba	
	G	^{im}šu-rin-na-kalam-ma-ke$_4$ níĝ-tab ak-a-ba	
8		**an ki-ta ba-da-bad-rá-a-ba**	when heaven had been separated from earth,
	A	an ki-ta ba-da-bad-r[á]-a-ba	
	B	[a]n ki-ta ba-ra-bad-rá-a-ba	
	C	an ki-ta ba-e-da-[ ]	
	D	an ki-ta [ ]	

	E	an ki-ta ba-da-bad-rá-a-[ ]	
	F	an [ ]-ta ba-da-bad-rá-a-[ ]	
	G	an ki-ta ba-ta-bad-rá-a-ba	
9		**ki an-ta ba-da-sur-ra-a-ba**	when earth had been delimited from heaven,
	A	ki an-ta ba-da-sur-ra-a-ba	
	B	[ ] an-ta ba-da-sur-r[a]-a-ba	
	C	ki an-na-ta ba-e-dè-[ ]	
	D	[ ] an-ta [ ]	
	E	[t]a ba-da-sur-ra-a-b[a]	
	F	[ ] an-ta ba-da-sur-ra-a-b[a]	
	G	ki an-ta ba-ta-su$_{13}$-rá-a-[b]a	
10		**mu-nam-lú-u$_{18}$-lu ba-an-ĝar-ra-a-ba**	(and) the name of mankind had been established on it;
	A	mu-nam-lú-u$_{18}$-lu ba-a[n]-ĝar-ra-a-ba	
	B	[m]u-na[m]-lú-u$_{18}$-lu ba-ĝar-ra-a-ba	
	C	mu-nam-lú-u$_{18}$-lu b[a]	
	D	[l]ú-u$_{18}$-lu [ ]	
	E	[n]am-lú-u$_{18}$-lu ba-an-ĝar-ra-a-b[a]	
	F	mu-nam-[ ]-u$_{18}$-lu ba-an-ĝar-ra-a-ba	
	G	mu-nam-lú-u$_{18}$-[ ] ba-an-ĝar-ra-a-ba	
11		**u$_{4}$ an-né an ba-an-de$_{6}$-a-ba**	at that time, when An had taken heaven,
	A	u$_{4}$ an-né an ba-an-de$_{6}$-a-ba	
	B	u$_{4}$ an-né an ba-an-de$_{6}$-a-b a	
	E	[a]n-né an bad-de$_{6}$-a-b[a]	
	F	[ ]-né an ba-an-de$_{6}$-a-ba	
	G	u$_{4}$ an-né an [ ]-a-ba	
	H	[ ] ba-a[n]	
12		**^{d}en-líl-le ki ba-an-de$_{6}$-a-ba**	Enlil had taken earth,
	A	^{d}en-líl-le ki ba-an-de$_{6}$-a-ba	
	B	^{d}en-líl-le ki ba-an-de$_{6}$-a-ba	
	E	[l]íl-le ki bad-de$_{6}$-a-b[a]	
	F	^{d}en-líl-le ki ba-an-de$_{6}$-a-ba	
	G	^{d}en-líl-le ki ba-an-[ ]	
	H	[ ] ki ba-an-de$_{6}$-[ ]	
13		**^{d}ereš-ki-gal-la-ra kur-ra saĝ-rig$_{7}$-bi-šè im-ma-ab-rig$_{7}$-a-ba**	(and) when they had given the Netherworld to Ereškigal as a present;

	A	^{d}ereš-ki-gal-la-ra kur-ra saĝ-rig$_7$-bi-šè / im-ma-ab-rig$_7$-a-ba	
	B	^{d}ereš-ki-gal-la-ra kur-ra saĝ-rig$_7$-bi-šè im-ma-ab-rig$_7$-a-b[a]	
	E	[ ]-gal-la-ra kur-ra saĝ-rig$_7$-bi-šè im-ma-a[b]	
	F	^{d}ereš-ki-gal-la-ra kur-ra saĝ-rig$_7$-bi-šè / im-ma-ab-rig$_7$-a-b[a]	
	G	^{d}ereš-ki-gal-la-šè kur-ra saĝ-r[ig$_7$]	
	H	[ ]-ra kur-ra saĝ-rig$_7$-bi-šè im-[ ]	
	I	^{d}ereš-ki-[ ]	
	J	[ ]-[g]a-⸢a⸣-[ ]	
14		**ba-u$_5$-a-ba ba-u$_5$-a-ba**	when he sailed away, when he sailed away,
	A	ba-u$_5$-a-ba ba-u$_5$-a-ba	
	B	ba-u$_5$-a-ba ba-u$_5$-a-ba	
	E	[ ]u$_5$-a-ba ba-u$_5$-a-ba	
	F	[ ]-a-ba ba-u$_5$-a-[ ]	
	G	[u]$_5$-a-ba [ ]	
	H	[ ]-ba ba-u$_5$-a-[ ]	
	I	ba-u$_5$-a-ba ba-u$_5$-a-ba	
	J	[ ]-[u$_5$⸣-a-ba	
15		**a-a kur-šè ba-u$_5$-a-ba**	when Father sailed away for the Netherworld,
	A	a-a kur-šè ba-u$_5$-a-ba	
	B	a-a kur-šè ba-u$_5$-a-ba	
	E	[ ] kur-šè ba-u$_5$-a-ba	
	F	[ ] kur-šè ba-u$_5$-[ ]	
	G	a-a k[ur-]-a-ba	
	H	[š]è ba-u$_5$-a-[ ]	
	I	a-a kur-šè ba-u$_5$-a-ba	
	J	[ ]-⸢u$_5$-a-ba	
16		**^{d}en-ki kur-šè ba-u$_5$-a-ba**	when Enki sailed away for the Netherworld:
	A	^{d}en-ki kur-šè ba-u$_{[5]}$-a-ba	
	B	^{d}en-ki kur-šè ba-u$_5$-a-ba	
	E	[b]a-u$_5$-a-ba	
	F	[ ] ba-u$_5$-[ ]	
	G	^{d}en-líl[sic] kur-[ ] ba-[ ]-a-ba	
	H	[ ] ba-u$_5$-[ ]	
	I	^{d}en-ki kur-šè ba-u$_5$-a-ba	
	J	[ ] [x-u$_5$-a-ba	

17		**lugal-ra tur-tur ba-an-da-ri**	for the king small things were put on board,
	A	lugal-ra tur-tur ba-an-da-ri	
	B	lugal-ra tur-tur ba-da-an-ri [-eš ^{erasure}]	
	E	[ ]-an-da-ri	
	F	lugal-ra tur-tur ba-an-[ ]	
	G	lugal-ra [tu]r ba-da-an-ri	
	H	[ ] tur-tur ba-an-da-r[i]	
	I	lugal-ra tur-tur ba-an-da-ri	
	J	[ ]-da-ri	
18		**^{d}en-ki-ra gal-gal ba-an-da-ri**	for Enki large things were put on board;
	A	^{d}en-ki-ra gal-gal ba-an-da-r[i]	
	B	^{d}en-ki-ra gal-gal ba-da-an-ri	
	E	[-a]n-[ ]	
	F	[ ]-ki-«ra» gal-gal ba-a[n]	
	G	^{d}en-ki-ra ga[l-ga]l ba-da-an-ri	
	H	[r]a gal-gal ba-an-da-[ ]	
	I	^{d}en-ki-ra gal-gal ba-an-da-ri	
	J	[ ]-da-ri	
19		**tur-tur-bi na$_4$-šu-kam**	the small things were like stones fitting in the hand,
	A	tur-tur-bi na$_4$-šu-[ ]	
	B	[tu]r-bi na$_4$-šu-kam	
	F	[tu]r-bi [ ]	
	G	tur-tur-bi na$_4$-šu-a-kam	
	H	[ ] na$_4$-šu-[ ]	
	I	tur-tur-bi na$_4$-šu-kam	
20		**gal-gal-bi na$_4$-gi-gu$_4$-ud-da-kam**	the large things were like stones which make the reed dancing;
	A	ga[l-ga]l-bi na$_4$-gi-gu$_4$-ud-[ ]	
	B	[ ] na$_4$-gi-gu$_4$-ud-da-a-kam	
	F	[ ]-gal-bi na[$_4$]-g[u$_4$]	
	G	gal-gal-bi na$_4$-gi$_4$-gu$_4$-ud-da-kam	
	H	[ ] na$_4$-gi-gu$_4$-ud-da-[ ]	
	I	gal-gal-bi na$_4$-<gi>-gu$_4$-ud-da-kam	
	J	[ ]-ud-da-kam	
21		**úr-^{ĝiš}má-tur-re-^{d}en-ki-ga-ke$_4$**	they covered $^{(l. 22)}$ the bottom of Enki's boat
	A	úr-^{ĝiš}má-tur-re-^{d}en-k[i]	
	B	[m]á-[tu]r-re-[^{d}]en-ki-ga-ke$_4$	

	F	[x x x]	
	G	úr-má-tur-ra-^{d}en-ki-g[a]	
	H	[ ]-re-^{d}en-ki-ga-[ ]	
	I	úr-^{ĝiš}má-tur-e-^{d}en-ki-ga-ke$_4$	
	J	[ ]-ga-ke$_4$	
22		**níĝ-bún-na-du$_7$-àm ì-šú-šú**	fitting like a turtle;
	A	níĝ-bún-na-du$_7$-à[m]	
	B	[ ]-na-[d]u$_7$-àm ì-šú-šú	
	F	[x x x]	
	G	níĝ-bún-na-du$_7$-àm mi-[ ]	
	I	[ ]-du$_7$-àm mi-šú-šú	
	J	[x x x]	
23		**lugal-ra a-^{ĝiš}má-saĝ-ĝá-ke$_4$**	for the king the water at the stem
	A	lugal-ra a-^{ĝiš}má-[ ]	
	B	lugal-ra a-$^{ĝi[š]}$[ ]-saĝ-ĝá-ke$_4$	
	C	[r]a [ ]	
	F	[x x x]	
	G	lu[gal]-ra a-$^{<ĝiš>}$má-saĝ-ĝá-[ ]	
	I	[ ] a-^{ĝiš}má-saĝ-ĝá-ke$_4$	
24		**ur-bar-ra-gin$_7$ téš mu-na-gu$_7$-e**	swallows up everything like a wolf;
	A	[ ]-bar-ra-gin$_7$ téš mu-na-[ ]	
	B	ur-bar-ra-gin$_7$ t[éš] mu-na-gu$_7$-e	
	C	ur-bar-ra-[ ]	
	G	[ ]-bar-ra-[erasure?]-gin$_7$ téš mu-un-n[a]	
	I	[ ] mu-na-gu$_7$-e	
25		**^{d}en-ki-ra a-^{ĝiš}má-eger-ra-ke$_4$**	for Enki the water at the stern
	A	[ ]-ra a-^{ĝiš}m[á]	
	B	^{d}en-ki-ra a-$^{[ĝiš]}$má-eger-ra-ke$_4$	
	C	^{d}en-ki-ra a-$^{<ĝiš>}$má-[-ke]$_4$	
	G	[-r]a a-$^{<ĝiš>}$má-eger-ra-[ke]$_4$	
	I	[ ] a-^{ĝiš}má-eger-ra-[ ]	
26		**ur-maḫ-gin$_7$ saĝ ĝiš im-ra-ra**	hammers like a lion.
	A	[x x x]	
	B	[m]aḫ-gin$_7$ saĝ ĝiš im-ra-ra	
	C	ur-maḫ-[ ]	
	G	[ ] saĝ ĝiš [i]m-[ ]	
	I	[ ]-maḫ<-gin$_7$> saĝ ĝiš [i]m-ra-[ ]	

Editions / Translations

Editions
1. A. Shaffer, *Sumerian Sources of Tablet XII of the Epic of Gilgameš*. Ann Arbor, MI: UMI, 1963.
2. Th. Jacobsen, The Descent of Enki. In: M.E. Cohen, D.C. Snell, D.B. Weisberg (eds), *The Tablet and the Scroll: Near Eastern Studies in Honor of William W. Hallo*. Bethesda: CDL Press, 1993; 120-123 (lines 1-26).
3. A. Gadotti, *'Gilgameš, Enkidu and the Netherworld' and the Sumerian Gilgameš Cycle*. Ann Arbor, MI: UMI, 2005.

Translations
1. J. van Dijk, Le Motif Cosmique dans la Pensée Sumérienne. *AcOr* 28 (1964) 17-20.
2. D. Wolkstein, S.N. Kramer, *Inanna, Queen of Heaven and Earth. Her Stories and Hymns from Sumer*. New York: Harper & Row Publishers, 1983; 3-9.
3. M. Dietrich, Die Kosmogonie in Nippur und Eridu. *JARG* 5 (1984) 160-161 (lines 1-16) [1244].
4. J. Bottéro, S.N. Kramer, Le Prologue de 'Gilgameš, Enkidu et l'Enfer'. In: J. Bottéro, S.N. Kramer, *Lorsque les Dieux faisaient l'Homme. Mythologie Mésopotamienne*. [Paris]: Gallimard, 1993; 478-479 (lines 1-13, in part).
5. R.J. Tournay, A. Shaffer, *L'Épopée de Gilgamesh*. Paris: CERF, 1994; 248-249 (lines 1-26).
6. W. Horowitz, *Mesopotamian Cosmic Geography*. Winona Lake: Eisenbrauns, 1998; 135-136 (lines 8-13).
7. A. George, *The Epic of Gilgamesh. The Babylonian Epic Poem and Other Texts in Akkadian and Sumerian*. London: Allen Lane, The Penguin Press, 1999; 175-179 (lines 1-26).
8. M.P. Streck, Die Prologe der sumerischen Epen. *OrNS* 71 (2002) 194-196.
9. C. Wilcke, Vom altorientalischen Blick zurück auf die Anfänge. In: E. Angehrn (ed.), *Anfang und Ursprung. Die Frage nach dem Ersten in Philosophie und Kulturwissenschaft*. Berlin, New York: Walter de Gruyter, 2007; 23 (lines 1-16).
10. ETCSL 1.8.1.4: Gilgameš, Enkidu and the Netherworld.

***

Comments

4-7. These lines may refer to the organization of the cult for the gods.

4-5. For the transcription in these lines of **níĝ-UL** as **níĝ-ul** and a translation as 'eternal things' – instead of **níĝ-du$_7$** 'the appropriate things'– , I have followed the argumentation of Bauer and Civil [1245].

[1244] The text of Dietrich is a literal translation into German of van Dijk's edition; therefore Dietrich's translation will not be discussed.

[1245] Bauer 2005, 26; Civil 2003, 289. Also Gadotti (2005, 42, 434) supported the transcription **níĝ-ul** instead of **níĝ-du$_7$** for **níĝ-UL**. Van Dijk 1967a, 231 had another opinion: '**nì-du$_7$** ist ein Schlüsselwort des sumerischen Denkens und bedeutet "das Passende, die von den Göttern be-

Van Dijk translated these lines as follows: '4. lorsque les fleurs s'épanouirent, conformément à l'ordre divin, 5. lorsque les fleurs furent plantées dans la terre, conformément à l'ordre divin'. These translations presumably are based on his interpretation of the first **UL** as **ul** 'bud' (in both lines) and a different transcription in line 5: **im-ma** instead of **mí-zi**.

5-6. ETCSL adds to the translation of these lines: 'for the first time', but that is not written in the Sumerian text.

6. **šú** is taken as the Akkadian *lêmu*, (inter alia): 'to take food or drink' [1246].

7. **níĝ-tab** = 1. *nappašu* (inter alia) 'air vent' [1247]
2. *našraptu* (inter alia) 'burning' [1248].

Only George has tried to combine both meanings of **níĝ-tab** in his translation: 'after the ovens of the land had been fired up with bellows'. The same intention has led to the present translation: 'to blow fire in the ovens', so that the locative ending of the complex **^{im}šu-rin-na-kalam-ma-ka̲** also has been expressed. Van Dijk translates **^{im}šu-rin-na** as 'tinnūr', which has been derived from the Akkadian *tinūru(m)* [1249].

8-9. According to van Dijk and Streck, the verbs in these lines are intransitive [1250]. In comparison with the present passive translation, intransitivity shifts the accent of the action to heaven and earth themselves. This is in contradiction to other myths, in which it is Enlil who separates **an** and **ki** [1251].

10. Streck (2002, 195): '(sich) die Menschheit einen Namen gemacht hatte', and at note 10: '**nam-lú-u$_{18}$-lu** Ergativ (beachte /**n**/ vor der Verbalbasis).' This implies an inversion of the usual Sumerian syntax: 'subject-object-verb' then has been changed into 'object-subject-verb'. But there may be a more likely explanation for the prefix-**n**. In the present case the /**n**/ before the stem represents the locative infix **ni**, referring to the place where the name of mankind was established, viz. on earth [1252].

---

stellten Ordnung"; **nì-du$_7$-e** ... **pa-è** bedeutet: "etwas im Rahmen der von den Göttern bestellten Ordnung *mirifice* verwirklichen".'

1246 *lêmu*: AHw, 543; CAD L, 126.

1247 CAD N I, 311; AHw 740: 'Luftloch'.

1248 CAD N II, 78; AHw 761: 'eine Brennform??'

1249 *tinūru(m)*: AHw, 1360: '(Brat-, Back-) Ofen'; CAD T, 420-421: '1. oven, kiln, 2. (a star or constellation)'.

1250 Van Dijk 1964, 19: 'que le ciel s'éloigna de la terre, que la terre descendit du ciel,'. Streck 2002, 195: 'der Himmel sich von der Erde getrennt hatte, die Erde sich vom Himmel gelöst hatte,'.

1251 Some examples: I. Biggs 1974: IAS 136 *iii*: 1'-3'; IAS 113 *ii*: 5-10; IAS 203 *ii*: 3'-5' (see also ch. 2.1.1b and Appendix Text editions, no. 1b); II. 'The Song of the Hoe' (ETCSL 5.5.4), ll. 3-5; see text no. 9 in this Appendix.

1252 The solution of Gadotti 2005, 436: 'The infix -an in the verbal chain **ba-an-gar-ra-a-ba** is not in my opinion to be interpreted as a pronominal infix of the third person, but rather as a phonetic bridge requested by the verb (**gar**, where the first consonant, /ĝ/ is nasalized)', may be considered as an argument *faute de mieux*.

Van Dijk translated this line as: 'als der Same der Menschheit gepflanzt wurde'[1253].

11-12. For the transcription **DU** = **de$_6$**: see Sallaberger 2005b.

At line 11 a new part of the introduction begins. According to Gadotti (2005, 47), the second section of the prologue starts at line 14. But there are arguments for the division as proposed by the present author: 1. the content of the first ten lines is a coherent unit; 2. the construction of line 11 which starts again with **u$_4$ ... -a-ba**, and 3. the separation of **an-ki** – now mentioned with the protagonists in active operation – is repeated.

13. The usual Sumerian expression for 'to give, to present with' is **saĝ-šè rig$_7$**. As rendered in this line, the construction can be explained as follows: the ***a**-ending in **kur-ra** is an anticipatory genitive, repeated in the ***bi** (not ***ga** [1254]) after **saĝ-rig$_7$**.

The subject in the prefix chain is represented by a ***b** before the stem: the collective 'they', being An and Enlil.

The interpretation of Shaffer and of Tournay & Shaffer is different. In view of their translations it is Enlil, who had given earth to Ereškigal as a (wedding) gift in the Netherworld [1255]. This implies that the ***b** before the stem should represent Enlil, which would be unique [1256].

17-18. The verb **ri** = *ramû* [1257] is here translated as 'to put on board'. If accompanied by the comitative ***da**, it can also mean 'to take along, to carry along' [1258]. In that case the

---

1253 Van Dijk 1971, 453. The origin of this interpretation can be seen in his article: van Dijk 1964, 18-19 and note 44. Van Dijk's trancription reads: **[nu]mun-nam-lú-u$_x$-lu ba-gar-ra-a-ba** [with translation: 'que la «semence» de l'humanité fut établie'] instead of **mu-nam-lú-u$_{18}$-lu ba-ĝar-ra-a-ba**. Indeed, **numun** means 'seed'.

1254 Streck 2002, 195, and note 11: ... **saĝ rig$_7$-ga-šè** ... 'ETCSL 1.8.1.4 liest hier das schwierige – und nicht übersetzte – **bi**.' First of all: Streck did not translate, nor comment on, the ***a** in **kur-ra**. Moreover, his reading **saĝ rig$_7$-ga-šè** is not very likely; for why should two different styles of writing be present in one sentence, viz. a doubling of ***g** in **saĝ rig$_7$-ga-šè**, and no doubling of the ***g** in **im-ma-ab-rig$_7$-a-ba** ?

1255 Tournay/Shaffer 1994, 249, note *b.*: 'Allusion à un mythe perdu, cf. Bottéro and Kramer 1993, 479', but Bottéro and Kramer do not speak about a lost myth *l.c.* Katz (2003, 440-441) shows evidence for an association of Enlil with the netherworld, and for a marriage of Enlil and Ereškigal: 1. an incantation against evil spirits names Enlil and Ereškigal as the parents of Namtar (*Udugḫul 360*); 2. Enlil is the father of Ninazu in 'Enlil and Ninlil'; in other texts Ereškigal is the mother of Ninazu; 3. the text GEN, ll. 12-13, 'which tells that the netherworld was given to Ereškigal by Enlil as a gift when he took the earth for himself'. However, the text GEN ll. 12-13 does not say that Enlil has given the netherworld to Ereškigal, but 'to Ereškigal they (= An and Enlil) had given the Netherworld as a present'.

1256 Very remarkable is the translation of Jacobsen 1993, 121: 'and when he (i.e. Enki) was given (as slave) to Ereshkigal as Netherworld dowry'. Apart from the fact that Enki has not yet been mentioned at all, this should be a unique mythological story.

1257 *ramû* II 'werfen', AHw, 952-953; *ramû* B '1. to set in place, to endow, imbue', CAD R, 133, with in the lexical part examples of the verbal form ...**-n-da-ri**. See also: Al-Fouadi 1969, 74, l. 84: **su-zi mu-un-da-ri** 'he is full of awesomeness' [= ETCSL 1.1.4, l. 84: 'he instils fear']; Wilcke 1969, 100, l. 86: ... **ni-ḫuš ḫé-em-da-ri** '[auf meinem Nest] lastet schreckliche Furcht ...' [= ETCSL 1.8.2.2., l. 86: 'Foreboding weighs [upon my nest]'].

1258 See e.g.: Wilcke 1969, 92, l. 14 and comments p. 136; 'Enki and the world order' [ETCSL 1.1.3], ll. 398, 400, 408; 'Enki's journey to Nibru' [ETCSL 1.1.4], l. 84; 'Nanna-Suen's journey to Nibru' [ETCSL 1.5.1], *passim*; 'A *tigi* to Enlil for Ur-Namma' [ETCSL 2.4.1.2], l. 6; 'A

translation should be: 'for the king small things were taken along, for Enki large things were taken along'. The possible interpretation of **tur-tur** and **gal-gal** [1259] may be more clear after a closer examination of the lines 17-22 together.

19-20. **na$_4$-šu-kam** 'like stone(s) of the hand'. The small things are stones, or are like stones, which are of a size to fit into the hand [1260]. The big things, on the contrary, are like stones that are used to flatten reeds. Nota bene: we still do not know what the exact nature of the small and the big things is. The comparison with some kinds of stones seems only an indication for the relative size of these things.

21-22. These lines form the clue to the lines 17-22. The small and the big things, that cover the bottom of Enki's boat, form a heap in such a way, that it looks like a turtle resting on the bottom [1261]. The metaphor of the turtle may represent the image of the future residence of Enki in the **abzu**: being covered like a turtle by his shield, Enki has the necessary protection to reside in his subterranean home [1262].

The same idea, putting things in a boat or on a raft, is found in '*The bilingual creation of the world by Marduk*', lines 17-18. These lines tell, that Marduk creates dry land by pouring soil on a raft [1263].

níĝ-bún-na-du$_7$-àm, literally: 'like a fitting turtle', has been translated as 'fitting like a turtle'. The full Sumerian expression for 'to be/make perfect, to complete' is **šu du$_7$**. If we suppose that **šu** was omitted in the text, then line 22 may be translated as: 'like a complete turtle'. Gadotti's translation of ll. 21-22 reads: '21. The keel of Enki's small boat 22. They overwhelmed like thrusting turtles,'. ETCSL has as translation of the lines 21-22: 'The keel of Enki's little boat was trembling as if it were being butted by turtles.' The plural of the verb **šú** 'to cover' in the line 22 (**ì-šú-šú**) refers to the small and the large things which are like certain small and large stones. The participle ***du$_7$** – as adjective at **níĝ-bún-na** – is singular, which implies that there is no question of more than one turtle, neither does the verb **šú** have any meaning like 'to tremble'.

In the opinion of Afanasieva, the lines 17-26 comprise only one metaphor: small and large stones spread around the keel of the boat and they produce 'aufwallende Kreise' on the surface of the water (ll. 17-22) [1264]. The effect of these waves should be described in the

---

praise poem of Šulgi' [ETCSL 2.4.2.05], l. 216; 'An **adab** (?) to Utu for Šulgi' [ETCSL 2.4.2.17], l. 4; 'A praise poem of Lipit-Eštar' [ETCSL 2.5.5.1], l. 34; 'An **adab** to An for Lipit-Eštar' [ETCSL 2.5.5.3], l. 36; 'A *šir-namšub* to Inana' [ETCSL 4.07.7], ll. 16-17.

1259 In his comments Shaffer writes about the stones: 'The "small stones"/"large stones" are figures of speech for hailstones.' (Shaffer 1963, 128).

1260 Thomsen 1984, 276-277, § 544: 'The enclitic copula may also be used in descriptions and comparisons, almost equivalent to **-gin$_7$** "like".' These stones are not to be confused with the **na$_4$-šu-gul-gul** 'lower millstone' (Stol 1979a, 89-92).

1261 The translation of the lines 21-22 by Jacobsen 1993, 121 is completely different: 'the oars of the small boat on which Enki was carried captive came down upon the turtles like goring (bulls)'. Jacobsen keeps following up this clue in the next lines, where he takes the 'oars' (**úr**) of line 21 also as subject of the lines 23-26.

1262 See also the more detailed discussion of this episode in ch. 2.1.7.

1263 Text CT 13 36: 17-18; see *amu,* CAD A II, 85. Also: Horowitz 1998, 130-131.

1264 Afanasieva 1998, 19-20. This scholar only gives an interpretation, no translation of these lines. Her interpretation 'aufwallende Kreise' (p. 20) most likely was inspired by the Sumerian verb **du$_7$-du$_7$** 'to whirl', which verb has this meaning only in its reduplicated form (ePSD); in line 22 **du$_7$** is not reduplicated.

next lines (23-26). The interpretation of Afanasieva, especially of the lines 21-22, is not very convincing; stones in the water that spread like turtles below the boat (thus: outside the boat) – 'um den Kiel herum, also unten am Boot' – , is an interpretation that is not in agreement with the Sumerian text.

23-26. In my opinion, these lines are metaphors for a tempestuous ocean. My interpretation is that Enki is on his way to his future residence, the Abzu, hence this journey over the primaeval ocean.

***

8. Enki and Ninmaḫ [1265]

Published sources

Siglum	*Signature*
A	SEM 116 + PBS I^1 4 (= PBS X^4 14)
C	K1711 + 2168 + 4896 + 4932
B	AO 7036 (TCL XVI 71)
D	composite version of ETCSL

1		**u$_4$-ri-a-ta u$_4$ an-ki-bi-ta ba-an-[...]**	From that day, the day when heaven and earth [....],
	A	[...]	
	C	u$_4$-ri-a-ta u$_4$ an-ki-bi-ta ba-an-[...]	
	C	*i-na u$_4$-mi ul-lu-ti / ša* an *u* ki-*tum x*[...]	
2		**ĝi$_6$-ri-a-ta ĝi$_6$ an-ki-bi-ta x-[....]**	from that night, the night when heaven and earth [....],
	A	[...]	
	C	ĝi$_6$-ri-a-ta ĝi$_6$ an-ki-bi-ta x-[...]	
	C	*i-na mu-ši ul-lu-ti* / [...] an *u* ki-*tum up*-[*ta-aṭ-ṭi-ru*]	
3		**[mu-ri-a-t]a mu nam a[n-ki-bi-da$^?$] / ba-t[ar-ra-ba$^?$]**	[from that year], the year when the fate [of heaven and earth was determined$^?$],
	A	[...]	
	C	[mu-ri-a-t]a mu nam a[n- ...] / ba-t[ar$^?$ ...]	
	C	*i-na ša-na-a-ti ul-la*-[*xxx*] / *šá ši-mat* an *u* ki-*tim* [*xxx*]	
4	A	**[diĝir-a]n-n[a]-ke$_{[4]}$-ne ba-tu-ud-da-a-ba**	after the gods of the heaven had been born,
	C	dim$_{10}$-me-er-šár-šár an-ki-a ba-tu-ud-d[a-eš-a-ba]	after all the gods in heaven and earth had been born,
	C	diĝir-meš gal-meš *ina* an-*e u* ki-*tim i'-a*[*l-du*]	the great gods in heaven and earth were born;
5	A	**^{d}amalu nam-NIR-PA-šè ba-tuku-a-ba**	after the mother goddesses had been taken as spouses,
	C-6	dim$_{10}$-me-er ama-^{d}inanna-ke$_4$-e-ne nam-dam-šè / ba-tuku-eš-a-ba	after the gods had taken the mother goddesses as spouses,
	C-6	diĝir-meš eš$_4$-dar-meš *ana aš-šu-ti i-ḫu-zu*	the gods took the mother goddesses as spouses;

1265 The translation of the bilingual text will be given only, if this deviates from the Old Babylonian version.

6	A	**^{d}amalu an-ki-a ba-ḫal-ḫal-la-a-ba**	after the mother goddesses had been distributed among heaven and earth,
	C-5	dim$_{10}$-me-er-šár-šár an-ki-a ba-ḫa-la-eš-a-b[a]	after the numerous gods had been distributed among heaven and earth,
	C-5	diĝir-meš gal-meš an *u* ki *ú-za-'i-z*[*u*]	the great gods distributed heaven and earth;
7	A	**^{d}amalu [... ù$^{?}$]-ma-a-peš$_{11}$ ù-tu-da-a-ba**	after the mother goddesses [...] had become pregnant, had given birth,
	C	dim$_{10}$-me-er ama-^{d}inanna-⌜ke$_{4}$-e-ne⌝ ĝìš bi-in-dug$_{4}$ / dim$_{10}$-me-er-e$^{?}$ [...]-tu-ud-da-eš-a-ba	the gods had intercourse with the mother goddesses; after the gods ... had brought forth,
	C	diĝir-m[eš] ⌜eš$_{4}$⌝-[dar-me]š *ir-ḫu-ma* / diĝir-meš [ ] *ú-al-li-du*	the gods had intercourse with the mother goddesses, the gods had ... brought forth;
8	A	**diĝir kurum$_{6}$-ma-bi [x...x]-bi-šè ba-ab-kéš-a**	after the gods ... because of ... of their food ration had set up an agreement,
	C	*missing*	
9	A	**diĝir-šár-šár kíĝ-ĝá al-su$_{8}$-ge-eš [diĝ]ir-tur-tur dú-lum / im-íl-íl-e-ne**	numerous gods are assigned to the work, (but only) a smaller group of gods is bearing the toil;
	B	[x] / [í]l-⌜e-ne	
	C	dim$_{10}$-m[e-] [xx] ⌞dim$_{10}$-me-er⌟-tur-tur / du$_{6}$-[x] [-n]e	
	C	diĝir-meš gal-[x] / *dul-lum* [ ]	
10	A	**diĝir íd <im>-dun-d[u]n-[ù]-ne saḫar-bi ḫa-ra-li im-dub-dub-bu-ne**	the gods dig out canals; its sand they heap up: Ḫarali.
	B	*missing*	
	C	dim$_{10}$-me-er íd dun-[ ] / saḫar-bi [ ]	
	C	diĝir-meš *na-ra-*[ ] / *ina e-pe-ri(-)ši-*[ ]	
11	A	**diĝir im ì-hur-re-[ne] zi-bé inim àm-ma-ĝar-re-ne**	The gods scratch the clay! They complain of their life.
	B	[ ] ⌜àm-ma-ĝar-re-ne	
	C	[ ]-eš-àm / [-m]i-ni-in-gi	
	C	[ ] *na-piš-ta-šú-nu* / [ ] GIR	

12	A	**u$_4$-ba ĝeštu$_2$-daĝal mud-diĝir-šár-šár-ĝál-ĝál**	At that time, the one with great wisdom, the creator among all the existing gods,
	B	[ ] diĝir-šár-šár-ĝal-ĝal	
	C	[k]e$_4$ mud-dim$_{10}$-me-er / [ ]-ke$_4$e-ne	
	C	[ ]*uz-ni ba-nu-ú* diĝir^{meš}-gal^{mes}	
13	A	**^{d}en-ki-ke$_4$ engur-buru$_3$ a-sur-ra ki diĝir-na-me šà-bi u$_6$ nu-um-me**	Enki, in the deep Engur, the subterranean water – a place the inside of which not a single god can observe –,
	B	[x] šà-bi / [ ] um [ ]	
	C	[ ]-ra ki diĝir-na-me / [-u]n-zu-àm	
	C	[*x*] *ru-qu-ú-ti* / [*l*]*a i-du-ú*	
14	A	**ki-nú-ni ì-nú ù-ku nu-um-zi-zi**	lay sleeping in his bed, and did not arise.
	B	[ ] um-z⌜i-zi]	
	C	[ ]-ab-zi-zi	
15	A	**diĝir ér-ra im-pà-pà-dè a-nir ĝál ì-ak im-me-ne**	The gods, continuously weeping, said: 'He caused the lament.'
	B	[ ] nir ĝál AK [ ]	
16	A	**lú-<ù>-ku-ra ì-nú-a-ra ki-nú-bi nu-um-zi-zi**	To the one who is lying asleep – he did not arise from his bed – ,
	B	[ ] nu-um-zi-zi	
17	A	**^{d}namma-ke$_4$ ama-palil ù-tud-diĝir-šár-šár-ra-ke$_4$-ne**	Namma, the primaeval mother, procreator of a multitude of gods,
	B	[ ] diĝir-šár-šár-ra-ke$_4$-e-ne	
18	A	**ér-ra-diĝir-re-e-ne dumu-ni-ir ba-ši-in-de$_6$**	has conveyed the weeping of the gods to her son:
	B	[ ]-ni-šè ba-ši-in-ku$_4$	
19	D	**[ù]-mu-un ši-nú-ù-nam ù-mu-un ši-ku-ku-na-nam**	'Lord, while you lie there, while you are sleeping indeed, lord,
	A	[ ]-⌞mu-un⌟ nú-ù-nam ù-⌞mu-x⌟ [ ]	
	B	[ ] ši-ku-ku-na-nam	
20	D	[ ] **/unu/$^{!}$** [**nu-um**]**-zi-zi**	[*while you lie in your*] dwelling [*and do not*] arise [*from your bed*],
	B	[x]-zi-zi	

21	D	**dìm-me-er šu-dím-dím-ma-zu [x] gú$^{?}$-bi im-tu$_{11}$-tu$_{11}$-ne**	the gods, who have approached you with a complaint, [ ] tired themselves out.
	A	⌜dìm⌝-me-er šè$^{!}$-dím-dím-ma g[ú-]	
	B	[x] gú-bi im-tu$_{11}$-tu$_{11}$-ne	
22	D	**du$_{5}$-mu-mu ki-nú-zu zi-ga [ì-bí]-ma-al-la-zu-ta/ na-áĝ-kù-zu ù-mu-e-kíĝ-ĝ[á]**	My son, arise from your bed! After you, with your wisdom, have let work your skill,
	A	du$_{5}$-mu-mu ki-nú-zu zi-ga [ ] / na-áĝ-kù-zu mu-[ ]	
	B	[ma-al-la-zu-ta / [x] ù-mu-e-kíĝ-[x]	
23	D	**kíĝ-sì-dìm-me-er-e-ne-ke$_{4}$ ù-mu-[e]-dím / *tir-ḫum*-bi ḫa-ba-tu-lu-n[e]**	(and) when you have created those who will take over the work of the gods, let them (= *the gods*) loose their basket.'
	A	kíĝ-sì-dìm-me-er-e-ne-ke$_{4}$ ù-mu-[ ] / *tir-ḫum*-bi ḫa-ba-tu-[x]	
	B	[ ]-dím *tir-ḫum* [ ] / [x]	
24	A/ D	**^{d}en-ki-ke$_{4}$ inim-ama-na-^{d}namma-ke$_{4}$ ki-nú-na ba-ta-zi**	Now Enki, at the word of his mother Namma, arose from his bed.
25	A/ D	**ḫal-an-kù niĝin$_{2}$ šà-kúš-ù-da-na TAR im-mi-ni-a[k]**	Roaming in Ḫalanku with a heavy heart, he has taken a decision.
26	A	**ĝeštu$_{2}$ ĝizzal$_{x}$ èn-tar [ ] nam-kù-zu mud-me-dím níĝ-nam-ma / se$_{12}$-en-se$_{12}$-šár im-ta-an-è**	The intelligent, understanding, investigative, [ ], wise one, creator of everything: he did emanate a birth goddess.
27	A	**^{d}en-ki-ke$_{4}$ da-né ba-ši-in-de$_{6}$ ĝeštu$_{2}$ ì-niĝin-e**	Enki brought her (= *the birth goddess*) at his side, (and) he directed the attention to her.
28	A	**^{d}en-ki-ke$_{4}$ mud-me-dím ní-te-a-na / šà-bi ĝeštu$_{2}$-ga$^{!}$ ù-mu-da$^{!}$-ni-de$_{5}$-ge**	When Enki had considered the character of the manifestations, which he created himself,
29	A	**ama-ni ^{d}namma-ra gù mu-un-na-dé-e**	he says to his mother Namma:
30	A	**ama-ĝu$_{10}$$^{!}$ mud mu-ĝar-ra-zu ì-ĝál-la-àm / zub-sìg-diĝir-re-e-ne kéš-da-ni**	'My mother, when the creature which you have suggested exists, fasten the carrier basket of the gods to him.
	C	am[a] [ ] / ki [ ]	
	C	*um-mi* [ ] / *it-ti* [ ]	

31	A	**šà-im-ugu-abzu-ka ù-mu-e-ni-in-šár**	When the birth goddesses (l. 32) have stirred for you in the clay on top of the abzu,
	C	ù-me-ni -[ ]	
32	A	**se$_{12}$-en-se$_{12}$-šár im mu-e-gir$_8$-gir$_8$-re-ne / za-e me-dím ù-mu-e-ni-ĝál**	they will pinch off clay for you. When you have moulded it into form,
	C	šà-tùr im ma [ ] / za-e ù-[ ]	the mother goddess .. clay .. / you...
	C	MIN *ṭi-id-da* [ ]	
33	A	**^{d}nin-maḫ-e an-ta-zu ḫé-ak-e**	let Ninmaḫ act as your companion.
34	A	**^{d}nin-imma$_3$ ^{d}šu-zi-an-na ^{d}nin-ma-da / ^{d}nin-šara$_2$ ^{d}nin-bara$_2$**	May Ninimma, Šuziana, Ninmada, Ninšara, Ninbara,
35	A	**^{d}nin-mug ^{d}mú-mú-du$_8$ ^{d}nin-⌜gùn⌝-na**	Ninmug, Mumudu and Ninguna
36	A	**tu-tu-a-zu ḫa-ra-gub-bu-ne**	assist you while you are giving birth.
37	A	**ama-ĝu$_{10}$ za-e nam-bi ù-mu-e-tar ^{d}nin-maḫ-e zub-sìg-bi ḫé-keše$_2$**	My mother, when you have determined their fate, let Ninmaḫ fasten their carrier baskets.'
38	B	[ ] **NI-dù nam-lú-[ulu$_3$]** [ ]	[ ] … mankind [ ]
	C	nit[a] [ ]	man [ ]
39	B	[**x] nam-lú-ulu$_3$ àm-**[ ]	[ ] mankind [ ]
	C	dùb-nir-ra-nit[a] šà-ta [ ]	the ejaculate of a man [ ] from the inside [ ]
	C	*ina ri-ḫu-ut z*[*i-ka-ri*] *il-da* [ ]	out of the sperm of a man offspring [ ]
40	B	[**pe]š$^?$ saĝ-e gù$^?$ àm-m[a- -sù$^?$**]	[pregn]ant$^?$; the first-born has screamed$^?$;
41	B	[ ] **dim$_4$ su-/unu/-RI^{su} dug-a à[m-**]	[ ] approached/was prudent; the afterbirth was [ ] in the jar;
42	B	**[ga-raš$^?$]^{sar}-e ĝiš-nu$_{11}$ mi-ni-in-íl nam-[maḫ$^?$-x]**	she made [the leek ?] raise his eyes; [ ]
43	B	[ ] **ság numun$_2$-e mi-ni-in-ri ù-tu na «bi» mu-de$_5$**	[ ] scattered; she has laid it down on the grass; giving birth was consecrated.
44	B	**^{d}en-ki-ke$_4$ kíĝ ni$_{10}$-ni$_{10}$-da ∅[1266] mi-ni-in-lá šà-bi ba-ḫúl**	Enki showed the work, while he completed it; their hearts rejoiced.

1266 ∅ = erasure

Editions / Translations

Editions

1. J. van Dijk, Le Motif Cosmique dans la Pensée Sumérienne. *AcOr* 28 (1964) 24-31.
2. C.A. Benito, '*Enki and Ninmaḫ*' and '*Enki and the World Order*'. Ann Arbor, Michigan: UMI, 1969; 20-81.
3. G. Pettinato, *Das altorientalische Menschenbild und die sumerischen und akkadischen Schöpfungsmythen.* Heidelberg: Carl Winter, Universitätsverlag, 1971; 69-73.
4. G. Komoróczy, Work and Strike of Gods. New light on the Divine Society in the Sumero-Akkadian Mythology. *Oikumene* 1 (1976) 9-37. [Comments on ll. 8-11 of 'Enki & Ninmaḫ': pp. 10-12.]
5. S.N. Kramer, J. Maier, *Myths of Enki, the Crafty God.* New York, Oxford: Oxford University Press, 1989; 31-37. [Comments pp. 211-215.]
6. W.G. Lambert, The Relationship of Sumerian and Babylonian Myth as Seen in Accounts of Creation. In: D. Charpin, F. Joannès (eds), *La Circulation des Biens, des Personnes et des Idées dans le Proche-Orient Ancien.* Paris: Éditions Recherche sur les Civilisations, 1992; 129-135.
7. J. Bottéro, S.N. Kramer, Enki et Ninmaḫ. In: J. Bottéro, S.N. Kramer, *Lorsque les Dieux Faisaient l'Homme. Mythologie Mésopotamienne.* [Paris]: Gallimard, 1993; 188-198.
8. W.H.Ph. Römer, Enki und Ninmach. In: W.H.Ph. Römer (ed.), mit einem Beitrag von D.O. Edzard, *Weisheitstexte, Mythen und Epen. Mythen und Epen I.* TUAT III/3. Gütersloh: Gütersloher Verlagshaus Gerd Mohn, 1993; 386-401.
9. H. Sauren, Nammu and Enki. In: M.E. Cohen, D.C. Snell, D.B. Weisberg (eds), *The Tablet and the Scroll. Near Eastern Studies in Honor of William W. Hallo.* Bethesda, Maryland: CDL Press, 1993; 198-208.
10. J. Klein, Enki and Ninmaḫ. In: W.W. Hallo (ed.), *The Context of Scripture, I: Canonical Compositions from the Biblical World.* Leiden, New York, Köln: Brill, 1997; 516-518.
11. W. Horrowitz, *Mesopotamian Cosmic Geography.* Winona Lake, Indiana: Eisenbrauns, 1998; 137-138.
12. M.P. Streck, Die Prologe der sumerischen Epen. *OrNS* 71 (2002) 197-198.

New-Assyrian bilingual text

R. Borger, Einige Texte religiösen Inhalts. II. 'Enki und Ninmaḫ' in Ninive. *OrNS* 54 (1985) 18-22.

Translations

1. Th. Jacobsen, The Birth of Man. In: *The Harps that Once... Sumerian Poetry in Translation.* New Haven and London: Yale University Press, 1987; 151-166.
2. C. Wilcke, Vom altorientalischen Blick zurück auf die Anfänge. In: E. Angehrn (ed.), *Anfang und Ursprung. Die Frage nach dem Ersten in Philosophie und Kulturwissenschaft.* Berlin, New York: Walter de Gruyter, 2007; 29-34.
3. ETCSL 1.1.2: Enki and Ninmaḫ.

***

Comments

1-3. The more usual formulation for the beginning of a myth is **u$_4$-ri-a** instead of **u$_4$-ri-a-ta**. Perhaps this formulation was influenced by the Akkadian *ina ūmi* ? [1267] A suggestion for the completion of the lines 1 and 2 was given by Streck [1268], by analogy with the lines 8 and 9 of 'Gilgameš, Enkidu and the Netherworld'.

For the Akkadian lines 1 and 2 of the Niniveh text, Streck supposes the verbal form *up-t[ar-ri-su]* 'getrennt wurden' [1269]. However, the Dt-stem of *parāsu* with a passive meaning has not been attested. In my opinion, a likely option for the completion of the Akkadian *UB-* is the 3rd pers. masc. plur. of the preterite of the Dt-stem of the verb *paṭāru*: *uptaṭṭirū* 'they were loosened; they fell apart' [1270].

5-7. Unlike the bilingual version (B), the Old Babylonian text (A) has no indication of the presence of more mother goddesses; both **^{d}amalu** and the verbal forms are singular, with the exception of the verbal form in line 6. **^{d}amalu** has been understood as a collective noun. The plural verbal form (line 6) may be explained as 'distributive', a consequence of the distribution of the mother goddesses among more places, as spouses of the gods in heaven and on earth.

5. **nam-NIR-PA-šè** most likely is a scribal error for **nam-dam-šè**; the standard expression is **nam-dam-šè tuku** 'to take as wife', as has been written in the bilingual version.

6. The Akkadian version suggests a reading ... **an-ki a-ba-ḫa-la-eš-a-ba** for the Sumerian line of the Niniveh text; the gods are the subject. In the OB text, the mother goddess is the absolutive. There is no example known in the Sumerian literature, where a mother goddess is distributed among heaven and earth. The meaning of this line may be, that the gods of both heaven and earth take the mother goddess as wife. Therefore the phrasing is: **an-ki-a ba-ḫal-ḫal-la-a-ba** instead of **an-ki a-ba-ḫal-ḫal-la-a-ba** [1271].

1267 A literature search has shown that no other literary texts begin like this. Three examples with **u$_4$-ri-ta** have been found: 'The Death of Gilgameš', segment F [ETCSL 1.8.1.3], lines 27 and 120; 'Lugalbanda in the mountain cave' [ETCSL 1.8.2.1], l. 429. In PSD B, 36^b some related formulations are mentioned: **[u$_4$ an k]i-ta bad-DU-ta**, TuM NF 3, 36 i:1 ('The Debate between Copper and Silver', Segment D [ETCSL 5.3.6], l. 1); **[...] ⌜ki⌝ an-ta bad-e-dè**, TuM NF 4, 79 rev. 4'.

1268 Streck 2002, 197. Line 1: **ba-an-[ba$_9$-rá / sur-ra-a-ba]**; line 2: **ba-[an- ba$_9$-rá / sur-ra-a-ba]**: '.. als [sich] Himmel und Erde ge[trennt hatten]'. The suggestion of Klein 1997a and ETCSL, to complete these lines with the verb **dím** 'to create', is less plausible: there is no example of a myth in which the *creation* of heaven and earth is mentioned.

1269 Streck 2002, 197, note 18.

1270 *parāsu*: AHw, 830-832; CAD P, 165-178. *paṭāru*: AHw, 849-851; CAD P, 286-303. The Dt-stem is passive at the D-stem: 'to be loosened, to fall apart'. Among the Akkadian equivalents for **bad** and **sur**, there is no likely option to complete the verbal form beginning with *UB*. The lexical list at *paṭāru* shows e.g. *ta-ár* **KUD** = *paṭāru* (CAD P 287a).

1271 A verbal form beginning with **a-ba** should have been a hapax in OB texts (Heimpel 1974), and seems less likely therefore.

7. It is supposed that in the OB text the sign **ḪA** = **peš₁₁** was used for the homophone **GIR** = **peš** 'to become/be pregnant' [1272].

8. Most probably this line (40% has been broken) tells us that the gods are obliged to provide for their meals by themselves [1273].

A compound expression **saĝ/zag/zú ... kéš** is very unlikely, because in those expressions 1) the nominal part has its place just before the verbal form, and 2) that nominal part is missing the suffix ***šè**. Therefore the translation of **kéš** was started from the basic meaning 'to bind'; the following meanings may be derived from it: 'to set up a binding agreement', 'to assign a person to a task' [1274]. The ***b** before the verbal stem does not allow a passive translation; it represents, in all likelihood, the collective of the gods as subjects.

9. For most scholars **diĝir-šár-šár** and **diĝir-tur-tur** represent the 'classical' opposition between the 'senior' or 'great' gods (= **diĝir-gal-gal**) and the 'younger' or 'minor' gods [1275]. Usually the epithet **diĝir-gal-gal** is used with the Anunna-gods; but they are not mentioned in this text. Therefore I stick to the literal meaning of **šár** 'numerous'.

10. ' ... dig out canals': **íd** also means 'river'. In the Akkadian version of *atra-ḫasīs*, the Igigi dig out the rivers the Tigris and the Euphrates [1276].

ḫa-ra-li has been interpreted as a place, which has come into being through the heaps of sand, caused by the digging by the gods. If the gods were heaping up the sand *in* Ḫarali [1277], an already existing place, I expect a locative ending behind **ḫa-ra-li**.

11. The meaning of **im ḫur** 'to scratch the clay' is in my opinion: the gods had to dig to such a depth, that in the end they reached the virgin soil, which consisted of clay [1278].

1272 In the literature there are some examples of the use of **ḪA / peš₁₁** instead of **peš**: 'Enlil and Ninlil' [ETCSL 1.2.1], l. 30; 'Enmerkar and the Lord of Aratta' [ETCSL 1.8.2.3], l. 546 (see also comments: Cohen 1973, 293-294. Mittermayer 2009, 214, reads in all texts at this line: **peš**); 'The lament for Nibru' [ETCSL 2.2.4], l. 34; 'A praise poem of Šulgi' [ETCSL 2.4.2.02], l. 246; 'A praise poem of Išme-Dagan' [ETCSL 2.5.4.01], l. A167; 'A *šir-namšub* to Ninurta' [ETCSL 4.27.07], l. 133. Pettinato (1971, 72; comment at l. 9) suggested to read **A-ḪA** = Ku'ara, also because in line 12 (in the present edition line 10) Ḫarali, of which Ku'ara should be the capital city, was mentioned. For this statement, however, no evidence has been found in the literature. About the position of Ku'ara in Mesopotamia, see e.g. Sjöberg and Bergmann 1969, 25 (no. 10), 80-81; Steinkeller 1980; Carroué 1991, 152, note 93; Steinkeller 1995a and 1995b; Bauer 1998, 504, 510.

1273 This idea is also found in other Sumerian texts: see the 'Appendix 1: Gods who (have to) work' at 'Enki & Ninmaḫ'.

1274 See *rakāsu*: AHw 945-947; CAD R , 91-105.

1275 Streck (2002, 197) translates **diĝir-šár-šár** as 'die mächtigen Götter'. Benito (1969, 47 *ad* 9) writes: '**diĝir-šár-šár** stands here, probably in opposition to **diĝir-tur-tur** and is therefore to be translated by 'the great gods' (not 'the numerous gods')'. He also founds this on some lexical texts: CT XXV 18: 3-4, and CT XI 30, I:71, with the equation **šár** = *ra-bu-u*. Cooper (1973, 583 note 9) supports the translation 'numerous' for **šár-šár**. The expression **diĝir-šár-šár** seems to have been used only in 'Enki & Ninmaḫ'.

1276 Lambert and Millard 1999, 42-43: tablet I: 20-26; tablet K 8562, *i*: 5-8.

1277 Komoróczy 1972 discusses Ḫarali and its localization in detail; further references at Sjöberg 1988, 174 *ad* line 4.

The expression **inim ĝar** (ePSD: 'to sue, to make a legal claim' [1279]) corresponds with the tendency of *atra-ḫasīs*, where the gods rebel against their tasks [1280].

12. **mud** = *banû* 'to create'. A translation like 'creator <u>of</u> the ... gods' is misleading. It might be interpreted as if Enki had created the majority of the gods, for which is no textual support. The context gives the clue: Enki is the one who *invents* things *for* the gods, usually in delicate situations. In that sense Enki may be called 'creator'.

13. The postposition ***e** at **^{d}en-ki-ke$_4$** cannot represent an ergative, because of the intransitive verb **nú** (l. 14). In this case it may be the deictic ***e**, in order to emphasize Enki. Or ***ke$_4$** may represent a weak spelling for ***kam** [1281], also meant to draw the attention to Enki. For **u$_6$ ... du$_{11}$ / e / di**: see Attinger [1282].

Compare the second part of this sentence with line 47 of 'Enki's Journey to Nippur': **šà-šeg$_9$-bar-ra lú igi nu-bar-re-dam** '(Nudimmud) lets nobody look into the midst of the Abzu.' [1283].

14. **ki-nú-ni** instead of the expected **ki-nú-na**: in the OB-time the suffixes after the pronomina possessiva are increasingly omitted [1284].

Thomsen remarks that the verb **ku** is always reduplicated and only found in the form **ù-ku-ku** [1285]. But there are also forms in the literature without the reduplication of **ku**, just as in this line [1286].

15. The usual construction is **ér pàd** instead of **ér-<u>ra</u> pàd**. The verb mostly combined with **a-nir** is **ĝar** instead of **ĝál** [1287].

1278 Streck 2002, 197, **diĝir im-ur$_5$-ur$_5$-re-ne** 'Die Götter mahlten (?Korn)'. The Sumerian text was copied from Benito 1969, 22, line 11.

1279 From some literature examples, e.g. 'The marriage of Martu' [ETCSL 1.7.1], lines 96, 103, 110, one may conclude that the 'object' of **inim ĝar** has the directive ending; therefore the transcription is **zi-bé**.

1280 The other possible transcription for **KA ĝar,** viz. **gù ĝar** 'to make noise; to shout' should led to the following translation: 'they (the gods) make noise'. This noise then comes from the panting of the gods during their work, or it is due to the noise of their work. In this way there is the antithesis: the noise (= activity) of the gods and the silence of Enki. But **gù ĝar** 'to make noise; to shout' seems to be a hapax. The correct verbal form should be: **àm-ma-ĝá-ĝá-ne**. Comparable 'corrupt' forms are also found in OBGT VI: 221-227. The Sumerian 'Story of the Flood' also contains many grammatical and lexical irregularities (Civil 1999, 139).

1281 See the edition of 'The Debate between Grain and Sheep', comments at ll. 1-2 *sub iii* (Appendix Text editions no. 6).

1282 Attinger 1993, 739-749, *sub* 5.3.197.

1283 Fouadi 1969, 71, 79 and 128 (comment *ad* 47). See also CAD A II: *apsû*, 194b in the lexical part for the translation of **šeg$_9$-bar** as Abzu.

1284 Krispijn: personal communication.

1285 Thomsen 1984, 309.

1286 Some examples with **ù-ku**: Castellino 1957, 20 (CBS 4560, rev. 1:36) [= 'The Death of Ur-Namma', ETCSL 2.4.1.1, l. 163]; van Dijk 1983, 117, l. 510 [= 'Ninurta's exploits: a *šir-sud* (?) to Ninurta', ETCSL 1.6.2, l. 509]; Sjöberg 1975, 194, l. 162 [= 'A hymn to Inana', ETCSL 4.07.3, l. 162]; Michalowski 1989, 50, l. 238 [= 'The lament for Sumer and Urim', ETCSL 2.2.3, l. 238]; Alster 1997, 19, no. 1.72 [= 'Proverbs: collection 1', ETCSL 6.1.01, segment B, l. 21]; Sefati 1998, 248, l. 46 [= 'A *kunĝar* to Inana', ETCSL 4.08.20, l. 46].

Van Dijk changed his original and correct translation [1288] afterwards into: 'Die Götter weinten und klagten: O, Sonnengott! (?)!', without any explanation for the new translation 'O, Sonnengott! (?)!' [1289]. Komoróczy translates: 'The gods shed their tears, "We fall into destruction" – they cried', but he left the possibility of another translation of the second part: 'The destruction was made (by him)' [1290]. Bottéro and Kramer combine the lines 15 and 16, thereby ignoring the dative construction – **lú-<ù>-ku-ra** – in line 16 [1291].

16. Van Dijk, Pettinato, Jacobsen and Römer suppose, that the gods of line 15 are also the subject in line 16, and they speak of 'gods who did not dare to disturb the sleeping Enki'. This part of the complete sentence (the lines 16-18) has been placed at the beginning, probably to emphasize once more the dogged persistence of Enki in his refusal to arise.

19. Instead of **nam** I expected the emesal-form **na-áĝ** because the goddess Namma is speaking. Most scholars consider **ù-mu-un** as part of the respective verbal forms. But in my opinion it is more likely that Namma opens with a form of address: **umun** (the emesal-form of **en** [1292]), 'Lord', and repeats this.

20. The context makes clear, that in this line Enki is tackled about his behaviour. This line shows some parallel with the lines 14 and 16, and therefore one may suppose that also the room in which Enki is lying is mentioned. Instead of the readings **TE BA** or **te-zu** [1293], I suggest the reading **TE-AB** = **unu**$_6$, or **TE-UNUG** = **unu**$_2$, both with the meaning 'dwelling'.

21. **šu-dím-dím** is supposed to be a phonetic spelling for **šu-dim**$_4$**-dim**$_4$. One of the meanings of **dim**$_4$ is: 'to approach an authority with a claim, a complaint' [1294]. The view of most scholars is: **šu dím** 'formed by hand', as if Enki or Namma were the creator of the gods with his or her own hands; there is no support for this in other mythological texts [1295].

My interpretation for **gú tu**$_{11}$ 'to smite the neck' is that the gods load their necks with the **zub-sìg** 'carrier basket' (lines 30 and 37); this has led to the translation: 'to tire oneself out'.

23. A search for the lemma **kíĝ .. sì / sig**$_{10}$ resulted in two attestations:

1287 Another example of the expression **a-nir ĝál** can be found in 'A man and his god' [ETCSL 5.2.4], line 121.

1288 Van Dijk 1964, 27, line 17: 'Les dieux répandaient des larmes: 'il a fait la misère', crièrent-ils'.

1289 Van Dijk 1971, 488.

1290 Komoróczy 1976, 16 and note 39. In an accompanying text, Komoróczy says: 'The god Enki, who in the epic stands at the head of the pantheon, (...)'. But nowhere the text of 'Enki & Ninmah' mentions this.

1291 Bottéro and Kramer 1993, 189: 'Et les dieux de gémir et de protester: « C'est la cause de notre malheur, Lui qui reste couché à dormir, sans se lever jamias! »'.

1292 Schretter 1990, 263-264, no. 469.

1293 **TE BA**: Benito 1969, 23 and ETCSL 1.1.2; **te-zu**: van Dijk, Pettinato.

1294 See CAD S, 133 *s.v. sanāqu*. In the lexical part: **šu-dim**$_4$ = *sanāqu šá* **ŠU**.

1295 See also my comment at line 12.

1. 'The advice of a supervisor to a younger scribe' [1296], l. 5: **gi al-gu$_4$-ud-da-gin$_7$ ì-gu$_4$-ud-dè-en kíĝ-ĝá bí-in-sig$_{10}$-ge-en** 'Like a springing reed, I leapt up and put myself to work'.
2. 'The Debate between Bird and Fish', l. 72: **làl-ḫar kug-zu kíĝ li-bí-ib-sig$_{10}$-ga** 'No skill has been expended on your holy shaping' [1297].

This results in the basic meanings of **kíĝ .. sì / sig$_{10}$**: 'to set to work, to exercise skill'. In the context of the present text, this has led to 'to take over the work'. Van Dijk translates 'remplaçants', and some scholars followed him in this respect [1298]. Kramer and Maier argue that ' "Servants" for the hapax legomenon *kin-sì* is a reasonable surmise; so, too, is "substitute" '. Klein wrote: 'a worker comparable (?) to the gods'; Römer wrote 'Arbeitsvertreter' [1299]. The new creatures are 'substitutes' of the gods in only one aspect: they have to do the heavy job of the gods.

The verbal form **ù-mu-[e]-dím** represents no imperative (so Benito: 'Fashion ...' and Lambert: 'Create ...').

While in the text SEM 116, line 9 clearly **dú-lum** was written, in this line 23 two tablets (PBS 10/4, 14 and AO 7036) show – remarkably – the Akkadian word *tir-ḫum*. Was the word *tirḫum* such an everyday word in the daily Akkadian life that it had suppressed the Sumerian word **dulum** even for the scribes, or did it rather have a negative sound and was it used for that reason in this Sumerian text? *tirḫum* is the *pirs*-form of the Akkadian verb *tarāḫum* 'to dig' [1300], and most probably it refers in this context to a kind of wicker basket or pannier used to transport the digged clay or sand.

The modal prefix ***ḫa** is considered to have a deontic function rather than an epistemic one [1301].

24. The ***e** postposition in **^{d}en-ki-ke$_4$** has been interpreted as the deictic ***e**, which emphasizes that finally Enki is listening to his mother.

25. For the reading **ḫal-an-kù**, see George [1302]. Ḫalanku is a synonym for Enki's abode, the **abzu** [1303]; it is also described as a room for Enki's deliberations [1304].

The traces of the last sign in this line are not in contradiction with **AK**. The only known compound **TAR** + verb is **TAR AK**. Attinger [1305] mentions this expression as a

---

[1296] ETCSL 5.1.3. See Vanstiphout 1997c, 590 note 1, for the published texts.

[1297] ETCSL 5.3.5. See Vanstiphout 1997d, 581 note 4, for references to the published material.

[1298] Van Dijk 1964, 27, note 69: '«remplaçants» est deviné'. Some scholars: Benito 1969, 36, comment 52-53, 'substitute'; Pettinato 1971, 71, comment 72 (line 25) 'Ersatz'; Lambert 1992a, 131, 'substitute'.

[1299] Kramer and Maier 1989, 212, note 15; Römer 1993b, 390; Klein 1997a, 517.

[1300] AHw 1324 *tarāḫu* I 'aufgraben?'; CAD T 203 *tarāḫu* 'to dig'. AHw 1348 *terḫu(m)* 'ein Bierkrug (zum Libieren)'; CAD T 425 *tirḫu (terḫu)* '(a vessel)'.

[1301] Civil 2000b, § 3, 31-35.

[1302] George 1993, 26 *ad* 193.

[1303] CAD A II, 194: lexical part at *apsû*; Green 1975, 160 ; 205.

[1304] 'A hymn to Nisaba', l. 42 (Hallo 1970b, 125, 129; ETCSL 4.16.1).

[1305] Attinger 2005, 255, nr. 5.479. **TAR AK** is found in 'A hymn to Inana' [ETCSL 4.07.3], l. 84. Attinger's suggestion for a translation is 'forcer, briser (?)'. In 'A hymn to Nanše' [ETCSL 4.14.1], in l. 141 we find: **saĝ níĝ PA KA níĝ <u>TAR ak</u>-a níĝ šab-ta ba-a**; this line is considered as untranslatable until now. Due to the context (according to Heimpel 1981, 115, *ad* 137-154, these lines refer to immoral acts) I suggest the following transcription for line 141: **saĝ níĝ sàg zú níĝ kud ak-a níĝ šab-ta ba-a**, with as tentative translation: 'he who has beaten a person, has

hapax. For **TAR AK** I suggest here: 'to take a decision' (literally: 'to make a decision'). Some scholars transcribe **TAR** as **ḫaš**, and suggest the verb **ra** [1306], which in their opinion might represent **ḫáš ráḫ** 'to slap the thigh'. The proposed expression **TAR AK** does not need an emendation, and its supposed meaning fits perfectly in the context.

26. A more literal translation of this sentence is: 'He – intellect, understanding, investigation, [ ], wisdom – creator of whatever manifestations: he did arise one who makes both **en**'s and a crowd come alive' [1307]. The verbal form also seems to indicate that it concerned only one **se$_{12}$-en-se$_{12}$-šár** (due to the Ø-morpheme). But as becomes clear from the lines 31-32, there is a plurality of the **se$_{12}$-en-se$_{12}$-šár**. In this line 26 only the concept of a **se$_{12}$-en-se$_{12}$-šár** is indicated, therefore rendered here as singular. Lambert [1308] based his interpretation of **se$_{12}$-en-^{sa7}sár** on line 32 of the bilingual version; this line offers **šà-tùr** as equivalent for **se$_{12}$-en-se$_{12}$-šár**. **šà-tùr** means, besides 'womb' also 'mother goddess' or 'birth goddess' [1309]. As Lambert says: ' (...) the seven (*in fact: eight; JL*) names in the lines 34-35 no doubt are these very goddesses.' These goddesses have also a task before the birth (see line 32); during the birth they have to assist (ll. 34-36); then they function as midwife. Therefore a translation of **se$_{12}$-en-se$_{12}$-šár** as 'mother goddess' seems to be inadequate here; I prefer 'birth goddess'.

Lambert is convinced, that **mud** in this line has to be translated as 'blood' [1310], most likely under the influence of the text of the Akkadian *atra-ḫasīs* epos. As already discussed at line 12, **mud** in this epos in connection with Enki means: to create, to invent. Every epithet of Enki emphasizes his intellectual skill and smartness, not his manual dexterity. A very well-known epithet of Enki is **^{d}nu-dím-mud**. Moreover: from whom or from where should this blood originate? The text does not tell us.

me-dím, in a narrower sense *binātu* 'limbs', but in a broader sense also: 'form, appearance, design' [1311].

28. For **ní-te-a-na**: see the comment of van Dijk [1312].

---

bitten him, he who has shared something from the exta inspection'. This means that we are not dealing here with a combination **TAR ak**, but with **KA .. TAR**, transcribed as **zú .. kud** 'to bite'.

1306 According to van Dijk and Pettinato. Jacobsen, Klein, and Römer have the same opinion, as appears from their translations ('to slap the thigh'). Jaques 2006, 563, ex. S723 translates: 'Dans le Ḫalanku, sa chambre (?) dans laquelle il échange des confidences, il s'y frappa la cuisse'. Benito writes **TAR**, but has no suggestion for the verb.

1307 A text, although from the first millennium, VAT 17019, tells us that man and king are created separately (Mayer 1987). The expression **se$_{12}$-en-se$_{12}$-šár** shows a Semitic syntax: V(S)O; compare e.g. the word **dub-sar** 'scribe' which has the usual Sumerian syntax (S)OV. Other Semitic/Akkadian influences in this text are: 1. the use of the word *tirḫum* in line 23, and 2. the use of the word **zub-sìg**, a loanword from the Akkadian *tupšikkum*.

1308 Lambert 1992a, 132 *ad* 26.

1309 CAD Š II, 145-146: *šassūru* A 'womb; mother goddess'. There is no lexical list, known at this moment, which preserves the equation **se$_{12}$-en-se$_{12}$-šár** = **šà-tùr**.

1310 Lambert 1992a, 131, l. 26: 'Being expert in wisdom, discernment and consultation, he produced skill of blood, bodies and creativity, the birth goddesses'; see also his comments on this line on pag. 132. Lambert has held this view (personal communication in a letter, february 2010).

1311 Via a search in ETCSL with the form **me-dím**.

1312 Van Dijk 1964, 28, note 75.

For the compound verb **ĝeštu$_2$ de$_5$**, and the correction of **ĝeštu$_2$-<u>ta</u>** (in text A) to **ĝeštu$_2$-<u>ga</u>**[!]: see Karahashi [1313].

The form **ù-mu-da-ni-de$_5$-ge** may be analysed as follows: **u-m-en-da-ni-en-**/*ḫamṭu-stem*/**-e** , in which ***en-da** refers to Enki and ***ni** represents the locative (**ĝeštu$_2$-<u>ga</u>**). The ***e** after the *ḫamṭu*-stem may have been written instead of a nominalizing suffix ***a** [1314].

Because of Lambert's translation of **mud** = blood and **me-dím** = body, his translation of this line seems rather curious: 'After Enki had in wisdom reflected upon his own blood and body'.

30. **mud mu-ĝar-ra-zu**: 'to create – it has been placed by you'; 'to place' represents here, in my opinion, the conceptual phase: 'to propose, suggest' [1315].

zub-sìg, used in the lines 30 and 37, seems to be a hapax in the Sumerian literature. It is a loanword from the Akkadian *tupšikkum* [1316].

Lambert – 'My mother, there is my/the blood which you set aside, (...)' – ends the discussion, giving an account for his interpretation of the lines 30-32, as follows: 'Thus the choice is between an interpretation of these lines which agrees with other creation traditions from its world and interpretations which of necessity rely too much on the translator's whims. There is nothing in these lines which opposes taking them to offer the conception of man's creation from clay mixed with divine blood, and stronger cases for other interpretations have not yet been offered.' [1317]. It is precisely this wish to harmonize creation traditions from different worlds – Sumerian *vs* Akkadian – that has led Lambert to his translation and interpretation of these lines of 'Enki and Ninmaḫ' [1318].

31. Van Dijk supposes, that we are dealing with the construction '**šà .. ak-a**' with the meaning 'in', but then the object is lacking [1319]. When **šár** is translated as 'to stir' instead of 'to mix', it appears that no object is necessary [1320].

1313 Karahashi 2000, 86, § 6.6.4, and ex. 3.

1314 The relative past prefix ***ù** occurs only in perfective verbal forms (Jagersma 2010, 521); moreover, the verb **de$_5$.g** belongs to the reduplication class (*marû*-stem **de$_5$-de$_5$**; Sallaberger 2005a, 233). The nominalizing suffix ***a** – like other word-final /**a**/ morphemes – may have undergone a change to ***e** (Black and Zólyomi 2007, 22). Van Dijk writes: **ù-mu-e-ri-ge**; Benito and ETCSL write: **ù-mu-ni-ri / de$_5$-ge**. Jaques 2006, 12, note 24: **... ĝeštu-ta ù-mu-ni-ri-ge**, with a translation that is hardly to defend gramatically: 'Après qu'Enki, le créateur, *aura* prêté l'oreille *au coeur de lui-même*'.

1315 Wilcke (2007, 30), omitting the translation of **ama-ĝu$_{10}$**, translates the first part as follows: 'An das Blut, das du *vergossen* hast und da es noch gibt, ...'

1316 AHw, 1371; CAD T, 476-479.

1317 Lambert (1992a, 134) writes this discussion after comparing four translations of the lines 30-32 (Benito; Kramer/Maier; van Dijk; Jacobsen), and his own interpretation; he repeats the idea of creation a 'new race by using blood' in a later publication (Lambert 2000, 1833).

1318 These different traditions were discussed in more detail in the chapters 4.3, 4.4.3, 4.5.2 and 4.7.2 *sub* Anthropogeny.

1319 Van Dijk 1964, 30, note 77 (line 32 at van Dijk is line 30 in the present edition): 'La sentence est à corriger (...): **šà ... ak.a** signifie: «dans»; alors, il manque l'objet. Tel quel, il est impossible de chercher l'objet dans la ligne qui précède. Si on veut chercher le complément direct dans **mud *mu.e.gar.ra.zu**, il faut éliminer **zub.sìg kešda.ì**. La solution la plus probable me semble être: à partir de la ligne 32 Enki donne ses ordres à Nammu. Le style de son discours a été emprunté aux rituels des incantations. L'expression **šà ... ak.a ù.mu.e.ni.ḫé** est fréquente dans

The form **ù-mu-e-ni-in-šár** may be explained as follows: **u-m-er-a-ni-en-**/*ḫamṭu-stem*/-∅, in which ***er-a** is 'for you' and ***ni** represents the locative [1321]. There is a discrepancy between the singular person-prefix ***en** in the verbal form of this line, and the plural suffix-form ***ene** in line 32, referring to **se$_{12}$-en-se$_{12}$-šár**. I suppose that also in line 31 the plural form of the subject was meant.

Frymer-Kensky (1987, 129), perhaps by mistake, mentions Ninsun as the mother of Enki.

32. An ergative element is lacking after **se$_{12}$-en-se$_{12}$-šár**. In view of their translation, some scholars [1322] seem to neglect the fact that Enki already had made appear the **se$_{12}$-en-se$_{12}$-šár** (l. 26).

The form **ù-mu-e-ni-ĝál** shows metathesis: the subject ***er** and the element ***ni** have changed places [1323]. The element ***ni** can be interpreted as a marker of causativity for the verb **ĝál** [1324].

34-35. Most scholars read in the second part of this line two times **^{d}nin-bara$_2$**; Lambert two times **^{d}nin-šara$_2$**; Jacobsen **^{d}nin-šara$_2$** and **^{d}nin-bara$_2$**, respectively. The copy of the text in UMBS I^1, 4 (pl. 6) shows as the first sign **šara$_2$**, and as the second sign **bara$_2$** [1325]. The copy of the same tablet in UMBS X^4 (pl. CIII) shows **bara$_2$** twice. For a discussion about the meaning of these goddesses, see Lambert [1326]. His reading for **^{d}ŠAR-ŠAR-GABA** is **^{d}mú-mú-du$_8$**, which is also used in this edition. His suggestion, that this is a variant spelling of the name of the dream deity Mamu is less likely. I propose: 'who opens what is growing', as a translation for the name of this goddess; then it will be clear why she is mentioned in this series: at the end of the gestation, she 'opens' for the baby the way from the womb to the outside. The names in the lines 34-35 are those of the birth goddesses; some of them may be responsible for a specific part of the birth process, e.g. Mumudu. Ninimma, 'Lady of the physiognomy', might be responsible for the features of the newborn.

36. The translation of Pettinato, 'Ninimma (...), die du geboren hast, mögen (...)', implies, that Namma is the mother of the goddesses who may help her. The text in question does not mention that Namma is the mother of these birth goddesses.

---

ces rituels. Le scribe de notre texte a simplement utilisé cette expression sans plus la comprendre. J'élimine **šà** dans la traduction.'

1320 Green 1975, 170-171: 'After you mix the 'heart' of the clay above the abzu, the sigensigdu will nip off pieces of clay.' 'You' as subject (= mother Namma) of the first part of this sentence does not fit with the verbal form. Most scholars have 'you' (referring to Namma) as subject of the first part of this sentence.

1321 Zólyomi 2000, 342.

1322 Pettinato, Hruška and Bottéro/Kramer translate, as if the **se$_{12}$-en-se$_{12}$-šár** had to be formed from the clay.

1323 Krispijn 2004, 80 ad nr. 6. A search via ETCSL showed many examples of this kind of metathesis.

1324 Zólyomi 2000, § 5, 347-353.

1325 See e.g. Mittermayer 2006: **šara$_2$** (nr. 409, p. 162), **bara$_2$** (nr. 217, p. 86).

1326 Lambert 1992a, 134-135.

The form **ḫa-ra-gub-bu-ne** requires some attention [1327]. My translation for **gub** is 'to assist'. Thomsen mentions [1328], that the verb **gub** + comitative ***da** is 'to stand by, to serve'. The comitative is absent is this line, but in the prefix chain /**er** + **a**/ (you + dative) can be analysed. Furthermore, the subject of **gub** is plural, viz. the goddesses mentioned in the lines 34-35. Therefore the verb should be **su₈.g** (= **súg**), the plural form of **gub**. Finally, the pronominal suffix /**ene**/ suggests, that the verb **gub** may be transitive here [1329], but an object is not present.

38-43. Because of the incompleteness of these lines, it is only possible to guess at the content. It seems that the normal procreation of man, the birth of a baby and the rituals belonging to it are described [1330].

40. The suggestion for the completion of the expression **gù + verb** with the verb **sù/sud** instead of **dè**, was inspired by some examples in the literature [1331]: **gù .. sù/sud** seems to be used exclusively in connection with newborn or young children, human or animal ones.

41. The first legible sign in this line is **PAP.PAP**; two transcriptions may be possible in this case: 1. **buluĝ₃** 'to grow', and 2. **dim₄** 'to check, to approach'. Due to the context, **buluĝ₃** 'to grow' seems less likely. Therefore I transcribed **dim₄**, translated as '... approached'. In case the compound verb **šu .. dim₄** was used, the translation should be '... was prudent'.

This line clearly twice reads **su** instead of **zu**, as the other scholars write.

I have written /**unu**/, because it is not possible to decide whether **TE-AB** or **TE-UNUG** has been copied. **te** = *ipu, šišītu* 'membrane, afterbirth' [1332]. In the lexical part at

---

1327 Also van Dijk (1964, 30 note 78) discusses this verbal form. He says: 'Si on traduit **gub** + dat. «servir», il faut lire **-gub-bu-uš**.'

1328 Thomsen 1984, 304.

1329 In the literature there are more examples (see below) with an, in my opinion, 'incorrect' stem, where **gub** has been used instead of the 'expected' **su₈.g**. See also Krecher 1967-1968, 11. Some examples: *1.* 'Gilgameš and Ḫuwawa' (version A, ETCSL 1.8.1.5; l. 83): **úr.ḫur.saĝ.ĝá.ka nam.ba.e.dè.gub.bu.ne** 'they should not have to wait at the foot of the hills'. *2.* 'Gilgameš and Ḫuwawa' (version B, ETCSL 1.8.1.5.1; l. 70): **gu.ru.ma mu.ni.ib.gub.bu.ne** '(they) stacked them in piles'. *3.* 'A prayer to Nanna for Rīm-Sîn' (Rīm-Sîn D, ETCSL 2.6.9.4; l. 33): **ki kug ki nam.til₃.la ĝiri₃.zu ḫé.ri.ib.gub.bu.ne** 'may they cause you to place your feet in holy places, places of life'. *4.* 'Sîn-idinnam and Iškur' (Sîn-idinnam E, ETCSL 2.6.6.5 l. 18): **mu.un.na.gub.bé.eš** 'they set ... up for him'. *5.* 'A hymn to Ḫendursaĝa' (Ḫendursaĝa A, ETCSL 4.06.1 l. 76): **bí.in.gub.bu.ne** 'they establish ...[him]'. *6.* 'A *šir-namšub* to Inana' (Inana G; ETCSL 4.07.7 l. 63): **ki.bi te.àm ki àm.gub.bu.ne** 'approach their place, the place where they are stationing'. *7.* 'The farmer's instructions' (ETCSL 5.6.3 l. 106): **ḫa.ra.gub.bu.uš** 'they should work at moving .. around'.

1330 In 'Enki and the World Order' (Benito 1969, ll. 394-401; ETCSL 1.1.3, ll. 395-402) the attributes of Aruru, 'the midwife of the Land', are described; leek and a vessel for the afterbirth belong to those attributes.

1331 'A love song of Išme-Dagan', l. 1 [ETCSL 2.5.4.10]; 'A *balbale* to Inana', l. 1 [ETCSL 4.08.a]; 'A *šir-gida* to Ninisina', l. 77 [ETCSL 4.22.1].

1332 *ipu*: AHw, 385; CAD I, 173; *šišītu*: AHw, 1250 (*šišītu* I); CAD Š III, 125.

šišītu we read also **uzu-TE+UNU-ak/gu** = *šišītu*. I suppose that in the present line **su** has been used instead of **uzu**, and that the complex **su-/unu/-RI^{su}** means: the afterbirth [1333].

A hymn [1334] mentions Ninisina, who puts the afterbirth in a jar (**dug-sila$_3$-ĝar-ra**); in 'Enki and the world order' (l. 399) Aruru/Nintur receives this jar [1335]. The **dug** in the present line is supposed to be a shortened spelling for this jar.

42. The completion with **ga-raš** is suggested by 'Enki and the world order', where Aruru/Nintur, as mother goddess and as midwife, receives **ga-raš** 'leek' [1336]. I suppose that one of the goddesses mentioned in the preceding lines is the subject in the lines 42 and 43. The ***e** after **[ga-raš$^{?}$]^{sar}-e** suggests a causative construction [1337]. The literal translation of the first part of this line should be: 'she made the leek raise the light'. Presumably the leek was used for clearing the eyes of the newborn, which are covered up by *vernix caseosa*, so that the baby can open his eyes; perhaps the leek was (also) used because it makes the eyes water. The application of the leek seems to be (part of) a purification ritual. The use of **ĝiš-nu$_{11}$** 'light' with respect to 'to see' can also be found in 'Enki and Ninmaḫ':

62. **gi$_4$-bi ĝiš-nu$_{11}$ gi$_4$-gi$_4$ lú u$_6$-e àm-ma-ni-in-dím**
63. **^{d}en-ki-ke$_4$ ĝiš-nu$_{11}$ gi$_4$-gi$_4$ lú u$_6$-e igi du$_8$-a-ni-ta**

62. Second, she created a man who, astonished, was gazing because the light was returning continuously;
63. Enki, after he has looked at the man who, astonished, was gazing because the light was returning continuously,
[64 ff.: decreed his fate ... etc.] [1338].

Probably the creation of a blind man has been described here, as was also assumed as a possibility by Falkenstein [1339]. Another example of the combination **ĝiš-nu$_{11}$ ... íl** is found in 'Enlil in the E-kur', line 4, in which **ĝiš-nu$_{11}$** also refers to the light of the eyes [1340].

1333 **RI** may be a scribal error for **ak** or **gu**. In the literature there are two examples where **unu$_2$-RI(-bàn-da)** has been written: 'The lament for Sumer and Urim', l. 320 [ETCSL 2.2.3]; 'The temple hymns', l. 305 [ETCSL 4.80.1]. Michalowski (1989, 96 *ad* 320) remarks, that it was 'not possible to define what an **unu$_2$-RI-bàn-da** was'.

1334 Stol 2000, 145 and note 212: text SRT 6 III 2 (dupl. 7:12) = 'A *šir-gida* to Ninisina' [ETCSL 4.22.1], l. 75. Römer (1969b, 295, comments 296) transcribes ***dug-sila$_3$-ĝar-ra-ke$_4$** as **baḫar$_2$-gé** (?) 'ein Töpfer (?)'; in ETCSL 4.22.1, l. 75, we read **baḫar$_2^{!}$-ra-ke$_4$** 'potter'.

1335 In this line it only reads **sìla-ĝar-ra**, without **dug**.

1336 ETCSL 1.1.3, ll. 395-402. To my knowledge there is no description or indication of the function of the leek at birth rituals. Stol (2000, 112) writes: 'The use of leeks (**garaš**) remains a mystery.'

1337 Attinger 1993, 197-199, ch. 3.2.3.12 (§129).

1338 Benito (1969, 39) translated line 62: 'The second one that she created was one who could see (even) with blinking (?) eyes'. In ETCSL 1.1.2 we read the translation of this line: 'Second, she fashioned one who turned back (?) the light, a man with constantly opened eyes (?)'.

1339 Falkenstein 1948, 165, note 7.

1340 Falkenstein1959a, 11; ETCSL 4.05.1. The lines 3 and 4 are: 3. **igi-íl-la-ni kur-re sá-sá** 4. **ĝiš-nu$_{11}$-íl-la-ni kur-šà-ga igi-ĝál** '3. his glance, seizing the mountain, 4. his look, seeing into the interior of the mountain.' [transcription and translation are of the present author]. For **DI-DI** = **sá-sá** = *kašādu*: see CAD K, 274 *ad kašādu*, lexical part.

43. The compound verb **na de$_5$.g** has also been used in 'Enki and the world order' when Enki presents Aruru/Nintur with birth attributes [1341]; in line 400 we read: **a-lá-kù-na-de$_5$-ga-ni šu ḫé-em-ma-da-an-ri** 'She carried off her holy consecrated water vessel'. The meaning of **na de$_5$.g** has been elucidated by Sallaberger [1342]. He states: 'Sometimes, Enki is named as the agent of this consecration, (..). The act of 'consecration' is achieved through the words of Enki' (p. 239). And further (p. 241): '(..) **na de$_5$** does not refer to a purification through washing or cleaning (cf. **luḫ**), but **na de$_5$** seems to indicate that the purity is achieved by putting an object in order so that its clarified essence emerges.'

Since in not one example with the compound verb **na de$_5$** the nominal part **na** has been written with a pronomen possessivum, I suppose the ***bi** in the present line to be a scribal error («**bi**»).

44. **lá** = (inter alia) *kullumu* 'to show' [1343]. Two examples with the expression **kíĝ lá** have been found in the literature. One is in line 132 in 'Enki & Ninmaḫ': **^{d}nin-maḫ kíĝ-ĝá-zu ḫé-bí-lá-lá** 'Ninmaḫ, may your work be shown!' [1344]. The other example is in 'The Debate between Hoe and Plough' (l. 53): **^{ĝiš}al im-ma kíĝ-ĝá lá-a** 'Hoe, burrowing in the mud' [1345]. The meaning 'to show' seems to fit best in this line of 'Enki & Ninmah'.

šà-bi: the ***bi** represents, in all likelihood, the gods, who suffered the toil until now.

***

1341 'Enki and the world order' [ETCSL 1.1.3], line 400.

1342 Sallaberger 2005a; especially pp. 239-241. A few examples may be added to those given by Sallaberger: 'Enki and the world order' [ETCSL 1.1.3] , l. 400; 'Inana and Šu-kale-tuda' [ETCSL 1.3.3], l. 82; 'The building of Ninĝirsu's temple' [ETCSL 2.1.7], ll. 784, 887.

1343 *kullumu*: AHw, 503-504; CAD K, 519-525. See also ePSD at **lá**.

1344 Benito 1969, 43: 'Ninmaḫ (?)!, may your work be restrained'; ETCSL 1.1.2: 'Ninmaḫ, may your work be ... '.

1345 ETCSL 5.3.1, l. 53.

8.1 Appendix 1: Gods who (have to) work

In this appendix attention will be paid to those parts of one Akkadian and some Sumerian compositions, which show that the gods have to work, and, in some cases, what solution has been invented for this problem. The relevant lines from 'Ninurta's exploits' ('Lugale'), 'a hymn to Nibru and Išme-Dagan', and '*atra-ḫasīs*' will be discussed [1346].

a. Ninurta's exploits [1347]

334	**u$_4$-bi-a a-silim ki-ta-du a-gàr-ra nu-um-dé**	On that day, no healthy water, welling up from the earth, did flow out over the arable land.
335	**ḫalba$_6$ du$_8$-du$_8$-du$_8$-ù u$_4$-zal-le-da-gin$_7$ kur-ra é-ri-a ba-ni-íb-íl-a**	When the ceaselessly accumulating ice, which looks like the dawn, had raised the wasteland in the mountains,
336	**diĝir-kalam-ma ba-súg-ge-eš-a**	when the gods of the Land had been assigned to a task,
337	**^{ĝiš}al ^{ĝiš}dusu-bi mu-un-lá-eš-a**	when they had taken charge of hoe and basket
338	**ur$_5$-ra-àm éš-gàr-bi ì-me-a**	– this indeed was their orderly duty –,
339	**kalam tar-bi-šè é-lú gù ba-an-dé**	there was talk of a community of men for the ploughing of the Land.
340	**^{íd}idigna nam-gu-la-ba an-šè u$_5$-bi nu-íl-e**	The Tigris, in its greatness, did not rise to its full height;
341	**kun-bi a-ab-ba-e li-[bí-lá-a] a-dug$_3$ nu-um-[ ]**	its mouth did not [reach] the sea, nor did it [carry$^?$] sweet water.
342	**nesaĝ$^?$ kar-re lú nu-gíd$^?$-e**	No one did transfer first-fruit offerings to the market-place.
343	**šà-ĝar ḫul-a níĝ nu-tu-ud**	The famine was hard, nothing was brought forth.
344	**íd-tur-tur-re šu-luḫ lú li-bí-in-ak saḫar nu-mu-da-an-zi-zi-i**	Nobody cleaned the little canals, the mud was not dredged up.

1346 The text KAR 4, in which the gods also have to work, was treated separately in this study (ch. 2.1.10 and Appendix Text editions no. 10).

1347 The Sumerian text has been copied – with some adaptations – from ETCSL 1.6.2: 'Ninurta's exploits'; lines 334-346. The text may originate from the Ur III-period, shortly after the time of Gudea (van Dijk 1983, 2-3). Manuscript H of Lugal-e [ISET 2, 23; Ni 4138] most likely dates from the Ur III period (Jagersma, personal communication).

345	**gana$_2$-zid-da a nu-šub-bé ég ak nu-ĝál-la**	Water was not let down on fertile fields, the making of ditches did not exist.
346	**kur-kur-re ab-sín-na nu-gub-bu še bir-a i-im-ak**	For the lands there is nothing standing in furrows: grain was sown scattered.

Editions / Translations

1. G. Pettinato, *Das altorientalische Menschenbild und die sumerischen und akkadischen Schöpfungsmythen.* Heidelberg: Carl Winter, Universitätsverlag, 1971; 91-96; [the present lines 334-346 are ll. 5-17 at Pettinato].
2. G. Komoróczy, Work and Strike of Gods. New Light on the Divine Society in the Sumero-Akkadian Mythology. *Oikumene* 1 (1976) 9-37. [this text: p. 13.]
3. J. van Dijk, *LUGAL UD ME-LÁM-bi NIR-ĜÁL. Le Récit Épique et Didactique des Travaux de Ninurta, de Déluge et de la Nouvelle Création.* Tome I. Introduction, Texte Composite, Traduction. Tome II. Introduction à la reconstruction du texte, inventaire des textes, partition, copies des originaux. Leiden: E.J. Brill, 1983; 31-33; 93-94.
4. W. Heimpel, The Natural History of the Tigris according to the Sumerian Literary Composition Lugal. *JNES* 46 (1987) 309-317.
5. Th. Jacobsen, The Birth of Man. In: *The Harps that Once... Sumerian Poetry in Translation.* New Haven, London: Yale University Press, 1987; 251-252.
6. J. Bottéro, S.N. Kramer, Ninurta le Preux. In: J. Bottéro, S.N. Kramer, *Lorsque les Dieux Faisaient l'Homme. Mythologie Mésopotamienne.* [Paris]: Gallimard, 1993; 352-353.
7. W.H.Ph. Römer, Aus »Lugal ud me-lám-bi nir-ĝal«. In: W.H.Ph. Römer, D.O. Edzard (eds), *Weisheitstexte, Mythen und Epen. Mythen und Epen I, TUAT III.* Gütersloh: Mohn, 1993; 434-448; for the lines 334-346: pp. 445-446.
8. ETCSL 1.6.2: Ninurta' exploits: a *šir-sud* (?) to Ninurta.

***

Comments

334-335. Van Dijk [1348] considers lines 334-335 as a reference to the deluge, 'non au déluge primordial connu par Enūma eliš et d'autres textes, généralement très anciens, mais au déluge qui a précédé la «seconde création», connu par les récits de Gilgameš, de la Bible, d'Atrahasīs, etc. ', and also: 'L'interprétation des lignes 334-337 (la traduction de **zal** par «fondre» est évidente) me semble avoir quelque importance pour la question de l'historicité du déluge.'

334. Van Dijk translates: 'Ce jour-là, l'eau salubre ne sortant plus de la terre, ne montait pas sur les champs', two times a negation, the first time even 'no more'. This is not in accordance with the Sumerian text.

[1348] Van Dijk 1983, 31; 33.

335. This line and the expression **u$_4$-zal-le-da-gin$_7$** have given rise to a great variety of translations [1349].

1. One of the transcriptions for **A-MUŠ$_3$-DI** is **ḫalba$_6$** 'ice'.
2. The three times repeated **du$_8$** [1350] refers, in my opinion, to ice that is continuously heaping up. For the composer of this text, who was living in Mesopotamia, the mountains with a possible ice-cap were situated to the (north)-east. This is also the place where the sun rises. The *simile* (because of the ***gin$_7$** in **u$_4$-zal-le-da-gin$_7$**) presumably does not refer to *melting* ice. The meaning of this sentence may be, that the ice-cap on the mountains glitters and reflects light, just as if the sun is rising. 'Dawn' is one of the meanings of **u$_4$-zal-le-da** [1351].
3. 'The ice has raised the wasteland in the mountains' means, that as a consequence of the presence of ice in the mountains no growth of plants and trees is possible; the result is 'wasteland'. The inhospitable area was increasing more and more in the mountain area.

339. With respect to the transcription and translation of **UN TAR**, this line has led to very different translations [1352]. The context of this episode shows that people were not present. Therefore the reading **UN** = **ùĝ** ('people') is not suitable [1353].

The present transcription for **UN TAR**, viz. **kalam tar**, seems to be a hapax [1354]. When we take the meaning of **tar** literally, viz. 'to cut', then 'to cut the Land' probably means: 'to plough'. In the context of this text it refers to the job of the gods in the Land,

---

1349 Pettinato: 'eiskaltes Wasser, das alles anfüllte, wie der aufleuchtende Tag, brachte aus dem Bergland Verwüstung.' Komoróczy: 'the cold water, flooding everything, just like the early morning, from the hills brought destruction.' Van Dijk: 'puisque, la glace entassée partout, le jour où elle commença à fondre, portait la destruction dans la Montagne.' Heimpel: 'They (*the waters* [*l. 334*]; *JL*) as well as (those of) the melting snowpack were 'carried' in the Kur in the waste land.' Jacobsen: '(Nay!) As ice long accumulating they (*the waters* [*l. 334*]; *JL*) rose in the mountains on the far side.' Bottéro/Kramer: 'Mais, transformée en glace accumulée, elle ravinait, en fondant, les montagnes.' Römer: 'Es (*das Wasser* [*l. 334*]; *JL*) wurde (ebenso) wie (das) des angehäuften Eises, das schmelzen geht, im Bergland auf das Ödland getragen.' ETCSL: 'The cold water (?) was piled up everywhere, and the day when it began to .... it brought destruction in the mountains.'

1350 **du$_8$-du$_8$-du$_8$-ù**: participle of the reduplicated *marû*-stem. ePSD: **du$_8$** 'to heap up, pile up'.

1351 Some examples in the literature, where **u$_4$ zal** has the meaning 'dawning of the day': 'Inana and Šu.kale.tuda' [ETCSL 1.3.3], l. 239 ; 'Gilgameš, Enkidu and the Netherworld' [ETCSL 1.8.1.4], l. 47 ; 'Lugalbanda and the Anzud bird' [ETCSL 1.8.2.2], ll. 44-45 ; 'Enmerkar and the lord of Aratta' [ETCSL 1.8.2.3], l. 308 ; 'The building of Ninĝirsu's temple' (Gudea, cylinders A and B) [ETCSL 2.1.7], l. 928.

1352 Pettinato: 'Das Volk in seinem ....; der Mensch wurde gerufen.' Van Dijk: '(alors), on avait le droit de faire appel à une maison pour le recrutement de travailleurs (?).' Jacobsen: '(the gods) poured on a man's field according to what they would have chopped off (of the ice).' Bottéro/Kramer: 'Car, afin d'assurer la production, on n'avait pas d'autres ouvriers à engager (?).' Römer: '(die Götter) *gossen* es in Übereinstimmung mit *dem, was sie (davon) abgetrennt* haben würden, *auf die Parzellen*[!] *der Menschen aus*.' ETCSL: 'People called on a household for the recruitment of workers.'

1353 The expression **UN/ùĝ tar** has been attested only a few times in the Sumerian literature: 'A praise poem of Šulgi; Šulgi D' [ETCSL 2.4.2.04], l. 187 'crushed people'; 'An **adab**[?] to Nergal for Šulgi[?]; Šulgi U' [ETCSL 2.4.2.21], ll. 21 and 23: no translation; 'A hymn to Ḫendursaĝa' [ETCSL 4.06.1], l. A51: no translation; idem, l. B13: 'crushing (?) of the people'. Attinger and Krebernik (2005, p. 65 *ad* 49) discussed **UN TAR** briefly, but rendered no translation.

1354 Also a search for **kalam kud/ḫaš** shows no results in the literature.

mentioned in the lines 336-338. The translation 'ploughing of the Land' has to be seen in a broad and general sense: besides the making of furrows also the digging of canals.

The very absence of man forces the gods to work. The work to be done by the gods is a point of discussion among them; in this line a solution for their problem has been suggested in a nutshell: **é-lú** 'a community of men'. In this composition, the creation itself of mankind is not described; it is only an allusion to such a creation.

346. The first part of this line also has led to varying translations [1355]. **kur-kur-re** does not mean people or Sumer; it points to all lands in general. Most translations neglect the locative in **ab-sín-na**.

***

[1355] Pettinato: 'In den Ländern wurden die Saatfurchen nicht gezogen'. Van Dijk: 'Comme les pays ne creusaient pas de sillons (avec la charrue)'. Jacobsen: 'In (all) the country no seed-furrow was drawn'. Bottéro/Kramer: 'On ne creusait pas de sillons'. Römer: 'Das Land Sumer *tut nicht Dienst* an den Furchen'. ETCSL: 'People did not work (?) in furrows'.

b. A hymn to Nibru and Išme-Dagan [1356]

^{d}en-líl	Enlil
^{d}nin$_x$ (EGI$_2$)-líl-bi	and Ninlil
an-na	
igi im-mi-in-ĝál-eš	fastened their eyes upon the heaven (l. 99).
ki-a	On earth
zag mi-ni-in-ĝar-re-eš	they had established a sanctuary.
u$_4$-bi-a	On that day,
an-gal-	when in the great heaven
ki-daĝal-ba	and on the broad earth
šà-bi mu-un-è-a	their intention has come to light,
diĝir ^{d}a-nun-na-	the Anunna gods
an-ki-ke$_4$-ne	of heaven and earth
kíĝ íb-gi$_4$-gi$_4$-ne	gave orders.
^{ĝiš}al	Hoe
^{ĝiš}dusu	(and) basket,
níĝ iri ki ĝá-ĝá	things for founding cities,
šu-ba «na»	
mu-un-ĝál	were in their hands (l. 113).

Editions / Translations

1. K. Oberhuber, Eine Hymne an Nippur (UET VI 118). *ArOr* 35 (1967) 262-270.
2. G. Komoróczy, Work and Strike of Gods. New Light on the Divine Society in the Sumero-Akkadian Mythology. *Oikumene* 1 (1976) 9-37. [this hymn: p. 13.]
3. M.-C. Ludwig, *Untersuchungen zu den Hymnen des Išme-Dagan von Isin.* Wiesbaden: Otto Harrassowitz, 1990; 100, 102, 148-149.
4. ETCSL 2.5.4.23: A Hymn to Nibru and Išme-Dagan, Segment A, ll. 97-114.

***

Comments

102. **zag** = *aširtu* 'sanctuary' [1357].

109. **kíĝ gi$_4$** = *šapāru* [1358], (inter alia) 'to give orders'.

110-114. There is a striking difference with other myths that tell about the gods who work. Indeed, the tools are the same (hoe and basket), but the aim is here the foundation of cities instead of digging out canals and rivers. Alas, after the line 114 the text is broken; the first legible text after these lines does not link up with them.

113. Ludwig is in uncertainty as to the Sumerian text in this line [1359]. This line will be understandable, if we accept the **na** as a scribal error; **ba** is the collective pronomen possessivum for the gods.

***

[1356] Copy of the cuneiform text: UET 6, 118: rev. iv 13-30.
[1357] *aširtu* I: AHw, 80; *aširtu* A: CAD A II, 436-439.
[1358] *šapāru*: AHw, 1170-1171; CAD Š I, 430-448.
[1359] Ludwig: **šu$^{?}$.b[a$^{?}$.n]a$^{?}$**; ETCSL: **šu$^{?}$ ⌜BA$^{?}$ NA$^{?}$⌝**.

c. *atra-ḫasīs*

From the Akkadian composition *atra-ḫasīs*, two passages are discussed, that refer to the duty of the gods and their solution for it [1360].

I:1	*i-nu-ma i-lu a-wi-lum*	When gods were 'man',
I:2	*ub-lu du-ul-la iz-bi-lu šu-up-ši-[i]k-ka*	they bore the corvée work, they carried the basket.
I:3	*šu-up-ši-ik i-li ra-bi-[m]a*	The basket of the gods was large, and
I:4	*du-ul-lu-um ka-bi-it ma-a-ad ša-ap-ša-qum*	the corvée work was heavy, extensive the hardship.
I:5	*ra-bu-tum* ^{d}*a-nun-na-ku si-bi-it-tam*	The great Anunnaku, their seven,
I:6	*du-ul-lam ú-ša-az-ba-lu* ^{d}*i-*⌜*gi-gi*⌝	let the Igigi bear the corvée work.

I:190	*[š]à-as-sú-ru* **lú-u$_{18}$-lu**^{a} *li-ib-ni-ma*	May the mother goddess create mankind, and
I:191	*šu-up-ši-ik ilim a-wi-lum li-iš-ši*	may man carry the basket of the gods.

Editions / Translations

Only a few publications are mentioned; in these articles there are references to previous translations and discussions.

1. G. Komoróczy, Work and Strike of Gods. New light on the Divine Society in the Sumero-Akkadian Mythology. *Oikumene* 1 (1976) 9-37. [Discussion about Atra-ḫasīs: pp. 18-19.]
2. C. Wilcke, Die Anfänge der akkadischen Epen. *ZA* 67 (1977) 153-216. [Discussion about Atra-ḫasīs: pp. 160-163.]
3. Th. Jacobsen, Inuma Ilu awīlum. In: M. De Jong Ellis (ed.), *Essays on the Ancient Near East in Memory of Jacob Joel Finkelstein.* Memoirs of the Connecticut Academy of Arts & Sciences, vol. 19. Hamden, Connecticut: Archon Books, 1977; 113-117.
4. W. von Soden, Die erste Tafel des altbabylonischen Atramḫasīs-Mythus. 'Haupttext' und Parallelversionen. *ZA* 68 (1978) 50-94.
5. J. Bottéro, S.N. Kramer, La Grande Genèse Babylonienne: de la Création de l'Homme au Déluge. In: J. Bottéro, S.N. Kramer, *Lorsque les Dieux Faisaient l'Homme. Mythologie Mésopotamienne.* [Paris]: Gallimard, 1993; 530, 537.
6. W.G. Lambert, A.R. Millard, *Atra-ḫasīs. The Babylonian Story of the Flood.* Winona Lake: Eisenbrauns, 1999; 42-43, 56-57, 146-147.
7. B. Alster, *ilū awīlum : we-e i-la*, 'Gods : Men' versus 'Men : God'. Punning and the Reversal of Patterns in the Atrahasis Epic. In: T. Abusch (ed.), *Riches Hidden in Secret Places. Ancient Near Eastern Studies in Memory of Thorkild Jacobsen.* Winona Lake, Indiana: Eisenbrauns, 2002; 35-40.
8. S. Loesov, **I-NU-MA-I-LU-A-WI-LUM**. *Babel und Bibel: Ancient Near Eastern, Old Testament and Semitic Studies* 1 (2004) 183-201.

[1360] The Akkadian text of these lines has been copied from Lambert and Millard 1999; 42, 56, with amendment of line I:190 (see Comments at this line).

9. B.R. Foster, *Before the Muses. An Anthology of Akkakdian Literature.* Third edition. Bethesda, Maryland: CDL Press, 2005; 229, 235.

***

Comments

I:1. The first line of this composition has been amply discussed [1361]. The present translation corresponds with the grammatical view of Kouwenberg [1362]: it is a nominal sentence with the nominative of a noun as a predicate. With respect to the interpretation, the view of Metzler [1363] seems to come as close as possible to the Akkadian text: *awīlum* has not been used metaphorically – as Moran and Streck suppose [1364] –, but metonymically. This interpretation has been rendered in the present edition with the aid of quotation marks.

I:2-6. The lines I:19-26, and the Assyrian recension of these lines (K 8562 i:5-9) [1365], tell us about the work the gods had to do before the presence of man: to dig out canals and rivers.

I:5. The present translation of *sibittam* corresponds to the translation and view of Lambert and Millard [1366]. Their conclusion is, that this 'adverbial accusative' [1367], normally used for the limiting of verbs, here limits a noun, viz. the Anunnaku [1368]. Any connection of

---

1361 A short survey of some of the recent translations: Komoróczy: 'When the gods, (like) men,'. Wilcke: 'Als Götter Mensch waren,'. Jacobsen: 'When Ilu was the boss'. Von Soden: 'Als die Götter (auch noch) Mensch waren,'. Bottéro/Kramer: 'Lorsque les dieux (faisaient) l'homme,'. Lambert/Millard: 'When the gods like men'. Foster; Alster: 'When gods were man,'. Westenholz (1996, 188-189) commented on the first line of *atra-ḫasīs*: 'A clear example of the cognitive problems caused by a metaphor is one of the cruxes of Akkadian narrative literature.' In a recent comment, Loesov says: '(...), a nominal clause *i-nu-ma i-lu a-wi-lum* would seem to be a *hapax legomenon* in terms of grammar, meaning, and narrative technique. (...): a temporal clause **inūma ilū awīlum* is impossible at the beginning of a narrative text and can hardly be an Akkadian temporal clause at all.' (Loesov 2004, 195-196). With respect to *awīlum*, Loesov refers to a locative theory, but not to its 'comparative' version (Loesov 2004, 196-197). His solution is: 'When gods – *in place of Man* – '. In other words: Loesov exchanges one *hapax legomenon* (line 1 as a nominal sentence, in his view) for another one, viz. *awīlum* represents a locative with meaning 'in place of'.

1362 Kouwenberg 2000, 41.

1363 Metzler 2002, 308, note 32.

1364 Moran 1987, 247; Streck 1999, 97 *ad* (112).

1365 Lambert and Millard 1999, 42-43.

1366 Lambert and Millard 1999, 146-147.

1367 For the 'adverbial accusative', see: Ungnad 1926, 25, § 19; Huehnergard 1997, 172-173, § 18.3; Lambert and Millard 1999, 146-147.

1368 Wilcke (1999, 75, note 21) and Metzler (2002, 509-510 and note 715) give a survey of the discussion with respect to *sibittam*. Some translations are: Komoróczy: 'the great Anunnakū forced the seven Igigū to make the work.' Wilcke (1977): 'wobei die großen Anunna siebenfach die Igigū die Fronarbeit verrichten ließen.' Wilcke (1999): 'Die großen Anunakkū-Götter ließen nämlich siebenfach Arbeit die Igigū-Götter tragen.' Von Soden: 'Die großen Anunnaku wollten die (nur) sieben Igigu die Mühsal tragen lassen.' Bottéro/Kramer: 'Car les grands Anunnaku, aux Igigu, imposaient une corvée septuple!' Lambert/Millard: 'The Seven great Anunnaki were

sibittam with the Igigi seems very unlikely, in view of the distance between these two words in the text. For a 'sevenfold corvée' for the Igigi, no evidence can be found in the text.

I:190. With respect to the reading *li-gim*$^{?}$*-ma*$^{?}$: Lambert and Millard [1369] have some doubts, but they argue that the suggestion that some parallel texts offer, viz. *lullâ*, is impossible because of the clear initial *li-*.

George and Al-Rawi have published a duplicate of this text, which reads: *sa-*⸢*as*⸣*-su-ru lullâ* (**lú-u$_{18}$-lu**)^{a} *li-ib-ni* [1370]. In their comments on this line they write [1371]: 'The SB text's *lullâ* replaces the difficult word in MS E iii:3' (Lambert: *li-gim*$^{?}$*-ma*$^{?}$*-a* [Lambert & Millard 1999, 56; JL]; von Soden: *li-id-di-a* [von Soden 1978, 62; JL]); there only *li* and *a* are certain (coll.).' The present translation makes use of this duplicate text.

***

making the Igigi suffer the work.' Foster: 'The great Anunna-gods, the seven, were burdening the Igigi-gods with forced labor.'

1369 Lambert and Millard 1999, 152 *ad* 190.

1370 George and Al-Rawi 1996, 168, line 69.

1371 George and Al-Rawi 1996, 187 *ad* 69.

8.2 Appendix 2: Another 'creation' of man ?

The Sumerian Flood Story

Signature	*Publication*	*Provenience*
CBS 10673 [1372]	Poebel PBS V/1, pls. LXXXVI-LXXXVII Copy of Poebel's hand copy in Kramer 1981, 150.	Nippur

Obverse
column i [*Approx. 36 lines missing*]

1'	**[...] im-ĝá-⸢ĝá⸣ [...]**	' [...] he/she/it places [...]
2'	**nam-lú-ulu$_3$-ĝu$_{10}$ ḫa-lam-ma-bi-a ga-ba-⸢ni]-[ib-gi$_4$-gi$_4$ $^?$]**	I want to reconsider the forsaking of my human people;
3'	**^{d}nin-tu-ra níĝ-dím-dím-ma-ĝu$_{10}$ sì-[sì-ga-bi-a$^?$] /ga-ba-ni-ib-gi$_4$-[gi$_4$]**	on behalf of Nintu I want to reverse the degrading of my creatures;
4'	**ùĝ ki-ùr-bi-ta ga-ba-ni-ib-gur-ru-dè**	I will let the people turn from its habitat.
5'	**iri^{ki}-me-a-bi ḫé-em-mi-in-dù / ĝissu-bi ní ga-ba-ab-dúb-bu**	Let them (*the people*) build their cities (and) I will calm down their shadow.
6'	**é-me-a šeg$_{12}$-bi ki-kù-ga ḫé-em-mi-in-šub**	Let them put down the bricks of temples on holy places.
7'	**ki-eš-bar$^!$-a ki-kù-ga ḫé-em-mi-ni-ib-ri**	Let them establish places of divination on holy places,
8'	**kù-a níĝ-izi-te-na si mi-ni-in-si-sá**	(and when) they have arranged there pure water that has quenched fire,
9'	**ĝarza me-maḫ šu mi-ni-ib-šu-du$_7$**	have completed the divine rules and exalted powers,
10'	**ki a im-ma-ab-dug$_4$ silim ga-mu-ni-in-ĝar**	(and) have irrigated the earth, (then) I shall establish well-being there.'
11'	**an ^{d}en-líl ^{d}en-ki ^{d}nin-ḫur-saĝ-ĝá-ke$_4$**	When An, Enlil, Enki (and) Ninḫursaĝa
12'	**saĝ-gíg-ga mu-un-dím-eš-a-ba**	had formed the black-headed people,
13'	**níĝ-gilim ki-ta «ki-ta» mu-lu-lu**	they made small animals crawl out of the earth,
14'	**máš anše níĝ-úr-limmu edin-na / me-te-aš bí-íb-ĝál**	they made goats, donkeys (and) all kinds of quadrupeds that are appropriate in the plain be there.

[1372] A picture of this tablet is available at CDLI no. P265876. The present transcription is based on this picture, Poeble's hand copy as published by Kramer (1981, 150), and on the transcription of Civil (1999, 140).

Editions / Translations

Editions
1. A. Poebel, Publications of the Babylonian Section IV/1: *Historical texts.* Philadelphia: The University Museum, 1914; 7-70.
2. G. Pettinato, *Das altorientalische Menschenbild und die sumerischen und akkadischen Schöpfungsmythen.* Heidelberg: Carl Winter Universitätsverlag, 1971; 97-100.
3. Th. Jacobsen, The Eridu Genesis. *JBL* 100 (1981) 513-529.
4. S.N. Kramer, The Sumerian Deluge Myth; reviewed and revised. *AnSt* 33 (1983) 115-121.
5. M. Civil, The Sumerian Flood Story. In: W.G. Lambert & A.R. Millard, *Atra-ḫasīs. The Babylonian Story of the Flood.* Winona Lake: Eisenbrauns, 1999; 138-145; 167-172.

Transcriptions and/or Translations
1. S.N. Kramer, The Deluge. In: J.B. Pritchard (ed.), *Ancient Near Eastern Texts Relating to the Old Testament.* Princeton, N.J.: Princeton University Press, 1950; 42-44.
2. S.N. Kramer, *A Flood*: The first 'Noah'. In: S.N. Kramer, *History Begins at Sumer: Thirty-Nine Firsts in Recorded History.* Philadelphia: University of Pennsylvania Press, 1981; 148-153.
3. Th. Jacobsen, The Eridu Genesis. In: Th. Jacobsen, *The Harps that Once... Sumerian Poetry in Translation.* New Haven, London: Yale University Press, 1987; 145-150
4. W.H.Ph. Römer, Die Flutgeschichte. In: W.H.Ph. Römer, D.O. Edzard (eds), *Weisheitstexte, Mythen und Epen. Mythen und Epen I, TUAT III.* Gütersloh: Mohn, 1993; 448-458.
5. J. Bottéro, S.N. Kramer, Le Récit du Déluge en Sumérien. In: J. Bottéro, S.N. Kramer, *Lorsque les Dieux Faisaient l'Homme. Mythologie Mésopotamienne.* [Paris]: Gallimard, 1993; 564-567.
6. ETCSL 1.7.4: The Flood Story.

***

Comments

Most scholars who have studied this text agree at least on one point: the undamaged fragment of column *i* opens with the speech of a god; the opinions only differ about which god is speaking. Some keep it neutral and speak of 'a god' (Civil, Pettinato); Poebel proposes 'Ninḫursaĝa or Enlil'; for others (Jacobsen, Römer) is Nintu speaking [1373]. The present author shares the opinion of Kramer and Bottéro/Kramer, who think that it is (part of) a speech of Enki. This is mainly based on line *i*: 3', where the god relates about 'my creatures'. This may refer to a general characteristic of Enki, who is also named **^{d}nu-dím-mud** 'Lord of manufacturing and creating'.

The most controversial points in the past discussion are two episodes described in column *i*. The first one is (line 2'), that mankind was supposed to be threatened by some catastrophe, sometimes called: destruction [1374]. The second one (lines 11'-12') is the 'creation' of the black-headed people.

[1373] The translation of line 3' shows that Nintu as speaking god has to be excluded.

[1374] Poebel, Civil, Kramer, Bottéro/Kramer.

2'-4' For the interpretation of the speech of the god, one of the key words in line 2' is **ḫa-lam**. One of the meanings of **ḫa-lam** is indeed 'to destroy' [1375]. Pettinato discusses this point as follows [1376]:

> 'Man hat auf Grund des Wortes **ḫa-lam** "Zerstörung" die genannten drei Zeilen auf eine der Sintflut vorangegangene Strafe bezogen. Eine solche Deutung ist jedoch sehr unwahrscheinlich, da die Menschen zu diesem Zeitpunkt noch nicht erschaffen waren. Das entscheidende Wort **ḫa-lam** heißt aber nicht nur "Zerstörung", sondern auch "Vergessenheit", "Verlassenheit", (...).'

About this supposed destruction Civil says [1377]:

> ' (...) the text begins with allusions to the destruction of man, although he is at this point newly created. (...) With what is preserved the translation given seems the only one possible. Was there, then, a first destruction of the human race prior to the one recorded in *Atra-ḫasīs* ? In the present state of knowledge it would be incautious positively to affirm it.'

Just like Pettinato, Römer and Jacobsen also did not believe in a 'destruction of man'. Römer translated **ḫa-lam** as 'Vergessen', Jacobsen as '(all) forgotten as they (= humankind) are'. In my opinion, line 2' refers to people, that after their creation (by Enki?) were left to fend for themselves. They became (nearly) forgotten by the gods, their behaviour was becoming more and more that of animals, and civilization was lacking. The situation in which people were, may be compared with that described in 'The Debate between Grain and Sheep', lines 20-25 [1378]:

20	**nam-lú-ulu$_3$-u$_4$-ri-a-ke$_4$-ne**	Mankind of those faraway days
21	**ninda gu$_7$-ù-bi nu-mu-un-zu-uš-àm**	had not got to know eating bread,
22	**túg níĝ mu$_4$-mu$_4$-bi nu-mu-un-zu-uš-àm**	had not got to know a garment (or) something to dress in!
23	**ùĝ ĝeš-gen$_6$-na-a šu-bi mu-un-ĝen**	The people moved on all fours;
24	**udu-gin$_7$ ka-ba ú mu-ni-ib-gu$_7$**	like sheep they ate grass with their mouths,
25	**a-mú-sar-ra-kam i-im-na$_8$-na$_8$-ne**	the water from the gardens they were drinking.

In short this is also told in 'How Grain came to Sumer' ll. 1-2 [1379]:

1	**ùĝ-e udu-gin$_7$ ú ka-ba mu-ni-ib-«ni»-[gu$_7$]**	Men used to eat grass with their mouths like sheep.
2	**ud-re-a ^{d}ezina$_2$ še gu nu-[zu]**	In those times, they did not know grain, barley or flax.

1375 ePSD: **ḫa-lam** '(to be) bad, evil; to forsake, forget; to destroy'.
1376 Pettinato 1971, 34.
1377 Civil 1999, 139.
1378 See the edition of this debate in the Appendix Text editions no. 6.
1379 ETCSL 1.7.6.

The text of 'The Sumerian Flood Story' does not go so far as to compare people's behaviour with that of animals, but line 3' clearly indicates that the level of their conduct was becoming lower and lower: they were becoming degraded [1380]. The god, presumably Enki, while speaking of 'my creatures', wanted to stop this process of degeneration, not so much for himself, but more out of consideration for the mother goddess Nintu. Apparently people were roaming all over the country just like wild animals. Enki's plan was to let them return from all those places.

The text in line 4' reads: **ki-ùr-bi-<u>ta</u>**. In general Civil does not believe in the correctness of the Sumerian used in this text [1381]. In this case he comments [1382]:

> 'The fact that in late Sumerian the contrast between **-ta**, **-šè**, and **-a** is often lost, makes the interpretation of this line (l. 40 in Civil's edition, l. 4' in my edition, JL) somewhat doubtful. If the original sense of **-ta** is to be retained, the line implies that mankind will leave the grounds where they live now, and where there are no buildings (i.e. nomadic life?), to move to the cities; if **-ta** is for **-šè**, it means that people will go back to their territories, which they left after a destruction, and where cities will now be (re)built. Without the missing part preceding line 37 (= 1' in my edition, JL), it is not possible to reach a conclusion.'

My interpretation of the text is based on the presumption that **-ta** is used in its original meaning, 'from ..away', because it is the wish of Enki – 'I want to reverse the degrading of my creatures' [l. 3'] – that people will change their habits, change their nomadic way of life, and leave their dwelling grounds, in order to start with the building – not *re*-building – of e.g. cities and temples. They cannot move to cities because there are as yet no cities at all.

ki-ùr, the Akkadian *duruššu*, has among other things the meaning 'habitation' [1383]. In the present context it has been translated as 'habitat'.

With respect to the verbal form in line 4' – in Civil's transcription [l. 40] **ga-ba-ni-ib-gur-ru-ne** –: this is an 'incorrect use of the suffix **-e-ne**, which can be only an object suffix referring to **un** "people" ', and which therefore is a Semitism, according to Civil [1384]. The transcription of the final **-NE** into **-dè** results in the *marû* 'infinitive' **gur-ru-dè** in a finite form: **ga-ba-ni-ib-gur-ru-dè** [1385].

5'-10' These lines make clear what Enki's plans are. In short: Enki wants the civilization of the uncivilized people and the development of a civilized society with due regard for the cults of the gods, as a condition for the well-being of the people and of Sumer.

5' First of all people have to build cities. This may indicate that until now mankind did not know what it was to live in a city, but led a nomadic way of life. This is beautifully

---

1380 Line 3': **sì** is interpreted in the sense of *sapānu*, (inter alia) 'to level, to smooth' [CAD S, 158]; Civil 1999, 167, *ad* 39.

1381 Civil 1999, 139.

1382 Civil 1999, 168 *ad* line 40.

1383 CAD D 198 *duruššu* 'basis, base, foundation, habitation'.

1384 Civil 1999, 168, comment at line 40.

1385 See Attinger 1993, 292, *ad* §190c for the verbal forms **ga** + *marû*-stem + **ed**.

confirmed by the second half of this line: 'I will calm down their shadow' [1386]. In other words: once people has settled in cities, their restless nomadic way of life will end and they can 'relax', as they have a permanent address now; then 'their shadow comes to rest'.

6'-9' These lines refer in more detail to the next duties of man: they have to build temples (l. 6') and places of divination (l. 7') in holy places, and install the cults for the gods (ll. 8'-9'). In these last two lines Civil takes the god who is speaking as the subject ('I will put...; I perfected...') [1387]. But in my opinion the god speaking refers to himself only in the second half of line 10': **silim ga-mu-ni-in-ĝar** 'I shall establish well-being there.'

7' **ki-eš-bar**: 'place of decision' [1388]; from the context it is clear that divine decisions are meant, which may be deduced from e.g. extispicy.

10' The last task of man is to cultivate the land. When all these assignments (ll. 5'-10') will have been performed, then there will be well-being for humankind.

11'-12' In the light of the above these lines may be completely understood. It is not a matter of *de novo* creation of man, but of bringing about human civilization. In the broken lines preceding line 1' it is probably stated that Enki had created hominid beings – see line 3': **níĝ-dím-dím-ma-ĝu$_{10}$** 'my creatures' – and even perhaps what was the reason for this creation. These creatures were not yet civilized, they even seemed to deteriorate. At Enki's instigation, but ultimately with the approval of An, Enlil and Ninḫursaĝa, this process was stopped. It may be clear that the gods do not act completely without self-interest: temple building and the installation of cults and rites imply that the gods will receive offerings. This selfish interest of the gods in creating people – as shown in some other texts edited in this study: to take over the tasks of the gods (e.g. working with hoe and basket, digging canals) and to provide them with their daily rations – was hidden and only indirectly hinted at in lines 6'-10'. When all these conditions had been fulfilled, and the earth also had been irrigated – to deliver the necessary food for the gods? –, only then would the god who is speaking (Enki$^?$) establish prosperity.

1386 The expression **ní dúb** means 'to relax'. In my opinion 'To calm down their shadow' can have two meanings here: 1. people will be safe in cities, so they can relax; 2. they can stop with their nomadic way of life; 'roaming around', as nomads do, means that one's shadow has no rest at all. Of course: both meanings are closely connected. Nice examples of **ní dúb** are found in the literature; to cite just two of them: 1. 'Ninurta's eploits' [van Dijk 1983, 75-76; ETCSL 1.6.2], l. 189. **^{ĝiš}erin abzu-a mú-a aga ĝissu daĝal-la** l. 190. **dumu ki ní dúb-bu-ĝu$_{10}$ nu-ba-tuš šu-ĝu$_{10}$ a-ba <ba>-gid$_2$-i-de$_3$** 'a cedar rooted in the **abzu**, a crown with broad shadow, my son, my security: he is not here any more; who will take me by the hand?' 2. 'A hymn to Inana' (Sjöberg 1975, 188, l. 115; ETCSL 4.07.3): **kaš$_4$ kar-re te-en-te-en ní dúb-bu-de$_3$ ^{d}inanna za-a-kam** 'to run, to escape, to quiet and to pacify are yours, Inanna'.

1387 In all verbal forms of the lines 4' - 10' the reference to subject or object should have resulted in a ***b** before the stem; sometimes this ***b** has been correctly written (the **ga**-forms in ll. 4' and 5'; the verbal forms in ll. 7' and 9'). In ll. 5', 6' and 8' this ***b** should have functioned as 'collective' referring to **ùĝ** in l. 4'. In l. 10' the ***b** should refer to **silim**. Here Civil's remark may be mentioned again: 'Most of the verbal forms (...) do not fit into the paradigms of standard Sumerian.' (Civil 1999, 139).

1388 **eš-bar** = *purussû*: AHw 882; CAD P 529-535.

'Black-headed people were formed' means: the now civilized beings are worthy to be called 'black-headed men'. These lines are not dealing with the *de novo* creation of humankind, as it is the case in the story of 'Enki and Ninmaḫ' [1389].

This text reminds us of 'The Debate between Grain and Sheep' [1390], especially lines 20-25, where man behaves like the animals, and lines 35-36, where the gods inspirited mankind to fulfil the needs of the gods. Here also mankind existed already, but in a primitive form, not yet appropriate to serve the gods; therefore the gods had to make people suitable.

12'-14' Civil comments on line 48 (my line 12') [1391]: 'The line shows that **saĝ-gi$_6$** is a designation of human beings in contrast with animals (…), and not an ethnic designation.' In this respect I disagree with Civil; the expression **saĝ-gíg** is always used in the Sumerian texts to designate the Sumerians. In my opinion lines 13'-14', describing the appearance of several kinds of animals, do not form a contrast with line 12', but only emphasize that the gods care for the black-headed people by giving them useful and serviceable animals.

My conclusion is that the development of people from uncivilized beings to the black-headed humans stems from the self-interest of the gods. This self-interest is now presented in a masked altruistic attitude.

***

1389 In 'The Song of the Hoe' it is stated that Enlil 'placed the first of humankind in a brickmould' (l. 19), but ultimately (l. 21) it appeared to be a black-headed one. Then the Anunna asked him for people of black-headed ones (l. 25).

1390 See the edition of 'The Debate between Grain and Sheep' in the Appendix Text editions no. 6.

1391 Civil 1999, 169.

9. The Song of the Hoe [1392]

1	**en-e níĝ-ul-e pa na-an-ga-àm-mi-in-è**	Then the lord has also let appear what was present in the bud!
2	**en nam-tar-ra-na šu nu-bal-e-dè**	The lord, who does not alter a fate decreed by him,
3	**^{d}en-líl numun-kalam-ma ki-ta è-dè**	Enlil, who makes the seed of the Land come out from the earth,
4	**an ki-ta bad-re$_6$-dè saĝ na-an-ga-àm-ma-an-sì**	has then set himself to separating heaven from earth,
5	**ki an-ta bad-re$_6$-dè saĝ na-an-ga-àm-ma-an-sì**	has then set himself to separating earth from heaven.
6	**uzu-è-a saĝ mú-mú-dè**	In order to let the first-born grow in Uzuea,
7	**dur-an-ki-ka bulug$_2$ nam-mi-in-lá**	he has then erected the axis at Duranki.
8	**^{ĝiš}al-e mu-un-ĝar u$_4$ al-è-e**	The hoe he used! And the sun rose.

The lines 9-17 have been omitted in this edition.

18	**uzu-è-a ^{ĝiš}al nam-mi-in-dù**	In Uzuea he indeed let the hoe enter;
19	**saĝ-nam-lú-ulu$_3$ ù-šub-ba mi-ni-in-ĝar**	he placed the first of humankind in a brick-mould.
20	**^{d}en-líl-šè kalam-ma-ni-šè ki mu-un-ši-in-dar-re**	In favour of his Land he (*the first one*) split open the earth towards Enlil.
21	**saĝ-gíg-ga-ni-šè igi-zi mu-un-ši-in-bar**	He (*Enlil*) looked with favour at his black-headed (*first*) one.
22	**^{d}a-nun-na mu-un-na-su$_8$-su$_8$-ge-eš**	The Anunna stepped forward to him (*Enlil*).
23	**šu-bi kiri$_3$-ba mu-un-ni$^{!}$-ĝál**	They greeted him respectfully.
24	**^{d}en-líl a-ra-zu-a mu-ni-in-ḫuĝ-e-ne**	They calmed Enlil with a supplication.

1392 Delnero (2006, 1961-2019) made a score of the texts of this Song, the result of which is reproduced here with some minor adaptations (see the Comments).

25	**ùĝ-saĝ-gíg-ga ^{ĝiš}al mu-un-da-ab-bé-ne**	They expressed before him their wish for a black-headed people.
26	**nin en ù-tu-da lugal ù-tu-da**	The lady who gave birth to the **en**, who gave birth to the king,
27	**^{d}nin-men-na-ke$_4$ tu-tu al-ĝá-ĝá**	Ninmena now establishes procreation.
28	**mas-su-an-ki-a en ^{d}nu-nam-nir-ra**	For the leader of heaven and earth, Lord Nunamnir,
29	**saĝ-zi saĝ-kal-la mu-ni-ib-še$_{21}$-še$_{21}$**	she (*Ninmena*) calls them, who are loyal and strong, by name for him;
30	**saĝ-bi gu-dili-àm mu-un-è-dè-a**	she, who forms a row of them for him,
31	**diĝir-re-e-ne-ra kurum$_6$-šè mu-un-dab$_5$-bé**	she makes them provide for the daily rations of the gods.
32	**^{d}en-ki-ke$_4$ ^{ĝiš}al-a-ni zà-mí ba-an-du$_{11}$**	Enki praised his (*Enlil's*) hoe.
33	**ki-sikil ^{d}nisaba eš-bar-ra ba-an-gub**	The young lady Nisaba noted down the decisions.
34	**^{ĝiš}al-mul ^{ĝiš}al-kù-ba šu mu-un-ne-ĝál**	They (*the black-headed people*) had the shining and holy hoe in their hands.

Editions / Translations

Editions

1. Th. Jacobsen, Sumerian Mythology: a Review Article. *JNES* 5 (1946) 134-138.
2. G. Pettinato, *Das altorientalische Menschenbild und die sumerische und akkadische Schöpfungsmythen.* Heidelberg: Carl Winter Universitätsverlag, 1971; 82-85.

Translations

1. J. van Dijk, Le Motif Cosmique dans la Pensée Sumérienne. *AcOr* 28 (1964) 23-24 (ll. 18-20).
2. S.N. Kramer, *Sumerian Mythology. A Study of Spiritual and Literary Achievement in the Third Millennium B.C. (Revised Edition).* Philadelphia: University of Pennsylvania Press, 1972; 51-53.
3. M.-L. Thomsen, *The Sumerian Language. An Introduction to its History and Grammatical Structure.* Copenhagen: Akademisk Forlag, 1984; 172, example 358 (ll. 1-5).
4. J. Bottéro, S.N. Kramer, L'Invention de la Houe et l'Origine des Hommes. In: J. Bottéro, S.N. Kramer, *Lorsque les Dieux Faisaient l'Homme. Mythologie Mésopotamienne.* [Paris]: Gallimard, 1993; 508-511.
5. G. Farber, The Song of the Hoe. In: W.W. Hallo (ed.), *The Context of Scripture, I: Canonical Compositions from the Biblical World.* Leiden, New York, Köln: Brill, 1997; 511-513.
6. H. Vanstiphout, ***Eduba***. *Schrijven en Lezen in Sumer.* Amsterdam: SUN, 2004; 82-90.

7. C. Wilcke, Vom altorientalischen Blick zurück auf die Anfänge. In: E. Angehrn (ed.), *Anfang und Ursprung. Die Frage nach dem Ersten in Philosophie und Kulturwissenschaft.* Berlin, New York: Walter de Gruyter, 2007; 25.
8. ETCSL 5.5.4: The Song of the Hoe.

***

Comments

The composition 'The Song of the Hoe' is one of the so-called Decad, a set of compositions belonging to the Sumerian school curriculum in the Old Babylonian period [1393].

1. Most scholars have translated **níĝ-UL** as 'what is fitting, appropriate, perfect', in other words: **UL** = **du**$_7$ [1394]. The other possible transcription for **níĝ-UL** is **níĝ-ul** [1395]. The ePSD mentions for **níĝ-ul**: '1. an everlasting thing; 2. joy'. Both these meanings might be more appropriate than 'what is fitting, appropriate, perfect' in the context of this text, the particular character of which has been emphasized by some scholars [1396]. But there is another meaning for **ul**, viz. bud. Because of the, in my opinion, clear parallel between the lines 1 and 3, the meaning 'bud' seems the most fitting one [1397].

The use of the prefixes ***na** + ***nga** might suggest that this text continues a preceding story [1398].

2. When the compound verb **šu bal** is used, the things changed are marked by an oblique case suffix, either a locative ending (as is the case in this line: **nam-tar-ra-na̲**), or a locative-terminative ending [1399] (= directive [1400]).

2-3. The verbal forms – **nu-bal-e-dè** and **è-dè** – are active [1401] *marû* participles, with an ergative ending ***e** [1402].

1393 Tinney 1999; Michalowski 2010.

1394 Exceptions: Thomsen (1984, 172, ex. 358), with transcription **níĝ-ul** and translation 'everything'; Jacobsen, (1946, 137) with translation 'the normal order'.

1395 See also the discussion of the transcription and the translation of **níĝ-UL** in the comment on the lines 4-5 of GEN in the Appendix Text editions no. 7 of this study.

1396 Kinnier Wilson 1965, 273; Wilcke 1972-1975 (Hacke); Farber 1997; Farber 1999a; Vanstiphout 2004, 82-85.

1397 See also Selz (1999, 510-511, note 154) for an elaborate discussion about **ul**.

1398 Civil 2000b, 37-38, *ad* 7. Marker of Reported Speech **na-**. In three OB catalogues (ETCSL 0.2.01, 0.2.11, and 0.2.12) 'The Song of the Hoe' is preceded by 1) 'A praise poem of Šulgi' (ETCSL 2.4.2.01), and 2) 'A praise poem of Lipit-Eštar' (ETCSL 2.5.5.1). These texts give no clue for the use of the prefixes ***na** + ***nga** in 'The Song of the Hoe' line 1. Edzard (2003a, ch. 12.12.2.5, p. 125) deals with connecting indicators. He gives the lines 1 and 4 of 'The Song of the Hoe' ('Creation of the Pickaxe', as he called it) as an example where 'connecting indicators' **na-nga** in the prefix chain are used. In a more recent article Edzard (2004a, 495, note 39) writes that the affirmative prefix ***na** may be used to attract the attention of the audience.

1399 Karahashi 2000, 149-150.

1400 Falkenstein 1978b, § 106, 118-129; Balke 2006, ch. 7, 171 ff; Krispijn 2004, 23, § 3.2.4.5.

1401 Most scholars (except Farber and Bottéro/Kramer) translate this part as being intransitive or passive.

1402 Edzard 2003a, 132, ch. 12.14.2; Krispijn 2004, § 3.7.3.3, 46-47.

3. In three texts **èš-ta** was written instead of **ki-ta** [1403]; this was not indicated by Delnero (2006, 1969-1970). In these lines of 'The Song of the Hoe', Enlil is the acting god, and **dur-an-ki** was mentioned (l. 7), a place associated with the temple area of Nippur [1404]. This might explain the use of **AB** = **èš** in some texts. Or could it be an abbreviated spelling for **ab-sin₂** 'furrow'? In connection with **numun** it seems more appropriate to choose **ki** in the final text (see also line 20).

4-5. Gadotti's transcription of **DU** in lines 4-5 is inconsistent: **re₆** and **de₆** (on pag. 43 and 44, respectively) [1405]. Moreover, Gadotti seems to ignore the /**dr**/ phoneme in the Sumerian verb **bad**ʳ [1406], and her transcription **BAD** = **ba₉** is not proven [1407].

For the transcription of **saĝ SUM** there are two possibilities: 1. **saĝ sum**, *ḫiāšu* or *ḫâšu*, 'to hasten' [1408]; 2. **saĝ sì.g**, *paqādu*, 'to take care of' [1409]. In view of the context, with Enlil as subject, it seems very unlikely that 'to hasten' is suitable, in spite of the doubt of several scholars about the seriousness of this text [1410]. The second possibility, **saĝ sì.g**, has been translated here as 'to set oneself to'.

6. The majority of the texts has **uzu-è-(a)**, only one has **uzu-mú-a**. Behrens reviews the names of Nippur, cited in MSL XI [1411]. In this list **uzu-è-(a)** is not mentioned, as opposed to **uzu-mú-a**. The name **uzu-mú-a** also appears in KAR 4, line 18 [1412].

7. **dur-an-ki**: besides the name of a sanctuary in Nippur, it is also a name for Nippur as the centre of the universe [1413]. For **bulug₂** 'axis': see PSD B, *sub* **bulug** A, § 4, 174-175.

In the individual Sumerian texts there are three different verbs used: **ĝar**, **íl** and **lá**. Although this last one has been used in the final text, I have tried to combine the meaning of the three different verbs in one English term: 'to erect' [1414].

8. **u₄ al-è-e**: The prefix **al-**, which has a stative meaning, is in general infrequently used. It is one of the many examples that point to the fact, that the text of 'The Song of the Hoe' shows a play with the syllable ***al** [1415]. **ĝišal-e**: the ending ***e** is the deictic **e**; its meaning has been expressed by the exclamation mark.

1403 The texts PBS X/2 16, SRT 19 and SEM 33.

1404 Bergmann 1964, 8-9 *ad* 9; Edzard 1987, 15; George 1993, 80 *ad* no. 218.

1405 Gadotti 2005: **ba₉-re₆-dè** on p. 43, while she wrote on p. 44: 'The "Song to the Hoe," however, employs **de₆** (line 4 and 5), "to bring, carry," (...).'

1406 Jagersma 2000.

1407 Borger 2010, 68 *ad* no. 113 **BAD**.

1408 AHw, 343: *ḫiāšum* 'hineilen'; CAD Ḫ, 146: *ḫâšu* A 'to move quickly, to rush to a goal'.

1409 AHw, 824: *paqādu(m)* 'übergeben, anvertrauen; betreuen; beauftragen'; CAD P, 115: *ad* 4: 'to take care of.... '

1410 See comment at line 8.

1411 Behrens 1978, 58-60.

1412 For an edition of KAR 4: see Appendix Text editions no. 10.

1413 George 1993, 80 *ad* nr. 218.

1414 The verb **lá** (ePSD): '1. (among others) to hang, to suspend; 2. to stretch out'.

1415 Kinnier Wilson 1965, 273; Wilcke 1972-1975 (Hacke); Farber 1997; Farber 1999a; Vanstiphout 2004, 82-85; Michalowski 2010.

18. One manuscript (TCL XVI 72) reads: ⌜**uzu-**⌝**mú-a saĝ nu-ĝá-ĝá-dè** 'In Uzumua, he who no one dares to oppose'.

Van Dijk reconstructs this line as follows: **uzu-è ^{ĝiš}al-a saĝ-nu ĝá-ĝá-dè**, 'Celui qui fit émerger la chair, afin de faire à l'aide de la hache l'individu humain', with commentary in brackets: '**nu** = **lú**! **saĝ-nu**: le premier homme?' [1416] The expression **saĝ-nu** has not been attested in the published Sumerian literature to the best of my knowledge.

20. Three texts read **kalam-ma-ni**, which was the final choice of Delnero (2006, 1977). As a consequence **kalam-ma-ni** was treated as subject, with a translation of this line e.g. in the ETCSL: 'His Land started to break through the soil towards Enlil'. In my opinion this seems very unlikely. There are other readings: **kalam-ma-na** (2x) and **kalam-ma-ni-šè** (4x). The last expression was chosen in my final score. Should the final reading be: **kalam-ma-na**, then the corresponding translation may be: 'In his Land he (*the first one*) split open the earth towards Enlil.'

In my opinion, the 'first of mankind' of line 19 is the subject in this line. 'People' are not yet present [1417]. Line 25 tells us, that the Anunna will ask for them! Moreover, in all the texts the sign **UN** is followed by ***ma**, confirming the reading **kalam** (Land) instead of **ùĝ** (people).

21. In this line again the *first-born* black-headed person is meant. All previous translations read instead 'black-headed *people*'.

For a discussion of the possible transcriptions of **MI**, see the comments at line 1 of the edition of NBC 11108 in this study. Here the transcription **gíg** is preferred because of the following ***ga**.

23. The literal translation of this line, which preferably may be reconstructed as **šu-bi kiri$_3$-ba mu-un-ni$^{!}$-ĝál** [1418] is: 'their hand was at their nose for him'. Unlike Delnero I read **kiri$_3$** instead of **kiri$_4$**.

25. The spelling **^{ĝiš}al** instead of **al** in the expression **al ... du$_{11}$/e/di** [1419] is only one example of the humorous use of the syllable /**al**/ in this text.

Pettinato and Bottéro/Kramer end their translations with this line; in view of these translations [1420], they consider this line as a 'concluding remark'. Probably they have been led to their translation by 1) the spelling **^{ĝiš}al** instead of **al** (only two texts read **al**); and 2) the spellings ***da-ab-bé** and ***da-bé**, which – in all likelihood – they supposed to represent the *marû*-stem of the verb **dab$_5$**.

1416 Van Dijk 1964, 23-24.

1417 Van Dijk, Farber and Bottéro/Kramer all suppose that 'people' are already present.

1418 Several texts have **mu-un-NE-ĝál**, others have **mu-un-ni-ĝál**. Analysis of the prefix chain of **mu-un-ni-ĝál** shows: **m-en-a-i**, whereby ***en + a** refers to Enlil, and the ***i** refers to the locative **kiri$_3$-ba**. If the resulting verbal form should be **mu-un-ne-ĝál**, with concomitant analysis of the prefix chain **m-enē-a-i**, then this has to be explained as if the gods raised their hands to their noses for themselves (***enē + a**). This possibility is not very likely.

1419 For **al ... du$_{11}$ / e / di**: see Attinger 1993, ch. 5.3.13, 429-438; especially ex. 208, p. 437: 'demander qqc'.

1420 Pettinato (1971, 84): 'sie lassen sein sumerisches Volk die Spitzhacke fassen.'
Bottéro/Kramer (1993, 509): '[Les Anunna] transmirent la Houe aux têtes-noires.'

26. Delnero (2006, 1980) wrote in the composite text the verbal form **ù-tu-dè-en**. However, in the score the form **ù-tu-da**, a *ḫamṭu* participle, is the most frequent one. Even if the verbal form **ù-tu-dè-en** might be a mistake for the *marû* participle **ù-tu-dè** (which would be a reverse of the scribal mistake as described by Delnero 2012, 138-139), then the *ḫamṭu* participle still stays the most frequent one. Therefore I gave the preference to the verbal form **ù-tu-da** in my final text.

27. Ninmena is included in the list of birth goddesses of **an** = *anum* [1421]. Possibly Ninmena is identical with **^{d}MEN** in the Fara god list (VAT 12760).

28-31. These lines form one sentence, with Ninmena as subject.

28. Farber and ETCSL neglect the ***ra** at the end of the complex; they suppose that **en nu-nam-nir** is the subject.

29. Ninmena is creating a queue of future working people for the gods (l. 30). Therefore the epithet **kal** has been translated as 'strong' instead of 'precious'.

30. Compare this line with line 211 of 'A Hymn to Inana' (ETCSL 4.07.3): **nitaḫ munus-bi gu-dili-a mi-ri-è-dè**: the men and women form a row for you.

34. Literal translation: 'the hand was at the shining hoe, at the holy hoe, for them'.

***

[1421] Litke 1998, 68; tablet II 21.

10. KAR 4 [1422]

Published sources

Siglum	*Signature*	*Publication*
A	VAT 9307	Landersdorfer 1917, T. I, II (photographs of the tablet); KAR 4 (copy); *obverse ll. 1-29; reverse ll. 1-22.*
B	K.4175 + Sm.57	PSBA X (June 1888), pl. I, II (join with C and D)
C	BM 82-3-23,146	RA 17, 189 (join with B)
D	BM 80-7-19,184	CT XVIII, plate 47 (join with B)
E	A.17634	Pettinato 1971, T. I (photo of tablet)

/	new line within one 'compartment'
|	the same line, but in the adjacent column
Pett.	line numbering of Pettinato

A obv.	*Pett.*		
1	A1	**u$_4$ an ki-ta tab gi-na bad-a-ta-eš-a-[ ]**	After heaven had been separated from earth, with which it was firmly joined,
2	A2	**^{d}ama-^{d}inanna-ke$_4$-e-ne ba-se$_{12}$-se$_{12}$-e-dè**	so that the mother goddesses could live there (i.e. in heaven),
3	A3	**u$_4$ ki ĝá-ĝá-e-dè ki dù-dù-a-ta**	after the earth had been built in order to be able to establish cult places there,
4	A4	**u$_4$ ĝiš-ḫur-ḫur an-ki-a mu-un-gi-na-eš-a-ba**	when the designs in heaven and on earth had been fixed,
5	A5	**e pa$_5$-re šu si-sá ĝá-ĝá-e-dè**	when the Tigris and the Euphrates had set out their banks (l. 6),
6	A6	**^{íd}idigna ^{íd}buranun gú-ne-ne ĝar-eš-a-ba**	so that dikes could be built and ditches could be cut in good order (l. 5),
	A6	*i* [ ]	
7	A7 A8	**an ^{d}en-líl ^{d}utu ^{d}en-ki /** **diĝir-gal-gal-e-ne**	An, Enlil, Utu and Enki, the great gods,
	A7 A8	^{d}*a*-[*x*] / diĝir^{meš}	

[1422] Edzard and Wilcke (1977, 86) claim to have a monolingual, Sumerian, Old Babylonian version from Isin (tablet IB 591) of the KAR 4 text. According to Farber (1999b, 120) this tablet contains with respect to KAR 4 'eine einsprachige sumerische Version, die jedoch einen stark stenographischen Charakter hat, der gerade knapp eine Identifizierung der Komposition und der Zeilen zuläßt.' Until now (summer 2013) the text of this tablet has not yet been published.

8	A9	**^{d}a-nun-na diĝir-gal-gal-e-ne**	the Anunna, the great gods,
	A9	*^{d}a-nun-na-ku* [ ]	
9	A10	**bara$_2$-maḫ ní-te mú-a mu-un-durun$_x$-<eš>$^{!!}$/**	remained in an exalted shrine, grown with awe.
	A11	**ní-te-a-ni šu mi-ni-íb-gi$_4$-gi$_4$**	He himself (*i.e.* Enlil) spoke there:
	A10	*ina* bara$_2$ *ṣi-ri* [ ] /	
	A11	*ú-ši-bu-ma i-na r*[*a-*]	
10	A12	**u$_4$ ĝiš-ḫur-ḫur an-ki-a mu-un-gi-na-eš-a-ba**	'Now that the designs in heaven and on earth have been fixed,
	A12	*ú-ṣu-rat* an *u* ki *ina* [ ]	
11	A13	**e pa$_5$ šu si-sá ĝá-ĝá-e-dè**	and the Tigris and the Euphrates have set out their banks $^{(l. 12)}$,
	A13	*i-ka ù pal-ga* [ ]	
12	A14	**^{íd}idigna ^{íd}buranun /**	so that dikes can be built and
	A15	**gú-ne-ne ĝar-eš-a-ba**	ditches can be cut in good order $^{(l. 11)}$:
	A14	*i-di-ig-lat ù pu-*[*x*] /	... [*after they*] *had dug out the*
	A15	*ú-ḫar-*[*x*]	*Tigris and the Eu*[*phrates*]:
	B	[ ] /	
		mu [x] /	
		uš tum [*x*] /	
		ú-kin [*x*]	
13	A16	**a-na-àm ḫé-en-bal-en-dè$^{!}$ (ZE$_2$)-en**	what can we change?
	A17	**/a-na-àm ḫé-en-dím-en-dè$^{!}$ (ZE$_2$)-en**	what can we create?
	A16	*mi-na-a i ni-pu-uš /*	*what should we do?*
	A17	*mi-na-a i ni-te-pu-uš*	*what shall we go to do?*
	B	a-na-àm ga-ab-dù-en-dè$^{!}$ (ZE$_2$)-en /	what do we want to do?
		a-na-àm ga-ab-dím-en-dè$^{!}$ (ZE$_2$)-en	what do we want to create?
		/ *mi-na-a i ni-pu-uš /*	*What should we do?*
		MIN *i ni-ib-ni*	*What should we create?*
14	A18	**^{d}a-nun-na diĝir-gal-gal-e-ne**	Anunna, great gods,
	A18	*^{d}a-nun-na-ku* diĝir^{meš}-gal^{meš}	
	B	^{d}a-nun-na diĝir-gal-gal-e-ne /	
		^{d}MIN diĝir^{meš}-gal^{meš}	
15	A19	**a-na-àm ḫé-en-bal-en-dè$^{!}$ (ZE$_2$)-en**	what can we change?
	A20	**/a-na-àm ḫé-en-dím-en-dè$^{!}$ (ZE$_2$)-en**	what can we create? '
	A19	*mi-na-a i ni-pu-uš /*	*What should we do?*
	A20	*mi-na-a i ni-ib-ni*	*What should we create? '*
	B	a-na-àm ḫé-bal-e-dè$^{!}$ (ZE$_2$)-en /	What do we want to change?

		a-na-àm mu-un-me-e / e-dè$^{!}$(ZE$_2$)-en / *mi-na-a i nu-uš-<bal>-kit /* MIN *i ni-ib-bi*	What have we decreed? ’ *What should we change?* *What should we decree? ’*
16	A21 A22	**diĝir-gal-gal-e-ne mu-un-su$_8$$^{!}$(SUR)-ge$^{!}$(RE)-eš-a /** **^{d}a-nun-na diĝir nam-tar-re**	The great gods, who were standing there, the Anunna, the gods who decree the fate,
	A21 A22 B	diĝir^{meš}-gal^{meš} *šu-ut iz-zi-zu /* *^{d}a-nun-na-ku mu-ši-im ši-ma-ti* diĝir-gal-e-ne su$_8$-ge-eš / ^{d}a-nun-na diĝir-nam-tar-ra / diĝir^{meš}-gal^{meš} *i-^{zi}ziz-zu /* *^{d}a-nun-na-ki /* *mu-ším* NAM^{meš}	
17	A23 A23	**min-na-ne-ne ^{d}en-líl-ra mu-un-na-ni-íb-gi$_4$-/gi$_4$** *ki-lal-lu-šu-nu ^{d}en-líl ip-pa-*[ ]	two of them answered Enlil:
18	A24 A24	**uzu-mú-a^{ki} dur-an-ki-ke$_4$** *i-na* uzu-mú-a^{ki} *ri-ki-is* an *u* ki^{ti}	‘In Uzumua, the bond of heaven and earth, *‘Let us* $^{(l.\ 19)}$ *in Uzumua, the bond of heaven and earth,*
19	A25 A25 E	**^{d}NAĜAR ^{d}NAĜAR im-ma-an-šum-en-dè$^{!}$ (ZE$_2$)-en** ^{d}NAĜAR ^{d}NAĜAR *i ni-iṭ-bu-ḫa* [ ]*-uḫ-ma*	we shall slaughter the gods Alla and Illa, *slaughter the gods Alla and Illa.*
20	A26 A26 E	**úš-úš-e-ne nam-lú-u$_x$-lu mú-mú-dè** *i-na da-me-šu-nu i ni-ib-na-a a-mi-lu-tu* [ ] úš-e-ne / saĝ ḫé-mú-mú / *ina* MUD-*šú-nu a-mi-lu-ta /* *i ni-ib-ni*	so that their blood makes mankind grow. *Let us, with the aid of their blood, create mankind.* Let their blood make the first-born grow. *Let us, with the aid of their blood, create mankind.*
21	A27 A27 E	**á-ĝiš-ĝar-ra-diĝir-e-ne éš-gàr-bí ḫé-a** *iš-kar* diĝir^{meš} *lu iš-kar-ši-na* á-ĝiš ĝar-ra-diĝir-ra- / né-kam éš-gàr-bi ḫé / *iš-kar i-lu /* *lu iš-kar-šú-nu*	Let the work assignment of the gods be its job (= *the job of mankind*):

22	A28	**u$_4$-da-rí-šè e-sur /**	to establish for ever the boundary ditches;
	A29	**gi-<na-e>-dè**	
	A28	*a-na* u$_4$^{meš} *da-ru-ti /*	
	A29	*mi-iṣ-ra a-na ku-un-ni*	
	E	[r]i-šè e-pa$_5$ /	
		[ ]	
23	A30	**^{ĝiš}al ^{gi}dusi šu-bí /**	to take in hand the hoe and the pannier;
	A31	**ĝá-ĝá-e-dè**	
	A30	*al-la ù tup-ši-ik-ka /*	
	A31	*a-na qa-ti-ši-na a-na ša-ka-ni*	
24	A32	**é-diĝir-gal-gal-e-ne /**	planning (l. 25) the houses of the great gods, befitting an exalted shrine,
	A33	**bara$_2$-maḫ-a túm-ma**	
	A32	*šub-tu* gal^{tu} *ša* diĝir^{meš} /	
	A33	*ša a-na pa-rak-ki ṣi-ri šu-*[ ]	
25	A34	**a-gàr-a-gàr-re ĝiš ḫur-ḫur-re**	(and) meadows,
	A34	*ú-ga-ru a-na ú-ga-ri uṣ-*[ ]	
26	A35	**u$_4$-da-rí-šè e-sur /**	to establish for ever the boundary ditches;
	A36	**gi-na-e-dè**	
	A35	*a-na* u$_4$^{meš} *da-ru-*[ ] /	
	A36	*mi-iṣ-ra a-na ku-u*[*n-*]	
27	A37	**e si sá-e-dè-«zé-en» /**	to put in order and to consolidate the ditches;
	A38	**gi-na-e-dè**	
	A37	*i-ka a-na šu-t*[*e-*] /	
	A38	*mi-iṣ-ra a-na* [ ]	
28	A39	**é-limmu-sud ú-ḫi-a /**	to increase all kinds of plants for the estates in the four corners;
	A40	**nun-nun-e-dè**	
	A39	[*x x x*] /	
	A40	[ ]	
29	A41	**šeĝ$_{14}$-šeĝ$_{14}$ [x]**	... the rains ... ;
	A41	[ ]	

A
rev.

1	A42	**ki-ùr-sur gi-na-e-dè /**	to establish the boundaries of the dwelling grounds: then it (= *mankind*) will pile up heaps of grain.
	A43	**gur$_7$ nam-mi-ni-íb-gur-gur-re**	
	A42	*mi-iṣ-ra a-*[*x*] /	
	A43	gur$_7$ *a-*[*x*]	

2	A44	*ḫi-e-pi* | *ḫi-e-pi*	*break*
3	A45	*ḫi-e-pi* | *ḫi-e-pi*	*break*
4	A46	*ḫi-e-pi* | *ḫi-e-pi*	*break*
5	A47	**a-šà-^{d}a-nun-na-ke$_4$-e-ne dùg-dùg-e-dè**	to make the fields of the Anunna fertile;
	A47	a-šà ^{d}*a-nun-na-ki a-na* [ ]	
6	A48	**ḫé-ĝál kalam-ma nun-nun-e-dè**	to increase abundance in the Land.
	A48	ḫé-ĝál *i-na* kur *a-na r*[*u-*]-⸢*i*⸣	
7	A49	**ezen-diĝir-e-ne šu-du$_7$-a**	The feast for the gods is completed,
	A49	*i-sin-ni* diĝir^{meš} *a-na* [*x*]-*li*	
8	A50	**a-sed dé-dé-da**	when cold water is poured out,
	A50	a^{meš} *ka-ṣu-ti a-na nu-qí-i*	
9	A51	**unu$_2$-gal-diĝir-e-ne bara$_2$-maḫ-a túm-ma**	and when the large abode of the gods is suitable for an exalted shrine.
	A51	*šub-tu* gal^{tu} *ša a-na* bara$_2$ *ṣi-ri šu-lu-kát*	
10	A52 / A53	**^{d}ul-le-ĝar-ra an-né-ĝar-ra / mu-ne-ne ì-pà-da**	Their names, Uleĝara and Aneĝara, are mentioned.'
	A52 / A53	^{d}ul-le-ĝar-ra an-né-ĝar-ra / *šu-me-šu-nu ta-sà-na-kàr*	*You shall mention their names: Uleĝara and Aneĝara.'*
11	A54 / A55	**gu$_4$ udu máš-anše ku$_6$ mušen ne-ta-a / ḫé-ĝál kalam-ma nun-nun-e-dè**	In order to increase abundance in the Land by means of cattle, sheep, wild animals, fishes and birds,
	A54 / A55	gu$_4$ udu *bu-la* ku$_6$^{meš} ù mušen^{meš} / ḫé-ĝál *i-na* kur *a-na du-še-e*	
12	A56 / A57	**^{d}en-ul ^{d}nin-ul / ka-kù-ga-a-né zur-zur-re**	Enul and Ninul, while their pure mouth is honouring,
	A56 / A57	^{d}en-ul ^{d}nin-ul *i-na pi-i-šú-nu / el-li uk-ta-an-nu*	
13	A58	**^{d}a-ru-ru nam-nin-a túm-ma**	(and) Aruru, who is befitting the **nin**-ship,
	A58	^{d}nin-diĝir^{meš} *ša be-lu-te šu-lu-*[ ]	
14	A59	**ĝiš-ḫur-gal-gal ní-ba mu-un-ḫur-ḫur-re-<ne>$^{!}$**	draw up their own great plans.
	A59	*i-na ra-ma-ni-šú-nu ú-ṣu-ra-te ra-*[ ]	
15	A60	**gašam gašam lú$^{!}$-IM lú$^{!}$-IM**	The wise one after the wise one, ignorant person after ignorant person,

	A60	*um-ma-nu a-na um-ma-ni nu-u'-ú a-na* [ ]	
16	A61	**še-gin$_7$ ní-bi ne ki-ta si$_{12}$-si$_{12}$ ki dím**	– like grain, that of itself becomes green from the earth and that adorns the earth,
	A61	*ki-ma še-im a-na* <*ra-*> *ma-ni-šu a-na bu-ni-i*]	*- like grain of itself for adornment,*
17	A62	**níĝ nu-kúr-ru mul-<an-na> da-ri-šè**	something that cannot be changed, just like an eternal star in the sky, –
	A62	*ša la ut-ta-ka-ru* mul-an^e *da-r*[*u-x*]	
18	A63	**u$_4$-gi$_6$-na-ta ezen-diĝir-e-ne /**	to make day and night the feast for
	A64	**šu-du$_7$-a**	the gods perfect,
	A63	*ur-ra ù mu-ú-ša /*	
	A64	*i-sin-ni* diĝir^{meš} *a-na šuk-lu-lim*	
	B	ĝar-ša-ta /	-----
		iš-tu «ši» ši-im-ta /	*After they have determined the fate,*
		i-ši-im-mu mim-ma /	*they have provided all that is pure.*
		el-la iš-ku-nu	
19	A65	**ní-te-a-ni ĝiš-ḫur-gal-gal-la /**	he draws up himself great plans.
	A66	**mu-un-ḫur-ḫur-re**	
	A65	*ina ra-ma-ni-šu-nu ú-ṣu-ra-te /*	*they draw up themselves great*
	A66	*ra-ab-ba-te uṣ-ṣi-ru*	*plans.*
	B	ní-bi-ta ĝiš-ḫur-/	They draw up themselves their
		gal-bi e-ne ḫur-ḫur-re /	great plans.
		ina ra-ma-ni-šú-nu /	
		ú-ṣu-ra-a-ti /	*The plans are greatly drawn up by*
		ra-biš uṣ-ṣu-ra	*themselves.*
20	A67	**an ^{d}en-líl /**	Where (l. 70) An, Enlil, Enki and
	A68	**^{d}en-ki ^{d}nin-maḫ /**	Ninmaḫ,
	A69	**diĝir-gal-gal-e-ne**	the great gods,
	A67	*^{d}a-nu ^{d}en-líl /*	
	A68	*^{d}é-a ù ^{d}nin-maḫ /*	
	A69	diĝir^{meš}-gal^{meš}	
	B	^{d}en-ki-ke$_4$ [x] /	
		diĝir-gal [x] /	
		^{d}a-nu-um [ ] /	
		ù [ ] /	
		diĝir [ ]	
	B	[ ] /	
		sur [ ] /	
		zi [ ] /	
		ur [ ]	

21	A70 A71	**ki nam-lú-u$_{x}$-lu ba-ni-in-dím-eš /** ^{d}**nisaba ki-bé nam-en-na an-gub**	have created mankind, there is Nisaba honoured in her **en**-ship.
	A70 A71	*a-šar a-mi-lu-tu ib-ba-nu-ú /* *^{d}nisaba i-na aš-ri šu-a-tu ku-un-na-at*	
	B	^{d}nisaba zà [mí] / ^{d}nisaba nam [ ] / *ina* [*x*] / *ú*-[ ]	Praise be to Nisaba. Nisaba …
22	A72 A73	**ad-ḫal** *mu-du-ú mu-da-a lu-kal-lim* **al-til igi-kár gaba-ri libir-ra/** **šu^{II}** *ki-din-^{d}30* ^{lú}**dub-sar-banda A** *su-ti-e* ^{lú}**dub-sar-lugal**	Secret. The one who is competent should show it only to someone who is also competent. It has been completed and collated from an old copy. The hand of Kidin-Sîn, junior scribe, son of Sutû, the king's scribe.
	B	*ni-ṣir-ti nam* [ ] / *šá ina aš-ri šak-nu ba-x*[ ]	
23	B	[ ] A [ ]	

Editions / Translations

Editions

1. S. Landersdorfer, *Die sumerischen Parallelen zur biblischen Urgeschichte.* Münster: Verlag der Aschendorffschen Buchhandlung, 1917; 62-76; Tafel I, II.
2. G. Pettinato, *Das altorientalische Menschenbild und die sumerischen und akkadischen Schöpfungsmythen.* Heidelberg: Carl Winter Universitätsverlag, 1971; 74-81.
3. W.G. Lambert, Review: Giovanni Pettinato: *Das altorientalische Menschenbild und die sumerischen und akkadischen Schöpfungsmythen.* Heidelberg: Carl Winter Universitätsverlag, 1971; *BSOAS* 35 (1972) 134-135.
4. J. Bottéro, S.N. Kramer, Le Récit Bilingue de la Création de l'Homme. In: J. Bottéro, S.N. Kramer, *Lorsque les Dieux Faisaient l'Homme. Mythologie Mésopotamienne.* [Paris]: Gallimard, 1993; 502-508.
5. K. Hecker, Ein zweisprachiger Schöpfungsmythos. In: K. Hecker et al. (eds), *Weisheitstexte, Mythen und Epen.* TUAT III, Lfg. 4: Mythen und Epen II. Gütersloh: Gütersloher Verlagshaus, 1994; 606-608.
6. For other editions: see Borger 1967-1975, HKL I, p. 96 *ad* KAR 4.

Translation

B.R. Foster, *Before the Muses. An Anthology of Akkadian Literature.* Third edition. Bethesda, Maryland: CDL Press, 2005; 491-493.

***

General remarks

1. The indications '*obverse*' and '*reverse*' refer to text A only [1423].
2. There are some deviations from the presentation of the text as done by Pettinato and Bottéro/Kramer:
   a. Pettinato counted every line with cuneiform signs; in doing so he has not taken into account, that the tablet has clearly been ruled: sometimes there is one line, sometimes there are two lines within one case. The present edition numbers the cases [1424].
   b. Pettinato has not distinguished between '*obverse*' and '*reverse*', but he has continued the numbering.
   c. Unlike Pettinato did, I have not reconstructed the missing Akkadian text.
   d. Bottéro and Kramer do pay attention to the ruling of tablet A, but they did not distinguish between '*obverse*' and '*reverse*'.
3. The Akkadian passages of A, and the texts of B-E, have been translated only when they deviate from the Sumerian main text A.
4. Lambert, Edzard and Wilcke, and Lambert and Millard have published their opinions about the supposed joins (B+C+D) [1425]. According to Lambert & Millard, this tablet contains the bilingual creation myth together with the Silbenalphabet (= 'Silbenvokabular A'); in the colophon this myth is given as the second in the series which begins with the bilingual Silbenalphabet [1426]. This conclusion depends on the interpretation of the catch line **dub-2-kám-ma me-me [kúr-kú]r** *i-li*. Probably the 'Akkadian' part of the text beginning with **me-me** should be more extensive; for the second and third column of the Old Babylonian version of the 'Silben-vokabular' contain 'Akkadian equivalents' to the entries in the first column, starting with names of gods (or *i-li* may represent all these gods who are called by name ?) [1427].
5. Edzard and Wilcke [1428] mention the discovery of a tablet fragment (IB 591) from the Old Babylonian period, in which each column is divided into three parts; the middle part contains a Sumerian version of the 'KAR 4' creation myth, while the left and right columns contain the 'Silbenvokabular A': on the left the Sumerian version, and on the right the Akkadian one. Up to the present this fragment has not yet been published.

It is evident that the Sumerian used in the KAR 4-text cannot be considered as 'perfect' Old Babylonian Sumerian. This might lead to the supposition, that in this case the Akkadian version is the primary one, which thereafter was translated into Sumerian. The deviations, mostly observed in the verbal forms, are discussed in the comments at the respective lines.

Pettinato comments on the text KAR 4 as a whole: 'Inhaltsmäßig ist er trotz des deutlichen Einflusses der akkadischen Tradition noch echt sumerisch' [1429]. The reaction of Kümmel on this statement was as follows [1430]:

[1423] A photograph of the tablet is shown in Landersdorfer 1917, Tafel I (= *obverse*) and II (= *reverse*).

[1424] In the present comments, Pettinato's and Bottéro/Kramer's numbers of the corresponding lines are given between brackets.

[1425] Lambert 1972, 134; Edzard and Wilcke 1977, 86 *sub* IB 591; Lambert and Millard 1999, 35.

[1426] Lambert and Millard 1999, 35.

[1427] Sollberger 1965, 22: 1a. [me-m]e; 1b. ^{d}g[u-la]; 1c. [^{d} ; 2a. [pap]-pap; 2b. ^{d}nin-ti; 2c. ^{d}[; etc. The interpretation 'Akkadian equivalents' is Sollberger's (p. 21).

[1428] Edzard and Wilcke 1977, 86 *sub* IB 591.

[1429] Pettinato 1971, 18.

> 'Pettinato setzt Orginalität und Priorität der sumerischen Fassung ohne Begründung voraus, obwohl die größere Zahl von Varianten und nicht nur Fehlern im sumerischen Text im Vergleich zur akkadischen Fassung an sich sowohl auf dessen lange Tradition wie ebenso auf relativ junge Textredaktion schließen lassen könnte. Unbestritten bleibt, daß altes Material darin wie in den nur sumerisch überlieferten Schöpfungstexten enthalten ist, das mindestens auf Quellen altbabylonischer Zeit zurückgeht [Mehr hat A. Falkenstein (1965a, p. 133 n. 72 [1431]) nicht festgestellt]. Ebenso offenbar ist aber die Kompilation bzw. Überlagerung in der Darstellung der Menschenschöpfung: Der Schöpfungsort Uzumua der Nippur-Tradition steht neben dem Motiv der Erschaffung des Menschen aus dem Blut erschlagener Götter, wie es sonst nur akkadisch überlieferte Texte seit Atraḫasīs kennen.'

Bottéro and Kramer have the following opinion about the myth, and especially about its time of origin [1432]:

> 'Le plus vieux de nos manuscrits figurait parmi la présumée Bibliotèque de Teglatphalassar I (1115-1077): la rédaction du mythe serait donc antérieure à la fin du IIe millénaire; et sans doute remontait-il plus haute encore, dans la mesure où l'on peut tirer argument du fait qu'il ignore le rôle créateur de *Marduk*.'

***

Structure of the composition

For a better understanding of the text, first of all the structure of the composition will be outlined with the aid of a short paraphrase of the content:

Obverse

1-2 Information about **an.ki** and mother goddesses
3-9 Occupations of the gods:
- 3-6 Earth has been built up; watercourses have been constructed.
- 7-9 The Anunna reside; Enlil starts to speak.

10-15 Enlil poses some questions to the Anunna.
16-29 Two Anunna gods answer Enlil.
- 16-17 Introduction to this answer.
- 18-29 Contents of this answer:
  - 18-20 Decision to create mankind.
  - 21-29 Tasks of mankind.

1430 Kümmel 1973-1974, 27-28.

1431 Falkenstein writes in this note: 'Da in dieser Komposition Enlil eine wichtige Rolle zufällt, Schauplatz des Geschehens Nippur ist, wird man trotz der sprachlichen Gestalt des sumerischen Textes mit einer Entstehung in der späten altbabylonischen Zeit rechnen müssen. Bei einem späteren Ansatz wäre kaum zu verstehen, dass die Hauptrolle nicht Marduk von Babylon zugeschrieben ist.'

1432 Bottéro and Kramer 1993, 503.

Reverse

1-10	Continuation of the answer to Enlil:
	1-9 Tasks of mankind (continuation)
	10 Names of the first men.
11-14	Plans of some gods with respect to animals.
15-19	Plans of men, which - when executed - are a feast for the gods.
20-21	Where mankind has been created, that is the place where Nisaba is revered.
22	Closing formula; colophon.

***

Comments

Obverse

1. As Pettinato has already concluded, this line is clearly corrupt. Jacobsen reads this line as follows [1433]: **u$_4$ an ki-ta tab-ge-na-bad-a-ta bà-a-[ba]** 'When heaven from earth – from the far removed trusty twin – had been parted'. According to Pettinato [1434], van Dijk supposed that in this composition two different traditions coincide. But van Dijk translates this line 1: 'le jour où le ciel et la terre furent fondés ensemble', and then continues: 'C'est un théologoumenon nouveau, car l'original portait à coup sûr: **u$_4$ an-ki-ta ... bad-a-ta...** "dès le jour où le ciel s'éloignait de la terre".' [1435].

Instead of the verbal form **bad-a-ta-eš-a-[ ]**, the one used in 'Gilgameš, Enkidu and the Netherworld' line 8 [1436], **(an ki-ta) ba-da-bad-rá-a-ba**, is more correct.

2. As **se$_{12}$** is already a plural form (of **ti.l** 'to live, to dwell'), the reduplication of it may be used to stress the plurality of the absolutive: *all* mother goddesses.

3. **ki ĝá-ĝá-e-dè** 'to place foundations'. Compare in this connection the lines 11-12 of the **zà-mì** hymns from Abū Ṣalābīḫ [1437]: **^{d}en-líl a-nun / ki mu-ĝar-ĝar** 'Enlil has assigned plots of land to the Anunna'. 'Plots of land' have to be understood here as 'cult places'. This led to the present translation of line 3: 'to establish cult places'.

4. Farber-Flügge discusses the term **ĝiš-ḫur** [1438].

5. The expression **šu si-sá ĝar** could not be found in the literature. Perhaps it is a contamination [1439] of two compound verbs: **šu si-sá** 'to put in order' and **šu ĝar** 'to carry out'.

1433 Jacobsen 1946, 143, note 24.

1434 Pettinato 1971, 79.

1435 Van Dijk 1964, 14.

1436 See the edition of this text: Appendix Text editions no. 7.

1437 Biggs 1974, 46.

1438 Farber-Flügge 1973, 181-191.

1439 **šu si-sá**: Sjöberg and Bergmann 1969, 51-52. **šu si-sá** = *šutēšuru* 'to put and keep in good order': CAD E, *sub ešēru* 12, 359-363. **šu ĝar** = *gamālu*: AHw 275: 'Vergeltbares tun, vergelten, schonen'; CAD G 21-23: '(among others) to be obliging, to perform a kind act, to act so as to please'; ePSD: 'to carry out [a task]'.

7-8. Falkenstein, Pettinato, Bottéro/Kramer and Foster suppose, that two groups of gods are mentioned here [1440]: on the one hand An, Enlil, Utu and Enki; on the other hand the Anunna. But in my opinion, it concerns only one group instead of two groups of gods: the great gods *are* the Anunna. The generally known Sumerian figure of speech has been used: the same subject is repeated in different wordings. In this case: the gods mentioned in line 7 are part of the Anunna [1441].

9. The original sequence of signs reads: **bara$_2$-maḫ ní-te mu-un-ki-tuš mú-a**, a clearly corrupt sentence. An emendation might be formulated as: **bara$_2$-maḫ ní-te mú-a mu-un-durun$_x$-<eš>**, where **mú-a** belongs to **ní-te** 'grown with awe', in a more free translation: 'which is all awe'. In the text BM 82-3-23, 146 *obv. i'*: 5'-6' [1442], we read: 5' **[g]ùr-ru**; 6' [ ] **gi$_4$**. This may correspond very well with the line 9 of the KAR 4-text *siglum* A. **ní gùr** is the more usual expression with meaning 'to be clad in awesome luminosity'.

The original 'verbal part' is in fact a noun: **ki-tuš (ki-KU)** 'dwelling place'. The Akkadian form *ú-ši-bu-ma* makes clear, that here the Sumerian form **durun$_x$** (= **KU-KU**, the plural form of **KU** = **tuš** 'to sit') has been meant. Therefore an emendation has been formulated with <**eš**>, because of the plural absolutive: the gods.

The second part of this line is the announcement of the direct speech (therefore the verb **gi$_4$-gi$_4$** represents the *marû*-stem). It is only one god, presumably Enlil – see therefore line 17, where some gods answer Enlil –, who is going to talk; the verbal form and **ní-te-a-ni** exclude a plural subject.

10. In line 10 then the direct speech starts. This makes the emendation as Pettinato does – the addition of line 15a <**^{d}en-líl-le gù mu-ne-dé**> – superfluous. The beginning of the

---

[1440] Falkenstein 1965a, 133; Pettinato 1971, 77, ll. 7-9; Bottéro/Kramer 1993, 503, ll. 7-8; Foster 2005, 492, ll. 7-8. Falkenstein, Pettinato: 'An, Enlil, Utu (und) Enki, die großen Götter, (*und*) die Anunna, die großen Götter'. Bottéro/Kramer: 'Alors, An, Enlil, Ninmaḫ (!) et Enki, les dieux majeurs, *ainsi que les autres* grands-dieux: les Anunna'. Bottéro/Kramer justify the replacement of Utu by Ninmaḫ (p. 503, note 3): 'Le copiste, distrait ou maladroit, sams doute, a introduit ici le nom d'*Utu*, dont la présence serait fort surprenante dans ce contexte; mais il a bien écrit *Ninmaḫ* en 38.' [l. 38: mistake of B/K; has to be l. 48; *JL*].
Foster: 'Anu, Enlil, Ninmaḫ, and Ea, the great gods, *and* the Anunna gods, the great gods,' with the same justification for the replacement of Utu by Ninmaḫ (l. 7) as that of Bottéro/Kramer.

[1441] It is not exactly known, which gods belong to the Anunna. According to Falkenstein, An is the father of the Anunna (Falkenstein 1965a, 129). Further he says: 'Die Identität der Anunna mit den Grossen Göttern ist zwar nicht durch die häufigen Belege **^{d}a-nun-na dingir-gal-gal-e-ne** erwiesen, da dies auch "die Anunna **und** die Grossen Götter" heissen könnte, wenn auch zu beachten ist, dass bei diesen Belegen in der akkadischen Übersetzung nie ein "und" zwischen den beiden Gliedern erscheint, wohl aber durch die Aussagen der Texte selbst, die "Anunna" und "Grosse Götter" als gleichwertige und vertauschbare Benennungen erkennen lassen. (...) Wenn die "Grossen Götter" ebenso wie gelegentlich die "schicksalentscheidenen Götter" von den Anunna differenziert werden, so mag dabei weniger die Vorstellung von einer rangmässigen Gliederung unter den Anunna vorliegen, als die Tendenz zur "parallelen Reihung" als einem literarischen Stilmittel.' (Falkenstein 1965a, 139, note 193). Edzard 1976-1980, 38, §3, identifies the Anunna with the great gods: 'Das häufigste Epithet der Anunna ist **dingir-gal-gal-e-ne** "die großen Götter".' He does not discuss the possibility of two groups of gods; but from his enumeration it appears, that the gods in line 7 of KAR 4 all belong to the Anunna.

[1442] This text is *siglum* C in the present 'KAR 4' edition (Meek 1920, 189).

direct speech in line 10, instead of in line 13 (Pettinato line A16), is the more likely, since otherwise the lines 10-12 would be a mere repetition of lines 4-6, a repetition which would be difficult to explain.

12. Pettinato's *ú-kin-n*[*u*] (line A15) has most likely to be read as *ú-ḫar-*[*x*] [1443].

13. The pronominal elements of the Sumerian verbal forms, both in text A and in text B, do not correspond to their Akkadian counterparts. Because of the frequent corrupt Sumerian verbal forms in these texts, preference is given to the Akkadian 'we' instead of the Sumerian 'you'. The Sumerian verbal form of text B also suggests 'we', due to the use of the prefix ***ga**.

Pettinato (line A17) did not read the Akkadian text correctly: Lambert says that in the second part *ni-te-pu-uš* has been written, and not *ni-ib-ni* [1444]. The Gt-stem of *epēšu* [1445] may have an ingressive meaning [1446].

The modal prefix ***ḫé** has been interpreted as having an epistemic function [1447].

15. Text B reads **mu-un-me-e-e-zé-en**. In a lexical list it is shown that **me-e** = **ME** = *qá-a-bu-um* [1448].

16. Like Pettinato (see his comment on l. 21), I have understood **sur** as an unorthographic spelling of **su$_8$-g**, which is confirmed by the text of source B. The use of **sur**, with ist final /**r**/, might be explained by the plural *ḫamṭu*-stem of **ĝen**, being (**e-**)**re$_7$** or **er** [1449].

16-17. The translations of Falkenstein, Pettinato and Bottéro/Kramer [1450] show, that they all suppose that **min-na-ne-ne** refers to two groups of gods: the great gods and the Anunna. As discussed in my comments on line 7-8, the Anunna *are* the great gods. Four Anunna-gods are mentioned (l. 7). Enlil has spoken now, and two of the Anunna answer him. Who these two are has not been mentioned, but in all likelihood they are Utu and Enki. For An is seldom to the fore; he is the father of the Anunna, but further hardly noticeably involved in the progress of the history.

Another argument to think of only two gods, and not of two groups of gods, is the use of the Akkadian *kilallūšunu*. In the CAD the meaning of *kilallān* is 'both, two, pair', and the examples given at this lemma leave no room for an interpretation of two groups; it concerns again and again two individual persons or cases [1451].

1443 *ḫerû*: CAD Ḫ, 175-176 'to dig (out, up)'; AHw, 341 'graben, ausbaggern; **D** ~ **G**'.

1444 Lambert 1972, 134b.

1445 AHw (*s.v. epēšu* p. 227) suggests, that a Gt-stem of *epēšu* does not occur, and refers to its Gtn-stem, with iterative meaning. On the other hand, according to the CAD (E, *epēšu* pp. 191-235, especially p. 192) a Gt-stem of *epēšu* does occur.

1446 Kouwenberg 2005, especially § 2.4 (pp. 86-88), and § 4 (p. 98): 'The functions (of the t-infix; *JL*) (…) reveal the Gt-stem as a straightforward marker of detransitivity, the very function it also has in other Semitic languages.'

1447 Civil 2000b, § 3.2, 32-34.

1448 See the lexical parts at *qabû*: AHw, 889 (*qabû* II), and CAD Q, 22.; MSL XIV, 128, the secondary branch of Proto-Ea nr 10.

1449 Thomsen 1984, 306 *s.v.* **ĝen**.

1450 Falkenstein 1965a, 133; Pettinato 1971, 78, l. 23; Bottéro/Kramer 1993, 504, ll. 16-17.

1451 *kilallān*, CAD K, 353-356.

19. Krebernik [1452] has given a plausible argumentation for the reading **ᵈNAĜAR ᵈNAĜAR** = **ᵈalla** and **ᵈilla**, instead of ᵈAlla-gods in this text, as was done by Hecker, Bottéro & Kramer, and Foster. Pettinato speaks of 'Lamga-Götter' [1453].

The pronominal element of the verbal form in the text reads ***-en-zé-en**, which according to the Akkadian text has to be emended as ***-en-dè-en** 'we'.

20. Lambert [1454] observes several 'serious misreadings' by Pettinato line 26 text D, (in my edition l. 20 text E). Instead of Pettinato's line 26 D: **[úš]-úš-e-ne**, Lambert proposes: 'for **úš.úš.e.ne** read **[mú]d.bi.e.ne**.' This suggestion has not been followed in the present edition, because in my opinion the cuneiform text does not give rise to such reading. Moreover, the simultaneous use of /**bi**/ and /**ene**/ is not very likely.

21, 23. The use of **NE**, transcribed as **bí**, after **éš-gàr** and **šu**, instead of **bi** (so in text E in line 21), is remarkable. It refers to **nam-lú-u$_x$-lu** (l. 20) [1455].

22. For the emendation of **gi-dè** as **gi-<na-e>-dè**: see *obverse* the lines 26 and 27, and *reverse* line 1.

22, 26. These two lines are (nearly) identical. However, could it be that the boundary ditches of line 22 are different from those mentioned in line 26? My supposition is, that the term in line 22 is used in a general way – perhaps referring to the boundary of Sumer (?) –, whereas in line 26 the boundary ditches of the temple domains of the gods in particular are meant.

25. **ĝiš ḫur**: Pettinato (l. 34) 'einreihen'; Bottéro/Kramer (l. 24) 'ajouter'. The primary meaning of **ĝiš ḫur**, however, is 'to design'.

27. The verbal form **si-sá-e-dè-zé-en** has been accepted by Pettinato (l. 37) without any comment. We are dealing in this line (as in *obv*. 22- 28 and *rev*. 1-6) with a verbal form that denotes an intention or a pupose [1456]. The gods are summing up the tasks of mankind; therefore, by analogy with former lines, I expect an infinitive as a verbal form: **si sá-e-dè**, as is also suggested by the Akkadian text.

Bottéro/Kramer (l. 27) translate: '{Ils délimiteront les champs}', which they consider as 'dittographie' [1457].

1452 Krebernik 2002, 292-294. Krebernik recants hereby his former suggestion to read **ᵈNAĜAR** = **ᵈE$_{10}$** in KAR 4 (Krebernik 1998-2001a, 74b).

1453 The value 'Lamga' for **NAĜAR** was already criticized by Lambert (1972, 135a), but Lambert did not suggest an alternative; he concludes his remark: 'The ancient lists and commentaries give five possible values to the sign **NAĜAR** when used for a god's name, and Lamga is the least probable in this text.'

1454 Lambert 1972, 134b.

1455 In CAD I/J, 245, in the lexical part of *iškaru* A, this line of KAR 4 is cited. There it was supposed that **NE** is the pronomen possessivum of the animate class, therefore an emendation **éš.gàr.ne.<ne>** was written.

1456 Attinger 1993, 305-308, ch. 3.2.9.3; Edzard 1967b, 41-44; Thomsen 1984, 266-267, § 524.

1457 Bottéro/Kramer 1993, 504, and note 1.

28. Due to the context, a more broad translation of **é** has been chosen: 'estate', instead of the usual 'house'.

Pettinato (l. 39, with comment) has chosen for **sud / sù** 'sprengen (vom Wasser)'; but in combination with **limmu**, the meaning of **sud^{r}** 'to be far' seems more probable.

The verb **nun** seems to be present (nearly) only in the fossilized expression **ḫi-nun** or **ḫé-nun**, the Akkadian *nuḫšu*: abundance, plenty, prosperity [1458]. The Akkadian equivalent for **nun** in this text is *rubbû* 'to increase' (see A *rev* 6). The lexical part of the adjective *rabû* mentions **nun** = /*rabû*/ in two lexical lists [1459].

Reverse

1. There is no break between this line and the line A *obv* 29, as can be inferred from the continuity in the accompanying 'Silbenalphabet A'.

ki-ùr-sur is probably a contamination of **ki-ùr** [1460] and **ki sur** ('to mark a boundary'). Pettinato asserts that **ki-ùr** belongs to the first column, but he does not support his statement with arguments [1461]. In my opinion it was written in the second column. Moreover, **ki-ùr** does not form part of the 'Silbenalphabet A'. Bottéro and Kramer (l. 31) seem to follow Pettinato in their translation of this line.

6. This line is included in CAD [1462], with the remark that this line was collated by W.G. Lambert, who supplied the Akkadian part: *ru*-[*bi*]-*i*, which I am not able to see on the photograph in Landersdorfer's book.

7-9. The Sumerian verbal forms are nominalized, unlike the Akkadian forms, which are constructed with *ana*, as they were in the preceding lines. There these *ana*-constructions corresponded to the Sumerian verbal constructions ending with ***e-dè**.

The Sumerian lines seem to me to form a culmination of the preceding lines, some kind of 'conclusion'. The service to the gods and the reverence for them are the most important things mankind has to take care of, and everything men do has to end in the cult.

10. In my opinion the answer to Enlil ends here [1463]. Bottéro and Kramer make an attempt to explain the names of the first men [1464]:

> ' (...): *Ullegarra* et *Annegarra*, parfaitement inconnus d'autre part. Ils paraissent bizarrement formés, d'un terme sumérien: gar(ra), «posé», «placé», «établi», et de

---

1458 *nuḫšu*: AHw, 801 'Fülle, Fruchtbarkeit'; CAD N II, 319-321 'abundance, plenty, prosperity'; ePSD *sub* **henun**.

1459 *rabû* adj.: AHw, 936; CAD R, 27.

1460 For the meaning of **ki-ùr** *durušu*: see van Dijk 1964, 47-48; Sjöberg 1973, 16 *ad* 4'; Civil 1999, 167-168, comments at line 40.
duruššu: AHw, 178-179 '(planiertes) Fundament'; CAD D, 198-199 'basis, base, foundation, habitation'.

1461 Pettinato 1971, 81, comment at line 42.

1462 CAD R, 38a, at the verb *rabû* A, in the lexical part.

1463 The end of the answer to Enlil has not been indicated by Pettinato with the aid of closing quotation marks. According to Bottéro and Kramer, the answer ends with line *rev*. 20 (their number 48).

1464 Bottéro and Kramer 1993, 507.

> deux pronoms akkadiens qui signifient: *ulle* < *ullû*, «ce qui est plus éloigné dans le lieu ou le temps», et *anne* < *annû*, «ce qui est plus rapproché». A-t-on voulu représenter ainsi un couple («celui qui a été mis-au-monde d'abord / ensuite») ? ou résumer l'antique lignée humaine entière: dans sa durée («ceux-là», les plus archaïques, et «ceux-ci», plus rapprochés), ou son extension géographique, des plus lointains aux plus voisins, sur la terre? Bien malin qui résoudra ce mystère!'

Foster comments on these names [1465]: 'These names may mean something like "establishing abundance/prosperity", but may also contain the Akkadian words for "no" and "yes" (...).'

These scholars may have overlooked, that **ul** in the first name has been written with the **diĝir**-determinative. Therefore, it is not a question of a strange combination of Sumerian and Akkadian terms. In both names two gods are named, viz. Ul [1466] and An, the last one as always without **diĝir**-determinative. The names mean literally: 'placed by Ul' (**ul** perhaps with the meaning 'bud') and 'placed by An'. An analogous construction is found in the Temple Hymns: **en-ul-e ĝar-ra** 'founded by the primaeval lord' [1467].

11-19. The interpretation of these lines in the present edition is different from those by Pettinato (ll. 54-66) and by Bottéro & Kramer (ll. 40-47). In my opinion, lines 11-14 belong together, as do lines 15-19.

In lines 11-14 is stated that the gods Enul, Ninul and Aruru are drawing plans to provide the Land with all kinds of animals. A man cannot create animals, gods can; but mankind has to take care of these animals.

Lines 15-19 tell that in addition every man, according to his own capacities (l. 15), has to draw plans by himself for the well-being of the gods (ll. 18-19). Men's activities have to be like grain, that becomes green of itself, grows up and embellishes the earth. Then this comparison goes further: the beauty of the greening and growing grain cannot be changed, just like the eternal stars in the sky. The task of men is self-evident, as is the growing etc. of grain.

12. In this line the Sumerian **zur-zur** is equated with the Akkadian *kutennû* 'to treat with honour' [1468]. But instead of **zur-zur** (as collated by Lambert), in my opinion on the base of the photograph and the copy of KAR 4 [1469], the reading **siskur$_2$** (= **siskur-siskur**) is not impossible. In that case the tentative translation might be: 'while their pure mouth is blessing' or 'while their pure mouth pronounces a benediction'.

14. The original text **mu un ní ba ḫur ḫur re** has to be emended, in accordance with the comment of Pettinato at this line (l. 59): **ní-ba mu-un-ḫur-ḫur-re-<ne>**, the extra <**ne**>, because of the plurality of the subject: Enul, Ninul and Aruru. The collective

---

1465 Foster 2005, 493, note 1. He has written the names as: Ullegarra and Annagarra.

1466 This god is possibly related to, or even identical to, one of the ancestors of Enlil, viz. **^{d}en-ul**, who is present in several god lists (see the chapter 3 in this study).

1467 Sjöberg and Bergmann 1969, 20, l. 65.

1468 CAD K, 540-542, *s.v. kunnû*; this line of KAR 4 is cited (collated by Lambert) *sub* 2. *kutennû*, p. 542.

1469 Landersdorfer 1917, Tafel II (photograph); Ebeling 1919, nr. 4 (copy).

pronomen possessivum ***ba** (in **ní-ba**) and also the Akkadian *-šunu* (in *ramanišunu*) point to it.

15-16. Pettinato (ll. 60-61) and Bottéro/Kramer (ll. 43-44) interpret these lines as if men grow like grain, rising up from the earth [1470].

At first sight the translation 'to adorn' for the Sumerian **dím** 'to create' may be not self-evident. For a better understanding of the equation **dím** = *bunnû*, made in this line, the explanation in the CAD is very clear [1471]. In summary: *banû* A, to be equated with **dím** 'to build', has also a D-stem, but which is only used in El-Amarna texts [1472]. The other verb *banû* (B) 'to grow' has indeed a D-stem, *bunnû* 'to beautify'. This verb refers, *inter alia*, to plants which are 'exceptionally well formed'. In this way the equation **dím** = *bunnû* may be understood.

17. The Sumerian text was emendated: **mul-<an-na>**, because in the Akkadian text we read **mul an**^{e} = *kakkab šamê.*

18. The text in B is missing in A. The Sumerian text of this line is very defective, and unclear to Pettinato [1473]. Probably the first part of this line, on the basis of the Akkadian text, should read: **nam ĝar-ra-eš-ša-ta** 'After they have determined the fate'. The counterpart of the rest of the Akkadian sentence (*mim-ma el-la iš-ku-nu*) is completely missing.

20. In this line, the gods who are responsible for and actively involved in the creation of men are mentioned. One of them is, not surprisingly, Ninmaḫ. In line *obverse* 7 however, besides An, Enlil and Enki, Utu has been mentioned instead of Ninmaḫ. This led Bottéro and Kramer to think of a scribal error [1474], and they emended this line by replacing Utu with Ninmaḫ. In line *obverse* 7 some of the Anunna-gods hold a conference. Why should the participants in this conference be the same as those, who afterwards execute the decisions, made during that meeting? Especially when it concerns the creation of mankind (l. 21), the presence of Ninmaḫ as a mother goddess may be expected. In my opinion, an emendation like that of Bottéro and Kramer is unnecessary.

In the translation of Hecker we read: 'An, Enlil, Ea und Damkina…': Damkina, the wife of Ea, as a 'translation' of Ninmaḫ?

21. For the equation **an-gub** = *kunnât*: see the lexical part at the verb *kunnû* [1475], which has, *inter alia*, the meaning 'to honour a deity'.

22. **igi kár** ('to examine') and **gaba-ri** ('to confront') are translated with one term: 'to collate'. This colophon suggests that there was an older copy of this text, in accordance with the discovery of the Old Babylonian fragment IB 591 [1476]

1470 Pettinato: 'hervorsprießen'; Bottéro and Kramer: 'pousser'.

1471 CAD B, 93b-94, sub *banû* B, after *bunnû* 6b.

1472 CAD B, 83, sub *banû* A, 5.: II 'to build (EA only)'.

1473 Pettinato 1971, 81 *ad* l. 64: '**nì-ša-ta** ist mir unverständlich'.

1474 Bottéro and Kramer 1993, 503, line 7, and note 3.

1475 CAD K, 540; Antagal VIII 180: **[g]ub.ba** = *kun-nu-*[*ú*].

1476 Edzard and Wilcke 1977, 86 *sub* IB 591.

Bottéro and Kramer translate line 22 as follows: 'C'est là une doctrine secrète: on n'en doit parler qu'entre compétences!' [1477] In other words: copy, collation and name of the scribe have been omitted.

Veldhuis discussed the meaning of the 'secret' in colophons [1478].

***

A general comment on the 'Silbenvokabular' which accompanies the text of KAR 4: Cavigneaux and Jacques (2010, 10) wrote that 'il semble indubitable que SV J b 25-34 recèle *in nuce* l'anthropogonie du mythe MC (= mythe de création; JL)'. However, this Silbenvokabular (see Sollberger 1965) does contain neither the anthropogeny of KAR 4 nor any other anthropogeny, even not in a nutshell.

***

1477 Bottéro and Kramer 1993, 505, ll. 51-52.
1478 Veldhuis 2010, 79-80.

Excursus

The animate *vs* inanimate class of an and ki

During the preparation of the translations and editions of the texts that are discussed in this study, it became clear that it was worthwhile having a closer look at the class – animate or inanimate – of the two protagonists in the primaeval cosmos: **an** and **ki**. In this respect a distinction should be drawn between **an** and **ki** forming a group, and **an** and **ki** as separate entities.

One way to determine the Sumerian idea about the class – animate or inanimate – of the subject or object is to analyse the verbal form, i.e. the pronominal element(s) of the prefix- or suffix-chain.

I **an** *and* **ki** *in their unity*

As was discussed elsewhere in this study [1479], the unit **an-ki** is present at the Beginning; from this primaeval unit the whole cosmos will be developed. In three texts we meet **an** and **ki**, joined together. Their unity has been expressed as **an-ki niĝin$_2$** (IAS 114, *i*:1'), **an ki téš-ba** (Ukg 15, *ii*: 2), or **an ki téš-bi-a** (NBC 11108, *obv.* 5).

1. In IAS 114, *i*: 1'-2' the attendant verbal expression is **NAM$_2$-LU$_3$** / **nam-ĝar**. The pronominal element ***b** before the *ḫamṭu*-stem [1480] may represent a singular inanimate subject, or a plural animate or inanimate subject (collectively expressed). However, there is no certainty about the singularity or plurality of the ergative, since 1. the collective pronomen ***b** has no pluralis pronominal suffix ***eš**, so that it is not possible to distinguish between a singular or plural subject; 2. the plurality of animate subjects seems not to be expressed in this text (see e.g. *i*: 3'-4'). Therefore a definitive answer about the class of **an** and **ki** cannot be given.

2. In Ukg 15, *ii*: 2 analysis of the verbal form **am$_6$-gi$_4$-gi$_4$** [1481] also shows a pronominal prefix ***b**, now before a reduplicated *ḫamṭu*-stem. The subjects are **an** and **ki**; the reduplication then refers to the plural subject or to an iterative action, and no definitive conclusion can be made about the class of **an** and **ki**.

3. In the third example in which **an** and **ki** form a unit – NBC 11108, *obv.* 5 – the verbal form **mu-lu** with the intransitive ***lu.g**, gives neither a clue with respect to the class of the absolutive [1482], nor with respect to the question of what that absolutive is: **an** alone or **an-ki** that has to be considered as one, singular entity. The next line **[nam-dam-š]è ⌜nu-ù⌝-tuku** is not helpful with respect to the class of **an** [1483]; the subject of this line, represented by the pronominal element ***en** or ***b**, can only be **an**. Due to the uncertainty of the pronominal element – ***en** or ***b** –, no conclusion is possible about the class – animate or inanimate – of **an**. Because the subject of line 6 must be **an**, this suggests that in line 5

1479 Ch. 2 and 4, *passim*.

1480 Analysis of the verbal form **nam-ĝar**: **na-(i/a)-m-b-/*ḫamṭu-stem*/-Ø**.

1481 Analysis of the verbal form: **a-m-b-/*ḫamṭu-stem*-*reduplicated*/-Ø**.

1482 Analysis of the verbal form **mu-lu.g**: **(i)-m-b-ni-/*hamṭu-stem*/-Ø**.

1483 Analysis of the verbal form **nu-ù-tuku**: **nu-i-en/b-/*ḫamṭu-stem*/-Ø**.

an is also the subject instead of **an-ki**. Moreover, it suggests that **an** and **ki**, although laying together, are understood as separate units.

II an

1. In Ukg 15, *ii*: 1 **an** is called **en** and **šul** 'youthful man'; these epithets point to the animate class of **an**. The verbal form – with the intransitive **gub** – gives no indication about the class of the subject. Line *ii:* 2 suggests that An 'as **en** and youthful man' is in a courting situation. Due to the designations **en** and **šul** (*ii*: 1) it may be concluded that An in this line was also considered as animate.

It has been argued [1484] that **an** is presumably the ergative of the lines *i*: 2-3. Analysis of the prefix elements of the verbal form **ḫa-mu-ni-se$_{11}$-se$_{11}$** [1485] may point to an ergative pronomen of the animate class. In this respect the pronominal suffix ***ē** of the verbal form **ḫa-mu-ak-e** is not distinctive. The moment of the cosmogony has not been indicated, but it may be tentatively concluded that this action of **an** took place after his intercourse/marriage with **ki**.

2. In the Barton cylinder *i*: 12-13, the verbal form – **am$_6$-dab$_6$-e** – does not reveal the class of the ergative **an**, because of the *marû*-form of the verb. On the other hand, analysis of the prefix-chain of the same verbal form – **am$_6$-dab$_6$-e** – in *i*: 14 shows: **a-m-b-da-b-/e/-ē**. This means that **an** has been considered as being inanimate. The situation may be called 'courting'.

3. In line 1 of the NBC 11108-tablet **an** in **an-uru$_{16}$-né** is the ergative; the prefix **mu-** of the verbal forms, with transitive verbs, may be analysed as ***(i)-m-en-**. Consequently, this **an** has to be of the animate class. There is a clear differentiation in this line between **an-uru$_{16}$** (the god An) and **an**/heaven which has been lit up.

Line 5 gives no clue to the class of the subject **an** because of the intransitive form of **lu.g**. In the following line (6), with the same subject, the analysis of the prefix-chain **nu-ù-** may result in ***nu-i-en-** or ***nu-i-b-**; in other words: **an** may be considered here as animate or inanimate, respectively.

In line 8 the ***b** before the *ḫamṭu*-stem points unambiguously to the inanimate class of **an**, where it concerns a point before marriage. The alternative, that the verbal form is a *marû*-stem (and that the class of **an** therefore cannot be determined), is less probable, because the usual order of tenses in such successive clauses is *ḫamṭu - marû* (line 8 *ḫamṭu* - line 9 *marû*) [1486].

Comparison of lines 1, 5-6 and 8-9 seems to indicate a contradiction. If line 1 refers to a time before the marriage of **an** and **ki**, then it is not clear why **an** here has to be considered as animate. An explanation might be that the story of the Beginning has been told more or less in the reverse sequence; then line 1 may refer to a moment immediately after the marriage of **an** and **ki**; lines 5-6 refer to a future marriage, but lines 8-9 mention a situation which seems even further away, i.e. earlier in the time scale.

4. In 'The Debate between Tree and Reed' the position of **an** is as follows. In line 5 the use of the dative ***ra** in **an-kù-ra** points to the animate class of **an**. The next line (6) shows

---

[1484] See the edition of Ukg 15 in the Appendix Text editions no. 2.

[1485] Analysis of **ḫa-mu-ni-se$_{11}$-se$_{11}$**: **ḫe-(i)-m-b-ni-en-/*ḫamṭu-stem-plur.*/-Ø**.

[1486] Streck 1998.

a ***b** before the *ḫamṭu*-stem, indicating that in this sentence **an** has been considered as inanimate; this may be the consequence of his epithet in this line, viz. **an-maḫ** 'exalted heaven'. In line 7 the pronominal element of the ergative in the prefix-chain has to be analysed as ***en**; therefore **an** is now of the animate class. However, in the Old Babylonian period, at the time when these texts of 'The Debate between Tree and Reed' were copied, the elements that may be used to differentiate between the two classes (animate and inanimate) were not consistently employed [1487]. Thus e.g. the ***ra** in line 5 may be not conclusive for the animate class of **an** in **an-kù-ra**.

5. In 'The Debate between Grain and Sheep' we meet **an** as the subject in line 2. Analysis of the prefix-chain (***i-m-b-**) of the concomitant verbal form **im-tu-dè-eš-a-ba** teaches us that **an** was considered inanimate.

6. In line 11 of 'Gilgameš, Enkidu and the nether world' the pronominal element ***en** in the prefix-chain [1488] points to the animate class of **an**.

III ki

1. In Ukg 15 *i*: 3 **ki** is considered to be of the inanimate class because of the use of the directive ***e**.

2. The Barton cylinder *i*: 12-13 shows, after analysis of the verbal form **am$_6$-dab$_6$-e** (**a-m-<u>b-da</u>-b-/e/-ē**), that **ki** belongs to the inanimate class.

3. In 'The Debate between Tree and Reed' all pronominal elements in the verbal forms of the transitive verbs with **ki** as ergative suggest the inanimate class of **ki**, with one exception in one text, viz. line 3 **ba-ni-in-du$_{11}$** [1489]. The pronominal element ***en** in this case refers to the animate class. However, in a duplicate text the inanimate class of **ki** was indicated in this line.

Conclusion

ki: The conclusion with respect to **ki** is that it always belongs to the inanimate class.

an: For **an** the situation has been summarized in the next table:

1487 Thomsen 1984, 147-148, § 289.

1488 Line 11: analysis of the prefix chain **ba-an-** is **(i/a)-ba-en-**.

1489 The text concerned is: AO 6715: **... ba-ni-i<u>n</u>-du$_{11}$**; text Ni 4463: **... ba-ni-i<u>b</u>-du$_{11}$**. Attinger 1993, 720-725, ch. 5.3.181; this sentence is mentioned in § 857 d 1^0, p. 725.

	before marriage/ during courting	marriage/ intercourse	after marriage/ intercourse
inanimate	Barton *i*:12-14 NBC 11108: 8	Tree/Reed: 6 Sheep/Grain: 2	
animate	Ukg *ii*: 1-2 (?) Tree/Reed: 5	Tree/Reed: 7	Ukg *i*: 2-3 (?) NBC 11108: 1 GEN: 11
unknown *)	NBC 11108: 5-6		

(?) (?) indicates that the attribution 'animate' may be inferred from the interpretation of the text, but that it cannot be proven by grammatical analysis.

*) 'Unknown' means: it cannot be determined to which class **an** belongs; it may be animate or inanimate.

The tentative conclusion with respect to the class of **an** may be as follows:

- The scribes seem to hesitate between an inanimate and animate class of **an** before or during courting, even during the intercourse.
- After intercourse/marriage An is the god of heaven, thus animate.

***

Appendix

God lists

1 *Early Dynastic period*

1.1 *Fara*

SF 1 [1490]

Obv.

	transliteration	transcription	name	'meaning' [1491]
I:1	**an**	**an**	An	
2	**^{d}en-E$_2$**	**^{d}en-líl**	Enlil	
3	**^{d}MUŠ$_3$**	**^{d}inanna**	Inanna	
4	**^{d}en-ki**	**^{d}en-ki**	Enki	
5	**^{d}ŠEŠ-KI**	**^{d}nanna**	Nanna	
6	**^{d}UD**	**^{d}utu**	Utu	

VI:25	**^{d}nin-LAK 777-DU$_6$**	**^{d}nin-utua**	Nin'utua	stud, breeding ram
26	**^{d}nin-ki**	**^{d}nin-ki**	Ninki	earth
27	**^{d}nin-KID**	**^{d}nin-líl**	Ninlil	
28	**^{d}nin-BULUG$_3$**	**^{d}nin-bulug$_3$**	Ninbulug	growth
---				
VIII:10	**^{d}ENGUR**	**^{d}namma**	Namma	

[1490] Krebernik 1986.

[1491] 'Meaning' = translation of the second part of the god's name, especially of the 'ancestor' gods.

SF 23-24 [1492]

	transliteration	transcription	name	'meaning'
v:17	**en-ki**	**en-ki**	Enki	
18	**nin-ki**	**nin-ki**	Ninki	
19	**en-E$_2$**	**en-líl**	Enlil	
20	**nin-KID**	**nin-líl**	Ninlil	
vi:1	**en-UḪ̮**	**en-ĝiriš$_x$**	Engiriš	butterfly
2	**nin-UḪ̮**	**nin-ĝiriš$_x$**	Ningiriš	"
3	**en-BULUG$_3$**	**en-bulug$_3$**	Enbulug	growth
4	**nin-BULUG$_3$**	**nin-bulug$_3$**	Ninbulug	"
5	**en-LAK 777-DU$_6$**	**en-utua**	En'utua	stud, breeding ram
6	**nin-LAK 777-DU$_6$**	**nin-utua**	Nin'utua	"
7	**en-gukkal**	**en-gukkal**	Engukkal	fat-tailed sheep
8	**nin-gukkal**	**nin-gukkal**	Ningukkal	"
9	**en-á**	**en-á**	En'a	power
10	**nin-á**	**nin-á**	Nin'a	"

[1492] Deimel 1923; Mander 1986, 108-109, nos. 1-14.

1.2 *Abū Ṣalābīḫ*

IAS 82-89 [1493]

	transliteration	transcription	name	'meaning'
1	[**an** ?]	[**an** ?]	[An ?]	
2	[^{d}**en-E$_2$** ?]	[^{d}**en-líl** ?]	[Enlil ?]	
3	[^{d}**nin-K**]**ID**	^{d}**nin-líl**	Ninlil	
4	[^{d}**e**]**n-k**[**i**]	^{d}**en-ki**	Enki	
5	[^{d}**ŠE**]**Š-K**[**I**]	^{d}**nanna**	Nanna	
6	⌜^{d}**MUŠ$_3$**⌝	^{d}**inanna**	Inanna	
-				
273	^{d}**en-ki**	^{d}**en-ki**	Enki	earth
274	^{d}**nin-ki**	^{d}**nin-ki**	Ninki	"
275	^{d}**en-E$_2$**	^{d}**en-líl**	Enlil	
276	^{d}**nin-KID**	^{d}**nin-líl**	Ninlil	
277	^{d}**en-UḪ**	^{d}**en-ĝiriš$_x$**	Engiriš	butterfly
278	^{d}**nin-UḪ**	^{d}**nin-ĝiriš$_x$**	Ningiriš	"
279	^{d}**en-BULUG$_3$**	^{d}**en-bulug$_3$**	Enbulug	growth
280	^{d}**nin-BULUG$_3$**	^{d}**nin-bulug$_3$**	Ninbulug	"
281	^{d}**en-LAK 777-DU$_6$**	^{d}**en-utua**	En'utua	stud, breeding ram
282	^{d}**nin-LAK 777-DU$_6$**	^{d}**nin-utua**	Nin'utua	"
283	^{d}**en-gukkal**	^{d}**en-gukkal**	Engukkal	fat-tailed sheep
284	^{d}**nin-gukkal**	^{d}**nin-gukkal**	Ningukkal	"
285	^{d}**en-á**	^{d}**en-á**	En'a	power
286	^{d}**nin-á**	^{d}**nin-á**	Nin'a	"
287	^{d}**en-an**	^{d}**en-an**	En'an	heaven
288	^{d}**nin-an**	^{d}**nin-an**	Nin'an	"

1493 Mander 1986, 24 (nos. 1-6), 29 (nos. 273-288).

2 *Old Babylonian period*

2.1 *Isin: IB 1552 + IB 1568 + fragments* [1494]

col. i

		name	'meaning'
1	**[ ] an**	An	
2	**[^{d}] IB**	Uraš	
3	**^{d}e[n]-líl**	Enlil	
4	**^{d}n[u-nam]-nir**	Nunamnir	
5	**^{d}n[in]-líl**	Ninlil	
6	**^{d}e[n]-ki**	Enki	
7	**^{d}ni[n]-ki**	Ninki	
8	**^{d}e[n]-ul**	En'ul	bud, fruit
9	**^{d}ni[n]-ul**	Nin'ul	"
10	**^{d}e[n]-x**	En-[ ]	
11	**^{d}n[i]n-x**	Nin-[ ]	
12	**^{d}en-[ ]**	En-[ ]	
13	**^{d}nin-[ ]**	Nin-[ ]	
14	**^{d}en-[ ]**	En-[ ]	
15	**^{d}nin-[ ]**	Nin-[ ]	
16	**^{d}en-da-ra**	Endara	syllabic for **dara$_3$** ibex ?
17	**^{d}nin-da-ra**	Nindara	"
18	**^{d}en-unu-kù-ga**	En'unukuga	pure city
19	**^{d}nin-unu-[kù-ga]**	Nin'unukuga	"
20	**^{d}en-men-[na]**	Enmenna	crown
21	**^{d}nin-men-[na]**	Ninmenna	"
22	**diĝir-maḫ**	Diĝirmaḫ	(a mother goddess)

1494 Wilcke 1987, 94. The cuneiform texts have not yet been published (spring 2012).

2.2 *Nippur: SLT 122-124* [1495]

1	**an**	An
2	*an-tum*	*Antum*
3	**^{d}IB**	Uraš
4	**^{d}en-líl**	Enlil
5	**^{d}nu-nam-nir**	Nunamnir
6	**^{d}nin-líl**	Ninlil
-	-	-
37	**^{d}IB**	Uraš
-	-	-
107	**^{d}namma**	Namma

[1495] Chiera 1929, 122-124; Peterson 2009, 14-16, Composite Text. The numbers refer to Peterson's composite text.

2.3 *Mari* [1496]

		name	'meaning'
1	**^{d}e[n-líl]**	Enlil	
2	**^{d}nin-[líl]**	Ninlil	
3	**^{d}en-[ki]**	Enki	earth
4	**^{d}nin-[ki]**	Ninki	"
5	**^{d}en-[kur]**	Enkur	mountain
6	**^{d}nin-k[u]r**	Ninkur	"
7	**^{d}en-šár**	Enšar	all
8	**^{d}nin-šár**	Ninšar	"
9	**[^{d}e]n-bulug$_3$**	Enbulug	growth
10	**[^{d}nin]-bulug$_3$**	Ninbulug	"
11	**[^{d}en]-mul**	Enmul	star
12	**^{d}n[i]n-[m]ul**	Ninmul	"
13	**^{d}en-[ul]**	En'ul	bud, fruit
14	**^{d}nin-ul**	Nin'ul	"
15	**^{d}en-da-šurim-ma**	Endašurimma	power of dung [1497]
16	**^{d}n[in]-da-šurim-ma**	Nindašurimma	"
17	**^{d}en-du$_6$-kù-ga**	Endukuga	holy hill
18	**^{d}nin-du$_6$-kù-ga**	Nindukuga	"
19	**^{d}en-ama-a-a-^{d}en-líl-lá**	Lords ancestors of Enlil	
20	**^{d}nin-ama-a-a-^{d}en-líl-lá**	Ladies ancestors of Enlil	
21	**^{d}en-me-šár-ra**	Enmešarra	all **me**
22	**^{d}nin-me-šár-ra**	Ninmešarra	"

1496 Lambert 1985a, 182. For the order 1-4: see Lambert 1985a, 183, *ad* 1-22 Enlil.

1497 CAD L 152: *le'û*, lexical part: **da DA** = / *le'û* /. Medicinal use of dung(?): see CAD K 28-29, *kabû* A, *sub* lexical part and b).

2.4 *TCL XV 10 or 'Genouillac list'* [1498]

		name	'meaning'
1	^{d}**en-ki**	Enki	earth
2	^{d}**nin-ki**	Ninki	"
3	^{d}**en-mul**	Enmul	star
4	^{d}**nin-mul**	Ninmul	"
5	^{d}**en-ul**	En'ul	bud, fruit
6	^{d}**nin-ul**	Nin'ul	"
7	^{d}**en-nun**	Ennun	prince
8	^{d}**nin-nun**	Ninnun	"
9	^{d}**en-kur**	Enkur	mountain
10	^{d}**nin-kur**	Ninkur	"
11	^{d}**en-kin-gal**	Enkingal	grandee, high official
12	^{d}**nin-kin-gal**	Ninkingal	"
13	^{d}**en-šár**	Enšar	totality
14	^{d}**nin-šár**	Ninšar	"
15	^{d}**en-ḫal** [1499]	Enḫal	secret, extispicy
16	^{d}**nin-ḫal**	Ninḫal	"
17	^{d}**en-bulug$_3$**	Enbulug	growth
18	^{d}**nin-bulug$_3$**	Ninbulug	"
19	^{d}**en-giriš (ḪIxŠE)**	Engiriš	butterfly
20	^{d}**nin-giriš**	Ningiriš	"
21	^{d}**en-da-šurim-ma**	Endašurimma	power of dung
22	< ^{d}**nin-da-šurim-ma** >	Nindašurimma	"
23	< ^{d}**en-amaš** >	En'amaš	sheepfold
24	^{d}**nin-amaš**	Nin'amaš	"
25	^{d}**en-du$_6$-kù-ga**	Endukuga	holy hill
26	^{d}**nin-du$_6$-kù-ga**	Nindukuga	"
27	^{d}**en-an-na**	En'anna	heaven
28	^{d}**nin-an-na**	Nin'anna	"
29	^{d}**en-u$_4$-ti-la**	En'utila	lifetime
30	^{d}**nin-u$_4$-ti-la**	Nin'utila	"

1498 De Genouillac 1930, pl. XXV-XXXI.

1499 **ḪAL** may also be transcribed as **buluḫ** 'aromatic tree'.

		name	'meaning'
31	^d^**en-me-šár-ra**	Enmešarra	all **me**
32	^d^**nin-me-šár-ra**	Ninmešarra	"
33	**an**	An	
34	**an-šár-gal**	Anšargal	great totality
35	^d^**en-uru-ul-la**	En'uru'ulla	city of olden times
36	^d^**uraš (IB)**	Uraš	
37	^d^**nin-***ì-li*	*Belet-ili*	
38	^d^**namma**	Namma	
39	^d^**ama-tu-an-ki**	Amatu'anki	the mother who has born **an-ki**
40	^d^**en-líl**	Enlil	

3 *Middle Babylonian period*

an = *anum*, tablet I: 1-28 [1500]

	Sumerian part	name	explanatory part	name
1	**an**	An	^{d}*a-nu-um*	*Anum*
2	**an**	An	*an-tum*	*Antum*
3	**an-ki**	**an-ki**	^{d}*a-nu-um u an-tum*	*Anum and Antum*
4	**an-IB**	An-Uraš	*id.*	*id.*
5	^{d}**nin-IB**	Nin-Uraš	*id.*	*id.*
6	**an-šár-gal**	Anšargal	*id.*	*id.*
7	^{d}**ki-šár-gal**	Kišargal	*id.*	*id.*
8	**an-šár**	Anšar	*id.*	*id.*
9	^{d}**ki-šár**	Kišar	*id.*	*id.*
10	^{d}**en-šár**	Enšar	*id.*	*id.*
11	^{d}**nin-šár**	Ninšar	*id.*	*id.*
12	^{d}*du-rí*	Duri	*id.*	*id.*
13	^{d}*da-rí*	Dari	*id.*	*id.*
14	^{d}*làḫ-ma*	Laḫma	*id.*	*id.*
15	^{d}*la-ḫa-ma*	Laḫama	*id.*	*id.*
16	^{d}**é-kur**	Ekur	*id.*	*id.*
17	^{d}**gá-ra**	Gara	*id.*	*id.*
18	^{d}*a-la-la*	Alala	*id.*	*id.*
19	^{d}*be-li-li*	Belili	*id.*	*id.*
20	^{d}*a-la-la*-**ALAM**	Alala-ALAM	*id.*	*id.*
21	^{d}*be-li-li*-**ALAM**	Belili-ALAM	*id.*	*id.*
22	^{d}**en-uru-ul-la**	En'uru'ulla	*id.*	*id.*
23	^{d}**nin-uru-ul-la**	Nin'uru'ulla	*id.*	*id.*
24	**21 en-ama-a-a**	21 ancestors	**an-na-ke$_4$ne**	of An
25	^{d}**nin-**$ì$*-li*	*Belet-ili*	**dam-an-na-ke$_4$**	spouse of An
27	^{d}**namma**	Namma	**ama-**^{d}**en-ki-ga-ke$_4$**	mother of Enki
28	^{d}**ama-ù-tu-an-ki**	Ama'utu'anki	^{d}**namma**	Namma

[1500] Litke's reconstructed list (1998, 20-24).

an = *anum*, tablet I: 96-138 [1501]

	transcription	transcription	name	'meaning'
96-97	**^{d}en-ki**	**^{d}nin-ki**	En/Ninki	earth
98-99	**^{d}en-ul**	**^{d}nin-ul**	En/Nin'ul	bud, fruit
100-101	**^{d}en-mul**	**^{d}nin-mul**	En/Ninmul	star
102-103	**^{d}en-udu**	**^{d}nin-udu**	En/Nin'udu	sheep
104-105	**^{d}en-gin**	**^{d}nin-gin**	En/Ningin	firmness (?)
106-107	**^{d}en-da**	**^{d}nin-da**	En/Ninda	power (?) [1502]
108-109	**^{d}en-bulug$_{3}$**	**^{d}nin-bulug$_{3}$**	En/Ninbulug	growth
110-111	**^{d}en-ḫal** [1503]	**^{d}nin-ḫal**	En/Ninḫal	secret, extispicy
112-113	**^{d}en-ug**	**^{d}nin-ug**	En/Ninug	lion
114-115	**^{d}en-ga-raš** [1504]	**^{d}nin-ga-raš**	En/Ningaraš	decision
116-117	**^{d}en-šár**	**^{d}nin-šár**	En/Ninšar	all
118-119	**^{d}en-nun**	**^{d}nin-nun**	En/Ninnun	prince
120-121	**^{d}en-kur**	**^{d}nin-kur**	En/Ninkur	mountain
122-123	**^{d}en-amaš**	**^{d}nin-amaš**	En/Nin'amaš	sheepfold
124-125	**^{d}en-kin-gal**	**^{d}nin-kin-gal**	En/Ninkingal	grandee
126-127	**^{d}en-kù-ĝál**	**^{d}nin-kù-ĝál**	En/Ninkugal	holiness
128-129	**^{d}en-an-na**	**^{d}nin-an-na**	En/Nin'anna	heaven, sky
130-131	**^{d}en-u$_{4}$-ti-la**	**^{d}nin-u$_{4}$-ti-la**	En/Nin'utila	lifetime
132-133	**^{d}en-da-šurim-ma**	**^{d}nin-da-šurim-ma**	En/Ninšurimma	power of dung [1505]
134-135	**^{d}en-du$_{6}$-kù-ga**	**^{d}nin-du$_{6}$-kù-ga**	En/Nindukuga	holy hill
136-137	**^{d}en-me-šár-ra**	**^{d}nin-me-šár-ra**	En/Ninmešarra	all **me**
138	**42-àm en-ama-a-a**	**^{d}en-líl-lá-ke$_{4}$**	42 ancestors of Enlil	

***

[1501] Litke 1998, 30-34. Some of Litke's transliterations have been converted into transcriptions: 102-103 **LU** into **udu**, and 104-105 **DU** into **gin**. Instead of **garaš$_{3}$** (**GA** + **KAS**) [nos 114-115] I read: **ga-raš**. The explanatory part only has **šu** (with the exception of line 138), indicating that the name should be repeated and that no explication was necessary. Therefore this part has been omitted in this table.

[1502] CAD L 152: *le'û*, lexical part: **da DA** = / *le'û* /.

[1503] **ḪAL** may also be transcribed as **buluḫ** 'aromatic tree'.

[1504] **ga-raš** might have substituted the **giriš** of earlier god lists (Volk 1998-2001, 367). For the meaning 'decision': see CAD P 530, *purussû* (lexical part). The transcription **ga-eš$_{8}$** – with the same meaning 'decision' – is also possible (ePSD *ad* **kaš**).

[1505] Medicinal use (?): see CAD K 28-29, *kabû* A, *sub* lexical part and b).

Bibliography

Bibliography

Abusch 1998 — T. Abusch, Ghost and God: Some Observations on a Babylonian Understanding of Human Nature. In: A.I. Baumgarten, J. Assmann, G.G. Stroumsa (eds), *Self, Soul and Body in Religious Experience.* Leiden-Boston-Köln: Brill, 1998; 363-383.

Afanasieva 1998 — V. Afanasieva, Rationales und Irrationales im antiken Denken oder aus dem Blickwinkel des Dichters. In: J. Prosecký, (ed.), *Intellectual Life of the Ancient Near East. Papers Presented at the 43rd Rencontre Assyriologique Internationale; Prague, July 1-5, 1996.* Prague: Academy of Sciences of the Czech Republic, Oriental Institute, 1998; 19-28.

Akkermans, Schwartz 2004 — P.M.M.G. Akkermans, G.M. Schwartz, *The Archaeology of Syria. From Complex Hunter-Gatherers to Early Urban Societies (ca. 16,000-300 BC).* Cambridge: Cambridge University Press, 2004^2.

Alberti 1985 — A. Alberti, A Reconstruction of the Abū-Ṣalābīkh God-list. *SEL* 2 (1985) 3-23.

Algaze 2005 — G. Algaze, *The Uruk World System. The Dynamics of Expansion of Early Mesopotamian Civilization. Second Edition.* Chicago, London: The University of Chicago Press, 2005.

Algaze 2007 — G. Algaze, The Sumerian Takeoff. In: E.C. Stone (ed.), *Settlement and Society. Essays Dedicated to Robert McCormick Adams.* Los Angeles: Cotsen Institute of Archaeology, University of California; Chicago, Illinois: The Oriental Institute of the University of Chicago, 2007; 343-368.

Algaze 2013 — G. Algaze, The End of Prehistory and the Uruk Period. In: H. Crawford (ed.), *The Sumerian World.* London and New York: Routledge, 2013; 68-94.

Allotte de la Fuÿe 1912 — F.M. Allotte de la Fuÿe, *Documents Présargoniques. Fasc 2.1.* Paris: Leroux, 1912.

Alster 1970 — B. Alster, 'Note brève' on **en-ki nun-ki**. *RA* 64 (1970) 189-190.

Alster 1971-1972 — B. Alster, 'Ninurta and the Turtle', UET 6/1 2. *JCS* 24 (1971-1972) 120-125.

Alster 1974 — B. Alster, *The Instructions of Suruppak. A Sumerian Proverb Collection.* Mesopotamia 2. Copenhagen: Akademisk Forlag, 1974.

Alster 1976a — B. Alster, On the Earliest Sumerian Literary Tradition. *JCS* 28 (1976) 109-126.

Alster 1976b — B. Alster, Early Pattern in Mesopotamian Literature. In: B.L. Eichler, with the assistance of J.W. Heimerdinger, Å.W. Sjöberg (eds), *Kramer Anniversary Volume: Cuneiform Studies in Honor of Samuel Noah Kramer.* AOAT 25. Kevelaer: Verlag Butzon & Bercker; Neukirchen-Vluyn: Neukirchener Verlag, 1976; 13-24.

Alster 1982 — B. Alster, Emesal in Early Dynastic Sumerian? What Is the UD.GAL.NUN-orthography? *ASJ* 4 (1982) 1-6.

Alster 1992 — B. Alster, Interaction of Oral and Written Poetry in Early

Mesopotamian Literature. In: M.E. Vogelzang, H.L.J. Vanstiphout (eds), *Mesopotamian Epic Literature: Oral or Aural?* Lewiston/Queenston/Lampeter: The Edwin Mellen Press, 1992; 23-69.

Alster 1997 B. Alster, *Proverbs of Ancient Sumer. The World's Earliest Proverb Collections.* Vols. I and II. Bethesda, Maryland: CDL Press, 1997.

Alster 1999 B. Alster, Tigris. In: K. van der Toorn, B. Becking, P.W. van der Horst (eds), *Dictionary of Deities and Demons in the Bible.* Second extensively revised edition. Leiden, Boston, Köln: Brill; Grand Rapids, Michigan / Cambridge, U.K.: W.B. Eerdmans Publishing Company, 1999; 870-871.

Alster 2002 B. Alster, *ilū awīlum : we-e i-la*, 'Gods : Men' versus 'Men : God'. Punning and the Reversal of Patterns in the Atrahasis Epic. In: T. Abusch (ed.), *Riches Hidden in Secret Places. Ancient Near Eastern Studies in Memory of Thorkild Jacobsen.* Winona Lake, Indiana: Eisenbrauns, 2002; 35-40.

Alster 2005a B. Alster, Nanše and her Fish. In: Y. Sefati *et al.* (eds), *'An Experienced Scribe Who Neglects Nothing'. Ancient Near Eastern Studies in Honor of Jacob Klein.* Bethesda, Maryland: CDL Press, 2005; 1-18.

Alster 2005b B. Alster, *Wisdom of Ancient Sumer.* Bethesda, Maryland: CDL Press, 2005.

Alster, Vanstiphout 1987 B. Alster, H. Vanstiphout, Lahar and Ashnan. Presentation and Analysis of a Sumerian Disputation. *ASJ* 9 (1987) 1-43.

Alster, Westenholz 1994 B. Alster, A. Westenholz, The Barton Cylinder. *ASJ* 16 (1994) 15-46.

Amiet 1953 P. Amiet, Les Combats Mythologiques dans l'Art Mésopotamien du Troisième et du Début du Second Millénaires. *Rev. Archéol.* 42 (1953) 129-164.

Amiet 1980 P. Amiet, *La Glyptique Mésopotamienne Archaïque.* Paris: Éditions du Centre National de la Recherche Scientifique, 1980.

Amiet 1995 P. Amiet, La Naissance des Dieux. *RB* 102 (1995) 481-505.

Andersson 2012 J. Andersson, *Kingship in the Early Mesopotamian Onomasticon 2800-2200 BCE.* Uppsala: Uppsala Universitet, 2012.

Artemov 2012 N. Artemov, The Elusive Beyond: Some Notes on the Netherworld Geography in Sumerian Tradition. In: C. Mittermayer, S. Ecklin, *Altorientalische Studien zu Ehren von Pascal Attinger.* **mu-ni u$_4$ ul-li$_2$-a-aš ĝa$_2$-ĝa$_2$-de$_3$**. OBO 256. Fribourg: Academic Press; Göttingen: Vandenhoeck & Ruprecht, 2012; 1-30.

Assmann, Frankfurter 2004 J. Assmann, D. Frankfurter, Egypt. In: S.I. Johnston (ed.), *Religions of the Ancient World. A guide.* Cambridge, Massachusetts, and London: The Belknap Press of Harvard University Press, 2004; 155-164.

Attinger 1993 P. Attinger, *Eléments de linguistique sumérienne. La construction de* **du$_{11}$/e/di** *«dire».* OBO Sonderband. Fribourg, Suisse: Éditions Universitaires; Göttingen: Vandenhoeck & Ruprecht, 1993.

Attinger 1998 P. Attinger, /n/ réflexif ? *NABU* 1998/41, 41-43.

Attinger 2001 P. Attinger, Review of S. Tinney, The Nippur Lament. Royal Rhetoric and Divine Legitimation in the Reign of Išme-Dagan of Isin (1953-1935 B.C.) *ZA* 91 (2001) 133-142.

Attinger 2005 P. Attinger, A propos de AK «faire» (II). *ZA* 95 (2005) 208-275.

Attinger, Krebernik 2005 P. Attinger, M. Krebernik, L'Hymne à Ḫendursaĝa (Ḫendursaĝa A). In: R. Rollinger (ed.), *Von Sumer bis Homer. Festschrift für Manfred Schretter zum 60. Geburtstag am 25. Februar 2004.* AOAT 325. Münster: Ugarit-Verlag, 2005; 21-104.

Baasten 2009 M. Baasten, Scheppen of Scheiden? Over de Interpretatie van Genesis 1. *Met Andere Woorden* 28 (2009) 3-14.

Bailleul-LeSuer 2012 R. Bailleul-LeSuer (ed., with new photography by A.R. Ressman), *Between Heaven and Earth. Birds in Ancient Egypt.* OIMP 35. Chicago: The Oriental Institute of the University of Chicago, 2012.

Baines, Málek 1993 J. Baines, J. Málek, *Atlas of Ancient Egypt.* New York: Facts on File Inc., 1993 (reprint of ed. 1980).

Balke 2006 T.E. Balke, *Das sumerische Dimensionalkasussystem.* AOAT 331. Münster: Ugarit-Verlag, 2006.

Baltrušaitis 1934 J. Baltrušaitis, *Art Sumérien Art Roman.* Paris: Librairie Ernest Leroux, 1934.

Barré 2001 M.L. Barré, 'Wandering about' as a *Topos* of Depression in Ancient Near Eastern Literature and in the Bible. *JNES* 60 (2001) 177-187.

Barton 1918 G.A. Barton, No. 1. The Oldest Religious Text from Babylonia. In: G.A. Barton, *Miscellaneous Babylonian Inscriptions.* New Haven: Yale University Press; London, Humphrey, Milford: Oxford University Press, 1918; 1-20.

Bauer 1972 J. Bauer, *Altsumerische Wirtschaftstexte aus Lagasch.* Rome: Biblical Institute Press, 1972.

Bauer 1982a J. Bauer, Leben in der Urzeit Mesopotamiens. *AfO Beiheft* 19 (1982) 377-383.

Bauer 1982b J. Bauer, Die Göttin ^{d}**kug-sù(-g)**. *Altorientalische Notizen* 19 (1982) 4-7.

Bauer 1998 J. Bauer, Der vorsargonische Abschnitt der mesopotamischen Geschichte. In: J. Bauer, R.K. Englund, M. Krebernik, *Mesopotamien. Späturuk-Zeit und Frühdynastische Zeit.* OBO 160/1. Freiburg, Schweiz: Universitätsverlag; Göttingen: Vandenhoeck & Ruprecht, 1998; 431-585.

Bauer 2005 J. Bauer, Gudea-Studien. In: Y. Sefati *et al.* (eds), *'An Experienced Scribe Who Neglects Nothing'. Ancient Near Eastern Studies in Honor of Jacob Klein.* Bethesda, Maryland: CDL Press, 2005; 19-28.

Becking, Korpel 2010 B. Becking, M.C.A. Korpel, To Create, to Separate or to Construct: an Alternative for a Recent Proposal as to the Interpretation of ברא in Gen 1:1-2:4a. *JHS* 10 (2010) 1-21.

Beckman 1996 G. Beckman (and H.A. Hoffner, Jr. ed.), *Hittite Diplomatic*

Texts. Atlanta, Georgia: Scholars Press, 1996.

Behrens 1978 — H. Behrens, *Enlil und Ninlil. Ein sumerischer Mythos aus Nippur*. Studia Pohl: Series Maior, 8. Rome: Biblical Institute Press, 1978.

Benito 1969 — C.A. Benito, *'Enki and Ninmaḫ'* and *'Enki and the World Order'*. Ann Arbor, Michigan: UMI, 1969.

Bergmann 1964 — E. Bergmann, Untersuchungen zu syllabisch geschriebenen sumerischen Texten. I. Altbabylonische Texte. *ZA* 56 (NF 22) (1964) 1-43.

Berlejung 1998 — A. Berlejung, *Die Theologie der Bilder. Herstellung und Einweihung von Kultbildern in Mesopotamien und die alttestamentliche Bilderpolemik*. OBO 162. Freiburg, Schweiz: Universitätsverlag; Göttingen: Vandenhoeck & Ruprecht, 1998.

Berlin 1979 — A. Berlin, *Enmerkar and Ensuḫkešdanna. A Sumerian Narrative Poem*. Philadelphia: The University Museum, 1979.

Bickel 1994 — S. Bickel, *La Cosmogonie Égyptienne Avant le Nouvel Empire*. OBO 134. Fribourg, Suisse: Éditions Universitaires; Göttingen: Vandenhoeck & Ruprecht, 1994.

Biggs 1966 — R.D. Biggs, The Abū Ṣalābīkh Tablets. A Preliminary Survey. *JCS* 20 (1966) 73-88.

Biggs 1974 — R.D. Biggs, *Inscriptions from Tell Abū Ṣalābīkh*. OIP 99. Chicago & London: The University of Chicago Press, 1974.

Biggs 1976 — R.D. Biggs, *Inscriptions from Al-Hiba-Lagash: The First and Second Seasons*. Bibliotheca Mesopotamica 3. Malibu, California: Undena Publications, 1976.

Biggs 1981 — R.D. Biggs, Ebla and Abu Salabikh: The Linguistic and Literary Aspects. In: L. Cagni (ed.), *La Lingua di Ebla. Atti del Convegno Internazionale (Napoli, 21-23 aprile 1980)*. Series Minor XIV. Napoli: Instituto Universitario Orientale, Seminario di Studi Asiatici, 1981; 121-133.

Biggs 1988 — R. Biggs, The Semitic Personal Names from Abu Salabikh and the Personal Names from Ebla. In: A. Archi (ed.), *Eblaite Personal Names and Semitic Name-Giving. Papers of a Symposium held in Rome, July 15-17, 1985*. ARES I. Roma: Missione Archeologica Italiana in Siria, 1988; 89-98.

Binsbergen, v., Wiggermann 1999 — W. van Binsbergen, F. Wiggermann, Magic in History. A Theoretical Perspective, and its Application to Ancient Mesopotamia. In: T. Abusch, K. van der Toorn (eds), *Mesopotamian Magic. Textual, Historical, and Interpretative Perspectives*. Groningen: STYX Publications, 1999; 3-34.

Black 1992 — J. Black, Some Structural Features of Sumerian Narrative Poetry. In: M.E. Vogelzang, H.L.J. Vanstiphout (eds), *Mesopotamian Epic Literature: Oral or Aural?* Lewiston/Queenston/ Lampeter: The Edwin Mellen Press, 1992; 71-101.

Black 2003 — J. Black, Sumerian Noises: Ideophones in Context. In: W. Sallaberger, K. Volk, A. Zgoll (eds), *Literatur, Politik und Recht in Mesopotamien. Festschrift für Claus Wilcke*. Wiesbaden: Harrassowitz Verlag, 2003; 35-52.

Black, Green 1998 — J. Black, A. Green (illustrations by T. Rickards), *Gods,*

Demons and Symbols of Ancient Mesopotamia. An Illustrated Dictionary. London: British Museum Press, 1998².

Black, Zólyomi 2007 — J. Black, G. Zólyomi, Introduction to the Study of Sumerian. In: J. Ebeling and G. Cunningham (eds), *Analysing Literary Sumerian. Corpus-based Approaches*. London, Oakville: Equinox Publishing Ltd., 2007; 1-32.

Boehmer 1957-1971 — R.M. Boehmer, Götterkämpfe in der Bildkunst. In: E. Ebeling, B. Meissner (eds), *Reallexikon der Assyriologie*, Band 3; Berlin: Walter de Gruyter, 1957-1971; 471-473.

Boehmer 1965 — R.M. Boehmer, *Die Entwicklung der Glyptik während der Akkad-Zeit*. Berlin: Walter de Gruyter & Co., 1965.

Boehmer 1980-1983 — R.M. Boehmer, Kopfbedeckung. B. In der Bildkunst. In: E. Ebeling, B. Meissner (eds), *Reallexikon der Assyriologie*, Band 6; Berlin, New York: Walter de Gruyter, 1980-1983; 203-210.

Boese 2010 — J. Boese, Die Blau'schen Steine und der 'Priesterfürst' im Netzrock. *AoF* 37 (2010) 208-229.

Borger 1967-1975 — R. Borger, *Handbuch der Keilschriftliteratur*. Band I-III. Berlin, New York: De Gruyter, 1967-1975.

Borger 1985 — R. Borger, Einige Texte religiösen Inhalts. II. 'Enki und Ninmaḫ' in Ninive. *OrNS* 54 (1985) 18-22.

Borger 1998 — R. Borger, Zum Emesal-Vokabular. In: M. Dietrich, O. Loretz, T.E. Balke (eds), **dubsar anta-men**. *Studien zur Altorientalistik. Festschrift für Willem H. Ph. Römer zur Vollendung seines 70. Lebensjahres mit Beiträgen von Freunden, Schülern und Kollegen*. AOAT 253. Münster: Ugarit-Verlag, 1998; 17-37.

Borger 2010 — R. Borger, *Mesopotamisches Zeichenlexikon. Zweite, revidierte und aktualisierte Auflage*. AOAT 305. Münster: Ugarit Verlag, 2010.

Bottéro 1982 — J. Bottéro, La Création de l'Homme et sa Nature dans le Poème d'Atraḫasīs. In: M.A. Dandamayev *et al*. (eds), *Societies and Languages of the Ancient Near East. Studies in Honour of I.M. Diakonoff*. Warminster: Aris & Phillips Ltd., 1982; 24-32.

Bottéro 1987 — J. Bottéro, *Mésopotamie. L'Écriture, la Raison et les Dieux*. Paris: Gallimard, 1987.

Bottéro 1995 — J. Bottéro, *Mesopotamia. Writing, Reasoning, and the Gods*. (Translated by Z. Bahrani and M. van de Mieroop). Chicago & London: The University of Chicago Press, 1995.

Bottéro 2001 — J. Bottéro, *Religion in Ancient Mesopotamia*. (Translated by T.L. Fagan). Chicago and London: The University of Chicago Press, 2001.

Bottéro, Kramer 1993 — J. Bottéro, S.N. Kramer, *Lorsque les Dieux Faisaient l'Homme. Mythologie Mésopotamienne*. [Paris]: Gallimard, 1993.

Brandon 1963 — S.G.F. Brandon, *Creation Legends of the Ancient Near East*. London: Hodder and Stoughton Ltd., 1963.

Bruschweiler 1987 — F. Bruschweiler, *Inanna. La Deésse Triomphante et Vaincue dans la Cosmologie Sumérienne. Recherche Lexicographique*. Leuven: Éditions Peeters, 1987.

Buren, v. 1935-1936 — E.D. van Buren, Entwined Serpents. *AfO* 10 (1935-1936) 53-65.

Buren, v. 1939-1940 — E.D. van Buren, Religious Rites and Ritual in the Time of Uruk IV-III. *AfO* 13 (1939-1940) 32-45.

Burrows 1935 — E.R. Burrows, *Ur Excavations: Texts. II: Archaic Texts.* London: The Trustees of The British Museum and of the University Museum, University of Pennsylvania, Philadelphia, 1935.

Cameron 1981 — D.O. Cameron, *Symbols of Birth and Death in the Neolithic Era.* London: Kenyon-Deane Ltd., 1981.

Caquot 1980 — A. Caquot, Schöpfungsmythen der Kanaanäer. In: *Die Schöpfungsmythen Ägypter, Sumerer, Hurriter, Hethiter, Kanaaniter und Israeliten. Mit einem Vorwort von Mircea Eliade.* Darmstadt: Wissenschaftliche Buchgesellschaft, 1980; 173-182.

Carroué 1991 — F. Carroué, Études de Géographie et de Topographie Sumériennes. II. À la Recherche de l'Euphrate au IIIe Millénaire. *ASJ* 13 (1991) 111-156.

Cassirer 1953 — E. Cassirer, *Language and Myth. Translated by S.K. Langer.* (A republication of the first English edition by Harper and Brothers 1946. Originally published in German: *Sprache und Mythos: ein Beitrag zum Problem der Götternamen*, 1925). New York: Dover Publications Inc., 1953.

Castellino 1957 — G. Castellino, Urnammu. Three Religious Texts. *ZA* 52 (NF 18) (1957) 1-57.

Cavigneaux 1978 — A. Cavigneaux, L'Essence Divine. *JCS* 30 (1978) 177-185.

Cavigneaux 1980-1983 — A. Cavigneaux, Lexikalische Listen. In: E. Ebeling, B. Meissner (eds), *Reallexikon der Assyriologie*, Band 6; Berlin, New York: Walter de Gruyter, 1980-1983; 609-641.

Cavigneaux 1995 — A. Cavigneaux, La Pariade du Scorpion dans les Formules Magiques Sumériennes (Textes de Tell Haddad V). *ASJ* 17 (1995) 75-99.

Cavigneaux, Jaques 2010 — A. Cavigneaux, M. Jaques, Peut-on Comprendre le Silbenvokabular? In: D. Shehata, F. Weiershäuser, K.V. Zand (eds), *Von Göttern und Menschen. Beiträge zu Literatur und Geschichte des alten Orients. Festschrift für Brigitte Groneberg.* Leiden, Boston: Brill, 2010; 1-15.

Cavigneaux, Krebernik 1998-2001 — A. Cavigneaux, M. Krebernik, Nin-ki. In: E. Ebeling, B. Meissner (eds), *Reallexikon der Assyriologie*, Band 9; Berlin, New York: Walter de Gruyter, 1998-2001; 445-447.

Cavigneaux, Rawi 1993 — A. Cavigneaux, F.N.H. Al-Rawi, Textes Magiques de Tell Hadad (Textes deTell Haddad II); I. Grand Texte contre Namtar. *ZA* 83 (1993) 176-195.

Cavigneaux, Rawi 2000 — A. Cavigneaux, F.N.H. Al-Rawi, *Gilgameš et la Mort. Textes de Tell Haddad VI avec un Appendice sur les Textes Funéraires Sumériens.* CM 19. Groningen: STYX Publications, 2000.

CDLI — Cuneiform Digital Library Initiative. Homepage: http://cdli.ucla.edu/.

CDLI Wiki — Abbreviations for Assyriology:

http://cdli.ox.ac.uk/wiki/abbreviations_for_assyriology.

Charpin 1986 D. Charpin, *Le Clergé d'Ur au Siècle d'Hammurabi (XIXe - XVIIIe Siècles av. J.-C.)*, Genève-Paris: Librairie Droz, 1986.

Charpin 1987 D. Charpin, Tablettes Présargoniques de Mari. *MARI* 5 (1987) 65-127.

Charpin 1994 D. Charpin, Inanna/Eštar, Divinité Poliade d'Uruk à l'Époque Paléo-Babylonienne. *NABU* 1994/39, 37-38.

Charvát 1981 P. Charvát, The Kish Evidence and the Emergence of States in Mesopotamia. *CA* 22 (1981) 686-688.

Charvát 1998 P. Chavát, *On People, Signs and States. Spotlights on Sumerian Society c. 3500 - 2500 B.C.* Prague: The Oriental Institute, Academy of Sciences of the Czech Republic, 1998.

Charvát 2002 P. Charvát, *Mesopotamia Before History.* London, New York: Routledge, 2002.

Charvát 2005 P. Charvát, *The Iconography of Pristine Statehood. Painted Pottery and Seal Impressions from Susa, Southwestern Iran.* Prague: Charles University Prague, Karolinum Press, 2005.

Charvát 2006 P. Charvát, Iconographie des premiers états en Mésopotamie. In: P. Charvát et al. (eds), *L'État, le Pouvoir, les Prestations et leurs Formes en Mésopotamie ancienne. Actes du Colloque assyriologique franco-tchèque. Paris, 7-8 novembre 2002.* Praha: Universita Karlova v Praze, 2006; 23-30.

Chiera 1929 E. Chiera, *Sumerian Lexical Texts from the Temple School of Nippur.* OIP XI. Chicago, Illinois: The University of Chicago Press, 1929.

Chiodi 1997 S.M. Chiodi, *Offerte «Funebri» nella Lagaš Presargonica.* Roma: Università degli Studi di Roma 'La Sapienza', 1997.

Chiodi 2008 S.M. Chiodi, 'AN.KI. Alle Origini del Mondo Divino'. *Rendiconti della Classe di Scienze Morali, Storiche e Filologiche, Nota Serie IX, Volume XIX, fascicolo 3*; Roma: Accademia dei Lincei, 2008; 517-539.

Civil 1966 M. Civil, Notes on Sumerian Lexicography, I. *JCS* 20 (1966) 119-124.

Civil 1972 M. Civil, Das Lugalbandaepos by Claus Wilcke. *JNES* 31 (1972) 385-387.

Civil 1974-1977 M. Civil, Enlil and Namzitarra. *AfO* 25 (1974-1977) 65-71.

Civil 1980 M. Civil, Les Limites de l'Information Textuelle. In: *L'Archéologie de l'Iraq du Début de l'Époque Neólithique à 333 avant notre ère: Perspectives et Limites de l'Interprétation Anthropologique des Documents. Paris, 13-15 juin 1978;* Colloques internationaux du Centre National de la Recherche Scientifique no 580. Paris: Éditions du Centre National de la Recherche Scientifique, 1980; 225-232.

Civil 1983 M. Civil, Enlil and Ninlil: The Marriage of Sud. *JAOS* 103 (1983) 43-66.

Civil 1985 M. Civil, On Some Texts Mentioning Ur-Namma. *OrNS* 54 (1985) 27-45.

Civil 1989 M. Civil, The Statue of Šulgi-ki-ur$_5$-sag$_9$-kalam-ma. Part One: The Inscription. In: H. Behrens, D. Loding, M.T. Roth (eds),

DUMU-E_2-DUB-BA-A. *Studies in Honor of Åke W. Sjöberg.* Philadelphia: The Samuel Noah Kramer Fund, 1989; 49-63.

Civil 1994 — M. Civil, *The Farmer's Instructions: A Sumerian Agricultural Manual.* Sabadell-Barcelona: AUSA, 1994.

Civil 1999 — M. Civil, The Sumerian Flood Story. In: W.G. Lambert, A.R. Millard, *Atra-ḫasīs. The Babylonian Story of the Flood.* Winona Lake, Indiana: Eisenbrauns, 1999; 138-145; commentary 167-172.

Civil 2000a — M. Civil, Ancient Mesopotamian Lexicography. In: J.M. Sasson (ed.), *Civilizations of the Ancient Near East.* Peabody, Massachusetts: Hendrickson Publishers, Inc., 2000; 2305-2314.

Civil 2000b — M. Civil, Modal Prefixes. In: J. Black (†), G. Zólyomi (eds), *Special Volume in Honor of Professor Mamoru Yoshikawa, 1. The Study of Diachronic and Synchronic Variation in Sumerian: Papers Presented at the 6th Meeting of the Sumerian Grammar Discussion Group, Oxford, 17th and 18th September 1999* [Organized by J. Black (†) and G. Zólyomi]. *ASJ* 22 (2000) 29-42.

Civil 2003 — M. Civil, Recension of G.J. Selz, *Altsumerische Wirtschaftsurkunden aus amerikanischen Sammlungen.* FAOS 15, 2. Stuttgart: F Steiner Verlag; 1993. In: *AuOr* 21 (2003) 288-289.

Civil 2007 — M. Civil, Early Semitic Loanwords in Sumerian. In: M.T. Roth *et al.* (eds), *Studies Presented to Robert D. Biggs, June 4, 2004.* Chicago: The Oriental Institute of the University of Chicago, 2007; 11-33.

Clifford 1994 — R.J. Clifford, *Creation Accounts in the Ancient Near East and in the Bible.* Washington D.C.: The Catholic Biblical Association of America, 1994.

Cohen 1973 — S. Cohen, *Enmerkar and the Lord of Aratta.* Ann Arbor, Michigan: UMI, 1973.

Cohen 1988 — M.E. Cohen, *The Canonical Lamentations of Ancient Mesopotamia.* Vol. I. Potomac, Maryland: Capital Decisions Ltd, 1988.

Cohen 1989 — M.E. Cohen, A Bilingual Šuilla to Ningeštinanna. In: H. Behrens, D. Loding, M.T. Roth (eds), **DUMU-E_2-DUB-BA-A**. *Studies in Honor of Åke W. Sjöberg.* Philadelphia: The Samuel Noah Kramer Fund, 1989; 79-85.

Collins 2002 — B.J. Collins, Necromancy, Fertility and the Dark Earth: The Use of Ritual Pits in Hittite Cult. In: P. Mirecki, M. Meyer (eds), *Magic and Ritual in the Ancient World.* Leiden-Boston-Köln: Brill, 2002; 224-241.

Collon 1997 — D. Collon, Moon, Boats and Battle. In: I.L. Finkel, M.J. Geller (eds), *Sumerian Gods and their Representations.* CM 7. Groningen: STYX Publications, 1997; 11-17.

Conti 1993 — G. Conti, Ninirigal, Mère de Kullab. *MARI* 7 (1993) 343-347.

Conti 1997 — G. Conti, Incantation de l'Eau Bénite et de l'Encensoir et Textes Connexes. *MARI* 8 (1997) 253-272.

Cooper 1972-1975 — J.S. Cooper, Heilige Hochzeit. B. Archäologisch. In: E. Ebeling, B. Meissner (eds), *Reallexikon der Assyriologie*, Band

4; Berlin, New York: Walter de Gruyter, 1972-1975; 259-269.

Cooper 1973a J.S. Cooper, Review of G. Pettinato, *Das altorientalische Menschenbild und die sumerischen und akkadischen Schöpfungsmythen. JAOS* 93 (1973) 581-585.

Cooper 1973b J.S. Cooper, Sumerian and Akkadian in Sumer and Akkad. *OrNS* 42 (1973) 239-246.

Cooper 1978 J.S. Cooper, *The Return of Ninurta to Nippur.* **an.gim dím.ma**. *AnOr* 52. Roma: Pontificium Institutum Biblicum, 1978.

Cooper 1983 J.S. Cooper, *The Curse of Agade.* Baltimore and London: The Johns Hopkins University Press, 1983.

Cooper 1989 J.S. Cooper, Enki's Member: Eros and Irrigation in Sumerian Literature. In: H. Behrens, D. Loding, M.T. Roth (eds), **DUMU-E$_2$-DUB-BA-A**. *Studies in Honor of Åke W. Sjöberg.* Philadelphia: The Samuel Noah Kramer Fund, 1989; 87-89.

Cooper 1992 J.S. Cooper, Babbling on Recovering Mesopotamian Orality. In: M.E. Vogelzang, H.L.J. Vanstiphout (eds), *Mesopotamian Epic Literature: Oral or Aural?* Lewiston/Queenston/ Lampeter: The Edwin Mellen Press, 1992; 103-122.

Cooper 1993 J.S. Cooper, Sacred Marriage and Popular Cult in Early Mesopotamia. In: E. Matsushima (ed.), *Official Cult and Popular Religion in the Ancient Near East.* Heidelberg: Universitätsverlag C. Winter, 1993; 81-96.

Crawford 2013 H. Crawford (ed.), *The Sumerian World.* London and New York: Routledge, 2013.

Crenshaw 2000 J.L. Crenshaw, The Contemplative Life in the Ancient Near East. In: J.M. Sasson (ed.), *Civilizations of the Ancient Near East*. Peabody, Massachusetts: Hendrickson Publishers, Inc., 2000; 2445-2457.

Cros 1910 G. Cros, publiées avec le concours de Léon Heuzey et François Thureau-Dangin, *Nouvelles Fouilles de Tello.* Paris: Leroux, 1910.

Deimel 1923 A. Deimel, *Die Inschriften von Fara. II: Schultexte aus Fara.* Leipzig: Hinrichs, 1923.

Deimel 1924 A. Deimel, *Die Inschriften von Fara. III: Wirtschaftstexte aus Fara.* Leipzig: Hinrichs, 1924.

Delnero 2006 P. Delnero, *Variation in Sumerian Literary Compositions: a Case Study Based on the Decad.* Dissertation. University of Pennsylvania, 2006.

Delnero 2007 P. Delnero, Pre-verbal /**n**/: Function, Distribution, and Stability. In: J. Ebeling, G. Cunningham (eds), *Analysing Literary Sumerian. Corpus-based Approaches.* London, Oakville: Equinox Publishing Ltd., 2007; 105-143.

Delnero 2012 P. Delnero, The Textual Criticism of Sumerian Literature. *JCS, Supplemental Series 3*. Boston: American Schools of Oriental Research, 2012.

Dietrich 1984 M. Dietrich, Die Kosmogonie in Nippur und Eridu. *JARG* 5 (1984) 155-184.

Dietrich 1991 M. Dietrich, Die Tötung einer Gottheit in der Eridu-Babylon-Mythologie. In: D.R. Daniels, U. Gleßmer, M. Rösel (eds),

Ernten, was man sät. Festschrift für Klaus Koch zu seinem 65. Geburtstag. Neukirchen-Vluyn: Neukirchener Verlag, 1991; 49-73.

Dietrich 1993a — M. Dietrich, Ea hat die Geschicke des Menschen festgesetzt. *MARG* 8 (1993) 1-24.

Dietrich 1993b — M. Dietrich, Persönliches Unheil als Zeichen der Gottesferne. Das Verhältnis zwischen Schöpfer und Geschöpf nach babylonischem Verständnis. *MARG* 8 (1993) 115-141.

Dietrich 1994 — M. Dietrich, '*Wir wollen die Menschheit schaffen!*'. Der göttliche Ursprung des Menschenwegs nach der sumero-babylonischen Mythologie. *MARG* 9 (1994) 41-54.

Dietrich 1995 — M. Dietrich, *ina ūmī ullûti* 'An jenen (fernen) Tagen'. Ein sumerisches kosmogonisches Mythologem in babylonischer Tradition. In: M. Dietrich, O. Loretz (eds), *Vom alten Orient zum alten Testament. Festschrift für Wolfram Freiherrn von Soden zum 85. Geburtstag am 19. Juni 1993.* AOAT 240. Kevelaer: Verlag Butzon & Bercker; Neukirchen-Vluyn: Neukirchener Verlag, 1995; 57-72.

Dietrich 2007 — M. Dietrich, Enki/Ea und El - Die Götter der Künste und Magie. In: T.R. Kämmerer (ed.), *Studien zu Ritual und Sozialgeschichte im alten Orient / Studies on Ritual and Society in the Ancient Near East.* Tartuer Symposien 1998-2004. Berlin, New York: Walter de Gruyter, 2007; 93-125.

Dijk, v. 1953 — J.J.A. van Dijk, *La Sagesse Suméro-Accadienne. Recherches sur les Genres Littéraires des Textes Sapientiaux.* Leiden: E.J. Brill, 1953.

Dijk, v. 1957-1971 — J. van Dijk, Gott. A. Nach sumerischen Texten. In: E. Ebeling, B. Meissner (eds), *Reallexikon der Assyriologie*, Band 3; Berlin: Walter de Gruyter, 1957-1971; 532-543.

Dijk, v. 1960 — J. J.A. van Dijk, *Sumerische Götterlieder. II. Teil.* Heidelberg: Carl Winter Universitätsverlag, 1960.

Dijk, v. 1964 — J. van Dijk, Le Motif Cosmique dans la Pensée Sumérienne. *AcOr* 28 (1964) 1-59.

Dijk, v. 1967a — J. van Dijk, Einige Bemerkungen zu sumerischen religionsgeschichtlichen Problemen. *OLZ* LXII (1967) 229-244.

Dijk, v. 1967b — VAT 8382. Ein zweisprachiges Königsritual. In: *Heidelberger Studien zum alten Orient. Adam Falkenstein zum 17. September 1966.* Wiesbaden: Otto Harrassowitz, 1967; 233-268.

Dijk, v. 1969 — J. van Dijk, Les Contacts Ethniques dans la Mésopotamie et les Syncrétismes de la Religion Sumérienne. In: S.S. Hartman (ed.), *Syncretism: Based on Papers Read at the Symposium on Cultural Contact, Meeting of Religions, Syncretism held at Åbo on the 8th-10th of September, 1966.* Stockholm: Almqvist & Wiksell, 1969; 171-206.

Dijk, v. 1971 — J. van Dijk, Sumerische Religion. In: J.P. Asmussen, J. Læssøe, in Verbindung mit C. Colpe (eds), *Handbuch der Religionsgeschichte.* Band 1. Göttingen: Vandenhoeck & Ruprecht, 1971; 431- 496.

Dijk, v. 1973 — J. van Dijk, Une Incantation Accompagnant la Naissance de l'Homme. *OrNS* 42 (1973) 502-507.

Dijk, v. 1976 — J. van Dijk, Existe-t-il un 'Poème de la Création' Sumérien? In: B.L. Eichler, with the assistance of J.W. Heimerdinger and Å.W. Sjöberg (eds.), *Kramer Anniversary Volume: Cuneiform Studies in Honor of Samuel Noah Kramer*. AOAT 25. Kevelaer: Verlag Butzon & Bercker; Neukirchen-Vluyn: Neukirchener Verlag, 1976; 125-133.

Dijk, v. 1983 — J. van Dijk, **LUGAL UD ME-LÁM-bi NIR-ĜÁL**. *Le Récit Épique et Didactique des Travaux de Ninurta, de Déluge et de la Nouvelle Création. Tome I. Introduction, Texte Composite, Traduction. Tome II. Introduction à la reconstruction du texte, inventaire des textes, partition, copies des originaux*. Leiden: E.J. Brill, 1983.

Dijk, v. 1998 — J.J.A. van Dijk, Inanna raubt den 'großen Himmel'. Ein Mythos. In: S.M. Maul, *Festschrift für Rykle Borger zu seinem 65. Geburtstag am 24. Mai 1994. tikip santakki mala bašmu...* CM 10. Groningen: STYX Publications, 1998; 9-37.

Dijk, v. *et al.* 1985 — J. van Dijk, A. Goetze, M.I. Hussey, *Early Mesopotamian Incantations and Rituals*. YOS, Babylonian Texts XI. New Haven and London: Yale University Press, 1985.

Dijk, v., Geller 2003 — J.J.A. van Dijk, M.J. Geller, *Ur III Incantations from the Frau Professor Hilprecht Collection, Jena*. Wiesbaden: Harrassowitz Verlag, 2003.

Doty 2000 — W.G. Doty, *Mythography. The Study of Myths and Rituals*. (Second edition). Tuscaloosa, Alabama: The University of Alabama Press, 2000.

Driel, v. 2000 — G. van Driel, Closer to Mesopotamia's Proto-History. *BiOr* 57 (2000) 493-509.

Dunham 1986 — S. Dunham, Sumerian Words for Foundation. I. **temen**. *RA* 80 (1986) 31-64.

Dupré 1975 — W. Dupré, *Religion in Primitive Cultures. A Study in Ethnophilosophy*. The Hague - Paris: Mouton, 1975.

Durand 1993 — J.-M. Durand, Le Mythologème du Combat entre le Dieu de l'Orage et la Mer en Mésopotamie. *MARI* 7 (1993) 41-61.

Durand 2008 — J.-M. Durand, La Religion Amorrite en Syrie à l'Époque des Archives de Mari. In: G. del Olmo Lete (ed.), *Mythologie et Religion des Sémites Occidentaux. Vol. I. Ébla, Mari*. OLA 162. Leuven, Paris, Dudley MA: Uitgeverij Peeters and Departement Oosterse Studies, 2008; 161-752.

Ebeling 1919 — E. Ebeling, *Keilschrifttexte aus Assur religiösen Inhalts. Erster Band*. Leipzig: J.C. Hinrichs'sche Buchhandlung, 1919.

Ebeling 1938 — E. Ebeling, Enmešarra. In: E. Ebeling, B. Meissner (eds), *Reallexikon der Assyriologie*, Band 2; Berlin, Leipzig: Walter de Gruyter, 1938; 396-397.

Edzard 1965 — D.O Edzard, Mesopotamien. Die Mythologie der Sumerer und Akkader. In: H.W. Haussig (ed.), *Wörterbuch der Mythologie. Band I. Götter und Mythen im vorderen Orient*. Stuttgart: Ernst Klett Verlag, 1965; 17-140.

Edzard 1967a D.O. Edzard, Panteon und Kult in Mari. In: J.-R. Kupper (ed.), *La Civilisation de Mari. XVe Rencontre Assyriologique Internationale Organisée par le Groupe François Thureau-Dangin (Liège, 4-8 juillet 1966).* Liège: Bibliothèque de la Faculté de Philosophie et Lettres de l'Université de Liège; Paris: Société d'Édition «Les Belles Lettres», 1967; 51-71.

Edzard 1967b D.O. Edzard, Das Sumerische Verbalmorphem /**ed**/ in den alt- und neusumerischen Texten. In: D.O. Edzard (ed.), *Heidelberger Studien zum alten Orient: Adam Falkenstein zum 17. September 1966.* Wiesbaden: Harrassowitz, 1967; 29-62.

Edzard 1976 D.O. Edzard, 'Soziale Reformen' im Zweistromland bis ca. 1600 v. Chr.: Realität oder literarischer Topos? In: J. Harmatta, G. Komoróczy (eds), *Wirtschaft und Gesellschaft im alten Vorderasien.* Budapest: Akadémiai Kiadó, 1976; 145-156.

Edzard 1976-1980 D.O. Edzard, [Igigū,] Anunna und. A. Nach sumerischen Quellen. In: E. Ebeling, B. Meissner (eds), *Reallexikon der Assyriologie*, Band 5; Berlin, New York: Walter de Gruyter, 1976-1980; 37-40.

Edzard 1984 D.O. Edzard, *Hymnen, Beschwörungen und Verwandtes aus dem Archiv L. 2769.* ARET V. Roma: Missione Archeologica Italiana in Siria, 1984.

Edzard 1987 D.O. Edzard, Deep-Rooted Skyscrapers and Bricks: Ancient Mesopotamian Architecture and its Imagery. In: M. Mindlin, M.J. Geller, J.E. Wansbrough (eds), *Figurative Language in the Ancient Near East.* London: School of Oriental and African Studies, University of London, 1987; 13-24.

Edzard 1987-1990 D.O. Edzard, Lugal-kiniš(e)-dudu. In: E. Ebeling, B. Meissner (eds), *Reallexikon der Assyriologie*, Band 7; Berlin, New York: Walter de Gruyter, 1987-1990; 146.

Edzard 1991 D.O. Edzard, Irikagina (Urukagina). In: P. Michalowski *et al.* (eds), *Velles Paraules. Ancient Near Eastern Studies in Honor of Miguel Civil on the Occasion of his Sixty-Fifth Birthday. AuOr* 9 (1991) 77-79.

Edzard 2003a D.O. Edzard, *Sumerian Grammar.* HdO 71. Leiden - Boston: Brill, 2003.

Edzard 2003b D.O. Edzard, Zum sumerischen Verbalpräfix **a(l)**-. In: W. Sallaberger, K. Volk, A. Zgoll (eds), *Literatur, Politik und Recht in Mesopotamien. Festschrift für Claus Wilcke.* Wiesbaden: Harrassowitz Verlag, 2003; 87-98.

Edzard 2003c D.O. Edzard, Enlil, Vater der Götter. In: P. Marrassini *et al.* (eds), *Semitic and Assyriological Studies, Presented to Pelio Fronzaroli by Pupils and Colleagues.* Wiesbaden: Harrassowitz Verlag, 2003; 173-184.

Edzard 2004a D.O. Edzard, Altbabylonische Literatur und Religion. In: D. Charpin, D.O. Edzard, M. Stol, *Mesopotamien. Die altbabylonische Zeit.* OBO 160/4. Fribourg: Academic Press; Göttingen: Vandenhoeck & Ruprecht, 2004; 481-640.

Edzard 2004b D.O. Edzard, Eine Jahrtausendwende vor 4000 Jahren? In: J.-W. Meyer, W. Sommerfeld (eds), *2000 v. Chr. Politische, wirtschaftliche und kulturelle Entwicklung im Zeichen einer Jahrtausendwende. 3. Internationales Colloquium der Deutschen Orient-Gesellschaft 4.-7. April 2000 in Frankfurt /Main und Marburg/Lahn.* Saarbrücken: SDV, Saarbrücker Druckerei und Verlag, 2004; 1-14.

Edzard, Wilcke 1977 — D.O. Edzard, C. Wilcke, 8. Vorläufiger Bericht über die Inschriftenfunde Frühjahr 1973, Frühjahr 1974, Herbst 1974. In: B. Hrouda (ed.), *Isin - Išān Baḥrīyāt I. Die Ergebnisse der Ausgrabungen 1973-1974.* München: Verlag der Bayerischen Akademie der Wissenschaften, 1977.

Egli 1982 — H. Egli, *Das Schlangensymbol. Geschichte, Märchen, Mythos.* Olten und Freiburg im Breisgau: Walter-Verlag, 1982.

Eliade 1954 — M. Eliade, La Terre-Mère et les Hiérogamies Cosmiques. In: O. Fröbe-Kapteyn (ed.), *Mensch und Erde. Eranos-Jahrbuch* 1953, Band XXII. Zürich: Rhein-Verlag, 1954; 57-95.

Eliade 1958 — M. Eliade, *Patterns in Comparative Religion.* Translated by R. Sheed. London and New York: Sheed and Ward, 1958.

Englund 1988 — R.K. Englund, Administrative Timekeeping in Ancient Mesopotamia. *JESHO* 31 (1988) 121-185.

Englund 1998 — R.K. Englund, Texts from the Late Uruk Period. In: J. Bauer, R.K. Englund, M. Krebernik, *Mesopotamien. Späturuk-Zeit und Frühdynastische Zeit.* OBO 160/1. Freiburg Schweiz: Universitätsverlag; Göttingen: Vandenhoeck & Ruprecht, 1998; 13-233.

Englund 2011 — R.K. Englund, Notes on **KID**$_a$. *Cuneiform Digital Library Notes* 2011:006.

Englund, Nissen 1993 — R.K. Englund, H.J. Nissen, unter Mitarbeit von P. Damerow, *Die lexikalischen Listen der archaischen Texte aus Uruk; unter Mitarbeit von Peter Damerow.* ATU 3. Berlin: Mann Verlag, 1993.

ePSD — The Electronic Pennsylvania Sumerian Dictionary. Homepage: http://psd.museum.upenn.edu/.

Espak 2010 — P. Espak, *The God Enki in Sumerian Royal Ideology and Mythology.* Tartu: Tartu University Press, 2010.

ETCSL — Electronic Text Corpus of Sumerian Literature. Homepage: http://etcsl.orinst.ox.ac.uk/.

Falkenstein 1942 — A. Falkenstein, Untersuchungen zur sumerischen Grammatik (Fortsetzung). *ZA* 47 (NF 13) (1942) 181-223.

Falkenstein 1947-1952 — A. Falkenstein, Eine Hymne auf Šūsîn von Ur. *WO* 1 (1947-1952) 43-50.

Falkenstein 1948 — A. Falkenstein, Review of S.N. Kramer, Sumerian Mythology. A Study of Spiritual and Literary Achievement in the Third Millennium B.C. *BiOr* 5 (1948) 163-167.

Falkenstein 1953 — A. Falkenstein, Sumerische Hymnen und Gebete. In: A. Falkenstein, W. von Soden, *Sumerische und akkadische Hymnen und Gebete.* Zürich, Stuttgart: Artemis-Verlag, 1953; 57-231.

Falkenstein 1957 — A. Falkenstein, *Die neusumerischen Gerichtsurkunden. 3. Teil: Nachträge und Berichtigungen, Indizes und Kopien.* München: Verlag der Bayerischen Akademie der Wissenschaften, 1957.

Falkenstein 1959a — A. Falkenstein, *Sumerische Götterlieder. I. Teil.* Heidelberg: Carl Winter Universitätsverlag, 1959.

Falkenstein 1959b — A. Falkenstein, Sumerische religiöse Texte. 1. Drei 'Hymnen' auf Urninurta von Isin. *ZA* 49 (1959) 80-150.

Falkenstein 1960 — A. Falkenstein, Sumerische religiöse Texte. 2. Ein Šulgi-Lied. *ZA* 50 (1960) 61-91.

Falkenstein 1962 — A. Falkenstein, Sumerische religiöse Texte. 3. Ein adab-Lied auf Inanna mit Bitte für Urninurta von Isin. *ZA* 52 (1962) 58-75.

Falkenstein 1965a — A. Falkenstein, Die Anunna in der sumerischen Überlieferung. In: H.G. Güterbock and Th. Jacobsen (eds), *Studies in Honor of Benno Landsberger on his Seventy-Fifth Birthday: April 21, 1965.* AS 16. Chicago, Illinois: University of Chicago Press, 1965; 127-140.

Falkenstein 1965b — A. Falkenstein, Fluch über Akkade. *ZA* 57 (1965) 43-124.

Falkenstein 1966 — A. Falkenstein, *Die Inschriften Gudeas von Lagaš. I. Einleitung. AnOr* 30. Roma: Pontificium Institutum Biblicum, 1966.

Falkenstein 1967 — A. Falkenstein, Zum sumerischen Lexikon. *ZA* 58 (1967) 5-15.

Falkenstein 1978a — A. Falkenstein, *Grammatik der Sprache Gudeas von Lagaš (zweite Auflage). 1: Schrift- und Formenlehre. AnOr* 28. Roma: Pontificium Institutum Biblicum, 1978.

Falkenstein 1978b — A. Falkenstein, *Grammatik der Sprache Gudeas von Lagaš (zweite Auflage). 2: Syntax. AnOr* 29. Roma: Pontificium Institutum Biblicum, 1978.

Farber 1987-1990 — G. Farber, **me (ĝarza,** *parṣu*). In: E. Ebeling, B. Meissner (eds), *Reallexikon der Assyriologie*, Band 7; Berlin, New York: Walter de Gruyter, 1987-1990; 610-613.

Farber 1997 — G. Farber, The Song of the Hoe. In: W.W. Hallo, K.L. Younger (eds), *The Context of Scripture, I: Canonical Compositions from the Biblical World.* Leiden, New York, Köln: Brill, 1997; 511-513.

Farber 1999a — G. Farber, 'Das Lied von der Hacke', ein literarischer Spaß? In: H. Klengel, J. Renger (eds), *Landwirtschaft im alten Orient. Ausgewählte Vorträge der XLI. Rencontre Assyriologique Internationale.* Berlin, 4.-8.7.1994. Berlin: Dietrich Reimer Verlag, 1999; 369-373.

Farber 1999b — G. Farber, Kleiner Leitfaden zum Silbenvokabular A. In: B. Böck, E. Cancik-Kirschbaum, T. Richter (eds), *Munuscula Mesopotamica. Festschrift für Johannes Renger.* AOAT 267. Münster: Ugarit-Verlag, 1999; 117-133.

Farber 2000 — W. Farber, Witchcraft, Magic, and Divination in Ancient Mesopotamia. In: J.M. Sasson (ed.), *Civilizations of the Ancient Near East.* Peabody, Massachusetts: Hendrickson Publishers, Inc., 2000; 1895-1909.

Farber, Farber 2003 — G. Farber, W. Farber, Von einem, der auszog, ein **gudu$_4$** zu werden. In: W. Sallaberger, K. Volk und A. Zgoll (eds), *Literatur, Politik und Recht in Mesopotamien. Festschrift für Claus Wilcke.* Wiesbaden: Harrassowitz Verlag, 2003; 99-114.

Farber-Flügge 1973 — G. Farber-Flügge, *Der Mythos 'Inanna und Enki' unter besonderer Berücksichtigung der Liste der* **me**. Studia Pohl 10. Rome: Biblical Institute Press, 1973.

Faulkner 1985 — R.O. Faulkner, *The Ancient Egyptian Pyramid Texts.* Warminster: Aris & Phillips, 1985.

Ferrara 1973 — A.J. Ferrara, *Nanna-Suen's Journey to Nippur.* Rome: Biblical Institute Press, 1973.

Ferrara 2006 — A.J. Ferrara, A Hodgepodge of Snippets: Some Thoughts on Narrative Now and Then. In: P. Michalowski, N. Veldhuis (eds), *Approaches to Sumerian Literature. Studies in Honour of Stip (H.L.J. Vanstiphout).* CM 35. Leiden, Boston: Brill, 2006; 47-66.

Finkelstein 1966 — J.J. Finkelstein, The Genealogy of the Hammurapi Dynasty. *JCS* 20 (1966) 95-118.

Focke 1999-2000 — K. Focke, Die Göttin Ninimma. II. Ninimma in der keilschriftlichen Überlieferung des 3. bis 1. Jts. v. Chr. *AfO* 46-47 (1999-2000) 92-110.

Foster 2005 — B.R. Foster, *Before the Muses. An Anthology of Akkadian Literature.* Third edition. Bethesda, Maryland: CDL Press, 2005.

Fouadi 1969 — A.-H. A. Al-Fouadi, *Enki's Journey to Nippur: The Journeys of the Gods.* Ann Arbor, Michigan: UMI, 1969.

Frankfort 1934 — H. Frankfort, Gods and Myths on Sargonic Seals. *Iraq* 1 (1934) 2-29.

Frankfort 1939 — H. Frankfort, *Cylinder Seals: A Documentary Essay on the Art and Religion of the Ancient Near East.* London: Macmillan & Co., 1939.

Frankfort 1955 — H. Frankfort, *Stratified Cylinder Seals from the Diyala Region.* OIP 72. Chicago, Illinois: The University of Chicago Press, 1955.

Frankfort 1970 — H. Frankfort, *The Art and Architecture of the Ancient Orient.* 4th ed. New Haven and London: Yale University Press - Pelican History of Art, 1970.

Frayne 1993 — D.R. Frayne, *Sargonic and Gutian Periods (2334-2113 BC). The Royal Inscriptions of Mesopotamia. Early Periods, vol. 2.* Toronto, Buffalo, London: University of Toronto Press, 1993.

Frayne 2008 — D.R. Frayne, *Presargonic Period (2700-2350 BC). The Royal Inscriptions of Mesopotamia. Early Periods, vol. 1.* Toronto, Buffalo, London: University of Toronto Press, 2008.

Frazer 1968 — J.G. Frazer, *Creation and Evolution in Primitive Cosmogonies and other Pieces.* London: Dawsons of Pall Mall, 1968. (Reprint of the 1935 edition).

Fritz 2003 — M.M. Fritz, '*... und weinten um Tammuz'. Die Götter Dumuzi-Ama'ušumgal'anna und Damu.* AOAT 307. Münster: Ugarit-Verlag, 2003.

Frymer-Kensky 1987 — T. Frymer-Kensky, The Planting of Man: A Study in Biblical Imagery. In: J.H. Marks, R.M. Good (eds), *Love & Death in the Ancient Near East. Essays in Honor of Marvin H. Pope.* Guilford, Connecticut: Four Quarters Publishing Company,

1987; 129-136.

Frymer-Kensky 1992 — T. Frymer-Kensky, *In the Wake of the Goddesses. Women, Culture, and the Biblical Transformation of Pagan Myth.* New York: The Free Press; Ontario: Maxwell Macmillan Canada, Inc., 1992.

Gadotti 2005 — A. Gadotti, *'Gilgameš, Enkidu and the Netherworld' and the Sumerian Gilgameš Cycle.* Ann Arbor, Michigan: UMI, 2005.

Galter 1983 — H.D. Galter, *Der Gott Ea/Enki in der akkadischen Überlieferung. Eine Bestandsaufnahme des vorhandenen Materials.* Graz: Graz Verlag für die Technische Universität Graz, 1983.

Gantzert 2011 — M. Gantzert, *The Emar Lexical Texts. Part 4 - Theoretical Interpretation.* Maastricht: Boekenplan, 2011.

Gelb 1960 — I.J. Gelb, Sumerians and Akkadians in their Ethno-Linguistic Relationship. In: *Aspects du Contact Suméro-Akkadien: IXe Rencontre Assyriologique Internationale (Genève, 20-23 juin 1960).* Genève: Musée d'art et d'histoire, 1960; 258-271.

Gelb 1981 — I.J. Gelb, Ebla and the Kish Civilization. In: L. Cagni (ed.), *La Lingua di Ebla. Atti del Convegno Internazionale (Napoli, 21-23 aprile 1980).* Napoli: Instituto Universitario Orientale di Napoli, Seminario di Studi Asiatici, 1981; 9-73.

Gelb 1992 — I.J. Gelb, Mari and the Kish Civilization. In: G.D. Young (ed.), *Mari in Retrospect. Fifty Years of Mari and Mari Studies.* Winona Lake, Indiana: Eisenbrauns, 1992; 121-202.

Gelb *et al.* 1989 — I.J. Gelb, P. Steinkeller, R.M. Whiting, Jr., *Earliest Land Tenure Systems in the Near East: Ancient Kudurrus. Plates.* OIP 104. Chicago, Illinois: The Oriental Institute of the University of Chicago, 1989.

Gelb *et al.* 1991 — I.J. Gelb, P. Steinkeller, R.M. Whiting, Jr., *Earliest Land Tenure Systems in the Near East: Ancient Kudurrus. Text.* OIP 104. Chicago, Illinois: The Oriental Institute of the University of Chicago, 1991.

Geller 1985 — M.J. Geller, *Forerunners to* **udug-ḫul**. *Sumerian Exorcistic Incantations.* Stuttgart: Franz Steiner Verlag Wiesbaden GmbH, 1985.

Geller 1990 — M.J. Geller, Taboo in Mesopotamia. A Review Article. *JCS* 42 (1990) 105-117.

Geller 1998 — M.J. Geller, Reflexives and Antipassives in Sumerian Verbs. *OrNS* 67 (1998) 85-106.

Genouillac, de 1930 — H. de Genouillac, *Textes Religieux Sumériens du Louvre. Textes Cunéiformes, Tome XV.* Paris: Librairie Orientaliste Paul Geuthner, 1930.

George 1992 — A.R. George, *Babylonian Topographical Texts. Or. Lov. Anal.* 40. Leuven: Departement Oriëntalistiek; Leuven: Uitgeverij Peeters, 1992.

George 1993 — A.R. George, *House Most High. The Temples of Ancient Mesopotamia.* Winona Lake, Indiana: Eisenbrauns, 1993.

George 1999 — A. George, *The Epic of Gilgamesh. The Babylonian Epic Poem and Other Texts in Akkadian and Sumerian.* London: Allen Lane, The Penguin Press, 1999.

George 2003 A.R. George, *The Babylonian Gilgamesh Epic. Introduction, Critical Edition and Cuneiform Texts.* Oxford: Oxford University Press, 2003.

George 2005 A.R. George, In Search of the **é.dub.ba.a**: The Ancient Mesopotamian School in Literature and Reality. In: Y. Sefati *et al.* (eds), *'An Experienced Scribe Who Neglects Nothing'. Ancient Near Eastern Studies in Honor of Jacob Klein.* Bethesda, Maryland: CDL Press, 2005; 127-137.

George 2006 A.R. George, Review of P. Lapinkivi, *The Sumerian Sacred Marriage in the Light of Comparative Evidence. (*SAAS 15, 2004). *BSOAS* 69 (2006) 315-317.

George, Rawi 1996 A.R. George, F.N.H. Al-Rawi, Tablets from the Sippar Library VI. Atra-ḫasīs. *Iraq* 58 (1996) 147-190.

Gibson 2007 McG. Gibson, Nippur - Sacred City of Enlil, Supreme God of Sumer and Akkad. Website Oriental Institute Chicago (september 2012): http://www.oi.uchicago.edu/research/projects/nip/nsc.html. Revision of an article in: *al-Rāfidān* 14 (1993) 1-18.

Glassner 1992 J.-J. Glassner, Inanna et les **ME**. In: M. de Jong-Ellis (ed.), *Nippur at the Centennial. Papers Read at the 35ᵉ Rencontre Assyriologique Internationale, Philadelphia, 1988.* Philadelphia: The Samuel Noah Kramer Fund, Babylonian Section, University Museum, 1992; 55-86.

Glassner 2003 J.-J. Glassner, *The Invention of Cuneiform. Writing in Sumer.* Translated and Edited by Z. Bahrani and M. van de Mierop. Baltimore, Maryland: The Johns Hopkins University Press, 2003.

Goblet d'Alviella 1912 E.F.A. Goblet d'Alviella, *De Wereldreis der Symbolen.* 's-Gravenhage: Maçonnieke Uitgevers-Maatschappij, 1912.

Goetze 1968 A. Goetze, An Old Babylonian Prayer of the Divination Priest. *JCS* 22 (1968) 25-29.

Grafman 1972 R. Grafman, Bringing Tiamat to Earth. *IEJ* 22 (1972) 47-49.

Green 1975 M.W. Green, *Eridu in Sumerian Literature.* Chicago, Illinois: University of Chicago, 1975.

Green 1977 M.W. Green, A Note on an Archaic Period Geographical List from Warka. *JNES* 36 (1977) 293-294.

Green 2000 A. Green, Ancient Mesopotamian Religious Iconography. In: J.M. Sasson (ed.), *Civilizations of the Ancient Near East.* Peabody, Massachusetts: Hendrickson Publishers, Inc., 2000; 1837-1855.

Green, Nissen 1987 M.W. Green, H.J. Nissen, unter Mitarbeit von P. Damerow und R.K. Englund, *Zeichenliste der archaischen Texte aus Uruk.* ATU 2. Berlin: Gebr. Mann Verlag, 1987.

Güterbock 1952 H.G. Güterbock, The Song of Ullikummi. Revised Text of the Hittite Version of a Hurrian Mythe (continued). *JCS* 6 (1952) 8-42.

Güterbock 1980-1983 H.G. Güterbock, Kumarbi. In: E. Ebeling, B. Meissner (eds), *Reallexikon der Assyriologie*, Band 6; Berlin, New York: Walter de Gruyter, 1980-1983; 324-330.

Haarmann, Marler 2008 — H. Haarmann, J. Marler, *Introducing the Mythological Crescent. Ancient Beliefs and Imagery Connecting Eurasia with Anatolia.* Wiesbaden: Harrassowitz Verlag, 2008.

Haber 2009 — E. Haber, The Stylistic Role of the Anticipatory Genitive Construction in Sumerian Literature. *RA* 103 (2009) 37-47.

Hall 1985 — M.G. Hall, *A Study of the Sumerian Moon-God, Nanna / Suen.* Ann Arbor, Michigan: UMI, 1985.

Hallo 1970a — W.W. Hallo, Antediluvian Cities. *JCS* 23 (1970) 57-67.

Hallo 1970b — W.W. Hallo, The Cultic Setting of Sumerian Poetry. In: A. Finet (ed.), *Actes de la XVIIe Rencontre Assyriologique Internationale, Université Libre de Bruxelles, 30 juin - 4 juillet 1969.* Ham-sur-Heure: Comité Belge de Recherches en Mésopotamie, 1970; 116-134.

Hallo 1976 — W.W. Hallo, The Royal Correspondence of Larsa: I. A Sumerian Prototype for the Prayer of Hezekiah? In: B.L. Eichler, with the assistance of J.W. Heimerdinger, Å.W. Sjöberg (eds), *Kramer Anniversary Volume: Cuneiform Studies in Honor of Samuel Noah Kramer.* AOAT 25. Kevelaer: Verlag Butzon & Bercker; Neukirchen-Vluyn: Neukirchener Verlag, 1976; 209-224.

Hallo 1983 — W.W. Hallo, Dating the Mesopotamian Past: The Concept of Eras from Sargon to Nabonassar. *Bull. Soc. Mes. Stud.* 6 (1983) 7-18.

Hallo 1985 — W.W. Hallo, Biblical Abominations and Sumerian Taboos. *JQR* 76 (1985) 21-40.

Hallo 1987 — W.W. Hallo, The Birth of Kings. In: J.H. Marks, R.M. Good (eds), *Love & Death in the Ancient Near East. Essays in Honor of Marvin H. Pope.* Guilford, Connecticut: Four Quarters Publishing Company, 1987; 45-52.

Hallo 1992 — W.W. Hallo, Royal Ancestor Worship in the Biblical World. In: M. Fishbane, E. Tov, with the assistance of W.W. Fields (eds), *'Shaᶜarei Talmon'. Studies in the Bible, Qumran, and the Ancient Near East Presented to Shemaryahu Talmon.* Winona Lake, Indiana: Eisenbrauns, 1992; 381-401.

Hallo, v. Dijk 1968 — W.W. Hallo, J.J.A. van Dijk, *The Exaltation of Inanna.* New Haven and London: Yale University Press, 1968.

Hallo, Moran 1979 — W.W. Hallo, W.L. Moran, The First Tablet of the SB Recension of the Anzu-Myth. *JCS* 31 (1979) 65-115.

Haul 2004 — M. Haul, **lú an-šè ba-e$_{11}$-dè**. Der Mythos von Etana um 2000 v. Chr. In: J.-W. Meyer, W. Sommerfeld (eds), *2000 v. Chr. Politische, wirtschaftliche und kulturelle Entwicklung im Zeichen einer Jahrtausendwende. 3. Internationales Colloquium der Deutschen Orient-Gesellschaft 4.-7. April 2000 in Frankfurt/Main und Marburg/Lahn.* Saarbrücken: SDV Saarbrücker Druckerei und Verlag, 2004; 237-262.

Hecker 1994 — K. Hecker, Ein zweisprachiger Schöpfungsmythos. In: K. Hecker *et al.* (eds), *Weisheitstexte, Mythen und Epen.* TUAT III, Lfg. 4: *Mythen und Epen II.* Gütersloh: Gütersloher Verlagshaus, 1994; 606-608.

Heimpel 1968	W. Heimpel, *Tierbilder in der sumerischen Literatur.* Studia Pohl 2. Roma: Pontificium Institutum Biblicum, 1968.
Heimpel 1974	W. Heimpel, *The Structure of the Sumerian Prefix Chain.* Unpublished manuscript. Berkeley, 1974.
Heimpel 1980-1983	W. Heimpel, Ku'ara. Stadt in Südbabylonien. In: E. Ebeling, B. Meissner (eds), *Reallexikon der Assyriologie*, Band 6; Berlin, New York: Walter de Gruyter, 1980-1983; 256-257.
Heimpel 1981	W. Heimpel, The Nanshe Hymn. *JCS* 33 (1981) 65-135.
Heimpel 1987a	W. Heimpel, Das untere Meer. *ZA* 77 (1987) 22-91.
Heimpel 1987b	W. Heimpel, The Natural History of the Tigris According to the Sumerian Literary Composition Lugal. *JNES* 46 (1987) 309-317.
Heimpel 1998-2001	W. Heimpel, Nanše. A. Philologisch. In: E. Ebeling, B. Meissner (eds), *Reallexikon der Assyriologie*, Band 9; Berlin, New York: Walter de Gruyter, 1998-2001; 152-160.
Heinrich 1936	E. Heinrich, mit einem Beitrag von M. Hilzheimer, *Kleinfunde aus den archaischen Tempelschichten in Uruk.* Berlin: Deutsche Forschungs-gemeinschaft, 1936.
Henshaw 1994	R.A. Henshaw, *Female and Male: The Cultic Personnel; The Bible and the Rest of the Ancient Near East.* Allison Park, PA: Pickwick Publications, 1994.
Hesse 2010	K.J. Hesse, Cultural Interaction and Cognitive Expressions in the Formation of Ancient Near Eastern Societies. In: P.J.J. Sinclair *et al.* (eds), *The Urban Mind. Cultural and Environmental Dynamics.* Studies in Global Archaeology 15. Uppsala: African and Comparative Archaeology, Department of Archaeology and Ancient History, Uppsala University, 2010; 61-87.
Hilprecht 1893-1896	H.V. Hilprecht, *Old Babylonian Inscriptions chiefly from Nippur. Part I, II.* Philadelphia: MacCalla, 1893 (I), 1896 (II).
Hockmann 2008	D. Hockmann, Die Warka-Vase - eine neue Interpretation. *AoF* 35 (2008) 326-336.
Hoffner 1990	H.A. Hoffner, Jr. (transl.), G.M. Beckman (ed.), *Hittite Myths.* Atlanta, Georgia: Scholars Press, 1990.
Hole 1993	F. Hole, Prehistoric Susa: Susa I Pottery. In: P.O. Harper, J. Aruz, F. Tallon (eds), *The Royal City of Susa. Ancient Near Eastern Treasures in the Louvre.* New York: The Metropolitan Museum of Art, 1993^2; 32-41.
Horowitz 1998	W. Horowitz, *Mesopotamian Cosmic Geography.* Winona Lake, Indiana: Eisenbrauns, 1998.
Hruška 1971	B. Hruška, Zur Geschichte der sumerischen Religion. Die Grenzen einer Methode. I. Der Weg vom Präanthropomorphismus zum Anthropomorphismus dargestellt von Helga Piesl. *ArOr* 39 (1971) 190-194.
Hruška 1974	B. Hruška, Review of Pettinato Giovanni, *Das altorientalische Menschenbild und die sumerischen und akkadischen Schöpfungsmythen. ArOr* 42 (1974) 272-276.
Hruška 1996	B. Hruška, Zum 'Heiligen Hügel' in der altmesopotamischen Religion. *WZKM* 86 (1996) 161-175.

Huehnergard 1997 J. Huehnergard, *A Grammar of Akkadian.* Atlanta, Georgia: Scholars Press, 1997 (corrected reprint 1998).

Hunger 1968 H. Hunger, *Babylonische und assyrische Kolophone.* AOAT 2. Kevelaer: Butzon & Bercker; Neukirchen-Vluyn: Neukirchener Verlag des Erziehungsvereins, 1968.

Jacobsen 1939 Th. Jacobsen, *The Sumerian King List.* AS 11. Chicago, Illinois: The University of Chicago Press, 1939.

Jacobsen 1946 Th. Jacobsen, Sumerian Mythology: A Review Article. *JNES* 5 (1946) 128-152.

Jacobsen 1961 Th. Jacobsen, Formative Tendencies in Sumerian Religion. In: G.E. Wright (ed.), *The Bible and the Ancient Near East. Essays in Honor of William Foxwell Albright.* London: Routledge & Kegan Paul; New York: Garden City, 1961; 267-278.

Jacobsen 1968 Th. Jacobsen, The Battle Between Marduk and Tiamat. *JAOS* 88 (1968) 104-108.

Jacobsen 1970 Th. Jacobsen, *Toward the Image of Tammuz and Other Essays on Mesopotamian History and Culture.* (W.L. Moran, ed.). Cambridge, Massachusetts: Harvard University Press, 1970.

Jacobsen 1973 Th. Jacobsen, Notes on Nintur. *OrNS* 42 (1973) 274-298.

Jacobsen 1975 Th. Jacobsen, Religious Drama in Ancient Mesopotamia. In: H. Goedicke, J.J.M. Roberts (eds), *Unity and Diversity. Essays in the History, Literature, and Religion of the Ancient Near East.* Baltimore and London: The Johns Hopkins University Press, 1975; 65-97.

Jacobsen 1976 Th. Jacobsen, *The Treasures of Darkness. A History of Mesopotamian Religion.* New Haven and London: Yale University Press, 1976.

Jacobsen 1977a Th. Jacobsen, Mesopotamia. In: H. Frankfort *et al.*, *The Intellectual Adventure of Ancient Man. An Essay on Speculative Thought in the Ancient Near East.* Chicago & London: The University of Chicago Press, 1977; 123-219 (reprint of the edition of 1946).

Jacobsen 1977b Th. Jacobsen, *Inuma Ilu awīlum.* In: M. de Jong Ellis (ed.), *Essays on the Ancient Near East in Memory of Jacob Joel Finkelstein.* Memoirs of the Connecticut Academy of Arts & Sciences, vol. 19. Hamden, Connecticut: Archon Books, 1977; 113-117.

Jacobsen 1981 Th. Jacobsen, The Eridu Genesis. *JBL* 100 (1981) 513-529.

Jacobsen 1984 Th. Jacobsen, *The Harab Myth.* SANE 2/3. Malibu, California: Undena Publications, 1984.

Jacobsen 1985a Th. Jacobsen, The Name Dumuzi. *JQR* 76 (1985) 41-45.

Jacobsen 1985b Th. Jacobsen, Ur-Nanshe's Diorite Plaque. *OrNS* 54 (1985) 65-72.

Jacobsen 1987 Th. Jacobsen, *The Harps that Once... Sumerian Poetry in Translation.* New Haven and London: Yale University Press, 1987.

Jacobsen 1989 Th. Jacobsen, The **lil$_2$** of **^{d}En-lil$_2$**. In: H. Behrens, D. Loding, M.T. Roth (eds), **DUMU-E$_2$-DUB-BA-A**. *Studies in Honor of*

Åke W. Sjöberg. Philadelphia: The Samuel Noah Kramer Fund, 1989; 267-276.

Jacobsen 1990 — Th. Jacobsen, Enuma Elisj. In: D. van der Plas, B. Becking, D. Meijer (eds), *De Schepping van de Wereld. Mythische Voorstellingen in het Oude Nabije Oosten.* Muiderberg: Dick Coutinho, 1990; 57-83.

Jacobsen 1993 — Th. Jacobsen, The Descent of Enki. In: M.E. Cohen, D.C. Snell, D.B. Weisberg (eds), *The Tablet and the Scroll. Near Eastern Studies in Honor of William W. Hallo.* Bethesda, Maryland: CDL Press, 1993; 120-123.

Jacobsen, Kramer 1953 — Th. Jacobsen, S.N. Kramer, The Myth of Inanna and Bilulu. *JNES* 12 (1953) 160-188.

Jagersma 2000 — B. Jagersma, Sound Change in Sumerian: the So-Called /**dr**/-Phoneme. *ASJ* 22 (2000) 81-87.

Jagersma 2010 — A.H. Jagersma, *A Descriptive Grammar of Sumerian.* Dissertation. Leiden: Faculty of the Humanities, Leiden University, 2010.

Jaques 2006 — M. Jaques, *Le Vocabulaire des Sentiments dans les Textes Sumériens. Recherche sur le Lexique Sumérien et Akkadien.* AOAT 332. Münster: Ugarit-Verlag, 2006.

Jean 1931 — C.-F. Jean, Noms Divins Sumériens. Listes des Élèves-Scribes de Nippur du 3e Millénaire environ avant J.-C. *RA* 28 (1931) 179-194.

Jestin 1937 — R.R. Jestin, *Tablettes Sumériennes de Šuruppak Conservées au Musée de Stamboul.* Paris: De Boccard, 1937.

Jestin 1972 — R.R. Jestin, Review of Giovanni Pettinato, *Das altorientalische Menschenbild und die sumerischen und akkadischen Schöpfungsmythen. BiOr* 29 (1972) 38-39.

Kämmerer, Metzler 2012 — T.R. Kämmerer, K.A. Metzler, *Das babylonische Weltschöpfungsepos Enūma elîš*. AOAT 375. Münster: Ugarit-Verlag, 2012.

Kaplan 1976 — M.F. Kaplan, Another Slaying of Tiamat? *IEJ* 26 (1976) 174-177.

Karahashi 2000 — F. Karahashi, *Sumerian Compound Verbs with Body-Part Terms*. Ann Arbor, Michigan: UMI, 2000.

Katz 2003 — D. Katz, *The Image of the Netherworld in the Sumerian Sources.* Bethesda, Maryland: CDL Press, 2003.

Katz 2005 — D. Katz, Death They Dispensed to Mankind. The Funerary World of Ancient Mesopotamia. *Historiae* 2 (2005) 55-90.

Katz 2006 — D. Katz, Review of P. Lapinkivi, *The Sumerian Sacred Marriage in the Light of Comparative Evidence. (*SAAS 15, 2004). *BiOr* 63 (2006) 103-107.

Katz 2007a — D. Katz, Enki and Ninḫursaĝa, Part One: The Story of Dilmun. *BiOr* 64 (2007) 568-589.

Katz 2007b — D. Katz, Sumerian Funerary Rituals in Context. In: N. Laneri (ed.), *Performing Death. Social Analyses of Funerary Traditions in the Ancient Near East and Mediterranean.* OIS 3. Chicago, Illinois: The Oriental Institute of the University of Chicago, 2007; 167-188.

Katz 2008 — D. Katz, Enki and Ninḫursaĝa, Part Two: The Story of Enki and Ninḫursaĝa. *BiOr* 65 (2008) 320-342.

Katz 2010 — D. Katz, The Naked Soul. Deliberations on a Popular Theme. In: J. Stackert, B.N. Porter, D.P. Wright (eds), *Gazing on the Deep. Ancient Near Eastern and Other Studies in Honor of Tzvi Abusch.* Bethesda, Maryland: CDL Press, 2010; 107-120.

Katz 2011 — D. Katz, Reconstructing Babylon: Recycling Mythological Traditions toward a New Theology. In: E. Cancik-Kirschbaum, M. v. Ess, J. Marzahn (eds), *Babylon. Wissenkultur in Orient und Okzident.* Berlin: De Gruyter, 2011; 123-134.

Kienast 1965 — B. Kienast, *Igigū* und *Anunnakkū* nach den akkadischen Quellen. In: H.G. Güterbock, Th. Jacobsen (eds), *Studies in Honor of Benno Landsberger on his Seventy-Fifth Birthday, April 21, 1965.* AS 16. Chicago, Illinois: The University of Chicago Press, 1965; 141-158.

Kienast 1976-1980 — B. Kienast, Igigū, Anunnakkū und. B. Nach akk. Quellen. In: E. Ebeling, B. Meissner (eds), *Reallexikon der Assyriologie*, Band 5; Berlin, New York: Walter de Gruyter, 1976-1980; 40-44.

Kienast 1985 — B. Kienast, Überlegungen zum 'Pantheon Babylonicum'. *OrNS* 54 (1985) 106-116.

Kinnier Wilson 1965 — J.V. Kinnier Wilson, Review of C.J. Gadd and S.N. Kramer, *Ur Excavations, Texts, VI: Literary and Religious Texts, First Part. (1963).* In: *JSS* 10 (1965) 271-274.

Kirk 1970 — G.S. Kirk, *Myth. Its Meaning and Functions in Ancient and Other Cultures.* Cambridge: Cambridge University Press; Berkeley and Los Angeles: University of California Press, 1970.

Klein 1981 — J. Klein, *Three Šulgi Hymns. Sumerian Royal Hymns Glorifying King Šulgi of Ur*. Ramat-Gan: Bar-Ilan University Press, 1981.

Klein 1989 — J. Klein, Building and Dedication Hymns in Sumerian Literature. *ASJ* 11 (1989) 27-67.

Klein 1990 — J. Klein, Šulgi and Išmedagan: Originality and Dependence in Sumerian Royal Hymnology. In: J. Klein and A. Skaist (eds), *Bar-Ilan Studies in Assyriology Dedicated to Pinḥas Artzi.* Ramat-Gan: Bar-Ilan University Press, 1990; 65-136.

Klein 1997a — J. Klein, Enki and Ninmaḫ. In: W.W. Hallo (ed.), *The Context of Scripture, I: Canonical Compositions from the Biblical World.* Leiden, New York, Köln: Brill, 1997; 516-518.

Klein 1997b — J. Klein, The Sumerian **me** as a Concrete Object. *AoF* 24 (1997) 211-218.

Klein 1998-2001 — J. Klein, Namtar. In: E. Ebeling, B. Meissner (eds), *Reallexikon der Assyriologie*, Band 9; Berlin, New York: Walter de Gruyter, 1998-2001; 142-145.

Klein 2001 — J. Klein, The Genealogy of Nanna-Suen and its Historical Background. In: T. Abusch *et al.* (eds), *Historiography in the Cuneiform World. Proceedings of the XLVth Rencontre*

Assyriologique Internationale. Bethesda, Maryland: CDL Press, 2001; 279-301.

Koch 2000 — H. Koch, Früheste Götterdarstellungen in Elam und Mesopotamien. In: S. Graziani (ed.), con la collaborazione di M.C. Casaburi e G. Lacerenza, *Studi sul Vicino Oriente Antico Dedicati alla Memoria di Luigi Cagni*. Napoli: Instituto Universitario Orientale, 2000; 585-605.

Koch 2007 — H. Koch, *Frauen und Schlangen. Die geheimnisvolle Kultur der Elamer in Alt-Iran*. Mainz am Rhein: Verlag Philipp von Zabern, 2007.

Komoróczy 1972 — G. Komoróczy, Das mythische Goldland Ḫarali im alten Vorderasien. *Acta Orient. Hung.* 26 (1972) 113-123.

Komoróczy 1973 — G. Komoróczy, «The Separation of Sky and Earth». The Cycle of Kumarbi and the Myths of Cosmogony in Mesopotamia. *ActAnt* 21 (1973) 21-45.

Komoróczy 1976 — G. Komoróczy, Work and Strike of Gods. New Light on the Divine Society in the Sumero-Akkadian Mythology. *Oikumene* 1 (1976) 9-37.

Kouwenberg 2000 — N.J.C. Kouwenberg, Nouns as Verbs: The Verbal Nature of the Akkadian Stative. *OrNS* 69 (2000) 21-71.

Kouwenberg 2005 — N.J.C. Kouwenberg, Reflections on the Gt-stem in Akkadian. *ZA* 95 (2005) 77-103.

Kramer 1940 — S.N. Kramer, *Lamentation over the Destruction of Ur*. AS 12. Chicago, Illinois: The University of Chicago Press, 1940.

Kramer 1944 — S.N. Kramer, The Death of Gilgamesh. *BASOR* 94 (1944) 2-12.

Kramer 1950 — S.N. Kramer, The Deluge. In: J.B. Pritchard (ed.), *Ancient Near Eastern Texts Relating to the Old Testament*. Princeton, New Jersey: Princeton University Press, 1950; 42-44.

Kramer 1956 — S.N. Kramer, *From the Tablets of Sumer. Twenty-five Firsts in Man's Recorded History*. Indian Hills, Colorado: The Falcon's Wings Press, 1956.

Kramer 1960 — S.N. Kramer, Sumero-Akkadian Interconnections: Religious Ideas. In: *Aspects du Contact Suméro-Akkadien. IXe Rencontre Assyriologique Internationale (Genève, 20-23 juin 1960). Genava* 8. Genève: Musée d'Art et d'Histoire, 1960; 272-283.

Kramer 1972 — S.N. Kramer, *Sumerian Mythology. A Study of Spiritual and Literary Achievement in the Third Millennium B.C. (Paperback edition of the revised edition 1961; originally published in 1944)*. Philadelphia: University of Pennsylvania Press, 1972.

Kramer 1981 — S.N. Kramer, *History Begins at Sumer: Thirty-Nine Firsts in Recorded History*. Philadelphia: University of Pennsylvania Press, 1981.

Kramer 1983 — S.N. Kramer, The Sumerian Deluge Myth; Reviewed and Revised. *AnSt* 33 (1983) 115-121.

Kramer 1985 — S.N. Kramer, BM 23631: Bread for Enlil, Sex for Inanna. *OrNS* 54 (1985) 117-132.

Kramer 1989 — S.N. Kramer, BM 100042: A Hymn to Šu-Sin and an Adab of Nergal. In: H. Behrens, D. Loding, M.T. Roth (eds), **DUMU-**

E_2-**DUB-BA-A**. *Studies in Honor of Åke W. Sjöberg*. Philadelphia: The Samuel Noah Kramer Fund, 1989; 303-316.

Kramer, Maier 1989 — S.N. Kramer, J. Maier, *Myths of Enki, the Crafty God*. New York, Oxford: Oxford University Press, 1989.

Krebernik 1984 — M. Krebernik, *Die Beschwörungen aus Fara und Ebla. Untersuchungen zur ältesten keilschriftlichen Beschwörungsliteratur*. Hildesheim - Zürich - New York: Georg Olms Verlag, 1984.

Krebernik 1986 — M. Krebernik, Die Götterlisten aus Fāra. *ZA* 76 (1986) 161-204.

Krebernik 1993-1997 — M. Krebernik, Muttergöttin. A. I. In Mesopotamien. In: E. Ebeling, B. Meissner (eds), *Reallexikon der Assyriologie*, Band 8; Berlin, New York: Walter de Gruyter, 1993-1997; 502-516.

Krebernik 1994 — M. Krebernik, Zur Einleitung der **zà-me**-Hymnen aus Tell Abū Ṣalābīḫ. In: P. Calmeyer *et al.* (eds), *Beiträge zur altorientalischen Archäologie und Altertumskunde. Festschrift für Barthel Hrouda zum 65. Geburtstag*. Wiesbaden: Harrassowitz, 1994; 151-157.

Krebernik 1998 — M. Krebernik, Die Texte aus Fāra und Tell Abū Ṣalābīḫ. In: J. Bauer, R.K. Englund, M. Krebernik, *Mesopotamien. Späturuk-Zeit und Frühdynastische Zeit*. OBO 160/1. Freiburg, Schweiz: Universitätsverlag; Göttingen: Vandenhoeck & Ruprecht, 1998; 237-427.

Krebernik 1998-2001a — M. Krebernik, **NAĜAR**. In: E. Ebeling, B. Meissner (eds), *Reallexikon der Assyriologie*, Band 9; Berlin, New York: W. de Gruyter, 1998-2001; 74-75.

Krebernik 1998-2001b — M. Krebernik, Ninlil (Mulliltu, Mullissu). Göttin, Gemahlin Enlils. In: E. Ebeling, B. Meissner (eds), *Reallexikon der Assyriologie*, Band 9; Berlin, New York: Walter de Gruyter, 1998-2001; 452-461.

Krebernik 1998-2001c — M. Krebernik, Nin-irigala. In: E. Ebeling, B. Meissner (eds), *Reallexikon der Assyriologie*, Band 9; Berlin, New York: Walter de Gruyter, 1998-2001; 386-387.

Krebernik 2002 — M. Krebernik, Geschlachtete Gottheiten und ihre Nahmen. In: O. Loretz, K.A. Metzler und H. Schaudig (eds), *Ex Mesopotamia et Syria Lux. Festschrift für Manfried Dietrich zu seinem 65. Geburtstag*. AOAT 281. Münster: Ugarit-Verlag, 2002; 289-298.

Krebernik 2003 — M. Krebernik, Drachenmutter und Himmelsrebe? Zur Frühgeschichte Dumuzis und seiner Familie. In: W. Sallaberger, K. Volk und A. Zgoll (eds), *Literatur, Politik und Recht in Mesopotamien. Festschrift für Claus Wilcke*. Wiesbaden: Harrassowitz Verlag, 2003; 151-180.

Krecher 1966 — J. Krecher, *Sumerische Kultlyrik*. Wiesbaden: Otto Harrassowitz, 1966.

Krecher 1967a — J. Krecher, Zum Emesal-Dialekt des Sumerischen. In: *Heidelberger Studien zum alten Orient. Adam Falkenstein zum 17. September 1966*. Wiesbaden: Otto Harrassowitz, 1967; 87-110.

Krecher 1967b J. Krecher, Die sumerischen Texte in 'syllabischer' Orthographie. *ZA* 58 (1967) 16-65.

Krecher 1967-1968 J. Krecher, Die pluralischen Verba für 'gehen' und 'stehen' im Sumerischen. *WO* 4 (1967-1968) 1-11.

Krecher 1978a J. Krecher, Sumerische Literatur. In: W. Röllig, in Verbindung mit E. Brunner-Traut *et al.* (eds), *Altorientalische Literaturen. Neues Handbuch der Literaturwissenschaft; Band I.* Wiesbaden: Akademische Verlagsgesellschaft Athenaion, 1978; 101-150.

Krecher 1978b J. Krecher, Sumerische Literatur der Fara-Zeit: Die **UD.GAL.NUN**-Texte (I). *BiOr* 35 (1978) 155-160.

Krecher 1986 J. Krecher, The Present Possibilities and Limitations of Our Understanding and Translating Sumerian Texts. *Sumer* 42 (1986) 44-47.

Krecher 1987a J. Krecher, Morphemless Syntax in Sumerian as Seen on the Background of Word-Composition in Chukchee. *ASJ* 9 (1987) 67-88.

Krecher 1987b J. Krecher, **DU** = **ku**$_x$(**-ř**) 'eintreten', 'hineinbringen'. *ZA* 77 (1987) 7-21.

Krecher 1992 J. Krecher, **UD.GAL.NUN** versus 'Normal' Sumerian: Two Literatures or One? In: P. Fronzaroli (ed.), *Literature and Literary Language at Ebla.* QdS 18. Firenze: Dipartimento di Linguistica, Università di Firenze, 1992; 285-303.

Krispijn 1990 Th.J.H. Krispijn, Beiträge zur altorientalischen Musikforschung. 1. Šulgi und die Musik. *Akkadica* 70 (1990) 1-27.

Krispijn 2004 Th.J.H. Krispijn, *Inleiding in de Sumerische Taalkunde.* Leiden: TCNO/Assyriologie/TCMA, Universiteit Leiden, 2004.

Krispijn 2005 Th.J.H. Krispijn, The Change of Official Sumerian in the City-State of Lagaš. *ASJ* 22 (2005) 153-175.

Kümmel 1973 H.M. Kümmel, Die Religion der Hethiter. Grundzüge und Probleme. In: U. Mann (ed.), *Theologie und Religionswissenschaft. Der gegenwärtige Stand ihrer Forschungsergebnisse und Aufgaben im Hinblick auf ihr gegenseitiges Verhältnis.* Darmstadt: Wissenschaftliche Buchgesellschaft, 1973; 65-85.

Kümmel 1973-1974 H.M. Kümmel, Bemerkungen zu den altorientalischen Berichten von der Menschenschöpfung. *WO* 7 (1973-1974) 25-38.

Lambert 1957-1971 W.G. Lambert, Göttergenealogie[1)]; Götterlisten[2)]. In: E. Ebeling, B. Meissner (eds), *Reallexikon der Assyriologie*, Band 3. Berlin: Walter de Gruyter, 1957-1971; 469-470[1)]; 473-479 [2)].

Lambert 1968 W.G. Lambert, Another Look at Hammurabi's Ancestors. *JCS* 22 (1968) 1-2.

Lambert 1970 W.G. Lambert, The Reading of the Name **uru.KA.gi.na**. *OrNS* 39 (1970) 419.

Lambert 1972 W.G. Lambert, Review of Giovanni Pettinato, *Das altorientalische Menschenbild und die sumerischen und akkadischen Schöpfungsmythen.* Heidelberg: Carl Winter, Universitätsverlag, 1971. *BSOAS* 35 (1972) 134-135.

Lambert 1974 — W.G. Lambert, Der Mythos im alten Mesopotamien, sein Werden und Vergehen. *ZRG* 26 (1974) 1-16.

Lambert 1975a — W.G. Lambert, The Cosmology of Sumer and Babylon. In: C. Blacker, M. Loewe (eds), *Ancient Cosmologies.* London: George Allen & Unwin Ltd., 1975; 42-65.

Lambert 1975b — W.G. Lambert, The Historical Development of the Mesopotamian Pantheon: A Study in Sophisticated Polytheism. In: H. Goedicke, J.J.M. Roberts (eds), *Unity and Diversity. Essays in the History, Literature, and Religion of the Ancient Near East.* Baltimore and London: The Johns Hopkins University Press, 1975; 191-200.

Lambert 1976 — W.G. Lambert, Review of Robert D. Biggs (ed.), *Inscriptions from Tell Abū Ṣalābīkh.* OIP 99. Chicago and London: University of Chicago Press, 1974. *BSOAS* 39 (1976) 428-432.

Lambert 1976-1980 — W.G. Lambert, Kalkal. In: E. Ebeling, B. Meissner (eds), *Reallexikon der Assyriologie*, Band 5. Berlin, New York: Walter de Gruyter, 1976-1980; 323.

Lambert 1980 — W.G. Lambert, The Theology of Death. In: B. Alster (ed.), *Death in Mesopotamia. Papers read at the XXVI[e] Rencontre Assyriologique Internationale.* Copenhagen: Akademisk Forlag, 1980; 53-66.

Lambert 1980-1983a — W.G. Lambert, Kosmogonie. In: E. Ebeling, B. Meissner (eds), *Reallexikon der Assyriologie*, Band 6. Berlin, New York: Walter de Gruyter, 1980-1983; 218-222.

Lambert 1980-1983b — W.G. Lambert, Laḫar. In: E. Ebeling, B. Meissner (eds), *Reallexikon der Assyriologie*, Band 6. Berlin, New York: Walter de Gruyter, 1980-1983; 431

Lambert 1981 — W.G. Lambert, Studies in **UD.GAL.NUN**. *OrAnt* 20 (1981) 81-97; Addendum: *idem*, p. 305.

Lambert 1984 — W.G. Lambert, The Section **AN**. In: L. Cagni (ed.), *Il Bilinguismo a Ebla. Atti del Convegno Internazionale (Napoli, 19-22 aprile 1982).* Napoli: Instituto Universitario Orientale, Dipartimento di Studi Asiatici, 1984; 393-401.

Lambert 1985a — W.G. Lambert, A List of Gods' Names Found at Mari. In: J.-M. Durand, J.-R. Kupper (eds), *Miscellanea Babylonica. Mélanges Offerts à Maurice Birot.* Paris: Éditions Recherche sur les Civilisations, 1985; 181-189.

Lambert 1985b — W.G. Lambert, The Pair Laḫmu - Laḫamu in Cosmology. *OrNS* 54 (1985) 189-202.

Lambert 1985c — W.G. Lambert, The Pantheon of Mari. *MARI* 4 (1985) 525-539. Addenda: *MARI* 6 (1990) 644.

Lambert 1986a — W.G. Lambert, The Reading of the Divine Name Šakkan. *OrNS* 55 (1986) 152-158.

Lambert 1986b — W.G. Lambert, Ninurta Mythology in the Babylonian Epic of Creation. In: K. Hecker, W. Sommerfeld (eds), *Keilschriftliche Literaturen. Ausgewählte Vorträge der XXXII. Rencontre Assyriologique Internationale, Münster, 8.-12.7.1985.* Berlin: Dietrich Reimer Verlag, 1986; 55-60.

Lambert 1987a — W.G. Lambert, Devotion: The Languages of Religion and

Love. In: M. Mindlin, M.J. Geller, J.E. Wansbrough (eds), *Figurative Language in the Ancient Near East*. London: School of Oriental and African Studies, University of London, 1987; 25-39.

Lambert 1987b W.G. Lambert, Goddesses in the Pantheon: A Reflection of Women in Society? In: J.-M. Durand (ed.), *La Femme dans le Proche-Orient Antique. Compte Rendu de la XXXIIIe Rencontre Assyriologique Internationale (Paris, 7-10 Juillet 1986)*. Paris: Éditions Recherche sur les Civilisations, 1987; 125-130.

Lambert 1987c W.G. Lambert, Recension of A.Tsukimoto, *Untersuchungen zur Totenpflege (kispum) im alten Mesopotamien.* AOAT 216. Kevelaer: Verlag Butzon & Bercker; Neukirchen-Vluyn: Neukirchener Verlag, 1985. In: *OrNS* 56 (1987) 403-404.

Lambert 1989 W.G. Lambert, Review of C.H. Gordon, G.A. Rendsburg, N.H. Winter (eds), *Eblaitica: Essays on the Ebla Archives and Eblaite Language*. Vol. I. Winona Lake, Indiana: Eisenbrauns, 1987. *BSOAS* 52 (1989) 115-116.

Lambert 1990 W.G. Lambert, Ancient Mesopotamian Gods. Superstition, Philosophy, Theology. *RHR* 207 (1990) 115-130.

Lambert 1991 W.G. Lambert, A New Look at the Babylonian Background of Genesis. In: H.-P. Müller (ed.), *Babylonien und Israel. Historische, religiöse und sprachliche Beziehungen.* Darmstadt, Wissenschaftliche Buchgesellschaft, 1991; 94-113.

Lambert 1992a W.G. Lambert, The Relationship of Sumerian and Babylonian Myth as Seen in Accounts of Creation. In: D. Charpin, F. Joannès (eds), *La Circulation des Biens, des Personnes et des Idées dans le Proche-Orient Ancien. Actes de la XXXVIIIe Rencontre Assyriologique Internationale (Paris, 8-10 juillet 1991).* Paris: Éditions Recherche sur les Civilisations, 1992; 129-135.

Lambert 1992b W.G. Lambert, The Language of ARET V 6 and 7. In: P. Fronzaroli (ed.), *Literature and Literary Language at Ebla.* QdS 18. Firenze: Dipartimento di Linguistica, Università di Firenze, 1992; 41-62.

Lambert 1992c W.G. Lambert, The Reading of **Uru-KA-gi-na** Again. *AuOr* 10 (1992) 256-258.

Lambert 1996 W.G. Lambert, *Babylonian Wisdom Literature.* Winona Lake, Indiana: Eisenbrauns, 1996 (reprint of the Oxford University Press edition 1963).

Lambert 1998 W.G. Lambert, Technical Terminology for Creation in the Ancient Near East. In: J. Prosecký (ed.), *Intellectual Life of the Ancient Near East. Papers Presented at the 43rd Rencontre Assyriologique Internationale. Prague, July 1-5, 1996*. Prague: Academy of Sciences of the Czech Republic, Oriental Institute, 1998; 189-193.

Lambert 2000 W.G. Lambert, Myth and Mythmaking in Sumer and Akkad. In: J.M. Sasson (ed.), *Civilizations of the Ancient Near East.* Peabody, Massachusetts: Hendrickson Publishers, Inc., 2000; 1825-1835.

Lambert 2008 W.G. Lambert, Mesopotamian Creation Stories. In: M.J. Geller, M. Schipper (eds), *Imagining Creation.* Leiden, Boston: Brill, 2008; 15-59.

Lambert 2010 W.G. Lambert, Time as a Cosmological Principle in Sumerian and Babylonian Thought. *Abstract 56th Rencontre Assyriologique Internationale. Barcelona, 26-30 July, 2010.*

Lambert 2011 W.G. Lambert, Babylon: Origins. In: E. Cancik-Kirschbaum, M. v. Ess, J. Marzahn (eds), *Babylon. Wissenkultur in Orient und Okzident.* Berlin: De Gruyter, 2011; 71-76.

Lambert, Millard 1999 W.G. Lambert, A.R. Millard, *Atra-ḫasīs. The Babylonian Story of the Flood.* With: M. Civil, *The Sumerian Flood Story.* Winona Lake, Indiana: Eisenbrauns, 1999 (reprint of the Oxford University Press edition 1969).

Lambert, Walcot 1965 W.G. Lambert, P. Walcot, A New Babylonian Theogony and Hesiod. *Kadmos* 4 (1965) 64-72.

Landersdorfer 1917 S. Landersdorfer, *Die sumerischen Parallelen zur biblischen Urgeschichte.* Münster: Verlag der Aschendorffschen Buchhandlung, 1917.

Landsberger 1915 B. Landsberger, *Der kultische Kalender der Babylonier und Assyrer.* Leipzig: Hinrichs, 1915.

Landsberger 1964-1966 B. Landsberger, Einige unerkannt gebliebene oder verkannte Nomina des Akkadischen. *WO* 3 (1964-1966) 48-79.

Langdon 1917 S. Langdon, *Sumerian Liturgical Texts.* PBS X, 2. Philadelphia, Pennsylvania: The University Museum, 1917.

Lapinkivi 2004 P. Lapinkivi, *The Sumerian Sacred Marriage in the Light of Comparative Evidence.* SAAS XV. Helsinki: The Neo-Assyrian Text Corpus Project of the University of Helsinki in co-operation with the Finnish Oriental Society, 2004.

Larsen 1987 M.T. Larsen, The Mesopotamian Lukewarm Mind. Reflections on Science, Divination and Literacy. In: F. Rochberg-Halton (ed.), *Language, Literature, and History: Philological and Historical Studies Presented to Erica Reiner.* New Haven, Connecticut: American Oriental Society, 1987; 203-225.

Larsen 1998 S. Larsen, *The Shaman's Doorway. Opening Imagination to Power and Myth.* Rochester, Vermont: Inner Traditions, 1998.

Lebrun 2000 R. Lebrun, From Hittite Mythology: The Kumarbi Cycle. In: J.M. Sasson (ed.), *Civilizations of the Ancient Near East.* Peabody, Massachusetts: Hendrickson Publishers, Inc., 2000; 1971-1980.

Leeming 2010 D.A. Leeming, *Creation Myths of the World: An Encyclopedia.* (2nd ed., 2 vols.). Santa Barbara, California - Denver, Colorado - Oxford, England: ABC-CLIO, LLC, 2010.

Legrain 1936 L. Legrain, *Ur Excavations, vol. III. Archaic Seal-Impressions.* London: The Oxford University Press; Philadelphia: The University Museum of the University of Pennsylvania, 1936.

Leick 1994 G. Leick, *Sex and Eroticism in Mesopotamian Literature.* London and New York: Routledge, 1994.

Lévi-Strauss 1962 C. Lévi-Strauss, *La Pensée Sauvage.* Paris: Librairie Plon, 1962.

Lévi-Strauss 1969 C. Lévi-Strauss, *Structural Anthropology.* (Translated from the French by C. Jacobson and B. Grundfest Schoepf). London: Allen Lane, The Penguin Press, 1969.

Lévi-Strauss 1977 C. Lévi-Strauss, *Structural Anthropology. Vol. II.* (Translated from the French by M. Layton). London: Allen Lane, Penguin Books Ltd., 1977.

Lisman 2005 J.J.W. Lisman, Kosmogonie in Sumerische Teksten uit het 3e en 2e millennium v. Chr. tot en met de Oudbabylonische Tijd. Leiden: [s.n.] 2005.

Litke 1998 R.L. Litke, *A Reconstruction of the Assyro-Babylonian God-Lists,* **AN**: *^{d}a-nu-um and* **AN**: *anu ša amēli.* New Haven: Yale Babylonian Collection, 1998.

Liverani 1993 M. Liverani (ed.), *Akkad, The First World Empire. Structure, Ideology, Traditions.* HANE/S-V. Padova: Sargon srl, 1993.

Loesov 2004 S. Loesov, **I-NU-MA-I-LU-A-WI-LUM**. *Babel und Bibel: Ancient Near Eastern, Old Testament and Semitic Studies* 1 (2004)183-201.

Long 1963 C.H. Long, *ALPHA. The Myths of Creation.* Atlanta, Georgia: Scholars Press, 1963.

Lovin, Reynolds 1985 R.W. Lovin, F.E. Reynolds, In the Beginning. In: R.W. Lovin, F.E. Reynolds (eds), *Cosmogony and Ethical Order. New Studies in Comparative Ethics.* Chicago and London: The University of Chicago Press, 1985; 1-35.

Ludwig 1990 M.-C. Ludwig, *Untersuchungen zu den Hymnen des Išme-Dagan von Isin.* Wiesbaden: Otto Harrassowitz, 1990.

Luginbühl 1992 M. Luginbühl, *Menschenschöpfungsmythen. Ein Vergleich zwischen Griechenland und dem Alten Orient.* Bern, Frankfurt a.M., New York, Paris, Wien: Peter Lang, 1992.

Lutz 1919 H.F. Lutz, *Selected Sumerian and Babylonian Texts.* PBS 1, no. 2. Philadelphia: Philadelphia University Museum, 1919.

Maisels 1999 C.K. Maisels, *The Emergence of Civilization. From Hunting and Gathering to Agriculture, Cities, and the State in the Near East.* London and New York: Routledge, 1999.

Mander 1986 P. Mander, *Il Pantheon di Abu-Ṣālabīkh. Contributo allo Studio del Pantheon Sumerico Arcaico.* Napoli: Instituto Universitario Orientale, Dipartimento di Studi Asiatici, 1986.

Mander 1995 P. Mander, Designs on the Fara, Abū-Ṣalābīkh and Ebla Tablets. *Annali (Istituto Universitario Orientale)* 55 (1995) 18-29.

Mander 2008 P. Mander, Les dieux et le Culte à Ébla. In: G. del Olmo Lete (ed.), *Mythologie et Religion des Sémites Occidentaux. Vol. I. Ébla, Mari.* OLA 162. Leuven, Paris, Dudley MA: Uitgeverij Peeters and Departement Oosterse Studies, 2008; 1-160.

Marchesi 2004 G. Marchesi, Who Was Buried in the Royal Tombs of Ur? The Epigraphic and Textual Data. *OrNS* 73 (2004) 153-197.

Margalit 1981 B. Margalit, The Ugarit Creation Myth: Fact or Fiction? *UF* 13 (1981) 137-145.

Martin *et al.* 2001 H.P. Martin *et al.*, *The Fara Tablets in the University of Pennsylvania Museum of Archaeology and Anthropology.* Bethesda, Maryland: CDL Press, 2001.

Matthews 1993 — R.J. Matthews, *Cities, Seals and Writing: Archaic Seal Impressions from Jemdet Nasr and Ur.* MSVO 2. Berlin: Gebr. Mann Verlag, 1993.

Maul 1997 — S.M. Maul, Die altorientalische Hauptstadt - Abbild und Nabel der Welt. In: G. Wilhelm (ed.), *Die orientalische Stadt: Kontinuität, Wandel, Bruch. 1. Internationales Colloquium der Deutschen Orient-Gesellschaft 9.-10. Mai 1996 in Halle/Saale.* Saarbrücken: Saarbrücker Druckerei und Verlag, 1997; 109-124.

Mayer 1987 — W.R. Mayer, Ein Mythos von der Erschaffung des Menschen und des Königs. *OrNS* 56 (1987) 55-68.

McCaffrey 2013 — K. McCaffrey, The Sumerian Sacred Marriage: Texts and Images. In: H. Crawford (ed.), *The Sumerian World.* London and New York: Routledge, 2013; 227-245.

McCarthy 2010 — A. McCarthy, 'Let Us Submit to the House of Kish': Glyptic and the Genesis of State-Level Identity in Sumer. In: P. Matthiae *et al.* (eds), *Proceedings of the 6th International Congress of the Archaeology of the Ancient Near East: 5 May - 10 May 2009, »Sapienza«, Università di Roma. Vol. 1: Near Eastern Archaeology in the Past, Present and Future. Heritage and Identity. Ethnoarchaeological and Interdisciplinary Approach, Results and Perspectives. Visual Expression and Craft Production in the Definition of Social Relations and Status.* Wiesbaden: Harrassowitz Verlag, 2010; 837-847.

Meek 1920 — T.J. Meek, Some explanatory lists and grammatical texts. *RA* 17 (1920) 117-206.

Metzler 2002 — K.A. Metzler, *Tempora in altbabylonischen literarischen Texten.* AOAT 279. Münster: Ugarit-Verlag, 2002.

Meyer, Sommerfeld 2004 — J.-W. Meyer, W. Sommerfeld (eds), *2000 v. Chr.. Politische, wirtschaftliche und kulturelle Entwicklung im Zeichen einer Jahrtausendwende. 3. Internationales Colloquium der Deutschen Orient-Gesellschaft 4.-7. April 2000 in Frankfurt/Main und Marburg/Lahn.* Saarbrücken: SDV, Saarbrücker Druckerei und Verlag, 2004.

Meyer-Laurin 2012 — V. Meyer-Laurin, Zum Alter des sumerischen Adverbiativs. In: C. Mittermayer, S. Ecklin (eds), *Altorientalische Studien zu Ehren von Pascal Attinger.* **mu-ni u_4 ul-li_2-a-aš $\hat{g}a_2$-$\hat{g}a_2$-de_3**. OBO 256. Fribourg: Academic Press; Göttingen: Vandenhoeck & Ruprecht, 2012; 215-241.

Michalowski 1981 — P. Michalowski, Carminative Magic: Towards an Understanding of Sumerian Poetics. *ZA* 71 (1981) 1-18.

Michalowski 1983 — P. Michalowski, History as Charter. Some Observations on the Sumerian King List. *JAOS* 103 (1983) 237-248.

Michalowski 1989 — P. Michalowski, *The Lamentation over the Destruction of Sumer and Ur.* Winona Lake, Indiana: Eisenbrauns, 1989.

Michalowski 1990a — P. Michalowski, Presence at the Creation. In: T. Abusch, J. Huehnergard, P. Steinkeller (eds), *Lingering over Words. Studies in Ancient Near Eastern Literature in Honor of William L. Moran.* Atlanta, Georgia: Scholars Press, 1990; 381-396.

Michalowski 1990b P. Michalowski, The Shekel and the Vizier. *ZA* 80 (1990) 1-8.

Michalowski 1991 P. Michalowski, Negation as Description: The Metaphor of Everyday Life in Early Mesopotamian Literature. In: P. Michalowski *et al.* (eds), *Velles Paraules. Ancient Near Eastern Studies in Honor of Miguel Civil on the Occasion of his Sixty-Fifth Birthday. AuOr* 9 (1991) 131-136.

Michalowski 1993a P. Michalowski, The Torch and the Censer. In: M.E. Cohen, D.C. Snell, D.B. Weisberg (eds), *The Tablet and the Scroll. Near Eastern Studies in Honor of William W. Hallo.* Bethesda, Maryland: CDL Press, 1993; 152-162.

Michalowski 1993b P. Michalowski, On the Early Toponomy of Sumer: A Contribution to the Study of Early Mesopotamian Writing. In: A.F. Rainey (ed.), *kinattūtu ša dārâti. Raphael Kutscher Memorial Volume.* Tel Aviv: Tel Aviv University, Institute of Archaeology, 1993; 119-133.

Michalowski 1996 P. Michalowski, Sailing to Babylon, Reading the Dark Side of the Moon. In: J.S. Cooper, G.M. Schwartz (eds), *The Study of the Ancient Near East in the Twenty-First Century. The William Foxwell Albright Centennial Conference.* Winona Lake, Indiana: Eisenbrauns, 1996; 177-193.

Michalowski 1998 P. Michalowski, The Unbearable Lightness of Enlil. In: J. Prosecký (ed.), *Intellectual Life of the Ancient Near East. Papers Presented at the 43rd Rencontre Assyriologique Internationale; Prague, July 1-5, 1996.* Prague: Academy of Sciences of the Czech Republic, Oriental Institute, 1998; 237-247.

Michalowski 1998-2001 P. Michalowski, Nisaba. A. Philologisch. In: E. Ebeling, B. Meissner (eds), *Reallexikon der Assyriologie*, Band 9. Berlin, New York: Walter de Gruyter, 1998-2001; 575-579.

Michalowski 2000 P. Michalowski, The Life and Death of the Sumerian Language in Comparative Perspective. In: J. Black (†), G. Zólyomi (eds), *Special Volume in Honor of Professor Mamoru Yoshikawa, 1. The Study of Diachronic and Synchronic Variation in Sumerian: Papers Presented at the 6th Meeting of the Sumerian Grammar Discussion Group, Oxford, 17th and 18th September 1999* [Organized by J. Black (†) and G. Zólyomi]. *ASJ* 22 (2000) 177-202.

Michalowski 2002 P. Michalowski, 'Round About Nidaba: On the Early Goddesses of Sumer. In: S. Parpola, R.M. Whiting (eds), *Sex and Gender in the Ancient Near East. Proceedings of the 47th Rencontre Assyriologique Internationale, Helsinki, July 2-6, 2001. Part II.* Helsinki: The Neo-Assyrian Text Corpus Project, Institute for Asian and African Studies, University of Helsinki, 2002; 413-422.

Michalowski 2003 P. Michalowski, A Man Called Enmebaragesi. In: W. Sallaberger, K. Volk, A. Zgoll (eds), *Literatur, Politik und Recht in Mesopotamien. Festschrift für Claus Wilcke.* Wiesbaden: Harrassowitz Verlag, 2003; 195-208.

Michalowski 2006 P. Michalowski, The Strange History of Tumal. In: P.

Michalowski, N. Veldhuis (eds), *Approaches to Sumerian Literature. Studies in Honour of Stip (H.L.J. Vanstiphout)*. CM 35. Leiden, Boston: Brill, 2006; 145-165.

Michalowski 2007 — P. Michalowski, The Lives of the Sumerian Language. In: S.L. Sanders (ed.), *Margins of Writing, Origins of Cultures*. OIS 2 (second printing). Chicago, Illinois: The Oriental Institute of the University of Chicago, 2007; 163-188.

Michalowski 2010 — P. Michalowski, Where's **AL**? Humor and Poetics in the Hymn to the Hoe. In: A. Kleinerman, J.M. Sasson (eds), *Why Should Someone Who Knows Something Conceal It? Cuneiform Studies in Honor of David I. Owen on His 70th Birthday*. Bethesda: CDL Press, 2010; 195-200.

Mittermayer 2005 — C. Mittermayer, *Die Entwicklung der Tierkopfzeichen. Eine Studie zur syro-mesopotamischen Keilschriftpaläographie des 3. und frühen 2. Jahrtausends v. Chr.* AOAT 319. Münster: Ugarit-Verlag, 2005.

Mittermayer 2006 — C. Mittermayer (unter Mitarbeit von Pascal Attinger), *Altbabylonische Zeichenliste der sumerisch-literarischen Texte*. OBO Sonderband. Fribourg: Academic Press; Göttingen: Vandenhoeck & Ruprecht, 2006.

Mittermayer 2009 — C. Mittermayer, *Enmerkara und der Herr von Arata. Ein ungleicher Wettstreit*. OBO 239. Fribourg: Academic Press; Göttingen: Vandenhoeck & Ruprecht, 2009.

Moran 1987 — W.L. Moran, Some Considerations of Form and Interpretation in *Atra-ḫasīs*. In: F. Rochberg-Halton (ed.), *Language, Literature and History: Philological and Historical Studies Presented to Erica Reiner*. AOS 67. New Haven, Connecticut: American Oriental Society, 1987; 245-255.

Mundkur 1978 — B. Mundkur, Notes on Two Ancient Fertility Symbols. *East and West* 28 (1978) 263-282.

Myers 2002 — J. Myers, *The Sippar Pantheon: A Diachronic Study*. Ann Arbor, Michigan: UMI, 2002.

Neumann 1981 — H. Neumann, Eine Inschrift des Königs Lugalkisalsi (VA 4855). *AoF* 8 (1981) 75-82.

Nissen 1985 — H.J. Nissen, Ortsnamen in den archaischen Texten aus Uruk. *OrNS* 54 (1985) 226-233.

Nissen 1988 — H.J. Nissen, *The Early History of the Ancient Near East 9000-2000 B.C.* Chicago and London: The University of Chicago Press, 1988.

Nissen 1993 — H.J. Nissen, Settlement Patterns and Material Culture of the Akkadian Period: Continuity and Discontinuity. In: M. Liverani (ed.), *Akkad, The First World Empire. Structure, Ideology, Traditions*. HANE/S-V. Padova: Sargon srl, 1993; 91-106.

Nissen 1999 — H.J. Nissen, *Geschichte Altvorderasiens*. München: R. Oldenbourg Verlag, 1999.

Nissen 2001 — H.J. Nissen, Cultural and Political Networks in the Ancient Near East during the Fourth and Third Millennia B.C. In: M.S. Rothman (ed.), *Uruk Mesopotamia & Its Neighbors. Cross-*

	Cultural Interactions in the Era of State Formation. Santa Fe: School of American Research Press; Oxford: James Currey, 2001; 149-179.
Nöldeke, Heinrich 1932	A. Nöldeke, E. Heinrich, *Vierter vorläufiger Bericht über die von der Notgemeinschaft der deutschen Wissenschaft in Uruk unternommen Ausgrabungen.* Berlin: Verlag der Akademie der Wissenschafte, 1932.
Nützel 1976	W. Nützel, The Climate Changes of Mesopotamia and Bordering Areas 14000 to 2000 B.C. *Sumer* 32 (1976) 11-25.
Oates 1960	J. Oates, Ur and Eridu, The Prehistory. In: M.E.L. Mallowan, D.J. Wiseman (eds), *Ur in Retrospect. In Memory of Sir C. Leonard Woolley.* (*Iraq* 22). London: British School of Archaeology in Iraq, 1960; 32-50.
Oberhuber 1967	K. Oberhuber, Eine Hymne an Nippur (UET VI 118). *ArOr* 35 (1967) 262-270.
Oberhuber 1991	K. Oberhuber, *Linguistisch-philologische Prolegomena zur altorientalischen Religionsgeschichte.* Innsbruck: Institut für Sprachwissenschaft der Universität Innsbruck, 1991.
O'Connor 2011	D. O'Connor, The Narmer Palette: A New Interpretation. In: E. Teeter (ed.), *Before the Pyramids. The Origins of Egyptian Civilization.* OIMP 33. Chicago, Illinois: The Oriental Institute of the University of Chicago, 2011; 145-152.
Ong 2004	W.J. Ong, *Orality and Literacy. The Technologizing of the Word.* London and New York: Routledge, 2004.
Orthmann *et al.* 1985	W. Orthmann *et al.*, *Der alte Orient. Propyläen Kunstgeschichte 18.* Berlin: Propyläen Verlag, 1985.
Oshima 2012	T. Oshima, When the Gods Made Us from Clay. In: A. Berlejung, J. Dietrich, J.H. Quack (eds), *Menschenbilder und Körperkonzepte im alten Israel, in Ägypten und im alten Orient.* Tübingen: Mohr Siebeck, 2012; 407-431.
Osten-Sacken, v.d. 1992	E. von der Osten-Sacken, *Der Ziegen-'Dämon'. ʿObed- und Urukzeitliche Götterdarstellungen.* AOAT 230. Kevelaer: Verlag Butzon & Bercker; Neukirchen-Vluyn: Neukirchener Verlag, 1992.
Osten-Sacken, v.d. 2009	E. von der Osten-Sacken, Schlange. C. In der Bildkunst. In: E. Ebeling, B. Meissner (eds), *Reallexikon der Assyriologie*, Band 12. Berlin, New York: Walter de Gruyter, 2009; 219-221.
Otten, Siegelová 1970	H. Otten, J. Siegelová, Die hethitischen Gulš-Gottheiten und die Erschaffung der Menschen. *AfO* 23 (1970) 32-38.
Pardee 1997	D. Pardee, The Baʿlu Myth. In: In: W.W. Hallo, K.L. Younger (eds), *The Context of Scripture. Vol. I. Canonical Compositions from the Biblical World. 1. Ugaritic Myths.* Leiden, New York, Köln: Brill, 1997; 241-274.
Parker 2000	S.B. Parker, The Literature of Canaan, Ancient Israel, and Phoenicia: an Overview. In: J.M. Sasson (ed.), *Civilizations of the Ancient Near East*. Peabody, Massachusetts: Hendrickson Publishers, Inc., 2000; 2399-2410.
Peterson 2009	J. Peterson, *Godlists from Old Babylonian Nippur in the University Museum, Philadelphia.* AOAT 362. Münster:

Ugarit-Verlag, 2009.

Pettinato 1971 G. Pettinato, *Das altorientalische Menschenbild und die sumerischen und akkadischen Schöpfungsmythen.* Heidelberg: Carl Winter, Universitätsverlag, 1971.

Pettinato 1982 G. Pettinato (ed.), in collaborazione con E. Arcari, A. Magi-Spinetti e G. Visicato, *Testi Lessicali Bilingui della Bibliotheca L. 2769. Parte I. Traslitterazione dei Testi e Ricostruzione del VE.* MEE 4. Napoli: Instituto Universitario Orientale di Napoli, Seminario di Studi Asiatici, 1982.

Pettinato 2000 G. Pettinato, Discesa di Enki agli Inferi? In: S. Graziani (ed.), con la collaborazione di M.C. Casaburi e G. Lacerenza, *Studi sul Vicino Oriente Antico Dedicati alla Memoria di Luigi Cagni.* Napoli: Instituto Universitario Orientale, Dipartimento di Studi Asiatici, 2000; 863-879.

Pfeiffer, Speiser 1935-1936 R.H. Pfeiffer, E.A. Speiser, One Hundred New Selected Nuzi Texts. *AASOR* 16 (1935-1936) 1-168.

Pientka-Hinz 2009 R. Pientka-Hinz, Schlange. A. In Mesopotamien. In: E. Ebeling, B. Meissner (eds), *Reallexikon der Assyriologie*, Band 12. Berlin, New York: Walter de Gruyter, 2009; 202-218.

Piesl 1969 H. Piesl, *Vom Präanthropomorphismus zum Anthropomorphismus. Entwicklungsstadien im altmesopotamischen Pantheon dargestellt am Präanthropomorphem* **kur** *und an der Hauptgottheit* **[d]en-lil$_2$**. Innsbruck: Institut für vergleichende Sprachwissenschaft der Leopold-Franzens-Universität Innsbruck, 1969.

Poebel 1914 A. Poebel, *Historical Texts.* University of Pennsylvania, The University Museum, Publications of the Babylonian Section IV/1. Philadelphia: The University Museum, 1914; 7-70.

Poebel 1923 A. Poebel, *Grundzüge der sumerischen Grammatik.* Rostock: Selbstverlag des Verfassers, 1923.

Pohl 1935 A. Pohl, *Vorsargonische und sargonische Wirtschaftstexte. TMH 5.* Leipzig: Hinrichs, 1935.

Pollock 2001 S. Pollock, The Uruk Period in Southern Mesopotamia. In: M.S. Rothman (ed.), *Uruk Mesopotamia & Its Neighbors. Cross-Cultural Interactions in the Era of State Formation.* Santa Fe: School of American Research Press; Oxford: James Currey, 2001; 181-231.

Pomponio 1983 F. Pomponio, Notes on the Lexical Texts from Abū Salābīkh and Ebla. *JNES* 42 (1983) 285-290.

Pomponio, Xella 1997 F. Pomponio, P. Xella, *Les Dieux d'Ebla. Étude Analytique des Divinités Éblaïtes à l'Époque des Archives Royales du IIIe Millénaire.* AOAT 245. Münster: Ugarit-Verlag, 1997.

Porada 1992 E. Porada, A Cylinder with a Storm God and Problems. In: D.J.W. Meijer (ed.), *Natural Phenomena: Their Meaning, Depiction and Description in the Ancient Near East.* Amsterdam, Oxford, New York, Tokyo: North-Holland, 1992; 227-243.

Postgate 2000 J.N. Postgate, Royal Ideology and State Administration in Sumer and Akkad. In: J.M. Sasson (ed.), *Civilizations of the*

Ancient Near East. Peabody, Massachusetts: Hendrickson Publishers, Inc., 2000; 395-411.

Powell 1984 M.A. Powell, Sumerian Cereal Crops. *BSA* 1 (1984) 48-72.

PSD See: Sjöberg 1984+

Quirke 2008 S. Quirke, Creation Stories in Ancient Egypt. In: M.J. Geller, M. Schipper (eds), *Imagining Creation.* Leiden, Boston: Brill, 2008; 61-86.

Rappaport 1999 R.A. Rappaport, *Ritual and Religion in the Making of Humanity.* Cambridge: Cambridge University Press, 1999.

Rawi, Black 1993 F.N.H. Al-Rawi, J.A. Black, A Rediscovered Akkadian City. *Iraq* 55 (1993) 147-148.

Renger 1967a J. Renger, Untersuchungen zum Priestertum in der altbabylonischen Zeit. 1.Teil. *ZA* 58 (1967) 110-188.

Renger 1967b J. Renger, Götternamen in der altbabylonischen Zeit. In: *Heidelberger Studien zum alten Orient. Adam Falkenstein zum 17. September 1966.* Wiesbaden: Otto Harrassowitz, 1967; 137-171.

Renger 1969 J. Renger, Untersuchungen zum Priestertum in der altbabylonischen Zeit. 2.Teil. *ZA* 59 (1969) 104-230.

Renger 1972-1975 J. Renger, Heilige Hochzeit. A. Philologisch. In: E. Ebeling, B. Meissner (eds), *Reallexikon der Assyriologie*, Band 4; Berlin, New York: Walter de Gruyter, 1972-1975; 251-259.

Renger 2004 J. Renger, Wirtschaftsformen in Mesopotamien zwischen dem dritten und zweiten Jahrtausend v. Chr. In: J.-W. Meyer, W. Sommerfeld (eds), *2000 v. Chr. Politische, wirtschaftliche und kulturelle Entwicklung im Zeichen einer Jahrtausendwende. 3. Internationales Colloquium der Deutschen Orient-Gesellschaft 4.-7. April 2000 in Frankfurt/Main und Marburg/Lahn.* Saarbrücken: SDV Saarbrücker Druckerei und Verlag, 2004; 141-154.

Richter 2004 T. Richter, *Untersuchungen zu den lokalen Panthea Süd- und Mittelbabyloniens in altbabylonischer Zeit. (2., verbesserte und erweiterte Auflage).* AOAT 257. Münster: Ugarit-Verlag, 2004.

Roaf 1996 M. Roaf, *Cultural Atlas of Mesopotamia and the Ancient Near East.* New York: Facts On File, Inc., 1996.

Roberts 1972 J.J.M. Roberts, *The Earliest Semitic Pantheon. A Study of the Semitic Deities Attested in Mesopotamia before Ur III.* Baltimore and London: The Johns Hopkins University Press, 1972.

Röllig 1976 W.Röllig, Über die Moral der Götter. In: B. Gladigow (ed.), *Religion und Moral.* Düsseldorf: Patmos, 1976; 84-98.

Römer 1965 W.H.Ph. Römer, *Sumerische 'Königshymnen' der Isin-Zeit.* Leiden: E.J. Brill, 1965.

Römer 1969a W.H.Ph. Römer, Religion of Ancient Mesopotamia. In: C.J. Bleeker, G. Widengren (eds), *Historia Religionum; Handbook for the History of Religions.* vol. I. *Religions of the Past.* Leiden: E.J. Brill, 1969; 115-194.

Römer 1969b W.H.Ph. Römer, Einige Beobachtungen zur Göttin Nini(n)sina auf Grund von Quellen der Ur III-Zeit und der altbabylonischen Periode. In: W. Röllig (ed.) unter Mitwirkung von M. Dietrich, *lišān mithurti. Festschrift Wolfram Freiherr von Soden zum 19. VI. 1968 gewidmet von Schülern und Mitarbeitern.* Kevelaer: Verlag Butzon & Bercker; Neukirchen-Vluyn: Neukirchener Verlag des Erziehungs-vereins, 1969; 279-305.

Römer 1988 W.H.Ph. Römer, Sumerische Hymnen. II. Eine Sammeltafel von Hymnen(fragmenten) m.B.a. Išmedagān von Isin und Nibru (mit Dupll.). *BiOr* 45, 24-60.

Römer 1990 W.H.Ph. Römer, Der Vater und sein nichtsnutziger Sohn. In: W.H.Ph. Römer, W. von Soden (eds), *Weisheitstexte, Mythen und Epen.* TUAT III/1: *Weisheitstexte.* Gütersloh: Gütersloher Verlagshaus Gerd Mohn, 1990; 77-91.

Römer 1993a W.H.Ph. Römer, Die Urzeit: Vor der Schöpfung (I), (II). In: W.H.Ph. Römer (ed.), mit einem Beitrag von D.O. Edzard, *Weisheitstexte, Mythen und Epen. Mythen und Epen I.* TUAT III/3. Gütersloh: Gütersloher Verlagshaus Gerd Mohn, 1993; 353-354 (I), 355-356 (II).

Römer 1993b W.H.Ph. Römer, Enki und Ninmach. In: W.H.Ph. Römer (ed.), mit einem Beitrag von D.O. Edzard, *Weisheitstexte, Mythen und Epen. Mythen und Epen I.* TUAT III/3. Gütersloh: Gütersloher Verlagshaus Gerd Mohn, 1993; 386-401.

Römer 1993c W.H.Ph. Römer, Der Prolog des Streitgespräches zwischen Holz und Rohr (Z. 1-29). In: W.H.Ph. Römer (ed.), mit einem Beitrag von D.O. Edzard, *Weisheitstexte, Mythen und Epen. Mythen und Epen I.* TUAT III/3. Gütersloh: Gütersloher Verlagshaus Gerd Mohn, 1993; 357-360.

Römer 1993d W.H.Ph. Römer, Aus »Lugal ud me-lám-bi nir-ĝal«. In: W.H.Ph. Römer (ed.), mit einem Beitrag von D.O. Edzard (eds), *Weisheitstexte, Mythen und Epen. Mythen und Epen I.* TUAT III/3. Gütersloh: Gütersloher Verlagshaus Gerd Mohn, 1993; 434-448.

Römer 1993e W.H.Ph. Römer, Die Flutgeschichte. In: W.H.Ph. Römer, D.O. Edzard (eds), *Weisheitstexte, Mythen und Epen. Mythen und Epen I.* TUAT III. Gütersloh: Mohn, 1993; 448-458.

Römer 1994 W.H.Ph. Römer, Der Mythos vom Königtum der Götter und Kumarbi, CTH 344. In: K. Hecker (ed.), *Weisheitstexte, Mythen und Epen. Mythen und Epen II.* TUAT III. Gütersloh: Gütersloher Verlagshaus, 1994; 828-830.

Römer 2010 W.H.Ph. Römer, *Die Zylinderinschriften von Gudea.* AOAT 376. Münster: Ugarit-Verlag, 2010.

Rosengarten 1967 Y. Rosengarten, *Répertoire Commenté des Signes Présargoniques Sumériens de Lagaš.* Paris: E. de Boccard, 1967.

Rosengarten 1977 Y. Rosengarten, avec la collaboration de A. Baer s.j., *Sumer et le Sacré. Le Jeu des Prescriptions (***me***), des Dieux, et des Destins.* Paris: Éditions E. de Boccard, 1977.

Rothman (ed.) 2001 M.S. Rothman (ed.), *Uruk Mesopotamia & Its Neighbors. Cross-Cultural Interactions in the Era of State Formation.*

Santa Fe, New Mexico: School of American Research Press; Oxford: James Currey, Ltd, 2001.

Roux 1969 — G. Roux, Ancient Iraq. In: T. Jones (ed.), *The Sumerian Problem*. New York, London, Sydney, Toronto: John Wiley & Sons, Inc., 1969; 134-138.

Rubio 1999 — G. Rubio, On the Alleged 'Pre-Sumerian Substratum'. *JCS* 51 (1999) 1-16.

Rubio 2005 — G. Rubio, On the Linguistic Landscape of Early Mesopotamia. In: W.H. van Soldt (ed.), in cooperation with R. Kalvelagen and D. Katz, *Ethnicity in Ancient Mesopotamia. Papers Read at the 48th Rencontre Assyriologique Internationale, Leiden, 1-4 July 2002*. Leiden: Nederlands Instituut voor het Nabije Oosten, 2005; 316-332.

Rubio 2011 — G. Rubio, Gods and Scholars: Mapping the Pantheon in Early Mesopotamia. In: B. Pongratz-Leisten (ed.), *Reconsidering the Concept of Revolutionary Monotheism*. Winona Lake, Indiana: Eisenbrauns, 2011; 91-116.

Ruijter, de 1979 — A. de Ruijter, *Een Speurtocht naar het Denken. Een Inleiding tot het Structuralisme van Claude Lévi-Strauss*. Assen: Van Gorcum, 1979.

Safar *et al.* 1981 — F. Safar, M.A. Mustafa, S. Lloyd, *Eridu*. Baghdad: Republic of Iraq, Ministry of Culture and Information, State Organization of Antiquities and Heritage, 1981.

Saggs 1978 — H.W.F. Saggs, *The Encounter with the Divine in Mesopotamia and Israel*. London: The Athlone Press, 1978.

Sallaberger 1993 — W. Sallaberger, *Der kultische Kalender der Ur III-Zeit*. Berlin, New York: Walter de Gruyter, 1993.

Sallaberger 1997 — W. Sallaberger, Nippur als religiöses Zentrum Mesopotamiens im historischen Wandel. In: G. Wilhelm (ed.), *Die orientalische Stadt: Kontinuität, Wandel, Bruch. 1. Internationales Colloquium der Deutschen Orient-Gesellschaft 9.-10. Mai 1996 in Halle-Saale*. Saarbrücken: SDV Saarbrücker Druckerei und Verlag, 1997; 147-168.

Sallaberger 1999 — W. Sallaberger, Ur III-Zeit. In: W. Sallaberger, A. Westenholz, *Mesopotamien. Akkade-Zeit und Ur III-Zeit*. Freiburg, Schweiz: Universitätsverlag; Göttingen: Vandenhoeck & Ruprecht, 1999; 121-390.

Sallaberger 2003-2006 — W. Sallaberger, Pantheon. A.I. In Mesopotamien. In: E. Ebeling, B. Meissner (eds), *Reallexikon der Assyriologie*, Band 10. Berlin, New York: Walter de Gruyter, 2003-2006; 294-308.

Sallaberger 2005a — W. Sallaberger, The Sumerian Verb **na** $\mathbf{de_5}$**(-g)** 'To Clear'. In: Y. Sefati *et al.* (eds), *'An Experienced Scribe Who Neglects Nothing'. Ancient Near Eastern Studies in Honor of Jacob Klein*. Bethesda, Maryland: CDL Press, 2005; 229-253.

Sallaberger 2005b — W. Sallaberger, 'bringen' im Sumerischen. Lesung und Bedeutung von $\mathbf{de_6}$ (**DU**) und $\mathbf{tum_2}$ (**DU**). In: R. Rollinger (ed.), *Von Sumer bis Homer. Festschrift für Manfred Schretter zum 60. Geburtstag am 25. Februar 2004*. Münster: Ugarit-Verlag, 2005; 557-576.

Sasson (ed.) 2000 — J.M. Sasson (ed.), *Civilizations of the Ancient Near East.* Peabody, Massachusetts: Hendrickson Publishers, Inc., 2000.

Sauren 1993 — H. Sauren, Nammu and Enki. In: M.E. Cohen, D.C. Snell, D.B. Weisberg (eds), *The Tablet and The Scroll. Near Eastern Studies in Honor of William W. Hallo.* Bethesda, Maryland: CDL Press, 1993; 198-208.

Schipper 2010 — M. Schipper, *In het begin was er niemand. Hoe het komt dat er mensen zijn.* Amsterdam: Uitgeverij Bert Bakker, 2010.

Schmandt-Besserat 1993 — D. Schmandt-Besserat, Images of Enship. In: M. Frangipane *et al.* (eds), *Between the Rivers and over the Mountains. Archaeologica Anatolica et Mesopotamica Alba Palmieri Dedicata.* Roma: Dipartimento di Scienze Storiche Archeologiche e Antropologiche dell'Antichità, Università di Roma «La Sapienza», 1993; 201-219.

Schmidt 1996 — B.B. Schmidt, *Israel's Beneficent Dead. Ancestor Cult and Necromancy in Ancient Israelite Religion and Tradition.* Winona Lake, Indiana: Eisenbrauns, 1996.

Schretter 1990 — M.K. Schretter, *Emesal-Studien. Sprach- und Literaturgeschichtliche Untersuchungen zur sogenannten Frauensprache des Sumerischen.* Innsbruck: Verlag des Instituts für Sprachwissenschaft der Universität Innsbruck, 1990.

Scurlock 2000 — J.A. Scurlock, Death and the Afterlife in Ancient Mesopotamian Thought. In: J.M. Sasson (ed.), *Civilizations of the Ancient Near East.* Peabody, Massachusetts: Hendrickson Publishers, Inc., 2000; 1883-1893.

Sefati 1998 — Y. Sefati, *Love Songs in Sumerian Literature. Critical Edition of the Dumuzi-Inanna Songs.* Ramat Gan: Bar-Ilan University Press, 1998.

Selz 1990 — G. Selz, Studies in Early Syncretism: The Development of the Pantheon in Lagaš. Examples for Inner-Sumerian Syncretism. *ASJ* 12 (1990) 111-142.

Selz 1992 — G.J. Selz, Enlil und Nippur nach präsargonischen Quellen. In: M. de Jong Ellis (ed.), *Nippur at the Centennial. Papers Read at the 35*[e] *Rencontre Assyriologique Internationale, Philadelphia, 1988.* Philadelphia: The Samuel Noah Kramer Fund, 1992; 189-225.

Selz 1995a — G. Selz, *Untersuchungen zur Götterwelt des altsumerischen Stadtstaates von Lagaš.* Philadelphia: The Samuel Noah Kramer Fund, 1995.

Selz 1995b — G.J. Selz, **maš-da-ři-a** und Verwandtes. Ein Versuch über **da-ři** 'an der Seite führen': ein zusammengesetztes Verbum und einige nominale Ableitungen. *ASJ* 17 (1995) 251-274.

Selz 1998 — G. Selz, **du$_{11}$-ga-ni-mu-gi-na**, aber steht die Lesung von **URU-KA-gi-na(-k)** fest? Ein Beitrag zum Problem des göttlichen Wortes in sumerischen Personennamen. *AoF* 25 (1998) 312-327.

Selz 1999 — G.J. Selz, Vom 'vergangenen Geschehen' zur 'Zukunftsbewältigung'. Überlegungen zur Rolle der Schrift in Ökonomie und Geschichte. In: B. Böck, E. Cancik-Kirschbaum, T.

Richter (eds), *Munuscula Mesopotamica. Festschrift für Johannes Renger.* AOAT 267. Münster: Ugarit-Verlag, 1999; 465-512.

Selz 2008 — G.J. Selz, The Divine Prototypes. In: N. Brisch (ed.), *Religion and Power. Divine Kingship in the Ancient World and Beyond.* OIS 4. Chicago, Illinois: The Oriental Institute of the University of Chicago, 2008; 13-31.

Seri 2012 — A. Seri, The Role of Creation in Enūma eliš. *JANER* 12 (2012) 4-29.

Seux 1987 — M.-J. Seux, La Création du Monde et de l'Homme dans la Littérature Suméro-Akkadienne. In: F. Blanquart, L. Derousseaux (eds), *La Création dans l'Orient Ancien. Congrès de l'ACFEB, Lille (1985).* Paris: Les Éditions du CERF, 1987; 41-78.

Shaffer 1963 — A. Shaffer, *Sumerian Sources of Tablet XII of the Epic of Gilgameš.* Ann Arbor, Michigan: UMI, 1963.

Sjöberg 1960 — Å. Sjöberg, *Der Mondgott Nanna-Suen in der sumerischen Überlieferung. I. Teil: Texte.* Stockholm: Almqvist & Wiksell, 1960.

Sjöberg 1961 — Å.W. Sjöberg, Ein syllabisch geschriebener Urnammu-Text. *OrSu* 10 (1961) 3-12.

Sjöberg 1970-1971 — Å.W. Sjöberg, Hymns to Meslamtaea, Lugalgirra and Nanna-Suen in Honour of King Ibbīsuen (Ibbīsîn) of Ur. *OrSu* 19-20 (1970-1971) 140-178.

Sjöberg 1973 — Å.W. Sjöberg, Miscellaneous Sumerian Hymns. *ZA* 63 (1973) 1-55.

Sjöberg 1974 — Å.W. Sjöberg, A Hymn to **^{d}lama-sa$_6$-ga**. *JCS* 26 (1974) 158-177.

Sjöberg 1975 — Å.W. Sjöberg, **in-nin šà-gur$_4$-ra**. A Hymn to the Goddess Inanna by the **en**-Priestess Enḫeduanna. *ZA* 65 (1975) 161-253.

Sjöberg 1977 — Å.W. Sjöberg, Miscellaneous Sumerian Texts, II. *JCS* 29 (1977) 3-45.

Sjöberg 1982 — Å.W. Sjöberg, Miscellaneous Sumerian Texts, III. *JCS* 34 (1982) 62-80.

Sjöberg 1984+ — Å.W. Sjöberg (ed.) with the collaboration of H. Behrens *et al.*, *The Sumerian Dictionary of the University Museum of the University of Pennsylvania.* Volume 1, A part I-III; volume 2: B. Philadelphia, Pennsylvania: The Babylonian Section of the University Museum, 1984 +.

Sjöberg 1988 — Å.W. Sjöberg, A Hymn to Inanna and her Self-Praise. *JCS* 40 (1988) 165-186.

Sjöberg 1996a — Å.W. Sjöberg, UET 7 no. 93. A Lexical Text or a Commentary? *ZA* 86 (1996) 220-237.

Sjöberg 1996b — Å.W. Sjöberg, The Ebla List of Animals MEE 4, no. 116. *WO* 27 (1996) 9-24.

Sjöberg 2000 — Å.W. Sjöberg, Early Dynastic Animal Names. R. Jestin, Tablettes Sumériennes de Šuruppak, no. 46, pl. 183. In: J. Marzahn, H. Neumann (eds) unter Mitarbeit von A. Fuchs, *Assyriologica et Semitica. Festschrift für Joachim Oelsner anläßlich seines 65. Geburtstages am 18. Februar 1997.* AOAT 252. Münster: Ugarit-Verlag, 2000; 407-417.

Sjöberg 2002 — Å.W. Sjöberg, In the Beginning. In: T. Abusch (ed.), *Riches Hidden in Secret Places. Ancient Near Eastern Studies in Memory of Thorkild Jacobsen.* Winona Lake, Indiana: Eisenbrauns, 2002; 229-247.

Sjöberg, Bergmann 1969 — A.W. Sjöberg, E. Bergmann, *The Collection of the Sumerian Temple Hymns.* Locust Valley, New York: J.J. Augustin Publisher, 1969.

Smith 1994 — M.S. Smith, *The Ugarit Baal Cycle. Vol. I. Introduction with Text, Translation and Commentary of KTU 1.1-1.2.* Leiden - New York - Köln: E.J. Brill, 1994.

Smith 2000 — M.S. Smith, Myth and Mythmaking in Canaan and Ancient Israel. In: J.M. Sasson (ed.), *Civilizations of the Ancient Near East.* Peabody, Massachusetts: Hendrickson Publishers, Inc., 2000; 2031-2041.

Smith, Pitard 2009 — M.S. Smith, W.T. Pitard, *The Ugarit Baal Cycle. Vol. II. Introduction with Text, Translation and Commentary of KTU/CAT 1.3-1.4.* Leiden - Boston: Brill, 2009.

Soden, v. 1964 — W. von Soden, Babylonische Göttergruppen: Igigu and Anunnaku. Zum Bedeutungswandel theologischer Begriffe. In: *Compte Rendu de l'Onzième Rencontre Assyriologique Internationale, Organisée à Leiden du 23 au 29 Juin 1962 par le Nederlands Instituut voor het Nabije Oosten.* Leiden: Nederlands Instituut voor het Nabije Oosten, 1964; 102-111.

Soden, v. 1978 — W. von Soden, Die erste Tafel des altbabylonischen Atramḫasīs-Mythus. 'Haupttext' und Parallelversionen. *ZA* 68 (1978) 50-94.

Soden, v. 1995 — W. von Soden, *Grundriss der akkadischen Grammatik. 3., ergänzte Auflage unter Mitarbeit von W.R. Mayer. AnOr* 33. Roma: Editrice Pontificio Istituto Biblico, 1995.

Soldt, v. 2005 — W.H. van Soldt (ed.) in cooperation with R. Kalvelagen and D. Katz, *Ethnicity in Ancient Mesopotamia. Papers read at the 48th Rencontre Assyriologique Internationale, Leiden, 1-4 July 2002.* Leiden: Nederlands Instituut voor het Nabije Oosten, 2005.

Sollberger 1956 — E. Sollberger, *Corpus des Inscriptions 'Royales' Présargoniques de Lagaš.* Genève: Droz, 1956.

Sollberger 1961 — E. Sollberger, Le Syllabaire Présargonique de Lagaš. *ZA* 54 (1961) 1-50.

Sollberger 1962 — E. Sollberger, The Tummal Inscription, *JCS* 16 (1962) 40-47.

Sollberger 1965 — E. Sollberger, A Three-Column *Silbenvokabular A.* In: H.G. Güterbock, Th. Jacobsen (eds), *Studies in Honor of Benno Landsberger on his Seventy-Fifth Birthday: April 21, 1965.* AS 16. Chicago, Illinois: The University of Chicago Press, 1965; 21-28.

Sollberger 1966 — E. Sollberger, *The Business and Administrative Correspondence under the Kings of Ur.* Locust Valley, New York: J.J. Augustin, 1966.

Sollberger 1967 E. Sollberger, The Rulers of Lagaš. *JCS* 21 (1967) 279-291.

Sommerfeld 1987-1990 W. Sommerfeld, Marduk. A. Philologisch. I. In Mesopotamien. In: E. Ebeling, B. Meissner (eds), *Reallexikon der Assyriologie*, Band 7; Berlin, New York: Walter de Gruyter, 1987-1990; 360-370.

Sparks 2007 K.L. Sparks, 'Enūma Elish' and Priestly Mimesis: Elite Emulation in Nascent Judaism. *JBL* 126 (2007) 625-648.

Spronk 1986 K. Spronk, *Beatic Afterlife in Ancient Israel and in the Ancient Near East.* AOAT 219. Kevelaer: Verlag Butzon & Bercker; Neukirchen-Vluyn: Neukirchener Verlag, 1986.

Staal 1986 F. Staal, Over het Cyclische en het Rechtlijnige Tijdsbegrip. In: F. Staal, *Over Zin en Onzin in Filosofie, Religie en Wetenschap.* Amsterdam: Meulenhoff Informatief, 1986; 51-76.

Steible, Behrens1982 H. Steible, unter Mitarbeit von H. Behrens, *Die altsumerischen Bau- und Weihinschriften.* FAOS 5/I: *Inschriften aus 'Lagaš'.* FAOS 5/II: *Kommentar zu den Inschriften aus 'Lagaš'. Inschriften ausserhalb von 'Lagaš'.* Wiesbaden: Franz Steiner Verlag, 1982.

Stein, Özbal 2007 G.J. Stein, R. Özbal, A Tale of Two *Oikumenai*: Variation in the Expansionary Dynamics of ʿUbaid and Uruk Mesopotamia. In: E.C. Stone (ed.), *Settlement and Society. Essays Dedicated to Robert McCormick Adams.* Los Angeles: Cotsen Institute of Archaeology, University of California; Chicago, Illinois: The Oriental Institute of the University of Chicago, 2007; 329-342.

Steiner 2005 G. Steiner, Gibt es ein vor-sumerisches Substrat im südlichen Mesopotamien? In: W.H. van Soldt (ed.), in cooperation with R. Kalvelagen and D. Katz, *Ethnicity in Ancient Mesopotamia. Papers Read at the 48th Rencontre Assyriologique Internationale, Leiden, 1-4 July 2002.* Leiden: Nederlands Instituut voor het Nabije Oosten, 2005; 340-355.

Steinkeller 1979 P. Steinkeller, Notes on Sumerian Plural Verbs. *OrNS* 48 (1979) 54-67.

Steinkeller 1980 P. Steinkeller, On the Reading and Location of the Toponyms **ÚR x Ú.KI** and **A.ḪA.KI**. *JCS* 32 (1980) 23-33.

Steinkeller 1984 P. Steinkeller, The Sumerian Verb **lug$_x$** (**LUL**). *SEL* 1 (1984) 5-17.

Steinkeller 1992 P. Steinkeller, Early Semitic Literature and Third Millennium Seals with Mythological Motifs. In: P. Fronzaroli (ed.), *Literature and Literary Language at Ebla.* QdS 18. Firenze: Dipartimento di Linguistica, Università di Firenze, 1992; 243-275, and pl. 1-8.

Steinkeller 1993 P. Steinkeller, Early Political Development in Mesopotamia and the Origins of the Sargonic Empire. In: M. Liverani (ed.), *Akkad, The First World Empire. Structure, Ideology, Traditions.* HANE/S V. Padova: Sargon srl, 1993; 107-129.

Steinkeller 1995a P. Steinkeller, The Location of Kuwara. *NABU* 1995/91, 82-83.

Steinkeller 1995b P. Steinkeller, A Rediscovered Akkadian City? *ASJ* 17 (1995) 275-281.

Steinkeller 1995c P. Steinkeller, Review of M.W. Green, H.J. Nissen,

Zeichenliste der Archaischen Texte aus Uruk; unter Mitarbeit von Peter Damerov und Robert K. Englund. Berlin: Mann Verlag, 1987. *BiOr* 52 (1995) 689-713.

Steinkeller 1995d — P. Steinkeller, Review: Altsumerische Verwaltungstexte aus Girsu/Lagaš by Joachim Marzahn. *JAOS* 115 (1995) 540-543.

Steinkeller 1999 — P. Steinkeller, On Rulers, Priests and Sacred Marriage: Tracing the Evolution of Early Sumerian Kingship. In: K. Watanabe (ed.), *Priests and Officials in the Ancient Near East. Papers of the Second Colloquium on the Ancient Near East - The City and its Life held at the Middle Eastern Culture Center in Japan (Mitaka, Tokyo), March 22-24, 1996.* Heidelberg: Universitätsverlag C. Winter, 1999; 103-137.

Steinkeller 2002 — P. Steinkeller, Archaic City Seals and the Question of Early Babylonian Unity. In: T. Abusch (ed.), *Riches Hidden in Secret Places. Ancient Near Eastern Studies in Memory of Thorkild Jacobsen.* Winona Lake, Indiana: Eisenbrauns, 2002; 249-257.

Steinkeller 2003 — P. Steinkeller, An Ur III Manuscript of the Sumerian King List. In: W. Sallaberger, K. Volk, A. Zgoll (eds), *Literatur, Politik und Recht in Mesopotamien. Festschrift für Claus Wilcke.* Wiesbaden: Harrassowitz Verlag, 2003; 267-292.

Steinkeller 2005 — P. Steinkeller, The Priestess **ÉGI-ZI** and Related Matters. In: Y. Sefati *et al.* (eds), *'An Experienced Scribe Who Neglects Nothing'. Ancient Near Eastern Studies in Honor of Jacob Klein.* Bethesda, Maryland: CDL Press, 2005; 301-310.

Steinkeller 2007 — P. Steinkeller, On Sand Dunes, Mountain Ranges, and Mountain Peaks. In: M.T. Roth *et al.* (eds), *Studies Presented to Robert D. Biggs, June 4, 2004.* AS 27. Chicago: The Oriental Institute of the University of Chicago, 2007; 219-232.

Steinkeller 2010 — P. Steinkeller, More on the Archaic Writing of the Name of Enlil / Nippur. In: A. Kleinerman, J.M. Sasson (eds), *Why Should Someone Who Knows Something Conceal It? Cuneiform Studies in Honor of David I. Owen on His 70th Birthday.* Bethesda: CDL Press, 2010; 239-243.

Steinmetzer 1922 — F.X. Steinmetzer, *Die babylonischen Kudurru (Grenzsteine) als Urkundenform.* Paderborn: Verlag von Ferdinand Schöningh, 1922.

Stevens 1989 — K.G. Stevens, Eine ikonographische Untersuchung der Schlange im vorgeschichtlichen Mesopotamien. In: L. de Meyer, E. Haerinck (eds), *Archaeologia Iranica et Orientalis. Miscellanea in Honorem Louis Vanden Berghe.* Gent: Uitgeverij Peeters, 1989; 1-32.

Stol 1979a — M. Stol, *On Trees, Mountains, and Millstones in the Ancient Near East.* Leiden: Ex Oriente Lux, 1979.

Stol 1979b — M. Stol, Review of S.D. Simmons, *Early Old Babylonian Documents*. YOS: Babylonian Texts, 14. New Haven: Yale University Press, 1978. *JCS* 31 (1979) 177-183.

Stol 1993-1997 — M. Stol, Mohrhirse. In: E. Ebeling, B. Meissner (eds), *Reallexikon der Assyriologie*, Band 8. Berlin, New York: Walter de Gruyter, 1993-1997; 348-349.

Stol 2000 M. Stol (with a chapter by F.A.M. Wiggermann), *Birth in Babylonia and The Bible. Its Mediterranean Setting.* CM 14. Groningen: STYX Publications, 2000.

Stolz 1999 F. Stolz, Sea. In: K. van der Toorn, B. Becking, P.W. van der Horst (eds), *Dictionary of Deities and Demons in the Bible.* Second extensively revised edition. Leiden, Boston, Köln: Brill; Grand Rapids, Michigan / Cambridge, U.K.: William B. Eerdmans Publishing Company, 1999; 737-742.

Streck 1998 M.P. Streck, The Tense Systems in the Sumerian-Akkadian Linguistic Area. *ASJ* 20 (1998) 181-199.

Streck 1999 M.P. Streck, *Die Bildersprache der akkadischen Epik.* AOAT 264. Münster: Ugarit-Verlag, 1999.

Streck 2002 M.P. Streck, Die Prologe der sumerischen Epen. *OrNS* 71 (2002) 189-266.

Strommenger 1962 E. Strommenger, (Aufnahmen von M. Hirmer) *Fünf Jahrtausende Mesopotamien. Die Kunst von den Anfängen um 5000 v. Chr. bis zu Alexander dem Großen.* München: Hirmer Verlag, 1962.

Strommenger 2008 E. Strommenger, Netzrock und Wolle oder die Macht der Quellen. *ZA* 98 (2008) 1-11.

Such-Gutiérrez 2003 M. Such-Gutiérrez, *Beiträge zum Pantheon von Nippur im 3. Jahrtausend. Teil I.* Roma: Università degli Studi di Roma « La Sapienza», Dipartimento di Studi Orientali, 2003.

Such-Gutiérrez 2005/6 M. Such-Gutiérrez, Untersuchungen zum Pantheon von Adab im 3. Jt. *AfO* 51 (2005-2006) 1-44.

Sweet 1994 R.F.G. Sweet, A New Look at the 'Sacred Marriage' in Ancient Mesopotamia. In: E. Robbins, S. Sandahl (eds), *Corolla Torontonensis. Studies in Honour of Ronald Morton Smith.* Toronto: TSAR, with the Assistance of the Centre for Korean Studies, University of Toronto, 1994; 85-104.

Szarzyńska 1987-1988 K. Szarzyńska, Some of the Oldest Cult Symbols in Archaic Uruk. *JEOL* 30 (1987-1988) 3-21.

Szarzyńska 1996 K. Szarzyńska, Archaic Sumerian Standards. *JCS* 48 (1996) 1-15.

Szarzyńska 2000 K. Szarzyńska, The Cult of the Goddess Inana in Archaic Uruk. *NIN* 1 (2000) 63-74.

Tallqvist 1938 K. Tallqvist, *Akkadische Götterepitheta: mit einem Götterverzeichnis und einer Liste der prädikativen Elemente der sumerischen Götternamen.* Helsingforsiae: Societas Orientalis Fennica, 1938.

Talon 2005 P. Talon, *The Standard Babylonian Creation Myth Enūma Eliš. Introduction, Cuneiform Text, Transliteration, and Sign List with a Translation and Glossary in French.* Helsinki: The Neo-Assyrian Text Corpus Project, 2005.

Thomsen 1984 M.-L. Thomsen, *The Sumerian Language. An Introduction to Its History and Grammatical Structure.* Copenhagen: Akademisk Forlag, 1984.

Tinney 1996 — S. Tinney, *The Nippur Lament. Royal Rhetoric and Divine Legitimation in the Reign of Išme-Dagan of Isin (1953-1935 B.C.).* Philadelphia: The Samuel Noah Kramer Fund, 1996.

Tinney 1999 — S. Tinney, On the Curricular Setting of Sumerian Literature. *Iraq* 61 (1999) 159-172.

Tobler 1950 — A.J. Tobler, *Excavations at Tepe Gawra II: Levels IX-XX.* Philadelphia: University of Pennsylvania Press, 1950.

Tonietti 2003 — M.V. Tonietti, É = *bītum* or = *ʾÀ*? About Ea in Early Dynastic Sources. In: P. Marrassini *et al.* (eds), *Semitic and Assyriological Studies, Presented to Pelio Fronzaroli by Pupils and Colleagues.* Wiesbaden: Harrassowitz Verlag, 2003; 666-679.

Toorn, van der 1996 — K. van der Toorn, *Family Religion in Babylonia, Syria and Israel. Continuity and Change in the Forms of Religious Life.* Leiden, New York, Köln: E.J. Brill, 1996.

Toorn, van der 2000 — K. van der Toorn, Theology, Priests, and Worship in Canaan and Ancient Israel. In: J.M. Sasson (ed.), *Civilizations of the Ancient Near East.* Peabody, Massachusetts: Hendrickson Publishers, Inc., 2000; 2043-2058.

Tournay, Shaffer 1994 — R.J. Tournay, A. Shaffer, *L'Épopée de Gilgamesh.* Paris: Les éditions du CERF, 1994.

Tsukimoto 1985 — A.Tsukimoto, *Untersuchungen zur Totenpflege (kispum) im alten Mesopotamien.* AOAT 216. Kevelaer: Verlag Butzon & Bercker; Neukirchen-Vluyn: Neukirchener Verlag, 1985.

Unger 1938 — E. Unger, Eridu. In: E. Ebeling, B. Meissner (eds) *Reallexikon der Assyriologie*, Band 2; Berlin, Leipzig: Walter de Gruyter, 1938; 464-470.

Ungnad 1926 — A. Ungnad, *Babylonisch-Assyrische Grammatik. Mit Übungsbuch (in Transkription).* Zweite, durchgesehene Auflage. München: C.H. Beck'sche Verlagsbuchhandlung, 1926.

Vanstiphout 1987a — H.L.J. Vanstiphout, Joins Proposed in Sumerian Literary Compositions. *NABU* 1987/87, 46-47.

Vanstiphout 1987b — H.L.J. Vanstiphout, *Enûma eliš*, tablet i: 3. *NABU* 1987/95, 52-53.

Vanstiphout 1990 — H.L.J. Vanstiphout, The Mesopotamian Debate Poems. A General Presentation (Part I). *ASJ* 12 (1990) 271-318.

Vanstiphout 1997a — H.L.J. Vanstiphout, The Disputation Between Ewe and Wheat. In: W.W. Hallo, K.L. Younger (eds), *The Context of Scripture. Vol. I. Canonical Compositions from the Biblical World.* Leiden, New York, Köln: Brill, 1997; 575-578.

Vanstiphout 1997b — H.L.J. Vanstiphout, The Disputation Between Summer and Winter. In: W.W. Hallo, K.L. Younger (eds), *The Context of Scripture. Vol. I. Canonical Compositions from the Biblical World.* Leiden, New York, Köln: Brill, 1997; 584-588.

Vanstiphout 1997c — H.L.J. Vanstiphout, The Dialogue Between a Supervisor and a Scribe. In: W.W. Hallo, K.L. Younger (eds), *The Context of Scripture. Vol. I. Canonical Compositions from the Biblical World.* Leiden, New York, Köln: Brill, 1997; 590-593.

Vanstiphout 1997d — H.L.J. Vanstiphout, The Disputation Between Bird and Fish. In: W.W. Hallo, K.L. Younger (eds), *The Context of Scripture.*

Vol. I. Canonical Compositions from the Biblical World. Leiden, New York, Köln: Brill, 1997; 581-584.

Vanstiphout 1997e — H.L.J. Vanstiphout, Why Did Enki Organize the World? In: I.L. Finkel, M.J. Geller (eds), *Sumerian Gods and Their Representations.* CM 7. Groningen: STYX Publications, 1997; 117-134.

Vanstiphout 2004 — H. Vanstiphout, **Eduba**. *Hoe Men Leerde Schrijven en Lezen in het Oude Babylonië. Een Bloemlezing van Literaire Teksten uit de Scholen van Sumer.* Amsterdam: Uitgeverij SUN, 2004.

Vanstiphout 2009 — H. Vanstiphout, Die Geschöpfe des Prometheus, or How and Why Did the Sumerians Create Their Gods? In: B.N. Porter (ed.), *What is a God? Anthropomorphic and Non-Anthropomorphic Aspects of Deity in Ancient Mesopotamia.* Transactions of the Casco Bay Assyriological Institute, vol. 2. Winona Lake, Indiana: Eisenbrauns, 2009; 15-40.

Veldhuis 1997 — N.C. Veldhuis, *Elementary Education at Nippur. The Lists of Trees and Wooden Objects.* Proefschrift Rijksuniversiteit Groningen. [s.n., s.i.], 1997.

Veldhuis 1999 — N. Veldhuis, The Poetry of Magic. In: T. Abusch, K. van der Toorn (eds), *Mesopotamian Magic. Textual, Historical, and Interpretative Perspectives.* Groningen: STYX Publications, 1999; 35-48.

Veldhuis 2004 — N. Veldhuis, *Religion, Literature, and Scholarship: The Sumerian Composition Nanše and The Birds, with a Catalogue of Sumerian Bird Names.* CM 22. Leiden, Boston: Brill-STYX, 2004.

Veldhuis 2010 — N. Veldhuis, The Theory of Knowledge and the Practice of Celestial Divination. In: A. Annus (ed.), *Divination and Interpretation of Signs in the Ancient World.* OIS 6. Chicago, Illinois: The Oriental Institute of the University of Chicago, 2010; 77-91.

Virolleaud 1911 — Ch. Virolleaud, XIX. Table des Matières de Deux Traités de Divination: l'Un Terrestre, l'Autre Astrologique, Accompagnée d'Instructions du Mage à Son Élève. *Bab.* 4 (1911) 109-113.

Volk 1998-2001 — K. Volk, Nin-giriš. In: E. Ebeling, B. Meissner (eds), *Reallexikon der Assyriologie*, Band 9. Berlin, New York: Walter de Gruyter, 1998-2001; 367.

Waetzoldt 1972 — H. Waetzoldt, *Untersuchungen zur neusumersichen Textilindustrie.* Roma: Centro per le antichità e la storia dell'arte del Vicino Oriente, 1972.

Waetzoldt 1980-1983 — H. Waetzoldt, Kopfbedeckung. A. Philologisch. In: E. Ebeling, B. Meissner (eds), *Reallexikon der Assyriologie*, Band 6. Berlin, New York: Walter de Gruyter, 1980-1983; 197-203.

Wang 2011 — X. Wang, *The Metamorphosis of Enlil in Early Mesopotamia.* AOAT 385. Münster: Ugarit-Verlag, 2011.

Wasilewska 2000 — E. Wasilewska, *Creation Stories of the Middle East.* London and Philadelphia: Jessica Kingsley Publishers Ltd, 2000.

Watrin 2004-2005 — L. Watrin, From Intellectual Acquisitions to Political Change:

Egypt-Mesopotamia Interaction in the Fourth Millennium BC. *KBN* 2 (2004-2005) 48-94.

Weeden 2009 — M. Weeden, The Akkadian Words for 'Grain' and the God Ḫaya. *WO* 39 (2009) 77-107.

Weidner 1924-1925 — E.F. Weidner, Altbabylonische Götterlisten. *AfK* 2 (1924-1925) 1-18; 71-82.

Weidner 1952-1953 — E. Weidner, Die Bibliothek Tiglatpilesers I. *AfO* 16 (1952-1953) 197-215.

Wellek, Warren 1984 — R. Wellek, A. Warren, *Theory of Literature.* (Third edition). San Diego, New York, London: Harcourt Brace & Company, 1984.

Westenholz 1974-1977 — A. Westenholz, Old Akkadian School Texts, *AfO* 25 (1974-1977) 95-110.

Westenholz 1975 — A. Westenholz, *Early Cuneiform Texts in Jena: pre-Sargonic and Sargonic Documents from Nippur and Fara in the Hilprecht-Sammlung vorderasiatischer Altertümer, Institut für Altertumswissenschaften der Friedrich-Schiller-Universität, Jena.* København: Munksgaard, 1975.

Westenholz 1988 — A. Westenholz, Personal Names in Ebla and in Pre-Sargonic Babylonia. In: A. Archi (ed.), *Eblaite Personal Names and Semitic Name-Giving. Papers of a Symposium held in Rome, July 15-17, 1985.* ARES 1. Roma: Missione Archeologica Italiana in Siria, 1988; 99-117.

Westenholz 1992 — J.G. Westenholz, The Clergy of Nippur: the Priestess of Enlil. In: M. de Jong-Ellis (ed.), *Nippur at the Centennial. Papers Read at the 35e Rencontre Assyriologique Internationale, Philadelphia, 1988.* Philadelphia: The Samuel Noah Kramer Fund, Babylonian Section, University Museum, 1992; 297-310.

Westenholz 1996 — J.G. Westenholz, Symbolic Language in Akkadian Narrative Poetry: The Metaphorical Relationship Between Poetical Images and the Real World. In: M.E. Vogelzang, H.L.J. Vanstiphout (eds), *Mesopotamian Poetic Language: Sumerian and Akkadian.* CM 6. Groningen: STYX Publications, 1996; 183-206.

Westenholz 1999 — A. Westenholz, The Old Akkadian Period: History and Culture. In: W. Sallaberger, A. Westenholz, *Mesopotamien. Akkade-Zeit und Ur III-Zeit.* Freiburg, Schweiz: Universitätsverlag; Göttingen: Vandenhoeck & Ruprecht, 1999; 17-117.

Westenholz 2000 — J.G. Westenholz, King by Love of Inanna - An Image of Female Empowerment? *NIN* 1 (2000) 75-89.

Westenholz 2010 — J.G. Westenholz, Heaven and Earth. Asexual Monad and Bisexual Dyad. In: J. Stackert, B.N. Porter, D.P. Wright (eds), *Gazing on the Deep. Ancient Near Eastern and Other Studies in Honor of Tzvi Abusch.* Bethesda, Maryland: CDL Press, 2010; 293-326.

Whittaker 2005 — G. Whittaker, The Sumerian Question: Reviewing the Issues. In: W.H. van Soldt (ed.) in cooperation with R. Kalvelagen and D. Katz, *Ethnicity in Ancient Mesopotamia. Papers read at the*

48th Rencontre Assyriologique Internationale, Leiden, 1-4 July 2002. Leiden: Nederlands Instituut voor het Nabije Oosten, 2005; 409-429.

Wickede, v. 1990 — A. von Wickede, *Prähistorische Stempelglyptik in Vorderasien*. München: Profil Verlag, 1990.

Wiggermann 1981-1982 — F.A.M. Wiggermann, Exit *Talim*! Studies in Babylonian Demonology, I. *JEOL* 27 (1981-1982) 90-105.

Wiggermann 1988 — F.A.M. Wiggermann, An Unrecognized Synonym of Sumerian **sukkal**, 'Vizier'. *ZA* 78 (1988) 225-240.

Wiggermann 1992 — F. Wiggermann, Mythological Foundations of Nature. In: D.J.W. Meijer (ed.), *Natural Phenomena: Their Meaning, Depiction and Description in the Ancient Near East*. Amsterdam, Oxford, New York, Tokyo: North-Holland, 1992; 279-306.

Wiggermann 1993-1997 — F.A.M. Wiggermann, Mischwesen. A. Philogisch. Mesopotamien. In: E. Ebeling, B. Meissner (eds), *Reallexikon der Assyriologie*, Band 8; Berlin, New York: Walter de Gruyter, 1993-1997; 222-246.

Wiggermann 1995 — F.A.M. Wiggermann, Extensions of and Contradictions to Dr. Porada's Lecture. In: E. Porada, *Man and Images in the Ancient Near East*. Wakefield, Rhode Island & London: Moyer Bell, 1995; 77-92.

Wiggermann 1996 — F. Wiggermann, Scenes from the Shadow Side. In: M.E. Vogelzang, H.L.J. Vanstiphout (eds), *Mesopotamian Poetic Language: Sumerian and Akkadian*. CM 6. Groningen: STYX Publications, 1996; 207-230.

Wiggermann 1997 — F.A.M. Wiggermann, Transtigridian Snake Gods. In: I.L. Finkel, M.J. Geller (eds), *Sumerian Gods and Their Representations*. CM 7. Groningen: STYX Publications, 1997; 33-55.

Wiggermann 1998-2001a — F.A.M. Wiggermann, Nammu. In: E. Ebeling, B. Meissner, (eds), *Reallexikon der Assyriologie*, Band 9; Berlin, New York: Walter de Gruyter, 1998-2001; 135-140.

Wiggermann 1998-2001b — F.A.M. Wiggermann, Nergal. A. Philologisch. In: E. Ebeling, B. Meissner (eds), *Reallexikon der Assyriologie*, Band 9; Berlin, New York: Walter de Gruyter, 1998-2001; 215-223.

Wiggermann 2000 — F.A.M. Wiggermann, Theologies, Priests, and Worship in Ancient Mesopotamia. In: J.M. Sasson (ed.), *Civilizations of the Ancient Near East*. Peabody, Massachusetts: Hendrickson Publishers, Inc., 2000; 1857-1870.

Wiggermann 2011 — F.A.M. Wiggermann, The Mesopotamian Pandemonium. A Provisional Census. *SMSR* 77 (2011) 298-322.

Wilcke 1969 — C. Wilcke, *Das Lugalbandaepos*. Wiesbaden: Otto Harrassowitz, 1969.

Wilcke 1972-1975 — C. Wilcke, Hacke. B. Philologisch. In: E. Ebeling, B. Meissner (eds) *Reallexikon der Assyriologie*, Band 4; Berlin, New York: Walter de Gruyter, 1972-1975; 33-38.

Wilcke 1973 — C. Wilcke, Sumerische literarische Texte in Manchester und Liverpool. *AfO* 24 (1973) 1-18.

Wilcke 1974 — C. Wilcke, Zum Königtum in der Ur III-Zeit. In: P. Garelli (ed.), *Le Palais et la Royauté (Archéologie et Civilisation). XIXe Rencontre Assyriologique Internationale; Paris, 29 juin - 2 juillet 1971.* Paris: Librairie Orientaliste Paul Geuthner, 1974; 177-232.

Wilcke 1977 — C. Wilcke, Die Anfänge der akkadischen Epen. *ZA* 67 (1977) 153-216.

Wilcke 1987 — C. Wilcke, Die Inschriftenfunde der 7. und 8. Kampagnen (1983 und 1984). In: B. Hrouda (ed.), *Isin - Išān Baḥrīyāt III. Die Ergebnisse der Ausgrabungen 1983-1984.* München: Verlag de Bayerischen Akademie der Wissenschaften, 1987; 83-120.

Wilcke 1989 — C. Wilcke, Genealogical and Geographical Thought in the Sumerian King List. In: H. Behrens, D. Loding, M.T. Roth (eds), **DUMU-E$_2$-DUB-BA-A**. *Studies in Honor of Åke W. Sjöberg.* Philadelphia: The Samuel Noah Kramer Fund, 1989; 557-571. [Corrections: C. Wilcke, *NABU* 1990/37, 28].

Wilcke 1990 — C. Wilcke, Ti'āmat-baštī. *NABU* 1990/36, 28.

Wilcke 1997 — C. Wilcke, Nanāja-šamḫats Rechtsstreit um ihre Freiheit. In: B. Pongratz-Leisten, H. Kühne, P. Xella (eds), *Ana šadî Labnāni lū allik. Beiträge zu altorientalischen und mittelmeerischen Kulturen. Festschrift für Wolfgang Röllig.* AOAT 247. Kevelaer: Verlag Butzon & Bercker; Neukirchen-Vluyn: Neukirchener Verlag, 1997; 413-429.

Wilcke 1998 — C. Wilcke, Zu 'Gilgameš und Akka'. Überlegungen zur Zeit von Entstehung und Niederschrift, wie auch zum Text des Epos mit einem Exkurs zur Überlieferung von 'Šulgi A' und von 'Lugalbanda II'. In: M. Dietrich, O. Loretz, T.E. Balke (eds), **dubsar anta-men**. *Studien zur Altorientalistik. Festschrift für Willem H. Ph. Römer zur Vollendung seines 70. Lebensjahres mit Beiträgen von Freunden, Schülern und Kollegen.* AOAT 253. Münster: Ugarit-Verlag, 1998; 457-485.

Wilcke 1999 — C. Wilcke, Weltuntergang als Anfang. Theologische, anthropologische, politisch-historische und ästhetische Ebenen der Interpretation der Sintflutgeschichte im babylonischen *Atram-ḫasīs*-Epos. In: A. Jones (ed.), *Weltende. Beiträge zur Kultur- und Religionswissenschaft.* Wiesbaden: Harrassowitz Verlag, 1999; 63-112.

Wilcke 2002 — C. Wilcke, Der Kodex Urnamma (CU): Versuch einer Rekonstruktion. In: T. Abusch (ed.), *Riches Hidden in Secret Places. Ancient Near Eastern Studies in Memory of Thorkild Jacobsen.* Winona Lake, Indiana: Eisenbrauns, 2002; 291-333.

Wilcke 2007 — C. Wilcke, Vom altorientalischen Blick zurück auf die Anfänge. In: E. Angehrn (ed.), *Anfang und Ursprung. Die Frage nach dem Ersten in Philosophie und Kulturwissenschaft.* Berlin, New York: Walter de Gruyter, 2007; 3-60.

Wilkinson 2002 — T.A.H. Wilkinson, Uruk into Egypt: Imports and Imitations. In: J.N. Postgate (ed.), *Artefacts of Complexity: Tracking the Uruk in the Near East.* Iraq Archaeological Reports - 5.

[Warminster]: Published for the British School of Archaeology in Iraq, 2002; 237-248.

Willemsen 1992 — H. Willemsen (red.), *Woordenboek Filosofie.* Assen-Maastricht: Van Gorcum, 1992.

Wolde, v. 2009a — E. van Wolde, Why the Verb ברא Does Not Mean 'to Create' in Genesis 1.1 - 2.4a. *JSOT* 34 (2009) 3-23.

Wolde, v. 2009b — E. van Wolde, *Terug naar het Begin. Waarom Genesis1,1 niet gaat over Gods schepping van hemel en aarde.* Nijmegen: Valkhof Pers, 2009.

Wolkstein, Kramer 1983 — D. Wolkstein, S.N. Kramer (Art compiled by E. Williams-Forte), *Inanna, Queen of Heaven and Earth. Her Stories and Hymns from Sumer*. New York *etc.*: Harper & Row Publishers, 1983.

Woods 2005 — C. Woods, On the Euphrates. *ZA* 95 (2005) 7-45.

Woods 2009 — C. Woods, At the Edge of the World: Cosmological Conceptions of the Eastern Horizon in Mesopotamia. *JANER* 9 (2009) 183-239.

Yadin 1971 — Y. Yadin, A Note on the Scenes Depicted on the ʿAin-Samiya Cup. *IEJ* 21 (1971) 82-85.

Yeivin 1971 — Z. Yeivin, A Silver Cup from Tomb 204a at ʿAin-Samiya. *IEJ* 21 (1971) 78-81.

Yoshikawa 1988 — M. Yoshikawa, For a Better Understanding of Sumerian Grammar. *BiOr* 45 (1988) 499-509.

Yoshikawa 1993 — M. Yoshikawa, Studies in the Sumerian Verbal System. *ASJ Supplementary Series* 1 (1993).

Zgoll 1997a — A. Zgoll, Inana als **nugig**. *ZA* 87 (1997) 181-195.

Zgoll 1997b — A. Zgoll, *Der Rechtsfall der En-ḫedu-Ana im Lied* **nin-me-šara**. AOAT 246. Münster: Ugarit-Verlag, 1997.

Zgoll 1998-2001 — A. Zgoll, Ningal. A.I. In Mesopotamien. In: E. Ebeling, B. Meissner (eds), *Reallexikon der Assyriologie*, Band 9; Berlin, New York: Walter de Gruyter, 1998-2001; 352-356.

Zgoll 2011 — A. Zgoll, Der Mythos Enlil und Ninlil. Vom Schrecken des Kanalbaus durch Stadt und Unterwelt. In: L. Vacín, **u_4 du_{11}-ga-ni sá mu-ni-ib-du_{11}**. *Ancient Near Eastern Studies in Memory of Blahoslav Hruška.* Dresden: Islet, 2011; 287-299.

Ziegler 2005 — N. Ziegler, Les Vaisseaux Sanguins et *Enûma eliš* VI: 5. *JMC* 5 (2005) 4-5.

Zimmern 1930 — H. Zimmern, Ein Zyklus altsumerischer Lieder auf die Haupttempel Babyloniens. *ZA* 39 (1930) 245-276.

Zólyomi 2000 — G. Zólyomi, Structural Interference from Akkadian in Old Babylonian Sumerian. In: J. Black (†), G. Zólyomi (eds), *Special Volume in Honor of Professor Mamoru Yoshikawa, 1. The Study of Diachronic and Synchronic Variation in Sumerian: Papers Presented at the 6th Meeting of the Sumerian Grammar Discussion Group, Oxford, 17th and 18th September 1999* [Organized by J. Black (†) and G. Zólyomi]. *ASJ* 22 (2000) 335-360.

***

Indices

General index

Deities, subjects

F

G

H

I

J

K

L

M

N

R

S

T

U

Y

Z

Index of Sumerian words

Index of Akkadian words

Index of texts

Sumerian texts

Akkadian texts

Other texts

***

Ugarit-Verlag

ALASPM **Abhandlungen zur Literatur Alt-Syrien-Palästinas und Mesopotamiens**
AOAT **Alter Orient und Altes Testament**
AVO **Altertumskunde des Vorderen Orients**
ELO **Elementa Linguarum Orientis**
FARG **Forschungen zur Anthropologie und Religionsgeschichte**
GMTR **Guides to the Mesopotamian Textual Record**
LOS **Lehrbücher orientalischer Sprachen**
MARG **Mitteilungen für Anthropologie und Religionsgeschichte**
UBL **Ugaritisch-Biblische Literatur**
UF **Ugarit-Forschungen**
Eikon – Beiträge zur antiken Bildersprache

Alter Orient und Altes Testament (AOAT) (lieferbare Bände)

43 N.P. Heeßel, *Babylonisch-assyrische Diagnostik.* 2000 (ISBN 3-927120-86-3), XII + 471 S., 2 Abb.; € 98,17

44 A. Ulbrich, *Kypris. Heiligtümer und Kulte weiblicher Gottheiten auf Zypern in der kyproarchaischen und kyproklassischen Epoche (Königszeit).* 2008. xii + 557 S. + 66 Tf. + 1 Karte (ISBN 978-3-934628-56-4) € 174,–

45 D. Wicke, *Vorderasiatische Pyxiden der Spätbronzezeit und der Früheisenzeit.* 2008. xii + 468 S. + 105 Tf. (ISBN 978-3-934628-74-8) € 148,–

46 A. Schuster-Brandis, *Steine als Schutz- und Heilmittel. Untersuchung zu ihrer Verwendung in der Beschwörungskunst Mesopotamiens im 1. Jt. v. Chr.* 2008. xvi + 490 S. + 38 Tf. (ISBN 978-3-934628-75-5) € 140,–

47 E. Fischer, *Ägyptische und ägyptisierende Elfenbeine aus Megiddo und Lachisch. Inschriftenfunde, Flaschen, Löffel.* 2007. xiv + 456 S. + 138 Tf. (ISBN 978-3-934628-89-2) € 148,–

48 H.-G. Buchholz / H. Matthäus, *Tamassos. Ein antiker Stadtstaat im Bergbaugebiet von Zypern.*

48/1 Band I: H.-G. Buchholz, *Die Nekropolen I, II und III*, unter Mitarbeit von Chr. Sandner-Behringer, mit Beiträgen von A. Destrooper-Georgiades, M. Egetmeyer, J.-S. Kühlborn, I. Michaelidou-Nikolaou und Karin Nys. 2010. xvi + 819 S. (ISBN 978-3-86835-004-3) € 196,–

49 Ch.H. Richter, *Parthische Pantoffelsarkophage. Untersuchungen zu einer Sargform Mesopotamiens im Vergleich mit Tonsärgen von Ägypten über den Mittelmeerraum bis Zentralasien.* 2011. XXII + 414 S. + 271 S/W-Tf. + 9 Farb-Tf. (ISBN 978-3-86835-027-2) € 192,–

50 I. Hrůša, *Die akkadische Synonymenliste malku = šarru. Eine Textedition mit Übersetzung und Kommentar.* 2010. viii + 570 S. + 44 Tf. (ISBN 978-3-86835-044-9) € 118,–

245 Fr. Pomponio / P. Xella, *Les dieux d'Ebla. Étude analytique des divinités éblaïtes à l'époque des archives royales du IIIe millénaire.* 1997 (ISBN 3-927120-46-4), VII + 551 S.; € 59,31

248 *Religion und Gesellschaft. Studien zu ihrer Wechselbeziehung in den Kulturen des Antiken Vorderen Orients.* 1997 (ISBN 3-927120-54-5), VIII + 220 S.; € 43,97

250 M. Dietrich / I. Kottsieper (Hrsg.), *„Und Mose schrieb dieses Lied auf". Studien zum Alten Testament und zum Alten Orient. Festschrift Oswald Loretz.* 1998 (ISBN 3-927120-60-X), xviii + 955 S.; € 112,48

251 Th.R. Kämmerer, *Šimâ milka. Induktion und Reception der mittelbabylonischen Dichtung von Ugarit, Emār und Tell el-ʿAmārna.* 1998 (ISBN 3-927120-47-2), XXI + 360 S.; € 60,33

252 J. Marzahn / H. Neumann (Hrsg.), *Assyriologica et Semitica. Festschrift für Joachim Oelsner anläßlich seines 65. Geburtstages am 18. Februar 1997.* 2000 (ISBN 3-927120-62-6), xii + 635 S. + Abb.; € 107,88

254 M. Jursa, *Der Tempelzehnt in Babylonien vom siebenten bis zum dritten Jahrhundert v.Chr.* 1998 (ISBN 3-927120-59-6), VIII + 146 S.; € 41,93

257 (2) Th. Richter, *Untersuchungen zu den lokalen Panthea Süd- und Mittelbabyloniens in altbabylonischer Zeit* (2., verb. und erw. Aufl.). 2004 (ISBN 3-934628-50-8); xx + 608 S.; € 88,–

258 S.A.L. Butler, *Mesopotamian Conceptions of Dreams and Dream Rituals.* 1998 (ISBN 3-927120-65-0), XXXIX + 474 S. + 20 Pl.; € 75,67

259 R. Rothenbusch, *Die kasuistische Rechtssammlung im Bundesbuch und ihr literarischer Kontext im Licht altorientalischer Parallelen.* 2000 (ISBN 3-927120-67-7), IV + 681 S.; € 65,10

260 T. Zewi, *A Syntactical Study of Verbal Forms Affixed by -n(n) Endings in Classical Arabic, Biblical Hebrew, El-Amarna Akkadian and Ugaritic.* 1999 (ISBN 3-927120-71-5), VI + 211 S.; € 48,06

261 H.-G. Buchholz, *Ugarit, Zypern und Ägäis – Kulturbeziehungen im zweiten Jahrtausend v.Chr.* 1999 (ISBN 3-927120-38-3), XIII + 812 S., 116 Tf.; € 109,42

262 (3) W.H.Ph. Römer, *Die Sumerologie. Einführung in die Forschung und Bibliographie in Auswahl* (dritte, aktualisierte Auflage). 2012 (ISBN 978-3-86835-079-1), XII + 262 S.; € 59,–

263 R. Rollinger, *Frühformen historischen Denkens. Geschichtsdenken, Ideologie und Propaganda im alten Mesopotamien am Übergang von der Ur-III- zur Isin-Larsa-Zeit.* (ISBN 3-927120-76-6) (i.V.)

264 M.P. Streck, *Die Bildersprache der akkadischen Epik.* 1999 (ISBN 3-927120-77-4), 258 S.; € 61,36

265 B.I. Faist, *Der Fernhandel des assyrischen Reichs zwischen dem 14. und 11. Jahrhundert v. Chr.* 2001 (ISBN 3-927120-79-0), XXII + 322 S. + 5 Tf.; € 72,09

266 O. Kaelin, *Ein assyrisches Bildexperiment nach ägyptischem Vorbild. Zu Planung und Ausführung der „Schlacht am Ulai".* 1999 (ISBN 3-927120-80-4), 150 S., Abb.; € 49,08

267 B. Böck / E. Cancik-Kirschbaum / Th. Richter (Hrsg.), *Munuscula Mesopotamica. Festschrift für Johannes Renger.* 1999 (ISBN 3-927120-81-2), XXIX + 704 S., Abb.; € 124,76

268 Y. Gong, *Die Namen der Keilschriftzeichen.* 2000 (ISBN 3-927120-83-9), VIII + 228 S.; € 44,99

269/1 M. Dietrich / O. Loretz, *Studien zu den ugaritischen Texten I: Mythos und Ritual in KTU 1.12, 1.24, 1.96, 1.100 und 1.114.* 2000 (ISBN 3-927120-84-7), XIV + 554 S.; € 89,99

270 A. Schüle, *Die Syntax der althebräischen Inschriften. Ein Beitrag zur historischen Grammatik des Hebräischen.* 2000 (ISBN 3-927120-85-5), IV + 294 S.; € 63,40

271/1 M.P. Streck, *Das amurritische Onomastikon der altbabylonischen Zeit I: Die Amurriter, die onomastische Forschung, Orthographie und Phonologie, Nominalmorphologie.* 2000 (ISBN 3-927120-87-1); 414 S.; € 75,67

272 R. Dittmann / B. Hrouda / U. Löw / P. Matthiae / R. Mayer-Opificius / S. Thürwächter (Hrsg.), *Variatio Delectat – Iran und der Westen. Gedenkschrift für Peter Calmeyer.* 2000 (ISBN 3-927120-89-8), XVIII + 768 S. + 2 Faltbilder; € 114,53

273 (2) J. Tropper, *Ugaritische Grammatik. Zweite, stark überarbeitete und erweiterte Auflage.* 2012 (ISBN 978-3-86835-069-2), XXII + 1068 S.; € 100,–

274 G.J. Selz (Hrsg.), *Festschrift für Burkhart Kienast zu seinem 70. Geburtstag dargebracht von Freunden, Schülern und Kollegen.* 2003 (ISBN 3-934628-91-X), xxviii + 733 S.; € 122,–

275 P. Gesche, *Schulunterricht in Babylonien im ersten Jahrtausend v. Chr.* 2001 (ISBN 3-927120-93-6), XXXIV + 820 S. + XIV Tf.; € 112,48

276 W.H.Ph. Römer, *Hymnen und Klagelieder in sumerischer Sprache.* 2001 (ISBN 3-927120-94-4); XI + 275 S.; € 66,47

277 C. Friedl, *Polygynie in Mesopotamien und Israel. Sozialgeschichtliche Analyse polygamer Beziehungen anhand rechtlicher Texte aus dem 2. und 1. Jahrtausend v.Chr.* 2000 (ISBN 3-927120-95-2), 325 S.; € 66,47

278/1 A. Militarev / L. Kogan, *Semitic Etymological Dictionary. Vol. I: Anatomy of Man and Animals*. 2000 (ISBN 3-927120-96-0), cliv + 425 S.; € 84,87

278/2 A. Militarev / L. Kogan, *Semitic Etymological Dictionary. Vol. II: Animal Names*. 2005 (ISBN 3-934628-57-5), xci + 415 S.; € 104,–

279 K.A. Metzler, *Tempora in altbabylonischen literarischen Texten*. 2002 (ISBN 3-934628-03-6), XVII + 964 S.; € 122,–

280 B. Huwyler / H.-P. Mathys / B. Weber (Hrsg.), *Prophetie und Psalmen. Festschrift für Klaus Seybold zum 65. Geburtstag*. 2001 (ISBN 3-934628-01-X), XI + 315 S., 10 Abb.; € 70,56

281 O. Loretz / K.A. Metzler / H. Schaudig (Hrsg.), *Ex Syria et Mesopotamia Lux. Festschrift für Manfried Dietrich zu seinem 65. Geburtstag*. 2002 (ISBN 3-927120-99-5), XXXV + 950 S., zahl. Abb.; € 138,–

282 F.T. Zeeb, *Die Palastwirtschaft in Altsyrien nach den spätaltbabylonischen Getreidelieferlisten aus Alalaḫ (Schicht VII)*. 2001 (ISBN 3-934628-06-0), XIII + 757 S.; € 105,33

283 R. Schmitt, *Bildhafte Herrschaftsrepräsentation im eisenzeitlichen Israel*. 2001 (ISBN 3-934628-05-2), VIII + 231 S.; € 63,40

284/1 D.M. Clemens, *Sources for Ugaritic Ritual and Sacrifice. Vol. I: Ugaritic and Ugarit Akkadian Texts*. 2001 (ISBN 3-934628-07-9), XXXIX + 1407 S.; € 128,85

285 R. Albertz (Hrsg.), *Kult, Konflikt und Versöhnung. Beiträge zur kultischen Sühne in religiösen, sozialen und politischen Auseinandersetzungen des antiken Mittelmeerraumes*. 2001 (ISBN 3-934628-08-7), VIII + 332 S.; € 70,56

286 J.F. Diehl, *Die Fortführung des Imperativs im Biblischen Hebräisch*. 2004 (ISBN 3-934628-19-2), XIV + 409 S.; € 78,–

288 A. Kassian / A. Korolëv † / A. Sidel'tsev, *Hittite Funerary Ritual šalliš waštaiš*. 2002 (ISBN 3-934628-16-8); ix + 973 S.; € 118,–

289 Z. Cochavi-Rainey, *The Alashia Texts from the 14th and 13th Centuries BCE. A Textual and Linguistic Study*. 2003 (ISBN 3-934628-17-6) xiv + 129 S.; € 56,–

290 O. Loretz, *Götter – Ahnen – Könige als gerechte Richter. Der „Rechtsfall" des Menschen vor Gott nach altorientalischen und biblischen Texten*. 2003 (ISBN 3-934628-18-4); xxii + 932 S.; € 128,–

292 A. Behrens, *Prophetische Visionsschilderungen im Alten Testament. Sprachliche Eigenarten, Funktion und Geschichte einer Gattung*. 2002 (ISBN 3-934628-21-4); xi + 413 S.; € 82,–

293 A. Hausleiter / S. Kerner / B. Müller-Neuhof (Hrsg.), *Material Culture and Mental Sphere. Rezeption archäologischer Denkrichtungen in der Vorderasiatischen Altertumskunde. Internationales Symposium für Hans J. Nissen, Berlin 23.–24. Juni 2000*. 2002 (ISBN 3-934628-22-2); xii + 391 S.; € 88,–

294 K. Kiesow / Th. Meurer (Hrsg.), *„Textarbeit". Studien zu Texten und ihrer Rezeption aus dem Alten Testament und der Umwelt Israels. Festschrift für Peter Weimar zur Vollendung seines 60. Lebensjahres*. 2002 (ISBN 3-934628-23-0); x + 630 S.; € 128,–

295 G.W. Vera Chamaza, *Die Omnipotenz Aššurs. Entwicklungen in der Aššur-Theologie unter den Sargoniden Sargon II., Sanherib und Asarhaddon*. 2002 (ISBN 3-93428-24-9); 586 S.; € 97,–

296 M.P. Streck / St. Weninger (Hrsg.), *Altorientalische und semitische Onomastik*. 2002 (ISBN 3-934628-25-7); vii + 236 S.; € 68,–

298 M. Krebernik / J. van Oorschot (Hrsg.), *Polytheismus und Monotheismus in den Religionen des vorderen Orients*. 2002 (ISBN 3-934628-27-3); v + 269 S.; € 76,–

299 W.G.E. Watson (ed.), "He unfurrowed his brow and laughed". Essays in Honour of Professor Nicolas Wyatt. 2007 (ISBN 978-3-934628-32-8); xvi + 410 S.; € 108,–

300 K. Löning (Hrsg.), *Rettendes Wissen. Studien zum Fortgang weisheitlichen Denkens im Frühjudentum und im frühen Christentum*. 2002 (ISBN 3-934628-28-1); x + 370 S.; € 84,–

301 J. Hahn (Hrsg.), *Religiöse Landschaften*. 2002 (ISBN 3-934628-31-1); ix + 227 S.; € 66,–

302 C.G. den Hertog / U. Hübner / St. Münger (Hrsg.), *Saxa Loquentur. Studien zur Archäologie Palästinas/Israels. Festschrift für Volkmar Fritz zum 65. Geburtstag*. 2003 (ISBN 3-934628-34-6); x + 328 S.; € 98,–

303 M.P. Streck, *Die akkadischen Verbalstämme mit ta-Infix*. 2003 (ISBN 3-934628-35-4); xii + 163 S.; € 57,–

304 L.D. Morenz / E. Boßhard-Nepustil, *Herrscherpräsentation und Kulturkontakte: Ägypten – Levante – Mesopotamien. Acht Fallstudien*. 2003 (ISBN 3-934628-37-0); xi + 281 S., 65 Abb.; € 68,–

305 (2) R. Borger, *Mesopotamisches Zeichenlexikon. Zweite, revidierte und aktualisierte Auflage*. 2010. XII + 736 S. (ISBN 978-3-86835-043-2) € 80,–

306 R. Dittmann / Chr. Eder / B. Jacobs (Hrsg.), *Altertumswissenschaften im Dialog. Festschrift für Wolfram Nagel zur Vollendung seines 80. Lebensjahres*. 2003 (ISBN 3-934628-41-9); xv + 717 S., Abb.; € 118,–

307 M. M. Fritz, *„... und weinten um Tammuz". Die Götter Dumuzi-Ama'ušumgal'anna und Damu*. 2003 (ISBN 3-934628-42-7); 430 S.; € 83,–

308 A. Zgoll, *Die Kunst des Betens. Form und Funktion, Theologie und Psychagogik in babylonisch-assyrischen Handerhebungsgebeten an Ištar*. 2003 (ISBN 3-934628-45-1); iv + 319 S.; € 72,–

309 W.H.Ph. Römer, *Die Klage über die Zerstörung von Ur*. 2004 (ISBN 3-934628-46-X); ix + 191 S.; € 52,–

310 Th. Schneider (Hrsg.), *Das Ägyptische und die Sprachen Vorderasiens, Nordafrikas und der Ägäis. Akten des Basler Kolloquiums zum ägyptisch-nichtsemitischen Sprachkontakt Basel 9.-11. Juli 2003*. 2003 (ISBN 3-934628-47-8); 527 S.; € 108,–

311 D. Kühn, *Totengedenken bei den Nabatäern und im Alten Testament. Eine religionsgeschichtliche und exegetische Studie*. 2005 (ISBN 3-934628-48-6), x + 514 S. + 42 S. mit Abb.; € 95,80

312 R. Hempelmann, *„Gottschiff" und „Zikkurratbau" auf vorderasiatischen Rollsiegeln des 3. Jahrtausends v.Chr.* 2004 (ISBN 3-934628-49-4); viii + 154 S. + 31 Tf.; € 55,–

313 R. Schmitt, *Magie im Alten Testament*. 2004 (ISBN 3-934628-52-4); xiii + 471 S.; € 94,–

314 St. Timm, *„Gott kommt von Teman ..." Kleine Schriften zur Geschichte Israels und Syrien-Palästinas*. Hrsg. von Claudia Bender und Michael Pietsch. 2004 (ISBN 3-934628-53-29); viii + 274 S.; € 63,–

315 B. Janković, *Vogelzucht und Vogelfang in Sippar im 1. Jahrtausend v.Chr.* Veröffentlichungen zur Wirtschaftsgeschichte Babyloniens im 1. Jahrtausend v.Chr. Band 1. 2004 (ISBN 3-934628-54-0); xx + 219 S.; € 56,20

316 Chr. Sigrist (Hrsg.), *Macht und Herrschaft*. 2004 (ISBN 3-934628-55-9) € 63,–

317 B. Burtea / J. Tropper / H. Younansardaroud (eds.), Studia Semitica et Semitohamitica. Festschrift für Rainer Voigt anläßlich seines 60. Geburtstages am 17. Januar 2004. 2005 (ISBN 3-934628-73-7); v + 539 S; € 98,50

318 M. Hutter / S. Hutter-Braunsar (Hrsg.), *Offizielle Religion, lokale Kulte und Individuelle Religiosität*. 2004 (ISBN 3-934628-58-3); 504 S. € 121,–

319 C. Mittermayer, *Die Entwicklung der Tierkopfzeichen. Eine Studie zur syro-mesopotamischen Keilschriftpaläographie des 3. und frühen 2. Jahrtausends v.Chr.* 2005 (ISBN 3-934628-59-1); 169 S. € 48,80

321 G.W. Vera Chamaza, *Die Rolle Moabs in der neuassyrischen Expansionspolitik. 2005* (ISBN 3-934628-61-3); viii + 203 S.; € 58,–

322 S. Bahyro, *The Shemihazah and Asael Narrative of 1 Enoch 6-11: Introduction, Tanslation and Commentary with Reference to Ancient Near Eastern and Biblical Antecedents*. 2005 (ISBN 3-934628-62-1); x + 295 S.; € 66,50

323 M. Novák / Fr. Prayon / A.-M. Wittke (Hrsg.), *Die Außenwirkung des späthethitischen Kulturraumes. Güteraustausch – Kulturkontakt – Kulturtransfer*. 2005 (ISBN 3-934628-63-X); viii + 496 S.; € 106,–

324 W.H. van Soldt, *The Topography of the City-State of Ugarit*. 2005 (ISBN 3-934628-64-8); vi + 253 S.; € 64,–

325 R. Rollinger (Hrsg.), *Von Sumer bis Homer. Festschrift für Manfred Schretter zum 60. Geburtstag am 25. Februar 2004*. 2005 (ISBN 3-934628-66-4); xi + 697 S.; € 128,50

326 U. Susanne Koch, *Secrets of Extispicy. The Chapter Multābiltu of the Babylonian Extispicy Series and Niṣirti Barûti Texts mainly from Aššurbanipal's Library.* 2005 (ISBN 3-934628-67-2); x + 630 pp + liv Tf.; € 119,–

327 H. Weippert, *Unter Olivenbäumen. Studien zur Archäologie Syrien-Palästinas, Kulturgeschichte und Exegese des Alten Testaments. Gesammelte Aufsätze. Festgabe zum 4. Mai 2003 herausgegeben von Angelika Berlejung und Michael Niemann.* 2005. x. + 522 S. (ISBN 3-934628-68-0); € 94,–

328 E.A. Braun-Holzinger / E. Rehm, *Orientalischer Import in Griechenland im frühen 1. Jahrtausend v.Chr.* 2005 (ISBN 3-934628-72-9), vi + 208 S. + 39 Tf.; € 63,60

329 M. Herles, *Götterdarstellungen in der 2. Hälfte des 2. Jahrtausends v. Chr. Das anthropomorphe Bild im Verhältnis zum Symbol.* 2006 (ISBN 3-934628-76-1), xiii+ 394 S., 145 Tf.; € 112,–

330 H.D. Baker / M. Jursa (eds.), *Approaching the Babylonian Economy. Proceedings of the START Project Symposium Held in Vienna, 1-3 July 2004.* Veröffentlichungen zur Wirtschaftsgeschichte Babyloniens im 1. Jahrtausend v.Chr. Band 2. 2005 (ISBN 3-934628-79-6), viii + 448 S.; € 86,50

331 Th.E. Balke, *Das sumerische Dimensionalkasussystem.* 2006 (ISBN 3-934628-80-X), x + 287 S.; € 68,–

332 M. Jaques, *Le vocabulaire des sentiments dans les textes sumériens. Recherches sur le lexique sumérien et akkadien.* 2006 (ISBN 3-934628-81-8) € 122,–

333 A. Zgoll, *Traum und Welterleben im antiken Mesopotamien. Traumtheorie und Traumpraxis im 3.-1. Jt. v. Chr. als Horizont einer Kulturgeschichte des Träumens.* 2006 (ISBN 3-934628-36-2), vi + 568 S.; € 96,–

334 T. Bar / E. Cohen (eds.), *Studies in Semitic and General Linguistics in Honor of Gideon Goldenberg.* 2007 (ISBN 3-934628-84-7), 387 S.; € 94,–

335 I. Márquez Rowe, *The Royal Deeds of Ugarit. A Study of Ancient Near Eastern Diplomatics.* 2006 (ISBN 3-934628-86-9), 336 S.; € 69,–

336 J. Luchsinger / H.-P. Mathys / M. Saur (Hrsg.), „ ... der seine Lust hat am Wort des Herrn!“ Festschrift für Ernst Jenni zum 80. Geburtstag. 2007 (ISBN 978-3-934628-87-8), xii + 466 S.; € 118,–

337 M. Hutter / S. Hutter-Braunsar (Hrsg.), *Pluralismus und Wandel in den Religionen im vorhellenistischen Anatolien. Akten des religionsgeschichtlichen Symposiums in Bonn am 19.–20. Mai 2005.* 2006 (ISBN 3-934628-88-5), 263 S; € 68,–

338 J. Kinnier Wilson, *Studia Etanaica. New Texts and Discussions.* 2007 (ISBN 978-3-934628-90-8), 100 S. + 15 Tf.; € 58,–

339 N. Brisch, *Tradition and the Poetics of Innovation. Sumerian Court Literature of the Larsa Dynasty (c. 2003–1763 BCE).* 2007 (ISBN 978-3-934628-91-5), xii + 303 S. + XVII Tf.; € 78,–

340 I.K.H. Halayqa, *A Comparative Lexicon of Ugaritic and Canaanite.* 2008. (ISBN 978-3-934628-95-3), 573 S.; € 118,–

341 J. Hackl, *Der subordinierte Satz in den spätbabylonischen Briefen.* 2007 (ISBN 978-3-934628-96-0), xiv + 171 S.; € 62,–

342 E.A. Braun-Holzinger, *Das Herrscherbild in Mesopotamien und Elam. Spätes 4. bis frühes 2. Jt. v.Chr.* 2007 (ISBN 978-3-934628-98-4), vi + 218 S. + 77 Tf.; € 78,–

343 M. Dietrich (Hrsg.), Orbis Ugariticus. Ausgewählte Beiträge von Manfried Dietrich und Oswald Loretz zu Fest- und Gedenkschriften. Anläßlich des 80. Geburtstages von Oswald Loretz. 2008 (ISBN 978-3-934628-99-1), xii + 343 S.; € 88,-.

344 R. Lamprichs, *Tell Johfiyeh. Ein archäologischer Fundplatz und seine Umgebung in Nordjordanien. Materialien zu einer Regionalstudie.* 2007 (ISBN 978-3-86835-000-5), xi + 787 S.; € 132,–

345 S. Kyung Huh, *Studien zur Region Lagaš. Von der Ubaid- bis zur altbabylonischen Zeit.* 2008 (ISBN 978-3-86835-003-6), xi + 916 S. + 3 Beilagen; € 146,–

346 J. Kutter, *nūr ilī. Die Sonnengottheiten in den nordwestsemitischen Religionen von der Spätbronzezeit bis zur vorrömischen Zeit.* 2008 (ISBN 978-3-86835-002-9), iv + 524 S. + 23 Tf.; € 92,–

347 Th. Schneider / K. Szpakowska (eds.), *Egyptian Stories. A British Egyptological Tribute to Alan B. Lloyd on the Occasion of His Retirement.* 2007. (ISBN 978-3-934628-94-6), x + 459 S.; € 112,–

348 K. Lämmerhirt, *Wahrheit und Trug. Untersuchungen zur altorientalischen Begriffsgeschichte.* 2010 (ISBN 978-3-86835-005-0), x + 907 S. € 130,–

349 L. d'Alfonso / Y. Cohen / D. Sürenhagen (eds.), *The City of Emar among the Late Bronze Age Empires. History, Landscape, and Society. Proceedings of the Konstanz Emar Conference, 25.–26.04. 2006.* 2008 (ISBN 978-3-86835-006-7) xviii + 315 S. € 78,–

350 I. Kottsieper / R. Schmitt / J. Wöhrle (eds.), *Berührungspunkte. Studien zur Sozial- und Religionsgeschichte Israels und seiner Umwelt. Festschrift für Rainer Albertz zu seinem 65. Geburtstag.* 2008 (ISBN 978-3-86835-008-1). xii + 676 S.; € 112,–

351 S. Aletta Kersken, *Töchter Zions, wie seid ihr gewandet? Untersuchungen zu Kleidung und Schmuck alttestamentlicher Frauen.* 2008 (ISBN 978-3-86835-007-4). xiv + 357 S.; € 76,–

352 Chr. Niedorf, *Die mittelbabylonischen Rechtsurkunden aus Alalaḫ (Schicht IV).* 2008. xii + 495 S. (ISBN 978-3-86835-009-8); € 96,–

353 L. El-Khouri, *Roman Settlements in the Region of Northwest Jordan. Archaeological Studies (Land-use and Landscape Development).* 2009 (ISBN 978-3-86835-010-4); 227 S. + Faltkarte; € 66,–

354 B. Morstadt, *Phönizische Thymiateria. Zeugnisse des Orientalisierungsprozesses im Mittelmeerraum. Originale Funde, bildliche Quellen, originaler Kontext.* 2008. 489 S. + 86 Taf. (ISBN 978-3-86835-011-1); € 110,–

356 B. Mofidi-Nasrabadi, *Aspekte der Herrschaft und der Herrscherdarstellungen in Elam im 2. Jt. v. Chr.* 2009 (ISBN 978-3-86835-013-5), x + 104 S. + 20 Tf.; € 54,–

357 Chr. Mountfort Monroe, *Trade, Tradition, and Transformation in the Eastern Mediterranean ca. 1350–1175 BCE.* 2009 (ISBN 978-3-86835-015-9), xviii + 362 S.; € 74,–

358 Kr. Kleber, *Tempel und Palast. Die Beziehungen zwischen dem König und dem Eanna-Tempel im spätbabylonischen Uruk.* Veröffentlichungen zur Wirtschaftsgeschichte im 1. Jahrtausend v.Chr. Band 3. 2008. xiv + 404 S. + 41 Tf. (ISBN 978-3-86835-016-6); € 91,–

359 Chr. Karrer-Grube / J. Krispenz / Th. Krüger / Chr. Rose / A. Schellenberg (Hrsg), *Sprachen – Bilder – Klänge. Dimensionen der Theologie im Alten Testament und in seinem Umfeld. Festschrift für Rüdiger Bartelmus zu seinem 65. Geburtstag.* 2009. viii + 409 S. (ISBN 978-3-86835-017-3); € 84,–

360/1 M. Dietrich / O. Loretz / J. Sanmartín, *Die keilalphabetischen Texte aus Ugarit, Ras Ibn Hani und anderen Orten. Dritte, erweiterte Auflage. The Cuneiform Alphabetic Texts from Ugarit, Ras Ibn Hani and Other Places. Third, enlarged Edition. KTU*^{3}. 2013. xxvi + 782 S. (ISBN 978-3-86835-014-2) im Druck

360/2 I. Kottsieper, *Wortregister, rückläufiger Index und Indices der Buchstabengruppen von KTU*^{3}*. Word-List, Reverse Index and Index of Letter-Clusters of KTU*^{3}. (in Vorbereitung)

361 I.J. de Hulster / R. Schmitt (eds.), *Iconography and Biblical Studies. Proceedings of the Iconography Sessions at the Joint EABS/SBL Conference, 22–26 July 2007, Vienna, Austria.* 2009. xii + 239 S. (ISBN 978-3-86835-018-0) € 84,–

362 J. Peterson, *Godlists from Old Babylonian Nippur in the University Museum, Philadelphia.* 2009. x + 140 S. + 27 Tf. (ISBN 978-3-86835-019-7) € 58,–

363 Ph. Clancier, *Les bibliothèques en Babylonie dans la deuxième moitié du I*^{er} *millénaire av. J.-C.* 2009. iv + 493 S. (ISBN 978-3-86835-020-3) € 88,–

364 M. Pietsch / Fr. Hartenstein (Hrsg.), *Israel zwischen den Mächten. Festschrift für Stefan Timm zum 65. Geburtstag.* 2009. x + 486 S. (ISBN 978-3-86835-025-8) € 96,–

365 A. Löhnert, *„Wie die Sonne tritt heraus!" Eine Klage zum Auszug Enlils mit einer Untersuchung zu Komposition und Tradition sumerischer Klagelieder in altbabylonischer Zeit.* 2009. xiv + 502 S. + XVI Tf. (ISBN 978-3-86835-026-5) € 96,–

366 M.J. Boda / J. Novotny (eds.), *From the Foundations to the Crenellations. Essays on Temple Building in the Ancient Near East and Hebrew Bible*. 2010. xxxviii + 629 S. (ISBN 978-3-86835-031-9) € 128,–

367 W. Meinhold, *Ištar in Aššur. Untersuchung eines Lokalkultes von ca. 2500 bis 614 v. Chr.* 2009. xii + 568 S. + 31 Tf. (ISBN 978-3-86835-032-6) € 110,–

368 R. Prentice, *The Exchange of Goods and Services in Pre-Sargonic Lagash*. 2010. xiv + 238 S. (ISBN 978-3-86835-033-3) € 78,–

371 J. Becker / R. Hempelmann / E. Rehm (eds.), *Kulturlandschaft Syrien. Zentrum und Peripherie. Festschrift für Jan-Waalke Meyer*. 2010. xxii + 662 S. (ISBN 978-3-86835-034-0) € 126,–

372 J. Vidal (ed.), *Studies on War in the Ancient Near East. Collected Essays on Military History*. 2010. vi + 198 S. (ISBN 978-3-86835-035-7) € 78,–

373 M. Müller, *Akkadisch in Keilschrifttexten aus Ägypten. Deskriptive Grammatik einer Interlanguage des späten zweiten vorchristlichen Jahrtausends anhand der Ramses-Briefe*. 2010. xxvi + 599 S. (ISBN 978-3-86835-037-1) € 96,–

374 Z. Cochavi-Rainey, *The Akkadian Dialect of Egyptian Scribes in the 14th and 13th Centuries BCE*. 2011. xiv + 290 S. (ISBN 978-3-86835-039-5) € 62,–

375 *Das babylonische Weltschöpfungsepos Enūma elîš*. Herausgegeben von Th. R. Kämmerer und K. A. Metzler. 2012. viii + 418 S. + XLVII Tf. (ISBN 978-3-86835-036-4) € 80,–

376 W.H.Ph. Römer, *Die Zylinderinschriften von Gudea*. 2010. x + 130 S. (ISBN 978-3-86835-040-1) € 54,–

377 M. Jursa, *Aspects of the Economic History of Babylonia in the First Millennium BC. Economic Geography, Economic Mentalities, Agriculture, the Use of Money and the Problem of Economic Growth* (with Contributions by J. Hackl, B. Janković, K. Kleber, E.E. Payne, C. Waerzeggers and M. Weszeli). Veröffentlichungen zur Wirtschaftsgeschichte im 1. Jahrtausend v. Chr. Band 4. 2010. xviii + 900 S. (ISBN 978-3-86835-041-8) € 136,–

378 Dominik Elkowicz, *Tempel und Kultplätze der Philister und der Völker des Ostjordanlandes. Eine Untersuchung zur Bau- und zur Kultgeschichte während der Eisenzeit I–II*. 2012. x + 238 S. + 77 Tf. (ISBN 978-3-86835-045-6) € 78,–

379 Chr. G. Frechette, *Mesopotamian Ritual-prayers of "Hand-lifting" (Akkadian Šuillas): An Investigation of Function in Light of the Idiomatic Meaning of the Rubric*. 2012. xx + 316 S. (ISBN 978-3-86835-046-3) € 78,–

380 P. Mommer / A. Scherer (Hrsg.), *Geschichte Israels und deuteronomistisches Geschichtsdenken. Festschrift zum 70. Geburtstag von Winfried Thiel*. 2010. VIII + 334 S. (ISBN 978-3-86835-047-0) € 72,–

381 R. Schmitt, *Der „Heilige Krieg" im Pentateuch und im deuteronomistischen Geschichtswerk. Studien zur Forschungs-, Rezeptions- und Religionsgeschichte von Krieg und Bann im Alten Testament*. 2011. xii + 248 S. (ISBN 978-3-86835-048-7) € 65,–

382 M. Dietrich / W. Mayer, *Der hurritische Brief des Dušratta von Mīttānni an Amenḫotep III. Text – Grammatik – Kopie. Englische Übersetzung des Textes von D. Durkin-Meisterernst*. 2011. xx + 308 S. + 16 Tf. (ISBN 978-3-86835-049-4) € 82,–

383 U. Bock, *„Von seiner Kindheit bis zum Erwachsenenalter". Die Darstellung der Kindheit des Herrschers in mesopotamischen und kleinasiatischen Herrscherinschriften und literarischen Texten*. 2012. xx + 439 S. (ISBN 978-3-86835-050-0) € 82,–

384 Sh. Zelig Aster, *The Unbeatable Light: Melammu and Its Biblical Parallels*. 2012. xviii + 430 S. (ISBN 978-3-86835-051-7) € 82,–

385 X. Wang, *The Metamorphosis of Enlil in Early Mesopotamia*. 2011. xx + 297 S. (ISBN 978-3-86835-052-4) € 64,–

386 O. Loretz, *Hippologia Ugaritica. Das Pferd in Kultur, Wirtschaft, Kriegführung und Hippiatrie Ugarits — Pferd, Esel und Kamel in biblischen Texten —. Mit einem Beitrag von M. Stol über Pferde, Pferdekrankheiten und Pferdemedizin in altbabylonischer Zeit*. 2011. 496 S. (ISBN 978-3-86835-055-5) € 89,–

387 A. S. Zernecke, *Gott und Mensch in Klagegebeten aus Israel und Mesopotamien. Die Handerhebungsgebete Ištar 10 und Ištar 2 und die Klagepsalmen Ps 38 und Ps 22 im Vergleich*. 2011 xvi+ 408 S. (ISBN 978-3-86835-058-6) € 75,–

388 Aurélie Carbillet, *La figure hathorique à Chypre (IIe–Ier mill. av. J.-C.)*. 2011. 372 S. + 90 Tf. (ISBN 978-3-86835-059-3) € 92,–

389 R. Jacob, *Kosmetik im antiken Palästina*. 2011. xiv + 418 S. (ISBN 978-3-86835-060-9) € 87,–

390 Acta Antiqua Mediterranea et Orientalia

390/1: Thomas R. Kämmerer (ed.): *Identities and Societies in the Ancient East-Mediterranean Regions. Comparative Approaches. Henning Graf Reventlow Memorial Volume*. 2011. VIII + 359 S. (ISBN 978-3-86835-062-3) € 75,–

391 M. Hutter / S. Hutter-Braunsar (Hrsg.), *Hethitische Literatur. Überlieferungsprozesse, Textstrukturen, Ausdrucksformen und Nachwirken. Akten des Symposiums vom 18. bis 20. Februar 2010 in Bonn*. 2011. 311 S. (ISBN 978-3-86835-063-0) € 72,–

392 G. Galil / A. Gilboa / A.M. Maeir / D. Kahn (eds.), *The Ancient Near East in the 12th–10th Centuries BCE: Culture and History. Proceedings of the International Conference held at the University of Haifa, 2–5 May, 2010*. 2012 xviii+647 S. (ISBN 978-3-86835-066-1) € 116,–

393 M. Maggio, *L'ornementation des dieux à l'époque paléo-babylonienne. Étude du matériel ayant appartenu aux dieux d'après des documents de la pratique. Réflexions sur le don, l'ornementation des statues divines et la conservation des objets précieux*. 2012. xiv + 245 S. (ISBN 978-3-86835-067-8) € 68,–

394 G. del Olmo Lete / J. Vidal / N. Wyatt (eds.), *The Perfumes of Seven Tamarisks. Studies in Honour of Wilfred G. E. Watson*. 2012. xix + 330 S. (ISBN 978-3-86835-068-5) € 86,–

395/1 W. Mayer, *Assyrien und Urarṭu I. Der Achte Feldzug Sargons II. im Jahr 714 v.Chr.* 2013. xvi + 189 S. (ISBN 978-3-86835-078-4) € 64,–

395/2 W. Mayer, *Assyrien und Urarṭu II. Die assyrisch-urartäischen Bilinguen*. 2013. xvi + 156 S. (ISBN 978-3-86835-091-3) im Druck

397 H.D. Baker / K. Kaniuth / A. Otto (Hrsg.), *Stories of Long Ago. Festschrift für Michael D. Roaf*. 2012. xxii + 745 S. (ISBN 978-3-86835-071-5) € 120,–

398 M. Sandowicz, *Oaths and Curses. A Study in Neo- and Late Babylonian Legal Formulary*. 2012. xiv + 542 S. + XLI Tf. (ISBN 978-3-86835-072-2) € 92,–

402 G. Minunno, *Ritual Employs of Birds in Ancient Syria-Palestine*. 2013. 165 S. (ISBN 978-3-86835-083-8) € 64,–

404 O. Loretz / S. Ribichini / W.G.E. Watson / J.Á. Zamora (eds.), *Ritual, Religion and Reason. Studies in the Ancient World in Honour of Paolo Xella*. 2013. xxxii + 677 S. (ISBN 978-3-86835-087-6) € 135,–

405 G. Goldenberg, *Further Studies in Semitic Linguistics*. 2013. 262 S. (ISBN 978-3-86835-089-0) € 69,–

407 Ernst Axel Knauf, *Data and Debates. Essays in the History and Culture of Israel and Its Neighbors in Antiquity. Daten und Debatten. Aufsätze zur Kulturgeschichte des antiken Israel und seiner Nachbarn*. Herausgegeben von Hermann Michael Niemann, Konrad Schmid und Silvia Schroer. 2013. xvi + 770 S. (978-3-86835-093-7) € 136,–

409 J.J.W. Lisman, *Cosmogony, Theogony and Anthropogeny in Sumerian Texts*. 2013. xiv + 422 S. (ISBN 978-3-86835-095-1) im Druck

Ugarit-Verlag, Rosenstr. 9, D–48143 Münster (www.ugarit-verlag.de)

Auslieferung / Distribution: Sigloch Distribution, Am Buchberg 8, 74572 Blaufelden (www.sigloch.de)

Distributor to North America: Eisenbrauns, Inc., POB 275, Winona Lake, Ind., 46590, USA (www.eisenbrauns.com)